Quantifying the Evolution of Early Life

Aims and Scope Topics in Geobiology Book Series

Topics in Geobiology series treats geobiology – the broad discipline that covers the history of life on Earth. The series aims for high quality, scholarly volumes of original research as well as broad reviews. Recent volumes have showcased a variety of organisms including cephalopods, corals, and rodents. They discuss the biology of these organisms-their ecology, phylogeny, and mode of life – and in addition, their fossil record – their distribution in time and space.
Other volumes are more theme based such as predator-prey relationships, skeletal mineralization, paleobiogeography, and approaches to high resolution stratigraphy, that cover a broad range of organisms. One theme that is at the heart of the series is the interplay between the history of life and the changing environment. This is treated in skeletal mineralization and how such skeletons record environmental signals and animal-sediment relationships in the marine environment.
The series editors also welcome any comments or suggestions for future volumes.

Series Editors
Neil H. Landman, landman@amnh.org
Peter Harries, harries@shell.cas.usf.edu

For other titles published in this series, go to
http://www.springer.com/series/6623

Marc Laflamme · James D. Schiffbauer
Stephen Q. Dornbos
Editors

Quantifying the Evolution of Early Life

Numerical Approaches to the Evaluation of Fossils and Ancient Ecosystems

Springer

Editors
Marc Laflamme
Smithsonian Postdoctoral Fellow
Department of Paleobiology
Smithsonian Institution, National Museum
of Natural History, Washington, DC
20013-7012, USA
laflammem@si.edu

Stephen Q. Dornbos
Department of Geosciences
University of Wisconsin-Milwaukee
Milwaukee, WI 53201-0413
USA
sdornbos@uwm.edu

James D. Schiffbauer
Department of Geosciences
Virginia Polytechnic Institute
and State University
4044 Derring Hall, Blacksburg
VA 24061, USA
jdschiff@vt.edu

ISBN 978-94-007-0679-8 e-ISBN 978-94-007-0680-4
DOI 10.1007/978-94-007-0680-4
Springer Dordrecht Heidelberg London New York

Library of Congress Control Number: 2011921512

Cover illustration: Small top images: cartoons on left and right by Marc Laflamme, center small image by J. William Schopf, large center image by James D. Schiffbauer

Printed on acid-free paper

Springer is part of Springer Science+Business Media (www.springer.com)

Dedication

When we began organizing the GSA topical session that formed the basis of this volume in 2008, Hans Hofmann was first on our list of proposed keynote speakers due to his longstanding dedication to quantitative and qualitative approaches to Precambrian Paleontology. We saw him as a flagship to bring in the crowds so as to ensure a well-attended session. After hearing the news of Hans' passing, we asked Brian Pratt to write up a dedication to our mutual friend. We are honoured to dedicate this book to our esteemed friend and mentor, Hans Hofmann.

— Marc Laflamme, James D. Schiffbauer, and Stephen Q. Dornbos

Hans J. Hofmann (1936–2010)

Hans J. Hofmann was one of Canada's most celebrated geoscientists. He conducted peerless research on Precambrian stromatolites, Proterozoic cyanobacterial and algal microfossils, Ediacaran macrofossils, and other such elements of the early rock record. He was the winner of several prestigious scientific prizes, including the Geological Association of Canada's Logan Medal and Billings Medal for paleontology, the Willet Green Miller Medal of the Royal Society of Canada, as well as the Charles Doolittle Walcott Medal of the U.S. Academy of Sciences.

My first encounter with Hans was when he came to McMaster University in the mid-1970s to give a talk on the remarkably well preserved microbial microfossils in cherts from the Belcher Islands. I remember that lecture like it was yesterday, including Hans' expressive, bearded face lighting up as he interjected "sex!" while describing the cyanobacteria preserved during cell division. I can imagine those few in the audience who knew that bacteria do not engage in sexual reproduction were too amused and mesmerized by the lecture to point this out. Published in 1976, these microfossils have made it into the cannon of Earth History by being immortalized in textbooks and review papers.

The late David R. Kobluk introduced us personally during that visit and shortly afterwards I asked him to examine samples of the tidal flat cyanobacteria I had collected on Bonaire, Netherlands Antilles. I also asked his advice on a sample of what was for me at the time a wholly mysterious sedimentary feature: cone-in-cone structure. No such thing existed in the teaching collection, and it did not look like anything figured in textbooks like Pettijohn's Sedimentary Rocks. He sent me a nice letter with a list of key references, and deposited the sample in the teaching collection of the Université de Montréal. Later, when I was a graduate student at McGill University, Hans would look me up when he visited and he even offered to take me on as a graduate student. We saw each other regularly at conferences after that and roomed together on trips to faraway lands. One of the more memorable ones was a field workshop in Mauritania where we examined the fabulous array of Neoproterozoic stromatolites exposed near Atar (Fig. 1). As I got to know him I was always struck by how low-key a person he was, yet always kind, gracious and good-humoured.

Hans may well have been the most scholarly scientist I have ever encountered. At his fingertips was a complete library of everything about the Precambrian fossil record, regardless of language or obscurity. His work was exacting and carefully referenced as befitting a systematic paleontologist, which he was, amongst other talents (Fig. 2). His several review papers, such as those in Earth's Earliest Biosphere (1983) and The Proterozoic Biosphere (1992), are especially valuable compilations summarizing the state-of-the-art and the literature.

Hans' seminal 1969 Geological Survey of Canada Paper 69-39 followed by his 1973 review paper showed the earth sciences community that stromatolites have great geological utility providing they were approached with rigour. This meant measuring everything that could be measured, in order to acquire a thorough,

Fig. 1 Hans Hofmann and the author against a backdrop of Cambro-Ordovician sandstones, Amojjiar Pass, Mauritania, November 1987

Fig. 2 One of Hans' more recent interests involved the Ediacaran rangeomorph fronds such as *Avalofractus abaculus* redrawn here from the specimen NFM F-754. Drawing by James D. Schiffbauer

quantitative understanding of their geometry. Hans made use of the computer in the early days of mainframes to show how stromatolites could accrete by varying different numerical growth factors. He also established that branching stromatolite morphology could best be appreciated by serially sectioning large samples and having the computer assemble three-dimensional reconstructions. By taking this approach he was able to question from a position of authority the stromatolite biostratigraphic paradigm that was still in vogue in the USSR at the time.

Hans' research on other subjects was similarly characterized by careful counting and painstaking measurement, and simple but meaningful graphing and statistical analysis. Among his many publications, his 1994 memoir on Mesoproterozoic microfossils in shales from Baffin Island is a tour de force in this regard. His paper in the same year in which he developed improved methods for plotting grain-shape data is another example of how he approached practical problems with common sense and arithmetical ability. It shows how enduring Hans' scientific ideas are that this paper is still cited by enthusiasts of sedimentary particles, just as his systematics continues to be referred to in taxonomic studies.

Ichnofossils were a long-standing interest of Hans, and another of his forays into morphometrics was to quantify the shape of meandering traces of early Cambrian age, enabled by improvements in image analysis two decades ago. With a wide range of measurements he conducted computer simulations to test for randomness in foraging behaviour of the tracemaker. Again, the goal was practical: what is the taxonomic value of meander patterns, something that needs to be determined before paleobiological and paleoecological interpretations can be made.

I suppose Hans' grandest numerical contribution—in terms of scale—is his proposal that Precambrian geological time scale should be subdivided using 100 m.yr. intervals called 'geons'. A sensible idea that has some adherents, it is a shame that Hans will never know if the community of earth scientists will eventually adopt this or not.

Hans was born in Germany and immigrated with his family to Canada after World War II. He received his university education at McGill University where he completed his doctoral thesis on Ordovician stratigraphy in the St. Lawrence Lowlands under Prof. T. H. Clark. He taught briefly at the University of Cincinnati and McMaster University before joining the Geological Survey of Canada in 1966. He then moved to the Université de Montréal where he taught for nearly 30 years before returning to McGill in 1998 as an emeritus professor. I am sure I speak for everyone when I say I learned so much from him, and it was a privilege to have known him personally.

Brian R. Pratt
Department of Geological Sciences
University of Saskatchewan
Saskatoon, Saskatchewan S7N 5E2
Canada

Foreword

For a long time paleontologists did not bother to explore Precambrian rocks. It seemed futile to hammer overcooked rocks in search of skeleton-less and undersized fossils of some hypothetical microorganisms that might have been. Then, the first revolution had arrived: fossils could be found in Precambrian rocks. And not just below the base of the Cambrian, but even in the Archean. That revolution gave birth to *Precambrian Paleontology*. It produced a myriad of *Science* and *Nature* papers, helped to pave the way for the new scientific discipline of *Astrobiology* (now also a name of a high-impact scientific journal), and redefined the scope of paleontology.

As highlighted in this book, the second revolution is taking place now. Marc Laflamme, Jim Schiffbauer, and Stephen Dornbos have put together an edited volume on *Quantifying the Evolution of Early Life: Numerical and Technological Approaches to the Evaluation of Fossils and Ancient Ecosystems* that represents a timely tribute to that second coming of the Precambrian fossils. There is a growing realization that Precambrian fossils are informative enough and abundant enough to allow for application of sophisticated approaches, which have been long thought reserved for the high quality fossil records (produced by mollusks, arthropods, and other mundane bio-mineralized junk that litters quantitative data of ordinary paleontology).

The three editors have assembled here an array of chapters from the leading researchers in the field, who illustrate the informative value of the Precambrian fossil record. Two main themes are emphasized in the book: numerical methods and technological approaches.

First, Precambrian fossils are preserved frequently enough to produce quantitative data that even the late Jack Sepkoski would have likely found impressive. This Phanerozoic-like quality of the Precambrian fossil record is being increasingly appreciated by many paleontologists. Multiple chapters included here show that numerical approaches are a fruitful strategy when dealing with early fossils. Precambrian rocks can yield quantitative paleontological data. While many challenges remain, it is clear by now that early fossils can be used to develop numerical databases that are as credible (and as controversial, of course) as those derived from the Phanerozoic fossil record.

Second, it is now clear that Precambrian organisms are often preserved well enough to be explored in minute detail using sophisticated analytical instrumentation, including nanotechnological approaches that are being continuously developed for materials science. As shown here by multiple authors, an amazing wealth of data is contained within enigmatic and often dubious fossils. Once advanced instrumentation is applied, what may have looked like a meaningless blob often reveals itself as an organized and biologically identifiable structure. Likewise, an ambiguous outline on a bedding plane may expose itself as a spectacular find, once dissected in three dimensions by powerful instrumentation. And what may have appeared morphologically suspect, may become transparently obvious when attacked with modern geochemical tools. Let us also remember the fact that researchers who harness mysteries of advanced technologies, and usually do not give a damn about paleontology, are likely to appreciate Precambrian fossils as a formidable target: an exciting biological mystery and a logistic challenge with unforeseen intellectual rewards.

There are caveats, of course, to this enthusiastic story. As highlighted, explicitly or implicitly, in multiple chapters of this book, quantitatively and technologically advanced studies of Precambrian fossils remain challenging. We lack full understanding of biological affinity and anatomical meaning of most Precambrian fossils. And thus, we cannot simply transfer neontological tools of biology and other disciplines—as many Phanerozoic paleontologists do, for better or worse—to study enigmatic fossils of the Precambrian. And that means also that while many quantitative and analytical tools of the Phanerozoic paleontology can be applied to Precambrian fossils, a comparable interpretative confidence may not be justified. To the credit of the authors and editors, these caveats are voiced explicitly in many chapters of this book.

But, forget about the caveats! This book is a powerful tribute to the past achievements and the future promise of the *Precambrian Paleontology*. There are rapid advances in sophisticated instrumentation. There is an unexpected wealth of quantitative paleontological data that are being recovered from Precambrian rocks. To use the most horrible of all clichés, but justifiably so, “the best is yet to come”.

Michał Kowalewski
Virginia Tech

Preface

Paleontologists strive to unravel the mysteries of ancient life. Although many of us are most comfortable with a rock hammer in our hand and a camera at our side, the tools at our disposal have increased exponentially in this digital age. Statistical software packages are easily accessible and can run effortlessly on most personal computers, allowing for an increasing amount of paleontological studies relying on the interpretation of large fossil datasets. Furthermore, recent advancements in material sciences have expanded the arsenal of paleontologically-friendly instrumentation available for researchers. As these tools become increasingly available at research facilities, paleontologists have begun exploring their capabilities for the examination of fossil materials. These technologies have ultimately allowed for increased analytical power whether you wish to observe and interpret the microstructure of fossils on the sub-μm scale or determine their chemical make-up. In essence, these fields are expanding at such a rapid rate that it becomes difficult to keep up with of all of the research potential available. Our goal when we first started brainstorming about this book was to expand the breadth of understanding of numerical approaches and novel technologies available to both young and seasoned paleontologists alike, all the while making new tools more comprehensible so that they can be easily incorporated into existing paleontological research programs. Because most if not all of these tools are applicable to fossil types spanning the entire geological record, we opted to concentrate on one of the most exciting timeframes in Earth history, the Proterozoic-Cambrian transition.

The Proterozoic-Cambrian transition marks a fundamental change in the fossil record as organisms achieve, for the first time, complex macroscopic morphologies ideally suited for numerical studies. Although detailed qualitative studies dominate this time interval, recent research trends have favored quantitative approaches due in part to the fact that many of these fossils are taxonomically problematic and therefore are difficult to study from a purely qualitative means. Taxon-free numerical approaches have opened the door to novel interpretations of these fossil communities, and have been instrumental in showcasing similarities while highlighting differences between these assemblages and the familiar fossils of the Phanerozoic. Of possibly greater importance has been the incorporation of advanced microbeam technologies, including environmental scanning electron microscopy (ESEM), focused ion beam electron microscopy (FIB-EM), transmission electron microscopy (TEM), confocal laser scanning microscopy and Raman spectroscopic imagery, to the study of these fossils. The very enigmatic nature of organisms from this timeframe lends itself quite well to the expansion of advanced technological methods into the study of, for instance, fossil ultrastructure and microchemistry. These technologies have already revealed a wealth of information concerning the fossil composition and as such the likely natural history of fossils which had remained unclassifiable until now—and, without a doubt, more pioneering data are certain to come as these technologies evolve.

We went to great lengths to assemble under one banner some of the most widely used numerical approaches and novel instruments so as to juxtapose the recent advances in these fields while making these topics accessible to numerically-and-technologically-philic or phobic scientists alike. We understand the difficulty and frustration in approaching a novel methodology from an outsider's perspective, and hope to facilitate the transition into these approaches by providing a detailed description of a wide range of techniques separated into individual chapters written both by seasoned paleontologists and boundary-pushing newcomers. The chapters individually focus on describing the strengths and limitations of each approach and ultimately apply the chosen mode of attack to a real dataset. Each chapter is accompanied by an extensive reference list for further reading, which we hope will guide the reader to a broader range of applications of the technique while additionally providing alternate views or applications of the topic. We hope this collection of papers proves to be the first stop for scientists wishing to expand or diversify their research programs to include numerical techniques and broaden their experiences with advanced instrumentation—potentially saving both time and money!

Quantifying the Evolution of Early Life is divided into two separate parts, although in many instances individual chapters will incorporate aspects of both. The first part focuses on numerical methods while the second concentrates on technological approaches. In the numerical methods part, Chapter 1 by Matthew Clapham (*Ordination methods and the evaluation of Ediacaran communities*) describes three of the most widely used ordination methods, principal components analysis (PCA), detrended correspondence analysis (DCA), and non metric multidimensional scaling (NMDS), in order to evaluate Ediacaran paleocommunities. In Chapter 2, John Huntley (*Exploratory multivariate techniques and their utility for understanding ancient ecosystems*) extends this approach by introducing readers to several well-known ordination tasks including PCA, NMDS, DCA, in addition to Principal Coordinates Analysis (PCO), Discriminant Analysis (DA), and Canonical Variate Analysis (CVA) in order to evaluate Proterozoic microfossil assemblages. Chapter 3 by Marc Laflamme and Michelle Casey (*Morphometrics in the study of Ediacaran fossil forms*) compares traditional morphometric techniques that employ multivariate statistics such as PCA, PCO, and NMDS with geometric morphometric methods employing landmarks in order to study the size and shape variations in Ediacaran fossils. Lindsey Leighton (*Analyzing predation from the dawn of the Phanerozoic*) in Chapter 4 extends landmark studies and other techniques for analyzing predation and drillhole stereotypy in order to evaluate the role of predation in the Cambrian radiation. Leighton employs several Cambrian datasets to discuss the strengths and limitations of predation studies in the early fossil record.

In Chapter 5, Andrew Bush, Richard Bambach, and Douglas Erwin (*Ecospace utilization during the Ediacaran radiation and the Cambrian eco-explosion*) summarize a theoretical ecospace technique that is highly applicable to the study of ecosystems through time. Their case study focuses on Ediacaran and early Cambrian faunas and how they compare to modern communities in terms of tiering, motility, and feeding strategies. Chapter 6 by Katherine Marenco and David Bottjer (*Quantifying bioturbation in Ediacaran and Cambrian rocks*) reviews and describes a number of established methods for quantitatively studying bioturbation, both from bedding plane surfaces and vertical outcrops or core sections, in addition to describing a new grid-based technique directly applicable to early life studies. In Chapter 7, Sara Pruss and Hannah Clemente (*Assessing the role of skeletons in Early Paleozoic carbonate production: insights from Cambro-Ordovician strata, western Newfoundland*) continue the trend of ecosystem analysis by highlighting a simple and effective point-counting method to evaluate the relative contribution of various skeletal components to carbonate successions, and presents an excellent case study of a Cambrian-Ordovician succession in western Newfoundland. Rounding out the numerical methods section in Chapter 8 is Amelinda Webb and Lindsey Leighton (*Exploring the ecological dynamics of extinction*), whose chapter reviews techniques applicable to exploring both the taxonomic and ecological dynamics of extinction events by using an Early Cambrian data set.

The second part deals with technological approaches applied to fossil studies. Jonathan Antcliffe and Martin Brasier (*Fossils with little relief: using lasers to conserve, image, and analyze the Ediacara biota*) present a review of laser scanning and imaging techniques specifically applied to the Ediacara biota in Chapter 9. The authors showcase how laser scanning and imaging techniques are instrumental in conservation efforts when dealing with rare fossils and have the possibility of increasing access to rare or protected fossils through electronic distribution of digital versions of specimens. Continuing with laser technologies in Chapter 10, William Schopf and Anatoliy Kudryavtsev (*Confocal laser scanning microscopy and Raman (and fluorescence) spectroscopic imagery of permineralized Cambrian and Neoproterozoic fossils*) provide a thorough introduction to confocal laser scanning microscopy and Raman (and fluorescence) spectroscopic imagery specifically applied to the study of early prokaryotic and metazoan life. These techniques allow for accurate reconstruction of permineralized fossils in three-dimensions in addition to analyzing the chemical composition of the fossils and the embedding matrix. Chapter 11, presented by Patrick Orr and Stuart Kearns (*X-ray microanalysis of Burgess Shale and similarly preserved fossils*), follows with a systematic overview of the physical principles underpinning SEM-based x-ray mapping (energy dispersive x-ray spectroscopy), offer a detailed discussion of sample preparation and appropriate analytical methodologies, paying special attention to the effects of beam working conditions (i.e., current and accelerating voltage) on analysis of 2D Burgess Shale-type fossils. Sebastian Willman and Phoebe Cohen (*Ultrastructural approaches to the microfossil record: assessing biological affinities by use of transmission electron microscopy*), in Chapter 12, offer a detailed explanation of how TEM enables high resolution studies of cross-sections of specimens, while focusing on what information can (and cannot!)

be extracted from microfossils with controversial biological affinities. Continuing with sub-surface imaging technologies in Chapter 13 are James Schiffbauer and Shuhai Xiao (*Paleobiological applications of focused ion beam electron microscopy (FIB-EM): an ultrastructural approach to the (micro)fossil record*), focusing on high resolution analysis using the FIB-EM which combines the best aspects of SEM (surface) and TEM (sub-surface) analysis into a single powerful workstation. The authors illustrate a method of sequential sectioning, or FIB-EM nanotomography, which permits detailed 3D fossil reconstructions at unsurpassed spatial scales. In Chapter 14, Christian Hallmann, Amy Kelly, Neal Gupta, and Roger Summons (*Reconstructing deep-time biology with molecular fossils*) shift the focus to geochemical techniques by providing an overview of the use of organic biomarkers in Precambrian molecular paleontology. They outline how to identify likely sample contamination and properly interpret data, all the while providing a theoretical and practical introduction into the field of organic geochemistry as applied to biomarkers. Finally with Chapter 15, Kathleen McFadden and Amy Kelly (*Carbon and sulfur stable isotopic systems and their application in paleoenvironmental analysis*) provide a review of carbon and sulfur stable isotope geochemical systems focusing on their applications in the study of ancient paleoenvironments. These systems are then employed to evaluate a case study of the single largest negative carbon isotopic excursion, the Neoproterozoic Shuram anomaly.

This collection is by no means a comprehensive review of either numerical or technological approaches to paleontological study, but rather serves as a starting point for delving into these types of approaches in one's own research. As enigmatic (or familiar) fossils from this timeframe and beyond are further studied, our numerical approaches must undoubtedly adapt accordingly. Similarly, as advanced instrumentation evolves, so does the methodology required for data collection—and for that matter, the type of data acquirable. Therefore, we feel strongly that continued discussion and collaboration between scientists—regardless of field of study—is not only a necessity, but will allow these modes of paleontological science to grow and develop.

This book would not have been possible without the diligent help from expert reviewers who ensured that these articles represented the highest quality of research. Julie Bartley, Kelly Batten-Hender, Dave Bottjer, Andy Bush, Nickolas Butterfield, Michelle Casey, Brad De Gregorio, Greg Dietl, Philip Donoghue, Doug Erwin, Robert Gaines, Murray Gingras, Thomas Hegna, Pincelli Hull, John Huntley, Emannuelle Javaux, Linda Kah, Richard Krause, Srinath Krishnan, Arnold Miller, Phil Novack-Gottshall, Jonathan Payne, Matthew Powell, Peir Pufahl, William Schopf, Thomas Servais, Carrie Tyler, Herbert Volk, Amelinda Webb, and Shuhai Xiao are to be commended for their insightful and timely reviews. We would also like to thank Tamara Welschot and Judith Terpos at Springer who first approached us to organize this volume and who supported our vision at every turn.

Marc Laflamme
James D. Schiffbauer
Stephen Q. Dornbos

Contents

Part II Technological Approaches

Contributors

Jonathan B. Antcliffe
Department of Earth Sciences, University of Bristol, Wills Memorial Building, Queen's Road, Bristol, BS81RJ, UK
Jon.antcliffe@bris.ac.uk

Richard K. Bambach
Department of Paleobiology, MRC-121, National Museum of Natural History, Smithsonian Institution, P.O. Box 37012, Washington, DC 20013-7012, USA
bambachr@si.edu

David J. Bottjer
Department of Earth Sciences, University of Southern California, Los Angeles, CA 90089, USA
dbottjer@usc.edu

Martin D. Brasier
Department of Earth Sciences, University of Oxford, South Parks Road, Oxford, OX13AN, UK
Martin.Brasier@earth.ox.ac.uk

Andrew M. Bush
Department of Ecology and Evolutionary Biology and Center for Integrative Geosciences, University of Connecticut, 75 North Eagleville Road, Storrs, CT 06269-3043, USA
andrew.bush@uconn.edu

Michelle M. Casey
Department of Geology and Geophysics, Yale University, New Haven, CT 06520-8109, USA
michelle.casey@yale.edu

Matthew E. Clapham
Department of Earth and Planetary Sciences, University of California, Santa Cruz, CA 95064, USA
mclapham@ucsc.edu

Hannah Clemente
Department of Geosciences, Clark Science Center, Smith College,
Northampton, MA 01063, USA
hclement@smith.edu

Phoebe A. Cohen
Department of Earth and Planetary Sciences, Harvard University,
26 Oxford Street, Cambridge, MA 02138, USA;
Massachusetts Institute of Technology, NASA Astrobiology Institute,
Cambridge, MA, USA
pacohen@fas.harvard.edu

Douglas H. Erwin
Department of Paleobiology, MRC-121, National Museum of Natural History,
Smithsonian Institution, P.O. Box 37012, Washington, DC 20013-7012, USA;
Santa Fe Institute, 1399 Hyde Park Road, Santa Fe, NM 87501, USA
erwind@si.edu

S. Neal Gupta
Department of Earth, Atmospheric and Planetary Sciences,
Massachusetts Institute of Technology, 77 Massachusetts Ave,
Cambridge, MA 02139, USA;
Indian Institute of Science Education and Research, Mohali, India
ngupta@ciw.edu

Christian Hallmann
Department of Earth, Atmospheric and Planetary Sciences,
Massachusetts Institute of Technology, 77 Massachusetts Ave,
Cambridge, MA 02139, USA
hallmann@mit.edu

John Warren Huntley
GeoZentrum Nordbayern, Friedrich-Alexander Universität
Erlangen-Nürnberg, Erlangen, Bavaria, Germany
jhuntley@vt.edu

Stuart L. Kearns
Department of Earth Sciences, University of Bristol, Queen's Road,
BS8 1RJ Bristol, UK
stuart.kearns@bristol.ac.uk

Amy E. Kelly
School of Earth and Space Exploration, Arizona State University,
85281-1404, Tempe, AZ, USA
kellya@alum.mit.edu

Anatoliy B. Kudryavtsev
Department of Earth and Space Sciences, Institute of Geophysics and Planetary Physics (Center for the Study of Evolution and the Origin of Life), and Molecular Biology Institute, University of California, Los Angeles, CA 90095, USA;
PennState Astrobiology Research Center, 435 Deike Building,
University Park, PA 16802, USA
Kudryavtsev@ess.ucla.edu

Marc Laflamme
Smithsonian Postdoctoral Fellow, Department of Paleobiology,
Smithsonian Institution, National Museum of Natural History,
Washington, DC 20013-7012, USA
laflammem@si.edu

Lindsey R. Leighton
Earth and Atmospheric Sciences Department, University of Alberta,
Edmonton, AB, Canada T6G 2E3
lleighto@ualberta.ca

Katherine N. Marenco
Department of Geology, Bryn Mawr College, Bryn Mawr,
PA 19010, USA
kmarenco@brynmawr.edu

Kathleen A. McFadden
ConocoPhillips, Houston, TX 77079-1100, USA
meikailing@yahoo.com

Patrick J. Orr
UCD School of Geological Sciences, University College Dublin,
Belfield, Dublin 4, Ireland
patrick.orr@ucd.ie

Sara B. Pruss
Department of Geosciences, Clark Science Center, Smith College,
Northampton, MA 01063, USA
spruss@smith.edu

James D. Schiffbauer
Department of Geosciences, Virginia Polytechnic Institute and State University,
4044 Derring Hall, Blacksburg, VA 24061, USA
jdschiff@vt.edu

J. William Schopf
Department of Earth and Space Sciences, Institute of Geophysics
and Planetary Physics (Center for the Study of Evolution and the Origin of Life),
and Molecular Biology Institute, University of California, Los Angeles,
CA 90095, USA;
PennState Astrobiology Research Center, 435 Deike Building,

University Park, PA 16802, USA
schopf@ess.ucla.edu

Roger E. Summons
Department of Earth, Atmospheric and Planetary Sciences,
Massachusetts Institute of Technology, 77 Massachusetts Ave,
Cambridge, MA 02139, USA
rsummons@mit.edu

Amelinda E. Webb
Department of Geology and Geophysics, Yale University,
New Haven, CT 06511, USA
amelinda.webb@yale.edu

Sebastian Willman
Department of Earth Sciences, Palaeobiology, Uppsala University,
Villavägen 16, Uppsala SE-752 36, Sweden
Sebastian.Willman@geo.uu.se

Shuhai Xiao
Department of Geosciences, Virginia Polytechnic Institute and State University,
4044 Derring Hall, Blacksburg, VA 24061, USA
xiao@vt.edu

Part I
Numerical Methods

Chapter 1
Ordination Methods and the Evaluation of Ediacaran Communities

Matthew E. Clapham

Contents

Abstract Analysis of paleocommunity data poses a challenge because of its multivariate nature, containing counts of many species in many samples. Comparison of the abundance of a single species among all samples provides only incomplete information, whereas attempting to consider every species is impractical. Ordination methods are analytical techniques that reduce the original multivariate dataset to a few important components by creating new synthetic variables designed to explain the maximum amount of original data variability. The ultimate goal is to order the samples along ecologically or environmentally meaningful gradients in order to interpret differences in community structure. This chapter describes three of the most widely-used ordination methods, principal components analysis (PCA), detrended correspondence analysis (DCA), and non-metric multidimensional scaling (NMDS), explaining the methodology of each and outlining their strengths and weaknesses for analysis of paleoecological data. The techniques are illustrated using Ediacaran paleocommunity data from Mistaken Point, Newfoundland. PCA relies on assumptions that are inappropriate for ecological data, such as the requirement that species abundances change in a linear fashion along the environmental

M.E. Clapham (✉)
Department of Earth and Planetary Sciences, University of California,
Santa Cruz, CA 95064, USA
e-mail: mclapham@ucsc.edu

M. Laflamme et al. (eds.), *Quantifying the Evolution of Early Life*, Topics in Geobiology 36,
DOI 10.1007/978-94-007-0680-4_1, © Springer Science+Business Media B.V. 2011

gradient, and is not well suited for community ordination. In contrast, DCA and NMDS both perform well with ecological data; DCA incorporates a more ecologically-realistic measure of distance between samples but some of the detrending methods have been criticized, whereas NMDS only assumes a monotonic relationship between compositional similarity and gradient distance. The two methods also have complementary strengths, with DCA typically better at extracting the primary gradient and NMDS better at resolving the overall pattern.

Keywords Paleoecology • Principal components analysis • Correspondence analysis • Detrended correspondence analysis • Non-metric multidimensional scaling

1.1 Introduction

Ecologists and paleoecologists are commonly faced with large datasets containing many variables from many samples (multivariate data), with the ultimate goal of revealing and interpreting ecologically significant patterns in those data. In paleoecology these datasets often take the form of an occurrence matrix of presence/absence or relative-abundance counts of perhaps 10 to >100 taxa (variables) from as many as 10–100s of collection sites. Analysis of between-sample similarities in a single variable (e.g., the abundance of a single taxon) is inadequate at capturing the full range of variability in the original dataset. However, the original dataset also contains a great deal of redundant information as, for example, several taxa can respond in a similar fashion to the same environmental gradient (Palmer 1993). Thus, it is almost always desirable to reduce the number of variables, typically to the two or three most important components, allowing the data to be displayed and interpreted more easily than by examining the dozens of variables in the original data.

A wide range of different analytical methods have been devised for reducing complex multivariate datasets so that they can be displayed in two or three dimensions (James and McCulloch 1990; Shi 1993). They are collectively called ordination techniques and in ecology and paleoecology they typically are used for gradient analysis – identifying environmental gradients and placing species in their correct position along those gradients (Bray and Curtis 1957). Environmental gradients, such as depth (more precisely, depth-related variables such as energy and substrate) in marine communities, exert a fundamental control on the spatial distribution of species and, as a result, are the primary influence on community composition (Cisne and Rabe 1978; Holland et al. 2001; Holland 2005). Other important gradients may include biogeographic variation, other environmental changes such as salinity (in estuarine environments, for example), temporal changes in community structure due to origination/immigration and extinction/emigration, or ecological succession as argued by Clapham et al. (2003).

This chapter will focus on three techniques for reconstructing environmental gradients and analyzing community structure: (1) Principal Components Analysis (PCA), (2) Correspondence Analysis (CA) and its relative Detrended Correspondence Analysis (DCA), and (3) Non-Metric Multidimensional Scaling (NMDS). Other methods (Polar Ordination and Principal Coordinates Analysis) will be discussed briefly but are less commonly applied by paleoecologists. These are all indirect gradient analysis methods, as they use the distribution of species among the sites to infer the original environmental gradients that structured those sites (Minchin 1987; Palmer 1993). In contrast, direct gradient analysis techniques (e.g., Canonical Correspondence Analysis) explicitly relate species composition to measured environmental parameters (Palmer 1993), but are less applicable to paleontological data where environmental parameters typically are not directly measured or even estimated. The key differences between these three indirect gradient analysis methods are the mechanisms by which they quantify sample dissimilarity and the assumptions inherent in relating dissimilarity to separation along the environmental gradient (Faith et al. 1987). The implications of those differences will be discussed in greater detail below, with the ultimate goal of answering the question: which ordination method should you use to analyze your data?

1.2 Dataset Summary

This chapter will use a dataset of relative abundances from Ediacaran communities at Mistaken Point (Table 1.1), previously used by Clapham et al. (2003), to explore the procedures, assumptions, and application of different multivariate techniques (see Chap. 3 for a more detailed description of the fossil sites). The occurrence matrix contains abundance counts of 16 taxa from nine fossiliferous bedding planes, including three outcrops of the well-known "E" surface. All abundances

Table 1.1 Occurrence matrix showing abundance of 16 taxa in nine Mistaken Point samples

Sample/Taxon	PC	BC	LMP	D	Ey	Eq	Ewc	G	SH
Aspidella	0	0	0	1	3	0	1	0	1
Bradgatia	0	1	0	76	226	15	14	55	10
Charnia	18	0	152	20	84	5	14	4	1
"Charnia" B	0	26	58	0	0	0	0	0	0
Charniodiscus	0	0	13	0	1,398	52	35	49	3
"Dusters"	0	0	4	8	612	28	35	5	0
Fractofusus "Spindle"	0	76	4	1,169	1,497	71	100	0	0
Hiemalora	0	0	1	0	21	1	0	0	0
"Holdfast Frond"	0	0	0	0	0	0	0	22	0
Ivesheadia ("*Ivesia*")	79	1	1	7	77	5	3	8	27
"Lobate Disc"	0	0	0	0	142	3	10	0	0
Hapsidophyllas "Network"	0	0	0	1	1	0	2	0	0
"Ostrich Feather"	0	0	62	0	0	0	0	0	0
Pectinifrons "Pectinate"	0	0	0	175	0	0	0	0	304
"Spoon Frond"	0	0	0	0	1	1	1	0	0
Thectardis "Triangle"	140	0	0	0	25	3	7	0	0
Total	237	104	295	1,457	4,087	184	222	143	346

were standardized to percent abundance by normalizing by the total number of specimens per sample, as many of the analyses are sensitive to variations in absolute abundance among sites that reflect sampling intensity rather than an actual ecological parameter. Most taxa are identified to genus level but several are polyphyletic form taxa: the "dusters" category likely encompasses two or more undescribed genera, while *Charnia* counts also include the recently-described genus *Beothukis*, which was not differentiated in the original data collection of Clapham et al. (2003). All of the ordination techniques will be performed on the same occurrence matrix to facilitate understanding of their methodology and enable comparison of their results. Most analyses shown here were performed with the free software package PAST (Hammer et al. 2001) but can be done with a variety of statistical programs.

1.3 Data Standardization

The first important choice to be made, even before deciding on the appropriate ordination technique, is whether to analyze the raw abundance counts or whether to standardize them. All of the analyses in this chapter will be performed after converting each raw abundance value to percent abundance, normalizing the data to eliminate the influence of variations in total sample size. Otherwise, it is possible that two samples may be erroneously grouped together because rare species in the large sample have a similar raw abundance to common species in the small collection. This standardization was chosen because sample size in the Mistaken Point dataset is a reflection of sampling intensity, rather than having ecological significance. This is the case in many, but not all, paleoecological datasets, and investigators will need to consider whether variations in total abundance have biological significance (in which case they should not convert the data to proportions) or reflect sampling intensity.

It is also possible to normalize species to their maximum abundance so that the species relative abundances all sum to a value of 1 (or 100%), or to perform a double standardization by both site and species totals (Faith et al. 1987). Standardization by species total abundance strongly weights rare species and reduces the influence of abundant taxa, which may not be appropriate unless the species occur at similar abundance or different trophic levels are present and the investigator wishes to account for the rarity of carnivores relative to primary producers, for example Jongman et al. (1995).

1.4 Ordination Methods

1.4.1 Principal Components Analysis (PCA)

Principal Components Analysis is one of the oldest techniques used in multivariate data ordination and was the first to be applied to ecological data (Goodall 1954).

It still enjoys some use in paleoecology (e.g., Rodriguez 2004; Botquelen et al. 2006; McKinney and Hageman 2006) despite having several attributes that are not ideally suited to certain types of ecological data (Gauch and Whittaker 1972; Fasham 1977; Minchin 1987). The most severe issue, discussed in more detail below, is the distortion induced by fitting a linear (Euclidean) distance model to non-linear ecological gradients (Gauch and Whittaker 1972; Minchin 1987). As a result, PCA should not be used for analysis of ecological (species count) data. It is instructive, however, to examine the analytical procedure used in PCA because it illustrates the general principles of ordination, in particular eigenanalysis-based multivariate ordination. PCA, like all ordination, seeks to summarize the original multivariate dataset, which can often contain dozens of species at multiple sites, in a smaller number of new, synthetic variables – the "principal components" in the method's name (James and McCulloch 1990). It is based on linear algebra, using a technique called eigenanalysis to rigidly rotate the original data matrix (i.e., changing the coordinate system without altering the relative position of the points) so that the new principal component axes account for the maximum amount of variability in the original data (Quinn and Keough 2002). The underlying mathematical basis is complex but the basics of PCA and other eigenanalysis methods can be explained graphically by considering a simplified example with only two variables (Fig. 1.1).

In this simple example, analysis of ecological data begins with a bivariate plot in which the axes are the abundance of species 1 and species 2, and the data points represent each sample in the study (Fig. 1.1a). The data are then "centered" by subtracting the mean species abundance from the abundance in each sample; this has the effect of moving the points so that the graph axes run through the mean of each species (Fig. 1.1b). The final step rotates the axes so that one axis, principal component 1, is aligned with the maximum variance of the sampled points (Fig. 1.1c). In fact, the new axes are simply obtained by a linear regression through the data points. Principal component 2 is created in a similar fashion and accounts for the maximum amount of remaining variance with the constraint that it is perpendicular to axis 1. This procedure can easily be conceptually generalized to

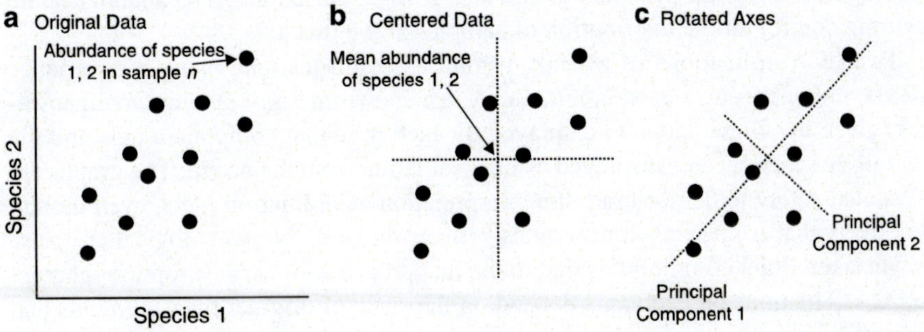

Fig. 1.1 Simplified two-variable example conceptually illustrating the process involved in PCA. (**a**) The abundance of each species is plotted on a bivariate (or multivariate) plot. (**b**) Data points are centered so that the axes pass through the mean point of each species. (**c**) Axes are rotated to align with the direction of maximum variability in the original data

multivariate datasets: principal component 1 is the direction of maximum variance in the multidimensional cloud of points and additional axes are generated to explain the maximum remaining variance while remaining perpendicular to all other axes.

In practice, PCA solutions are not derived graphically but are instead calculated using a linear algebra technique involving spectral decomposition of an association matrix to extract the eigenvectors and eigenvalues (these two terms will be explained below). The association matrix defines the interrelationships between the samples based on the covariance or correlation between variables (covariance and correlation are mathematical metrics of the degree to which changes in two variables are associated) (Quinn and Keough 2002). It is important to note that PCA with the covariance matrix will often give significantly different results from PCA with a correlation matrix; however, because PCA is not appropriate for ecological data analysis, the details are not important for this chapter. In brief, the correlation matrix is calculated from standardized data, removing the effects of differences in the variances of the original variables (Quinn and Keough 2002). As such, it is useful for analyzing variables measured in different units or with different scales, whereas the covariance matrix (sensitive to differences in variance) is more appropriate when differences in the variances have biological significance.

Regardless of the choice of association matrix, PCA will produce a final plot with several principal component axes ("eigenvectors") each explaining a certain amount of the variance present in the initial data (quantified by the axis "eigenvalue"). Because the principal component axes are linear regressions through the data cloud, each axis can be defined by a linear combination of the analyzed variables. For example,

$$\mathrm{PC1} = c_1X_1 + c_2X_2 + \ldots + c_nX_n$$

The coefficients c_n are termed "loadings" and quantify the importance of each variable in influencing the sample's position along the principal component axis. In the case of community analysis, they indicate the degree to which a sample's position is influenced by the abundance of each species (X_n). When the loading of a species on a particular principal component is high, variations in its abundance are a strong control on the distribution of samples along that axis.

Two PCA ordinations of Mistaken Point assemblages, one using a covariance matrix and one using a correlation matrix, are shown in Fig. 1.2 with the eigenvalues (given the Greek letter λ) displayed by each principal component axis and the loadings of each species displayed as lines radiating from the origin. The graphs are presented solely to demonstrate the interpretation of ordination plots, even though it is likely that neither result accurately reflects the underlying ecological structure of Mistaken Point communities due to the inappropriate linear assumption inherent to PCA. The units on each axis depend on the scale of original measurements and are not of importance for data interpretation. Note that the amount of variance explained by each principal component is low (no more than 40%) – a result that is typical of ecological data ordination. Explained variance decreases with successive principal components so only the first two, or sometimes three, axes are typically

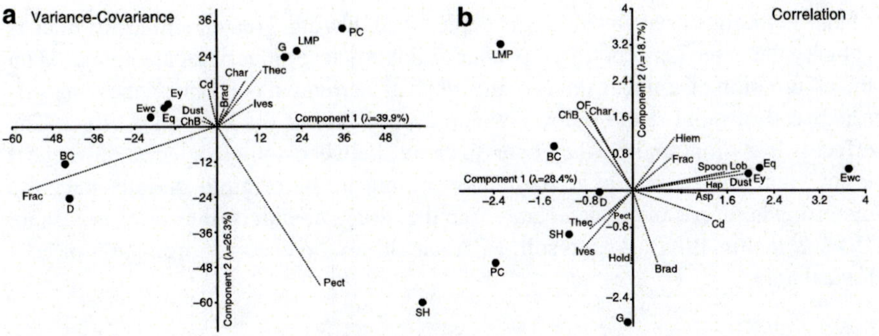

Fig. 1.2 Principal Components Analysis (PCA) ordination of Mistaken Point data (only PC1 and PC2 are shown for simplicity). (**a**) Ordination using the variance-covariance matrix. Samples are indicated by *black circles* and taxon loadings are indicated by *dotted lines* with the length of the *line* indicating the degree of loading on each axis. (**b**) Ordination using the correlation matrix

interpreted in ecological studies, although it is possible that lower order axes may also contain important information. In the covariance analysis (Fig. 1.2a), *Fractofusus* ("spindles") and *Pectinifrons* ("pectinates") have high loadings (indicated by longer lines radiating from the origin) on axes 1 and 2, primarily because they account for 57% of the variance in the original data. This indicates that variations in their abundance are the primary controls on the position of samples along those axes – for example, principal component 1 is essentially the abundance of *Fractofusus*, from high abundance at negative PC1 scores to low abundance at positive PC1 scores. Variations in the abundance of rare species (*Aspidella*, lobate discs, *Hapsidophyllas* ("networks"), "holdfast fronds", etc.) make essentially no contribution to the structure of the ordination plot. In contrast, all species contribute equally to the correlation PCA due to the effects of variable standardization (small variations in the length of the loadings are an artifact of displaying only two axes of a multidimensional diagram), which gives as much weight to the abundant *Fractofusus* as it does to *Hiemalora*, only present on three surfaces and never accounting for more than 0.5% of the community.

As mentioned above, it is likely that neither PCA ordination accurately represents the actual ecological structure of Mistaken Point communities. That error arises because PCA requires the assumption of an underlying linear relationship between variables (Euclidean distance) – a severe problem that results in a distorted representation of the original ecological gradient when applied to community data (Gauch and Whittaker 1972; Minchin 1987). Many studies have documented curved unimodal variations in taxon abundance along ecological gradients (depth, altitude, etc.) and a non-linear relationship between compositional similarity and distance between samples along the gradient, with the rate of decrease in similarity reduced at greater distances (Minchin 1987). Because PCA attempts to fit this non-linear relationship to an underlying assumption of linearity, the resulting PCA ordination is distorted and likely does not reflect the actual ecological gradients. One common distortion is the characteristic "horseshoe" shape, in which the ends of the gradient curve inwards are depicted as having greater similarity than is actually the case. This contrasts to other ordination techniques that are not based on the assumption of a linear relationship, such as detrended correspondence analysis (discussed in more detail below), which do not exhibit the horseshoe effect. The effect is less pronounced when beta diversity (between-sample species difference) is low, but when multiple gradients are present in the original sample space the distortion is severe and undetectable, and the results cannot be interpreted (Fasham 1977; Minchin 1987). As a result, PCA should not be used for ordination of ecological data.

1.4.2 Correspondence and Detrended Correspondence Analysis (CA/DCA)

The recognition of faults in PCA methodology when applied to community data (e.g., Gauch and Whittaker 1972) led to the development of new techniques for ecological ordination. Correspondence analysis, the underlying mechanism for DCA, is an eigenanalysis technique similar to principal components analysis (CA can also be calculated through an iterative process called reciprocal averaging) (Hill and Gauch 1980). Whereas PCA conceptually seeks to rotate axes through the multidimensional cloud of sample points to maximize the variance explained, CA can be thought of as rotating the axes simultaneously through both species and sample points to maximize the correspondence between the two. In matrix algebra terminology, this simultaneous axis rotation means that eigenvectors for species and samples are extracted jointly and allows samples and species to be plotted directly on the same axis instead of separately determining loadings of species onto sample axes. Correspondence analysis performs the same matrix decomposition techniques as PCA but on the original data after data transformation to yield chi-square distances rather than on an association matrix of correlation or covariance between sites (Chardy et al. 1976; Faith et al. 1987). As a result, many features of the output of CA are analogous to those in PCA, only differing due to the use of chi-square distance rather than Euclidean distance (Kenkel and Orloci 1986; Faith et al. 1987). The sum of CA eigenvalues measures the lack of independence between species and samples in the original data set (the overall chi-squared statistic divided by total frequency – called "inertia"); in PCA they measure the total variance in the original data set. As in PCA, successive axes account for progressively less inertia, with the first axis explaining a relatively high proportion of the lack of independence between species and samples.

This chi-square transformation of the original data matrix is a key aspect of CA and is one reason why it is more suitable for ordination of ecological data (Faith et al. 1987). Chi-square distance does not assume a linear relationship between sample dissimilarity and ecological distance, an assumption inherent in PCA that is rarely met by ecological data (Faith et al. 1987; Minchin 1987). As a result, CA correctly ordinates samples from the ends of the environmental gradient along the primary axis, avoiding the "horseshoe effect" common in PCA. The chi-square distance has other limitations, however, and is not robust to variations in the underlying structure of the ecological gradient when taxon abundances deviate from a unimodal normal distribution (Faith et al. 1987), a relatively common feature in natural ecosystems (Oksanen and Minchin 2002). It tends to overweight species with low abundance, exaggerating the distance of samples containing more rare taxa (Minchin 1987), although different software packages contain options for downweighting their importance (Hill and Gauch 1980; Holland et al. 2001; Scarponi and Kowalewski 2004). Chi-square distance also does not reach a constant maximum when samples with no compositional overlap are compared, instead varying depending on the relative occurrence of common and rare species in those samples (Minchin 1987).

Despite the limitations of the chi-square distance, correspondence analysis often performs well at reconstructing the primary ecological gradient (Gauch et al. 1981; Olszewski and West 1997), although there are two potential artifacts of the CA process that may hamper interpretation. First, samples are often compressed at either end of the gradient relative to the middle (Hill and Gauch 1980) because taxon breadths along the gradient are artificially truncated and narrowed at the edge of the study area (Peet et al. 1988). Second, and more significantly, CA does not produce additional axes that are independent of prior axes (although they must not be linearly correlated, they can have a non-linear relationship) and the second axis often appears as a quadratic function of the first (due to projection of a multidimensional dataset into two dimensions) – termed the "arch effect" (Hill and Gauch 1980). Two separate techniques have been devised to correct for these artifacts in CA: detrending to eliminate the arch effect on higher-order axes, and rescaling to reduce compression at the ends of axes. The resulting ordination is called detrended correspondence analysis (DCA) (Hill and Gauch 1980).

Detrending uses a running window to divide the primary axis into segments; all points within a segment are centered by subtracting the mean axis 2 score in that segment from each axis 2 value (Fig. 1.3) (Hill and Gauch 1980). This technique

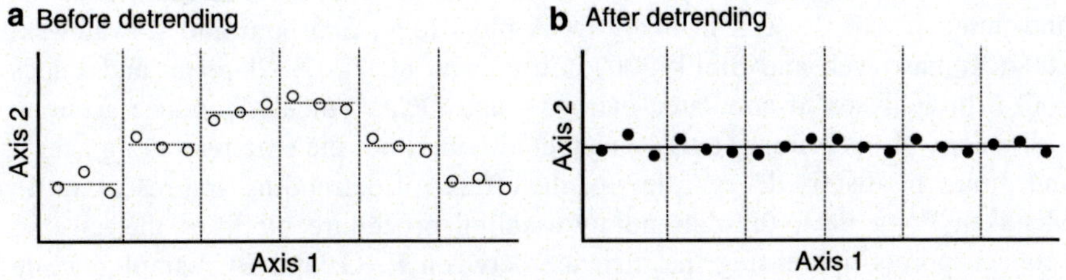

Fig. 1.3 Example illustrating the detrending procedure used in DCA. (**a**) Axis 1 is divided into segments (*dashed lines*) and the original points in each section (*open circles*) are centered by subtracting the mean axis 2 value for that segment (*dotted line*). (**b**) The resulting detrended points are shown by *filled circles*. The actual detrending procedure is more complicated because it uses overlapping running segments and is performed in conjunction with rescaling (Modified from Hill and Gauch (1980))

removes any arch effect (by distorting the ordination plot), regardless of whether the arch is an artifact or a real feature of the ecological structure (Kenkel and Orloci 1986; Minchin 1987; Wartenberg et al. 1987; Olszewski and West 1997). Detrending can also be sensitive to the number of segments in some cases (Jackson and Somers 1991), although the default value (26 segments) often improves the ordination (Peet et al. 1988; Knox 1989).

DCA also uses a mathematical rescaling process to remove the effects of gradient compression near the ends of axes, assumed to be an artifact (Hill and Gauch 1980). It requires that the abundances of all taxa are normally distributed along the gradient with equal variances (i.e., that species composition changes at the same rate along the gradient) and adjusts the position of species along the axis to maintain a constant within-sample variance of species scores at each point (Hill and Gauch 1980; Peet et al. 1988). This has the benefit of rescaling the axis so that distance can be directly interpreted in terms of compositional change (Peet et al. 1988) but has been criticized because of its potentially unjustified assumption of constant rates of taxon turnover (Wartenberg et al. 1987). Species abundances along a gradient are not always Gaussian (Oksanen and Minchin 2002) and different species response curves may have different variances (Holland 2005).

The choice of whether or not to detrend the data has been intensely debated (Minchin 1987; Wartenberg et al. 1987; Peet et al. 1988; Knox 1989) and CA (Cisne and Rabe 1978; Olszewski and West 1997) and DCA (Holland et al. 2001; Scarponi and Kowalewski 2004; Holland 2005; Zuschin et al. 2007; Clapham and James 2008) have both been used in paleoecological data analysis. Despite the arbitrary and perhaps unfounded assumptions inherent to DCA (Wartenberg et al. 1987), the detrending and rescaling process may yield more accurate and interpretable ordination results, especially of long ecological gradients (Peet et al. 1988; Knox 1989) or in the case of rapid ecological turnover (Kenkel and Orloci 1986), although in other cases they produce distortions in the underlying gradient (Kenkel and Orloci 1986; Minchin 1987). In particular, DCA ordination is often twisted so that residual variation is actually spread over axes 2 and 3, resulting in the characteristic wedge shape, where variation on axis 2 is maximal near the mid-point of axis 1, seen in many DCA plots (e.g., Scarponi and Kowalewski 2004; Tomasovych and Siblik 2007; Zuschin et al. 2007; Clapham and James 2008). In analyses of simulated data, CA and DCA typically produce accurate ordinations of the primary axis (Gauch et al. 1981) but the structure along axes 2 and 3 may be distorted, regardless of the effects of detrending and rescaling. In Mistaken Point data, the detrending/rescaling procedure modifies the spacing between points, increasing the distance between E, G, and BC samples while bringing SH and PC samples closer to the main cluster, but does not greatly alter the relative position of the samples (Fig. 1.4). The main difference is in the position of the PC sample, which is likely difficult to ordinate due to its low species count (only three taxa). Overall, CA and DCA both appear to result in a reasonable ordination of Mistaken Point surfaces.

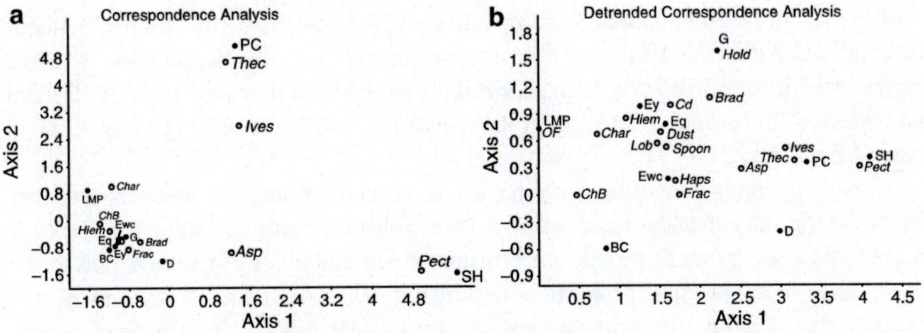

Fig. 1.4 Correspondence Analysis (**a**) and Detrended Correspondence Analysis (**b**) ordinations of Mistaken Point data (only the first two axes are shown for simplicity). Sample positions are shown by *black filled circles* and species positions by *gray filled circles*

1.4.3 Non-Metric Multidimensional Scaling (NMDS)

Eigenanalysis ordination methods such as PCA or DCA are metric techniques that stipulate a proportional relationship between compositional change and distance along the gradient (Kenkel and Orloci 1986; Minchin 1987). As discussed above, ordination results may be distorted when the actual relationship deviates from the idealized assumptions. Another family of techniques compares between-sample similarity using quantitative distance metrics (discussed below) rather than taking an eigenanalysis approach. This section will focus primarily on non-metric multidimensional scaling, the most widespread of these techniques, but it is useful to briefly mention two other distance-based ordination methods: polar ordination and principal coordinates analysis (PCoA).

Polar ordination (also called Bray-Curtis ordination) is so named because it projects all samples onto an axis using their relationship to two selected reference points ("poles"), as calculated by the chosen distance metric (Bray and Curtis 1957; Cisne and Rabe 1978; Beals 1984). In the original formulation of Bray and Curtis (1957), the two data points with the maximum separation were chosen as the reference points; later workers proposed a variety of methods for selecting reference points (Beals 1984). Higher-order axes can be constructed from a matrix of residual distances (i.e., the remaining distance not accounted for by the first axis), producing axes that are perpendicular and uncorrelated to the primary axis (Beals 1984). The comparison of all samples to only two reference points, rather than analyzing all intersample differences, may seem like a limiting constraint but polar ordination appears to perform well at ecological ordination (Gauch and Whittaker 1972; Beals 1984).

Principal Coordinates Analysis (PCoA) attempts to maximize the linear correlation between actual intersample distances (using a user-chosen distance metric) and those distances projected onto a two- or three-dimensional space (Gower 1966). It is sometimes called "metric multidimensional scaling" because it is similar to non-metric multidimensional scaling but uses a parametric (or "metric") linear correlation. Although PCoA is able to use an ecologically-appropriate distance metric, the linear correlation imparts similar distortion to that seen in PCA (Podani and Miklós 2002) and PCoA generally performs poorly at ordinating ecological data (Minchin 1987).

Non-metric multidimensional scaling differs markedly from those metric approaches by assuming only that the relationship between distance and sample dissimilarity is monotonic (i.e., increasing rank-order distance corresponds to increased rank-order dissimilarity) rather than proportional (Kruskal 1964; Fasham 1977; Kenkel and Orloci 1986; Clarke 1993). In this manner, rank-order NMDS is analogous to non-parametric classical statistics. NMDS uses this rank-order agreement between the calculated solution and the original dataset, a value called "stress," to assess the quality of ordination results. Analysis begins with a random configuration of points (or a user-specified configuration, such as the results of another ordination method) and performs an iterative optimization procedure to refine the solution until the stress reaches a minimum (Clarke 1993). The algorithm may reach a local stress minimum rather than identifying the globally best solution (Fasham 1977; Kenkel and Orloci 1986; Minchin 1987; Clarke 1993), but this is rare when the starting configuration is random and can generally be overcome by selecting several random starting points and comparing the final stress to confirm that the optimal solution has been reached (Clarke 1993). The lowest stress indicates the best rank-order agreement between calculated solution and original matrix, with a stress of zero indicating perfect rank-order agreement and stress values less than 0.1 corresponding to a good representation that can reliably be interpreted (Clarke 1993). Higher stress values can still be interpreted, but with greater caution, although stresses greater than 0.2 indicate a poor to nearly random ordering of points (Clarke 1993).

Another fundamental difference of NMDS is the ability to select from a variety of similarity or distance coefficients, whereas PCA is based on Euclidean distance and DCA is constrained to use chi-square distance to calculate sample dissimilarity (Faith et al. 1987). This is a significant benefit because Euclidean (and perhaps also chi-square) distance often perform poorly with ecological data (Faith et al. 1987). A myriad of similarity or distance metrics have been proposed, each with benefits and drawbacks for specific data types (Faith et al. 1987; Shi 1993). Studies with various simulated ecological gradients have suggested that the Bray-Curtis, Kulczynski, and Relativized Manhattan (i.e., each species standardized to percent abundance within a sample and normalized to its maximum abundance in any sample) coefficients produce the closest rank-order and linear correlation between compositional dissimilarity and ecological distance along the gradient under most conditions (Faith et al. 1987). Although some coefficients perform better than others with simulated ecological data, there is no objective guide to choosing a distance metric (see Faith et al. (1987) for a detailed review of several distance measures). The Bray-Curtis coefficient was designed for interpretation of ecological data (Bray and Curtis 1957) and performs well on simulated datasets (Faith et al. 1987). Because it enjoys widespread use and is intuitive – simply calculate the absolute value of the difference in abundance of a species (X_i–X_j) between two samples divided by the total abundance of that species in the two samples (X_i+X_j), and sum the results for all species – it is a good choice for use in NMDS. It is by no means the only possibility, as other metrics perform well and yield robust solutions, but Bray-Curtis similarity will be used in the NMDS ordination in this chapter. Other distance metrics, such as the Kulczynski coefficient (Faith et al. 1987), yield very similar solutions.

In contrast to the eigenanalysis methods, NMDS simply arranges the points in *n*-dimensional space (where *n* is a number of dimensions, usually two or three, chosen by the user) rather than extracting multiple orthogonal axes that explain decreasing amounts of variance (Clarke 1993). This has two major implications for viewing and interpreting NMDS ordinations. First, axes in an NMDS ordination have no specific meaning and the data plot can be rotated, translated, or scaled without altering the results, as long as the relative position of the points remains unchanged (Clarke 1993). For example, the ordination is unchanged after rotating points so that the direction of maximum variability is horizontal to facilitate visual comparison with DCA or PCA plots (e.g., Tomasovych and Siblik 2007). Second, as the NMDS algorithm specifically arranges the samples to fit in *n*-dimensional space, the best solution for two dimensions will not necessarily be a two-dimensional projection of the best solution for three dimensions (Clarke 1993). This contrasts with eigenanalysis methods, where many eigenvectors are calculated and a plot of axis 1 vs. axis 2 will simply be a two-dimensional projection of the multivariate data cloud. Although it is rare to require more than three dimensions, how should the number of dimensions be chosen? The iteration will yield a better solution (i.e., lower stress) as additional dimensions are included because the original data-set is multivariate, until the number of dimensions is one less than the number of samples. However, a plot of stress against the dimensionality (called a "scree plot," available in some software packages) will often exhibit a distinct break in slope, providing a rough guideline for the minimum number of dimensions that should be analyzed (Kruskal 1964). It is also reasonable to interpret one extra dimension if the ordination results in additional structure (Kruskal 1964). One caveat: the resulting ordination plot may be distorted when the specified dimensionality is greater than the actual number of dimensions in the original environmental gradient (Austin 1976).

NMDS was adopted slowly due to its intensive computational requirements but the calculations are now routine and it is used widely in paleoecological analysis (Olszewski and West 1997; Bonuso et al. 2002; Clapham et al. 2003; Dominici and Kowalke 2007; Tomasovych and Siblik 2007; Zuschin et al. 2007). Many studies have shown that NMDS produces more accurate ordinations than CA/DCA, for simulated ecological data at least (Fasham 1977; Kenkel and Orloci 1986; Minchin 1987), although each technique has individual strengths and weaknesses. In particular, NMDS is less susceptible to distortion when beta diversity is high and when there are multiple underlying gradients with different beta diversities (Fasham 1977). The non-metric technique may be less adept than the eigenanalysis techniques at delineating clusters within the data, but may perform better at optimizing relative distance between samples even when there are clusters (Kenkel and Orloci 1986). As discussed more below, NMDS may be of reduced use when sample size is small;

in analyses with fewer than 30 samples, use of rank-order may result in significant loss of information (Anderson 1971). Overall, NMDS is a robust ordination technique that performs excellently with ecological data.

Figure 1.5 shows NMDS ordination plots for Mistaken Point data using the Bray-Curtis similarity coefficient, based on 2D and 3D dimensionality. As discussed above, the Bray-Curtis coefficient was designed for comparison of ecological samples and produces good results with simulated ecological data (Faith et al. 1987). Three different low-stress two-dimensional solutions (arbitrarily chosen from many) yield virtually the same arrangement of points, differing slightly in the position of the three E surface subsamples (Fig. 1.5a–c). In contrast, the three randomly-selected 3D solutions are highly erratic (the substantial difference in arrangement does not result from only viewing two dimensions) (Fig. 1.5d–f). The stress value is lower, but recall that stress is always reduced when extra dimensions are added (Kruskal 1964). The unpredictable performance of the 3D solution may indicate that the original environmental gradient was two-dimensional (Austin 1976), but could instead result from the small number of samples and low species richness within each sample. Replicate random analyses did not converge upon a repeated solution of lowest stress, suggesting that there was no single configuration of points that best fit the criteria, likely because there were few samples with few shared taxa.

1.5 Comparison and Interpretation of Results

These results raise an obvious question – which ordination yields the "correct" ecological representation of Mistaken Point communities? As with all natural datasets, there is no objective way of testing the accuracy of the ordination results because the positions of samples along the underlying environmental gradients are unknown. Analyses of simulated ecological data suggest that DCA and NMDS should yield the most accurate and robust results (Fasham 1977; Kenkel and Orloci 1986; Minchin 1987) and, with the exception of the placement of the PC surface, they produce very similar ordination of the data (Fig. 1.6). Note that the NMDS ordination has been rotated from the plots in Fig. 1.5 to highlight the similarity with the DCA plot, and recall that rotations such as this do not alter the results of the ordination. Two primary gradients can be observed: one from samples with abundant *Fractofusus* to samples with no *Fractofusus* and also containing more *Charniodiscus* and *Bradgatia* (vertical axis) and one from samples with abundant *Pectinifrons* to samples with abundant *Charnia* and "*Charnia*" B (horizontal axis). Placement of the PC surface is erratic with different methods, almost certainly because that assemblage only contains three taxa (*Thectardis*, *Ivesheadia*, and *Charnia*). DCA ordination places it close to the SH surface because of their shared abundance of *Ivesheadia*, but its position in the NMDS plot closer to the frond-rich LMP community (with *Charnia*), E surface (with *Thectardis*), and G surface seems intuitively more reasonable.

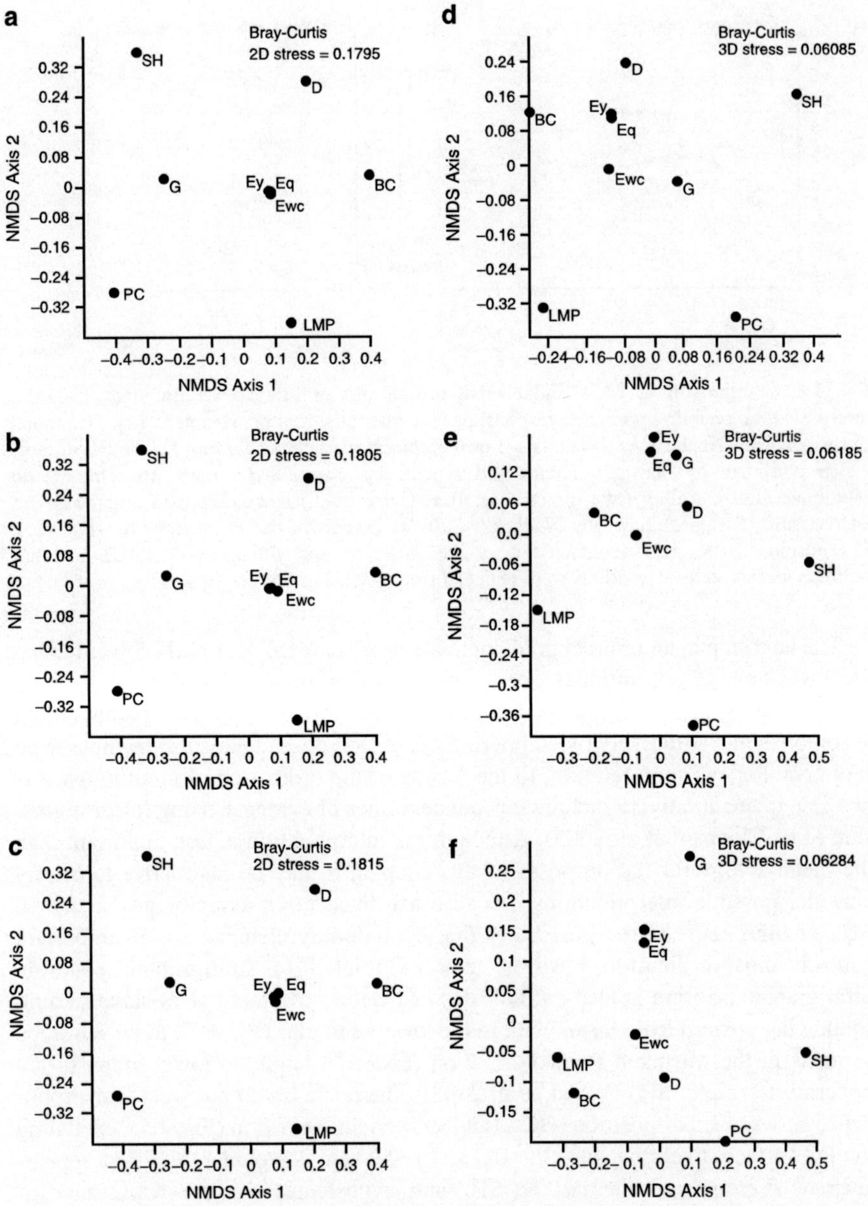

Fig. 1.5 Non-Metric Multidimensional Scaling ordinations of Mistaken Point data. **(a–c)** Three results of a two-dimensional ordination (arbitrarily chosen) using the Bray-Curtis coefficient. **(d–f)** Three random results of a three-dimensional ordination using the Bray-Curtis coefficient. Note the instability of the three-dimensional solution compared to the stable two-dimensional results. As discussed in the text, this may indicate that the underlying ecological gradient is two-dimensional or may be a limitation of the small dataset

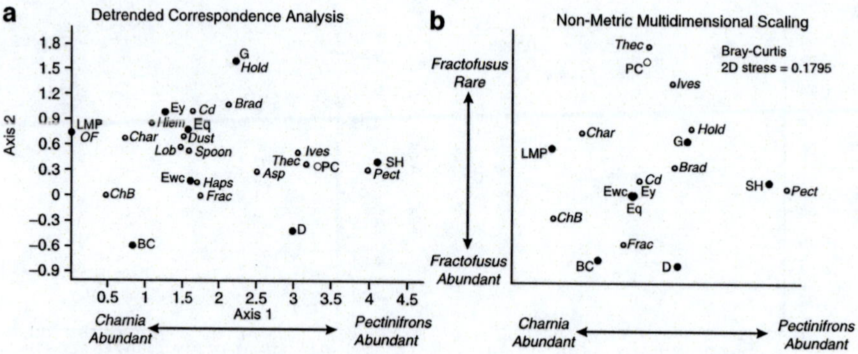

Fig. 1.6 Comparison of DCA and NMDS ordinations of Mistaken Point data, showing interpreted taxonomic gradients responsible for the position of samples. (**a**) Detrended Correspondence Analysis, with samples shown by *black filled circles* (except for the PC surface, which is shown by an *open circle*) and species by *gray filled circles*. (**b**) Non-Metric Multidimensional Scaling (two dimensions, Bray-Curtis coefficient) rotated to align with the DCA results. Samples shown by *black filled circles* except for the PC surface (*open circle*). The positions of selected abundant species are shown by *gray filled circles*; PAST does not ordinate species so their positions were obtained from NMDS ordination in R

The next step in an indirect gradient analysis such as DCA or NMDS is inferring the environmental, evolutionary, ecological, or other gradients that may have been responsible for generating the taxon gradients identified above. This is most often accomplished qualitatively by comparing other lines of evidence, for example from sedimentological interpretation, to the results of the ordination plot, although it is possible to quantitatively include independent lines of evidence using linear regression (e.g., Clapham et al. 2003). Although the interpretation is less important than the methodology for the purposes of this chapter, it may be instructive to briefly consider possible interpretations. Because Mistaken Point assemblages span >10 Myr of time, there is the possibility that evolutionary changes are an important control; thus, ordination position may be related to stratigraphic position. Stratigraphic position is also an indicator of paleoenvironment at Mistaken Point, shallowing upward from basin plain in the lowest samples (PC, BC) to toe-of-slope deposits in the Mistaken Point Formation (LMP, D, E, G) to lower slope in the uppermost sample (SH) (Wood et al. 2003). There is a broad but weak correspondence between axis 1 position (DCA) and stratigraphic position (Fig. 1.6), excluding the PC surface, likely because the BC and LMP samples predate the first appearance of *Pectinifrons* whereas the SH sample postdates the last appearance of *Fractofusus*. This implies that evolutionary changes in the Mistaken Point biota were one of the probable controls on community composition, as expected for samples spanning such a long time interval. Clapham et al. (2003) argued that each fossil surface may also reflect a snapshot of a different stage within an ecological succession trend from *Pectinifrons* and *Fractofusus*-dominated early successional communities to frond-dominated late successional communities. Mistaken Point communities have the temporal resolution to preserve a near-instantaneous snapshot of the community at a point in time; although the snapshots occur randomly throughout the stratigraphic section. Ordination techniques are ideally suited to take a jumbled series of snapshots and organize them into a gradient, whether that gradient is related to ecological succession or some other factor. The taxonomic gradient inferred for ecological succession can be observed along the vertical axis 2 in the DCA and NMDS plots, although it is modified by evolutionary changes in the biota along axis 1 (Fig. 1.6). Other ecological characteristics are at least consistent with expectations of a succession model (Clapham et al. 2003) but more surfaces may be needed to strengthen or falsify the link to ecological succession.

1.6 Conclusion

Quantitative ordination techniques are ideally suited for analyzing multivariate community ecology data. Principal Components Analysis (PCA) uses Euclidean distance to relate species dissimilarity to distance along the gradient, requiring the inappropriate assumption of a linear species response. Because that assumption is almost always invalid for community data, PCA ordinations are distorted, often severely, into a horseshoe shape or worse, usually making them uninterpretable. Although PCA is not suited for ecological ordination, it is still quite appropriate for data in morphometric or paleobiogeographic analyses, for example, where the dissimilarity-distance response is closer to linear (Chaps. 2 and 3). Detrended Correspondence Analysis (DCA) and Non-Metric Multidimensional Scaling (NMDS) can both yield accurate ecological ordinations. NMDS performs slightly better on simulated ecological data but DCA may be better at extracting the primary gradient. Detrending often results in a more accurate ordination of simulated data than in unaltered Correspondence Analysis (CA), but has also been criticized for its ad hoc methodology and assumptions. Thus, the best approach for analyzing your own paleocommunity data depends on the ultimate goal of the study. If you wish to identify the primary environmental gradient and reconstruct sample and species placement along that gradient, DCA is an excellent approach. If you are instead interested in the overall relationship among samples, NMDS may be the best option. Since DCA and NMDS ordinations have very complementary strengths, the most robust approach is to apply both methods; the overall result should be similar and gradients or clusters resolved by both DCA and NMDS are likely to be a good reflection of the real underlying ecological structure.

Acknowledgments Rodrigo Sala and Marc Laflamme assisted with data collection. Fieldwork in 2000 and 2001 in the Mistaken Point Ecological Reserve was carried out under Scientific Research Permits granted by the Parks and Natural Areas Division, Department of Tourism, Culture, and Recreation, Government of Newfoundland and Labrador. The original data collection was supported by a Natural Sciences and Engineering Research Council of Canada (NSERC) grant (to Guy M. Narbonne) and by an NSERC postgraduate scholarship (to Clapham). Thoughtful reviews from John Huntley, Richard Krause, and Amelinda Webb helped improve many aspects of this contribution.

References

Anderson AJB (1971) Ordination methods in ecology. J Ecol 59:713–726
Austin MP (1976) Performance of four ordination techniques assuming three different non-linear species response models. Vegetatio 33:43–49
Beals EW (1984) Bray-Curtis ordination: an effective strategy for analysis of multivariate ecological data. Adv Ecol Res 14:1–55
Bonuso N, Newton CR, Brower JC et al (2002) Statistical testing of community patterns: uppermost Hamilton Group, Middle Devonian (New York State: USA). Palaeogeogr Palaeoclimatol Palaeoecol 185:1–24
Botquelen A, Gourvennec R, Loi A et al (2006) Replacements of benthic associations in a sequence stratigraphic framework, examples from the Upper Ordovician of Sardinia and Lower Devonian of the Massif Armoricain. Palaeogeogr Palaeoclimatol Palaeoecol 239:286–310
Bray JR, Curtis JT (1957) An ordination of the upland forest communities of southern Wisconsin. Ecol Monogr 27:325–349
Chardy P, Glemarec M, Laurec A (1976) Application of inertia methods to benthic marine ecology: practical implications of the basic options. Estuar Coast Mar Sci 4:179–205
Cisne JL, Rabe BD (1978) Coenocorrelation: gradient analysis of fossil communities and its applications in stratigraphy. Lethaia 11:341–364
Clapham ME, James NP (2008) Paleoecology of Early-Middle Permian marine communities in eastern Australia: response to global climate change in the aftermath of the late Paleozoic ice age. Palaios 23:738–750
Clapham ME, Narbonne GM, Gehling JG (2003) Paleoecology of the oldest-known animal communities: Ediacaran assemblages at Mistaken Point, Newfoundland. Paleobiology 29:527–544
Clarke KR (1993) Non-parametric multivariate analyses of changes in community structure. Aust J Ecol 18:117–143
Dominici S, Kowalke T (2007) Depositional dynamics and the record of ecosystem stability: early Eocene faunal gradients in the Pyrenean foreland, Spain. Palaios 22:268–284
Faith DP, Minchin PR, Belbin L (1987) Compositional dissimilarity as a robust measure of ecological distance. Vegetatio 69:57–68
Fasham MJR (1977) A comparison of nonmetric multidimensional scaling, principal components and reciprocal averaging for the ordination of simulated coenoclines and coenoplanes. Ecology 58:551–561
Gauch HG, Whittaker RH (1972) Comparison of ordination techniques. Ecology 53:868–875
Gauch HG, Whittaker RH, Singer SB (1981) A comparative study of nonmetric ordinations. J Ecol 69:135–152
Goodall DW (1954) Objective methods for the classification of vegetation. Aust J Bot 2:304–324
Gower JC (1966) Some distance properties of latent root and vector methods used in multivariate analysis. Biometrika 53:325–338
Hammer Ø, Harper DAT, Ryan PD (2001) PAST: paleontological statistics software package for education and data analysis. Palaeontol Electronica 4:1–9
Hill MO, Gauch HG (1980) Detrended correspondence analysis: an improved ordination technique. Vegetatio 42:47–58
Holland SM (2005) The signatures of patches and gradients in ecological ordinations. Palaios 20:573–580
Holland SM, Miller AI, Meyer DL et al (2001) The detection and importance of subtle biofacies within a single lithofacies: the Upper Ordovician Kope Formation of the Cincinnati, Ohio region. Palaios 16:205–217
Jackson DA, Somers KM (1991) Putting things in order: the ups and downs of detrended correspondence analysis. Am Nat 137:704–712

James FC, McCulloch CE (1990) Multivariate analysis in ecology and systematics: panacea or Pandora's box. Annu Rev Ecol Syst 21:129–166

Jongman RHG, ter Braak CJF, Van Tongeren OFR (1995) Data analysis in community and landscape ecology. Cambridge University Press, Cambridge

Kenkel NC, Orloci L (1986) Applying metric and nonmetric multidimensional scaling to ecological studies: some new results. Ecology 67:919–928

Knox RG (1989) Effects of detrending and rescaling on correspondence analysis: solution stability and accuracy. Vegetatio 83:129–136

Kruskal JB (1964) Multidimensional scaling by optimizing goodness of fit to a nonmetric hypothesis. Psychometrika 29:1–27

McKinney FK, Hageman SJ (2006) Paleozoic to modern marine ecological shift displayed in the northern Adriatic Sea. Geology 34:881–884

Minchin PR (1987) An evaluation of the relative robustness of techniques for ecological ordination. Vegetatio 69:89–107

Oksanen J, Minchin PR (2002) Continuum theory revisited: what shape are species responses along ecological gradients? Ecol Model 157:119–129

Olszewski TD, West RR (1997) Influence of transportation and time-averaging in fossil assemblages from the Pennsylvanian of Oklahoma. Lethaia 30:315–329

Palmer MR (1993) Putting things in even better order: the advantages of canonical correspondence analysis. Ecology 74:2215–2230

Peet RK, Knox RG, Case JS et al (1988) Putting things in order: the advantages of detrended correspondence analysis. Am Nat 131:924–934

Podani J, Miklós I (2002) Resemblance coefficients and the horseshoe effect in principal coordinates analysis. Ecology 83:3331–3343

Quinn GP, Keough MJ (2002) Experimental design and data analysis for biologists. Cambridge University Press, Cambridge

Rodriguez J (2004) Stability in Pleistocene Mediterranean mammalian communities. Palaeogeogr Palaeoclimatol Palaeoecol 207:1–22

Scarponi D, Kowalewski M (2004) Stratigraphic paleoecology: bathymetric signatures and sequence overprint of mollusk associations from upper Quaternary sequences of the Po Plain, Italy. Geology 32:989–992

Shi GR (1993) Multivariate data analysis in palaeoecology and palaeobiogeography – a review. Palaeogeogr Palaeoclimatol Palaeoecol 105:199–234

Tomasovych A, Siblik M (2007) Evaluating compositional turnover of brachiopod communities during the end-Triassic mass extinction (Northern Calcareous Alps): removal of dominant groups, recovery and community reassembly. Palaeogeogr Palaeoclimatol Palaeoecol 244:170–200

Wartenberg D, Ferson S, Rohlf FJ (1987) Putting things in order: a critique of detrended correspondence analysis. Am Nat 129:434–448

Wood DA, Dalrymple RW, Narbonne GM et al (2003) Paleoenvironmental analysis of the late Neoproterozoic Mistaken Point and Trepassey formations, southeastern Newfoundland. Can J Earth Sci 40:1375–1391

Zuschin M, Harzhauser M, Mandic O (2007) The stratigraphic and sedimentologic framework of fine-scale faunal replacements in the middle Miocene of the Vienna basin (Austria). Palaios 22:285–295

Chapter 2
Exploratory Multivariate Techniques and Their Utility for Understanding Ancient Ecosystems

John Warren Huntley

Contents

Abstract Elucidating the ecology and evolution of the early biosphere is a critical area of research in geobiology. Indeed, the number of described species and data sets from early ecosystems has increased rapidly in the last few decades. Such datasets lend themselves well to more complex multivariate analyses. The purpose of this contribution is to serve as a primer for commonly used multivariate ordinations. I review data characterization, as it is critical to selecting the proper ordination. Then I discuss principal components analysis, principal coordinates analysis, non-metric multidimensional scaling, detrended correspondence analysis, and canonical variate analysis. For each ordination I discuss the purpose of the procedure, how it manipulates the data, which types of data are appropriate, and examples of how these procedures have been used to better understand the ecology and evolution of ancient life.

Keywords Multivariate Ordinations • Proterozoic • Ecology • Evolution

J.W. Huntley (✉)
GeoZentrum Nordbayern, Friedrich-Alexander Universität Erlangen-Nürnberg, Erlangen, Bavaria, Germany
e-mail: jhuntley@vt.edu

M. Laflamme et al. (eds.), *Quantifying the Evolution of Early Life*, Topics in Geobiology 36,
DOI 10.1007/978-94-007-0680-4_2,

2.1 Introduction

Let us begin with a consideration of the most commonly applied dichotomy to the history of life, those histories before and after the dramatic events of the Early Cambrian. People have considered the Phanerozoic (>540 mya) fossil record for hundreds of years (albeit sometimes in ways very different from the way many view it today (da Vinci 1939) owing to the broad geographic coverage of Phanerozoic-aged fossil-bearing rocks, the conspicuous size of many Phanerozoic fossils, and because Phanerozoic fossils are often quite similar to modern organisms. Indeed, vast amounts of data are available for this latest chapter in the history of life in the published literature and in large databases from which workers have gleaned much about macroevolutionary and macroecological patterns and processes (Alroy et al. 2008, 2001; Bambach 1983; Bambach et al. 2007; Boyce 2005; Briggs et al. 1992; Foote 1995; Huntley and Kowalewski 2007; Jablonski 1993; Jernvall et al. 1996; Madin et al. 2006; Payne et al. 2009; Raup 1972; Sepkoski et al. 1981; Thomas et al. 2000).

Our rich knowledge of the Phanerozoic has often been contrasted with that of older Eons (Proterozoic and Archean, or collectively with the Hadean known informally as the Precambrian). This apparent discrepancy was of course a thorn in the side of Darwin when he considered the seemingly rapid appearance of so many phyla, given that nothing was known at the time about fossils in older strata. Large and complex organisms seemed to appear very rapidly with no precursors, and geologists set out to address this issue. The seeds of Precambrian paleontology were sown from the 1860s through the early decades of the twentieth century (not without mistakes and setbacks), though it was not until the 1960s that the search for knowledge of early life began to produce fruit (Schopf 2000, 2009). The decades since have been witness to a rapid growth in the amount of data pertaining to the evolution and ecology of early life.

A cursory survey of the published literature through the Science Citation Index Expanded database reveals that Precambrian paleontology is indeed a vital and thriving discipline and confirms Schopf's (2000) assertion of rapid growth. A search of research articles using the keywords "Fossils" and "Precambrian" demonstrates that there has been a dramatic increase during the past 20 years in our knowledge of Precambrian life and early ecosystems (Fig. 2.1a, b). Searches utilizing proper, and higher-resolution, geochronological terms (Archean, Paleoproterozoic, Mesoproterozoic, and Neoproterozoic) reveal similar patterns of rapid growth, though with few pre-1990 citations (Fig. 2.1c). This discrepancy is likely an artifact of recent advances in geochronological methods producing higher resolution correlations. It should also be noted that though the SCIE database contains references back to 1900 it has the most complete coverage from 1990 to present; however the post-1990 increase in publication activity should not be interpreted as a spurious result of a biased database. The amount of literature published for each Era (as well as for the lone Eon, Archean) increased over the last 20 years. The rate of increase seems to be greatest for the youngest Era, the Neoproterozoic. Older Eras seem to

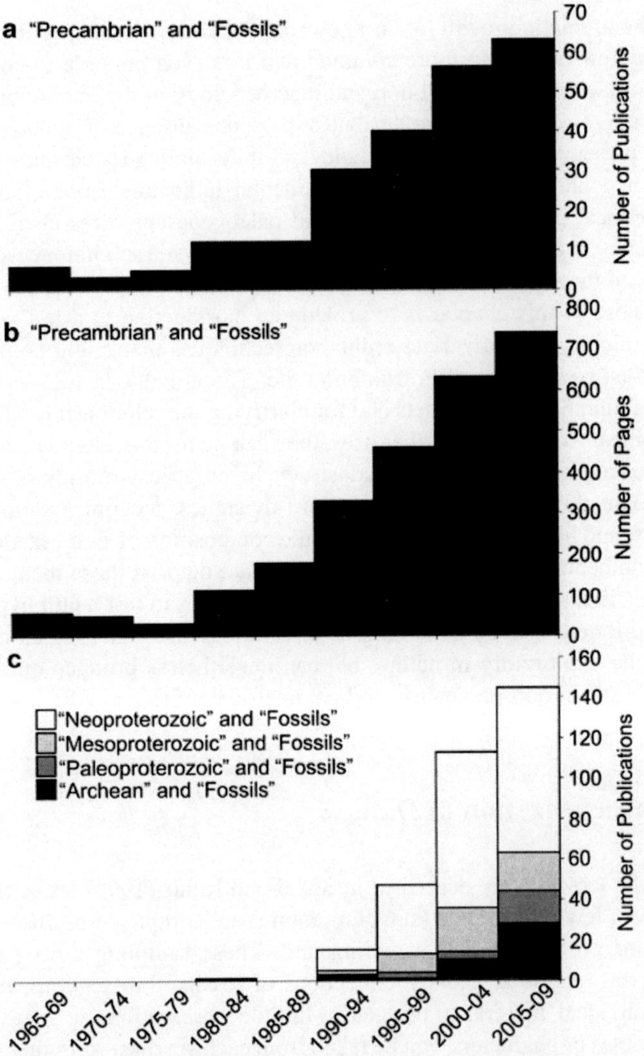

Fig. 2.1 Time series results from Science Citation Index Expanded database search for article citations between 1900 and present. (**a**) Number of articles published with the topics "Precambrian" and "Fossils". (**b**) Number of pages published with the topics "Precambrian" and "Fossils". (**c**) Number of articles published with the following sets of topics: "Archean" and "Fossils", "Paleoproterozoic" and "Fossils", "Mesoproterozoic" and "Fossils", and "Neoproterozoic" and "Fossils". Database accessed on 19 Sept 2009

display less dramatic growth. This pattern could be driven by the relative amount rock available from each Era. The notable exception, of course, is the Archean. Though there are fewer Archean rocks, the prospect of finding evidence of the oldest life is indeed tempting and perhaps attracts more researchers per unit rock than other times in the Precambrian.

Given the dramatic growth in our knowledge of early life, large paleobiological datasets are now becoming more common and it is even possible to conduct "big picture" analyses of macroevolution and macroecology in the Precambrian. These datasets contain many important attributes: presence/absence or abundance counts of species, presence/absence of morphological traits among species, mode of fossil preservation, taphonomic indices, environmental indicators from lithologic data and geochemical proxies, paleolatitude and paleogeographic reconstruction data, etc. Such datasets can be built from both original field and laboratory work as well as constructed through data collection from the published literature.

The purpose of this chapter is to provide an introduction to a few of the more common exploratory multivariate ordination techniques that might be particularly useful for the investigation of Precambrian life. These methods are useful for simplifying multidimensional datasets and for clarifying the relationships among large set of variables. Among many other uses, they can be used to elucidate community or paleobiogeographical structure, construct an empirical morphospace, aid in detecting extinction selectivity and recovery dynamics, recognize subtle shifts in morphology, and identify changing taxonomic composition of samples along one or more environmental gradients. It should be pointed out that these methods are not confirmatory in that one cannot use them by themselves to test a null hypothesis of significant differences among the samples or variables under investigation. They are fundamentally exploratory in nature, but can nevertheless produce outcomes that can then be used in non-parametric and parametric confirmatory statistical testing.

2.2 Characterization of Data

Following the terminology described in Sokal and Rohlf (1994) let us first define and describe a few terms. Data (singular: datum) are comprised of observations or measurements on the smallest sampling unit. These sampling units can include such objects as individual fossils, collections of fossils, thin sections, rock hand-samples, individual mineral grains, etc. Multiple observations or measurements, termed variables or characters, can be taken from each smallest sampling unit. Such variables could include species names, morphological characters, size data, taphonomic scores, mode of preservation, mineral identification, roundness of sediment grains, paleocurrent direction, stable isotope ratios, etc.

These observations and measurements are taken from a subset, or sample, of the total number of observations existing anywhere within space or time, or the population. For instance, there are likely many billions of acritarchs in Mesoproterozoic-aged rocks on Earth. The totality of these acritarchs makes up the population (or sampling universe). Clearly it is not feasible, nor advisable, to collect and observe all of these acritarchs; therefore, a sample of a few tens of thousands of individuals from various workers will have to suffice for characterizing acritarchs during the Mesoproterozoic. This is a primary function of statistical methods; to analyze samples and make meaningful statements with some degree of certainty about the population. Were we able to observe the population of Mesoproterozoic acritarchs, we would have no need for confirmatory statistical tests since we would know with 100% certainty the descriptive statistics (e.g., mean, median, mode, skewness, kurtosis, etc.) of any variable we may choose.

Variables can be classified into several different categories. It is important to be aware of the specific nature of the variables in your dataset as these may govern which ordination should be used. As is common with classification schemes there are various approaches for categorizing types of variables (Stevens 1946, though see Velleman and Wilkinson 1993; Sokal and Rohlf 1994; Reyment 1991). I will not discuss at length the arguments and multiple points of view; rather I will adopt the scheme that is in my opinion the most suitable for the investigation of early life and ecosystems. To quote Stevens (1946) from his landmark paper on the topic, "Any particular scale, sensory or physical, may be objected to on the grounds of bias, low precision, restricted generality, and other factors, but the objector should remember that these are relative and practical matters and that no scale used by mortals is perfectly free of their taint" (1946, p. 680). We will consider four main categories of variables: measurement, ranked, attributes, and derived.

By definition, **measurement variables** are expressed in a numerically ordered fashion (Sokal and Rohlf 1994). We can further sub-divide these variables based on continuity and the nature of zero in the scale. Theoretically, there is an infinite number of potential values between any two fixed points in the case of *continuous variables* (Sokal and Rohlf 1994). The width of an acritarch vesicle is an example of a continuous variable. Practically, it is impossible to obtain the true value of continuous variables. Our measurements must serve as an approximation of the true value. To return to our acritarch example, we may measure a vesicle diameter of 15 μm under a light microscope. This is a reasonable approximation of acritarch size, but we would have more precision by measuring the acritarch under an SEM, perhaps obtaining a value of 15.015 μm. Since there is an infinite number of possibilities between two points along the continuous variable scale, our accuracy is in part a function of our chosen methodology for data collection. One must make trade-offs between accuracy and resources (such as time and money) to determine what is necessary to collect the data needed for the question at hand. *Discontinuous variables* (also known as discrete or meristic) are expressed as whole numbers; values between these whole numbers are not possible (Sokal and Rohlf 1994). The number of individual *Chuaria* vesicles on a bedding plane or the number of quartz grains from a thin-section point count are examples of discontinuous variables. The accuracy of this type of measure is contingent upon our ability to count, and not necessarily by the accuracy of calipers or other instruments, as would be the case for continuous variables. We can also characterize measurement variables by the nature of zero in the scale. *Ratio scale* measurements have a constant interval size as well as a true zero. For example, a measure of temperature in degrees Kelvin is a ratio scale measure. The difference between 1 and 2 K is the same as the difference between 240 and 241 K. In addition 0 K is a true value of zero. *Interval scale* measurements also have a constant interval size, but do not have a true zero. A corresponding example is measuring temperature in degrees centigrade. The difference between 1°C and 2°C is the same as 240°C and 241°C. However, 0°C is an arbitrary zero value as negative values in degrees C are possible. Time of day and directional data (e.g. N 20° E) are also examples of interval scale data as they have zeros with arbitrary values.

Ranked variables cannot be measured *per se*, but are ranked in a meaningful order. The difference between two adjacent ranks is not necessarily equal or proportional to the difference between two other adjacent ranks on this type of scale. A ranking scheme wherein the quality of preservation of a fossil specimen is ranked is an example of a ranked variable. With such a ranking scheme, low values might represent low taphonomic alteration and high values represent high alteration. These scores are quantitatively meaningful inasmuch as they indicate relative values along a qualitative scale. Another example of a ranked variable is Moh's scale of hardness. Moh's scale is based on the relative hardness of specific minerals, though the differences between adjacent ranks are not necessarily equal.

Attribute variables cannot be measured. Coding in a database for such variables is an attempt to quantify qualities for the sake of data analysis. Examples of attribute variables are colors (e.g., red=0, blue=1, yellow=2, etc.), differing growth forms (e.g., as a result of sexual dimorphism or ecophenotypic plasticity), or species, etc.

Derived variables are based on two or more independently-measured variables. These are often expressed as ratios, percentages, and rates. Care should be taken when using derived variables. Much of the time derived variables can be considered redundant and should not be used in multivariate ordinations if one or more of their parent variables are also included in the ordination. If the parent variables are not included in the ordination, one should take care to properly transform the variable prior to analysis (e.g., logit-transformation for proportional data; this transformation is used because proportional data are typically not normally distributed and are bounded by the values of 0 and 1).

2.3 Common Multivariate Ordinations

The following sections will review various multivariate ordinations including Principal Components Analysis, Principal Coordinates Analysis, Non-metric Multidimensional Scaling, Detrended Correspondence Analysis, and Canonical Variate Analysis. These methods differ in purpose and in the types of data required for their proper use. The purpose and data requirements for each method are outlined in each section.

2.3.1 Principal Components Analysis

Principal components analysis (PCA) is one of the most commonly used ordinations for projecting the hyperspace of many variables to two or three dimensions.

PCA calculates new "hypothetical" variables or principal components as combinations of the original variables; therefore each observation is assigned a principal component score (PC score) for each of the newly-derived principal components. The scales on principal component axes are unit-less, but can be related to the original variables.

PCA works by extracting eigenvectors and eigenvalues from a matrix of variance–covariance (or correlation) values among the variables in a given data set. Principal components are defined by a linear equation based on the original variables. The coefficients for each original variable in this linear equation are called *eigenvectors* or *loadings*. Therefore each principal component will be defined as some combination of all the original variables through a simple linear equation. The magnitude of the eigenvector (whether positive or negative) will serve as a quick reference as to the importance of different variables in a particular principal component. These equations can also be used to calculate the principal component scores for each observation (Reyment 1991; Hammer and Harper 2006). The PCA ordination will also produce a measure of the variance for each principal component called an *eigenvalue* or *latent root*. The sum of eigenvalues for all principal components is equal to the variance amongst the original variables. By dividing the eigenvalue for a given principal component by this sum we can determine the relative amount of variance explained by that principal component (Reyment 1991; Hammer and Harper 2006). The matrix algebra involved in PCA is beyond the scope of this contribution, but see Jolliffe (2002) and McCune et al. (2002) for more specific descriptions of the procedure.

PCA essentially identifies the direction of maximum variance in the multivariate hyperspace of the data set. A commonly used visual depiction of PCA posits that once the axis of maximum variance is identified the data undergo an orthogonal rotation wherein the distances between the data points in the hyperspace are preserved. This image is useful for understanding the end result of PCA, but is not an accurate description of the actual procedure. This axis of maximum variance is known as the first principal component. Additional principal components are identified in the process of the ordination and are all at 90° to one another in the hyperspace. Each subsequent principal component accounts for the majority of the variance unaccounted for by previous principal components. The number of principal components produced by the ordination can, in theory, be equal to the number of original variables. The expectation in a highly structured data set is that the large majority of the variance will be accounted for with the first few axes.

The data matrix should be constructed such that observations occupy rows and variables occupy columns (as it should for all ordinations). A proper use of PCA requires continuous measurement variables (though Jolliffe (2002) relaxes this assumption when it is used as a purely descriptive technique) and more observations than variables. Interpretation of PCA works best when the dataset displays a multivariate normal distribution. There should be some correlation amongst the variables for PCA to be useful. If none of the variables are correlated, then the data hyperspace will take the form of a hypersphere with no possibility of diagnosing a direction of maximum variance.

One must consider the variances of the original variables when performing PCA. Variables with higher variances can artificially dominate the first principal component and confound interpretations. If variables are measured in the same units and have similar variances then the PCA should be performed on the *variance–covariance matrix* (a matrix consisting of covariance values between the original variables). A dataset characterized by variables in different units (e.g., mm^2 and K) and/or divergent variances (e.g., morphometric measurements of bottlenose dolphins and blue whales) needs to be standardized in order to prevent spurious results. This is accomplished by instead performing PCA on the *correlation matrix* (conversely, a matrix of correlation values between the original variables; Reyment 1991; Hammer and Harper 2006).

I culled data appropriate for PCA from the acritarch meta-analysis reported in Huntley et al. (2006a). The data were collected from the peer-reviewed literature of acritarch occurrences ranging in age from Paleoproterozoic to Cambrian. The original dataset contains a wide variety of variables including body size (continuous variable), presence/absence of 31 morphological characters (attribute variables), and best estimate age of deposition (derived variable from the maximum and minimum age estimates) among others. In this example we will consider the following variables from species descriptions (all in μm): minimum reported vesicle diameter, maximum reported vesicle diameter, process length, and process width. These are continuous variables and there are no missing values. Typically, one should have no missing values for PCA, however some software packages will substitute the mean value of the variable for the missing datum. The data matrix is therefore comprised of four variables and 62 observations (acritarch occurrences) ranging in age from Mesoproterozoic to Cambrian. This analysis is focused on acanthomorphic (process-bearing) acritarchs and overlooks sphaeromorphs. The measurements were log-transformed. Log transformation is important because of the assumption of PCA that variables display a linear relationship with one another; however a pattern of allometric growth (change in shape with growth) is more common. Log transformation will linearize the relationships between body size dimensions. PCA was performed in PAST 1.94b on a variance–covariance matrix (Hammer et al. 2001; Hammer and Harper 2006) (Fig. 2.2).

The eigenvectors for the PCA of acritarch body size data are presented in Table 2.1. These values are the coefficients in the linear equations that define PC scores. For example, to calculate the PC1 score of a particular variable one would solve the following equation: PC1 score=(–0.51*minimum)+(–0.56*maximum)+(–0.40*process length)+(–0.51*process width). We can see from the eigenvalues that in this example PC1 accounts for 61.7% of the variance (Table 2.2).

Once the ordination has been performed and the observations have been assigned PC scores, one must determine how many principal components are actually useful. One method for doing so is a qualitative examination of the scree plot, or descending curve plotting the percentage of variance (or log of the eigenvalue may be plotted as well) for each principal component (Fig. 2.2c). The scree plot from the acritarch PCA indicates that most variance is explained by the first three principal components. The curve begins to flatten after the third PC and little explanatory

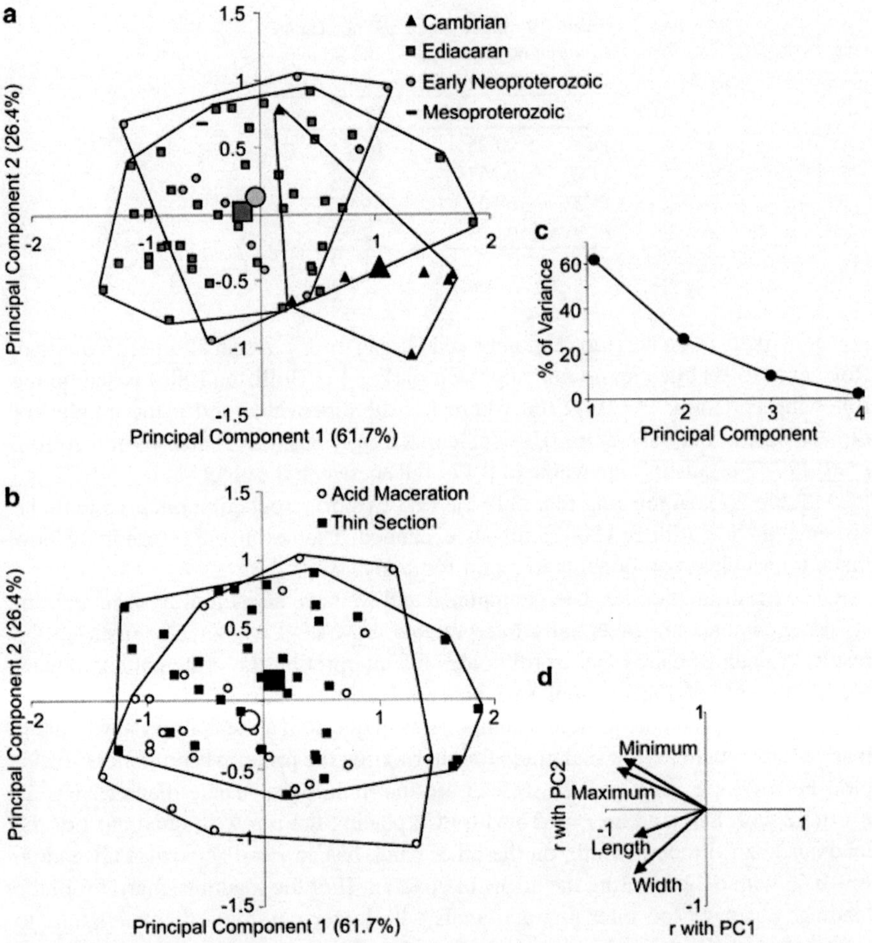

Fig. 2.2 Principal components analysis of acritarch body size metrics. (**a**) Scatter plot of PC1 vs. PC2 with convex hulls classified by age group. (**b**) Scatter plot of PC1 vs. PC2 with convex hulls classified by sample preparation technique. The large symbols in scatter plots (**a**) and (**b**) represent the centroid value for the respective groups. (**c**) Scree plot for percent of variance explained by the principal components. (**d**) Loadings chart for PC1 vs. PC2

Table 2.1 Eigenvectors of PCA of acritarch body size variables

	Minimum diameter	Maximum diameter	Process length	Process width
PC1	–0.51	–0.56	–0.40	–0.51
PC2	0.49	0.40	–0.23	–0.73
PC3	0.19	0.06	–0.88	0.44
PC4	–0.68	0.72	–0.11	–0.02

Table 2.2 Eigenvalues and percentage of variation explained by the principal components

	Eigenvalue	% Variance
PC1	0.625	61.7
PC2	0.267	26.4
PC3	0.101	9.9
PC4	0.019	2.0

power is likely to be derived (Hammer and Harper 2006). As an alternative method, Reyment (1991) cites an *ad hoc* approach outlined by Jolliffe (1986) based on the following equation: $l^*=0.7(\Sigma l)/p$ where l is the eigenvalue and p the number of dimensions. In our case $\Sigma l=1.0127$ (calculated from Table 2.2) and $p=4$; therefore $l^*=0.177$. The cut-off eigenvalue of 0.177 falls between the eigenvalues of PC2 and PC3 (Table 2.2), suggesting that only the first two principal components should be considered. See Jolliffe (2002) for an expanded discussion on determining how many to principal components to retain for analysis.

Once the ordination has been completed and we have an idea of how many principal components are potentially interpretable, how shall we actually interpret the results? Loadings charts are useful guides for interpreting the principal component space in terms of the original variables. A loadings chart plots the correlation coefficients between the principal components in question and the original variables. It serves as a qualitative visual guide for interpreting the principal component scatter plot. For example, in Fig. 2.2d we can see that minimum vesicle diameter has an $r=-0.82$ with PC1 and an $r=0.52$ with PC2 placing the arrow in quadrant I of the loadings chart. Process width, on the other hand, has an $r=-0.71$ with PC1 and an $r=-0.66$ with PC2, locating the arrow in quadrant II of the loadings chart. From the loadings chart we can infer that acritarchs with larger minimum diameters tend to fall toward quadrant I and acritarchs with wider processes tend to fall toward quadrant II. However, there is a more conspicuous pattern that we must first address. All of the original variables are significantly and negatively correlated with PC1. As is often the case, the first principal component is a strong indicator of overall body size (whether by a positive or negative correlation). Indeed, many times PC1 scores are ideal indices of body size as they are derived from many measures of size. In this case, a negative PC1 value indicates larger vesicle diameters and process dimensions while positive PC1 values indicate overall smaller body size. Subsequent principal components are often reflective of differences in shape (PC1 having removed most of the variance related to size). It is possible to remove the factor of body size before using PCA in order to focus primarily on differences in shape (see Chap. 3).

Equipped with an understanding of the relationship between the principal component space and the original variables one can now begin to address important questions. Perhaps we want to know if acritarch size or shape changed through time or if taphonomy and sample preparation techniques can bias size estimates. By classifying the observations by group (e.g., age or sampling technique) we can identify patterns and potentially test hypotheses through non-parametric or parametric confirmatory statistical techniques. Figure 2.2a presents the results of the PCA with the observations grouped by age and distinguished by convex hulls (a convex polygon which encloses the minimum area required to contain all data points of a given group). The large symbol for each grouping variable represents the centroid, or multivariate mean value. We can see that, on average, Cambrian acanthomorphic acritarchs were smaller (positive PC1 values) than their Precambrian counterparts and, for a given size, had shorter and narrower processes. These assessments can be rigorously tested via non-parametric tests (e.g., bootstrapping (Kowalewski et al. 1998; Efron 1981) and randomization (Huntley et al. 2006a; Huntley et al. 2008)) and parametric tests (e.g., MANOVA) (for more information about hypothesis testing in morphometrics see Chap. 3).

Precambrian acritarchs are typically preserved in one of two ways: compressions in shales or three-dimensional preservation in cherts. These differing modes of preservation necessitate very different modes of sample preparation which can bias estimates of size and morphology. Shale specimens are macerated in HCl, thus dissolving the non-organic content and liberating compressed vesicles. Chert specimens are thin-sectioned to reveal the acritarchs. Thin-sectioning will randomly cut across acritarchs; therefore, one will rarely measure the full diameter of an acritarch, but will, on average, underestimate size. Perhaps, then, we can predict that macerated specimens will display a larger body size and display less variation than thin-section specimens. Figure 2.2b suggests, as reflected by the relative position of the centroids, that acid maceration specimens are, indeed, slightly larger (more negative PC1 score) than thin-section specimens, but *F*- and *t*-tests show that the variance and mean values of size for the two groups are statistically indistinguishable ($p=0.95$ and $p=0.32$, respectively; $\alpha=0.05$). In this limited data set, there is therefore no indication that taphonomy and sample preparation techniques, which are difficult to disentangle in this case, significantly affect our perception of acanthomorphic acritarch body size. Of course one should keep in mind that this is merely a limited example, and not a broad statement about acritarchs in general. A more rigorous study would require one to address other potentially confounding factors such as the exclusion of sphaeromorphs, the age structure of each group, and secular trends in size (Huntley et al. 2006a).

There is much potential for the use of PCA in the study of early life and ecosystems, though primarily for morphometrics of organisms with similar baupläne, given the suggested constraint to continuous variables. Laflamme et al. (2004) applied PCA to clarify the taxonomy of the Ediacaran frond *Charniodiscus*; first in an ordination examining the differences among two species co-occurring on bedding planes in the Mistaken Point Formation of Newfoundland and then in multiple ordinations based on different combinations of variables including global occurrences of the genus (all ordinations confirmed the distinct morphologies of *C. procerus* and *C. spinosus*).

Laflamme et al. (2004) provide a good example of how one should consider principal component scores from different ordinations separately and not mix their results. Principal component scores are derived solely from the original variables in a given ordination; therefore scores from different ordinations are not comparable. However, the data matrices used in multiple ordinations by Laflamme et al. (2004) were comprised of various combinations of continuous measurement variables (linear body size measurements), a discontinuous measurement variable (number of primary branches in the frond), and an attribute variable (presence/absence of a terminal spine). A principal coordinate analysis (see below) would probably have been a more suitable ordination for this dataset of mixed variable types, but would most likely produce similar results (see Chap. 3).

For more information and discussion about PCA see Hammer and Harper (2006); Jolliffe (2002); Reyment (1991) and references therein. Hammer et al. (2001) provided a very user-friendly and, perhaps just as important, free software called PAST (Paleontological Statistics). This freeware is compatible with Excel spreadsheets and can perform many of the statistical operations commonly used by paleobiologists including PCA. For those interested in having more flexibility in programming their procedures, SAS and R are commonly used software packages. The supplementary information section of Reyment (1991), authored by L.F. Marcus, provides the code for principal components analysis written for SAS. R software, like SAS, is also a highly flexible and powerful platform; however, unlike SAS, R is freeware and available for download at http://www.r-project.org.

2.3.2 Principal Coordinates Analysis

Principal coordinates analysis (PCO) is another eigen-based method for reducing dimensionality by projecting multidimensional datasets to a smaller number of dimensions. To better understand the difference between PCA and PCO, it is instructive to distinguish *R*-mode analyses from *Q*-mode analyses. *Q*-mode analyses focus on the association between individual observations in a dataset while *R*-mode analyses address the relationships among the variables. PCA is an *R*-mode analysis of a matrix based on similarities (correlation or covariance). PCO is a *Q*-mode ordination based on measures of dissimilarity (with the flexibility to select from a broad family of dissimilarity coefficients, see Hammer and Harper (2006)) (Gower 1966; Reyment 1991). In PCO, the multivariate data do not undergo a rigid rotation (as associated with PCA), but the ordination does aim to preserve the relative distances between data points.

Principal coordinates analysis has less stringent assumptions about the data than PCA. PCO will work with continuous variables, discontinuous variables, ranked variables, attribute variables, or any combination thereof; though there should be no missing values. PCO will sometimes return negative eigenvalues which are typically associated with coordinates which explain very little variance. However, one should be wary of an ordination that produces large numbers of negative eigenvalues (Hammer and Harper 2006). As PCO produces only an estimate of the original distances between points (i.e., similar objects are found closer to one another in the projection), interpreting the ordination in terms of the original variables can be problematic.

However, due to the loosened assumptions about variables in PCO, one has more freedom with the analysis. An empirical morphospace can be produced based on a broader range of characters, and, thus, more dissimilar observations can be included (perhaps organisms of varying baupläne) than in PCA.

Principal coordinates analysis is not one of the more commonly used ordinations in ecology and paleoecology. Wills (1998) coupled PCO with cladistic analyses of modern and Cambrian priapulids to address evolutionary relationships and secular trends in occupied empirical morphospace. Specifically, Wills addressed Gould's assertion (1989) that morphological disparity peaked in the Cambrian and was subsequently greatly reduced. Even though there were more priapulid families in the Cambrian than in the Recent, Wills (1998) found that Recent priapulids actually occupy slightly more morphospace than their Cambrian forbearers. Moreover, the morphospaces occupied by the Cambrian and Recent groups were predominantly adjacent rather than overlapping.

For more information about PCO see Gower (1966), Hammer and Harper (2006), and Reyment (1991). PAST offers users the ability to perform PCO and the freedom to choose from a large number of similarity indices (Hammer et al. 2001; Hammer and Harper 2006). Reyment (1991) provides a SAS code for PCO written by L.F. Marcus. PCO is also sometimes referred to as metric multidimensional scaling (mMDS).

2.3.3 *Non-Metric Multidimensional Scaling*

Non-metric multidimensional scaling (nMDS or MDS) is a numerical technique whose function is to produce a "map" from a dissimilarity index data matrix. Kruskal and Wish (1978) presented the example of performing nMDS on a matrix of distances between ten US cities to produce an ordination that successfully reproduced the relative positions and distances between the cities. nMDS is different from an ordination like PCA where the axis of maximum variance is identified and the data hyperspace is rotated accordingly. nMDS is an iterative process wherein configurations of the data points are produced sequentially with the goal of preserving the original distances between the points. nMDS calculates a stress value for each iteration by comparing the ranked distances between observations in the original data with the ranked distances between nMDS scores from the ordination. Low stress values (~0.10 or less) indicate a close agreement between the ordination and the original data (Hammer and Harper 2006). This procedure runs until it reaches a minimum stress value. As nMDS is an iterative mathematical technique, multiple runs of this procedure will produce multiple results with differing stress values. I will borrow an illustration of nMDS from Kruskal and Wish (1978). One can imagine a landscape with hills and valleys that represents an infinite number of solutions for the nMDS procedure. The vertical dimension, or elevation, of this landscape represents the stress value. Hills represent solutions with high stress values and valleys represent low stress solutions. The nMDS procedure is like a sky diver who parachutes onto a random point in the landscape; this point represents the initial solution of the nMDS. The sky diver's goal is to reach the lowest point in the landscape, so she begins to walk down slope. With each step the sky diver finds herself at a lower elevation; in the same way each iteration of the nMDS produces a solution with a lower stress value. As this landscape could potentially have multiple hills and valleys, a single nMDS procedure could settle in a local minimum and not the global minimum (the lowest point in the landscape). For this reason software packages will often perform multiple runs of the nMDS procedure with different starting configurations to maximize the likelihood of approaching the global minimum. Moreover, the stress of an nMDS can vary based on the number of dimensions to which the data are reduced. This is another difference between nMDS and PCA. One defines the number of dimensions to which the data are reduced *a priori* with nMDS while the number of dimensions produced in PCA is a function of the number of variables. It is important to examine the effect of the number of dimensions on stress value. Some software packages do this automatically while others require the user to define the number of dimensions. Shepard plots are visual guides for conveying the amount of stress in an nMDS solution. These scatter plots compare the ranked distances between each data point from the original data matrix with the ranked distances between each point from the ordination. A perfect linear correlation indicates zero stress and lower r values indicate higher stress (Fig. 2.3).

Similar to PCO, interpreting nMDS axes in relation to the original variables can be problematic as the ordination axes do not necessarily correspond to the maximal directions of variance and are not likely to be orthogonal to one another. One can subject the nMDS scores to a PCA (as the nMDS scores are continuous variables) to sidestep this problem.

nMDS is commonly used in paleoecological studies and is well-suited to the vicissitudes of data collection from the fossil record. One can use any combination of continuous, discontinuous, ranked, and/or attribute variables with nMDS which also accommodates missing observations in the data matrix. Let us continue to use the acritarch body size data from the PCA example for nMDS. I subjected the four log-transformed body size variables to nMDS using PAST 1.94b with a Euclidian similarity measure (Hammer et al. 2001). PAST gives the option to reduce the data to either two or three dimensions while other software packages have different options. PC-ORD automatically calculates solutions for one to six dimensions (McCune and Mefford 2006) and SAS allows the user to define the number of dimensions (see Huntley et al. (2006b) for a SAS code for nMDS). PAST reported a stress value of 0.098 for the two-dimensional ordination and a stress value of 0.024 for the three-dimensional ordination. These stress values are quite low and indicate that both ordinations were successful in reconstructing the dissimilarities between the variables in the original data. Such low stress values are not surprising because we are considering a small number of continuous variables. One might not necessarily expect stress values this low with larger and more complicated data matrices of mixed variable types. Due to its lower stress value, let us continue the example with the ordination of three dimensions. The distribution of nMDS scores appears similar to the PC scores in Fig. 2.2 when categorized by age (Fig. 2.3a). The tight distribution

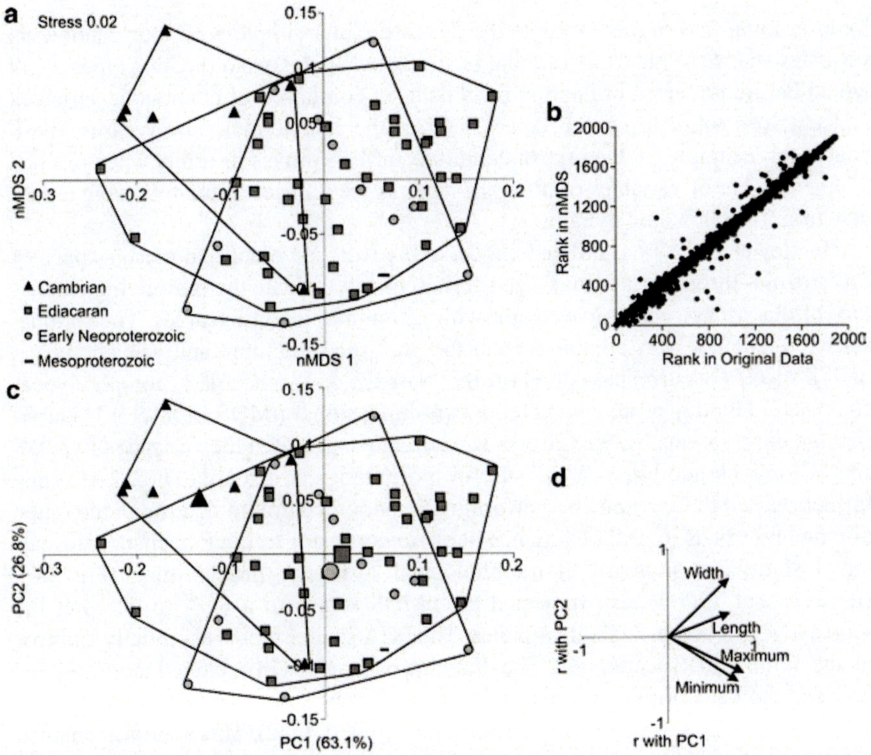

Fig. 2.3 Non-metric multidimensional scaling of acritarch body size metrics (scaled to three dimensions). (**a**) Scatter plot of nMDS1 vs. nMDS2 with convex hulls classified by age group. (**b**) Shepard plot for the nMDS ordination. The close association of data points with the line x=y indicates that there is a high agreement between the distances between observations in the original data set and the distances between observations in the ordination. This close association indicates a low stress value. (**c**) Results of PCA of nMDS scores (MDS1, MDS2, and MDS3): scatter plot of PC1 vs. PC2 with convex hulls classified by age group. Though similar in appearance, the nMDS values were slightly rotated by the PCA. Note how these PCA results are nearly a mirror image of those in Fig. 2.2a; though PC1 and PC2 explain slightly more variation in Fig. 2.3c than in Fig. 2.2. (**d**) Loadings chart for PC1 vs. PC2

of points about the x=y line in the Shepard plot confirms the low stress value (Fig. 2.3b). One should be suspect of ordinations with high stress values, but the stress values in this example are low and we can proceed with the assumption that the ordination accurately describes the data. As discussed earlier, one should not relate nMDS scores to the original variables; rather one should first subject the continuous nMDS scores to a principal components analysis and then proceed with interpreting the ordination space (see below for a broader example from my own work). One can see that the PCA results (Fig. 2.3c) are quite similar to the nMDS ordination (Fig. 2.3a). Though they appear identical, the PCA did rotate the nMDS values slightly. Using the loadings chart as a guide (Fig. 2.3d) we can interpret the morphospace and draw conclusions that are similar to the PCA example (Fig. 2.2).

Keep in mind that in this example the data are rather simple with four continuous variables and we might expect similar results between the two ordination types. PCA would be the preferred ordination for a data set composed of continuous variables (since it does not distort the distances between the data points); however I use these data in this example for the sake of continuity. nMDS proves its utility when one has a large number of variables of different types as we will see in the following examples from the published literature.

Huntley et al. (2006a) utilized nMDS to construct an empirical morphospace of Proterozoic- through Cambrian-aged acritarchs to elucidate the morphological history of this group, which is wrought with taxonomic inconsistencies. Their global acritarch database was compiled from the published literature and was composed of 778 species occurrences coded for the presence or absence of 31 morphological characters. Huntley et al. conducted a two-dimensional nMDS in SAS 9.1, classified the observations by age bin *a posteriori*, and calculated the variance of nMDS scores for each age bin as a measure of morphological disparity (Fig. 2.4). They further refined these results by performing a randomization to determine the range of temporal trends in nMDS variance likely to occur due to random chance, revealing a significant increase in morphological disparity through time (Fig. 2.4). Huntley et al. (2006) also subjected the nMDS scores to a PCA to interpret the scores in terms of the original variables. The PCA scatter plot is essentially a mirror image of the nMDS scatter plot. The dramatic expansions in occupied morphospace between the Paleoproterozoic and middle/late Cambrian is primarily driven by the diversification(s) of process-bearing acritarchs (Fig. 2.4a–d). This approach enabled Huntley et al. (2006) to address questions in a way not possible with a purely taxic approach. They concluded that the first major expansion of occupied morphospace preceded the first burst of taxic diversity by a half billion years (a pattern common amongst many Phanerozoic eukaryotes) and the occupation of morphospace through time was broadly correlated with major geological and biological events such as Snowball Earth, the first appearance of the large Ediacara organisms, and the Cambrian Explosion.

Fig. 2.4 nMDS of Paleoproterozoic- to Cambrian-aged acritarchs. (**a**) Occupied morphospace in the Paleoproterozoic. (**b**) Occupied morphospace in the middle and late Cambrian. (**c**) PCA of nMDS scores for all species occurrences. (**d**) Loadings chart for PC1 vs. PC2 of nMDS scores. (**e**) Temporal trend of nMDS variance (variance of nMDS1 plus variance of nMDS2). Inset is result of 1,000 iteration randomization of nMDS variation. The *center line* represents the mean value of randomization for each bin, and the *upper* and *lower lines* represent the 95% confidence intervals. nMDS variance values which fall outside the confidence intervals display either higher or lower morphological disparity than expected by random chance. Huntley et al. (2006) interpreted the overall trend as displaying a significant increase in morphological disparity. The *dashed lines* in (**a–c**) represent the convex hull of the total realized morphospace of all acritarchs considered in the study. The *vertical dashed line* in (**e**) represents the first occurrence of Ediacara organisms (Modified from Huntley et al. (2006))

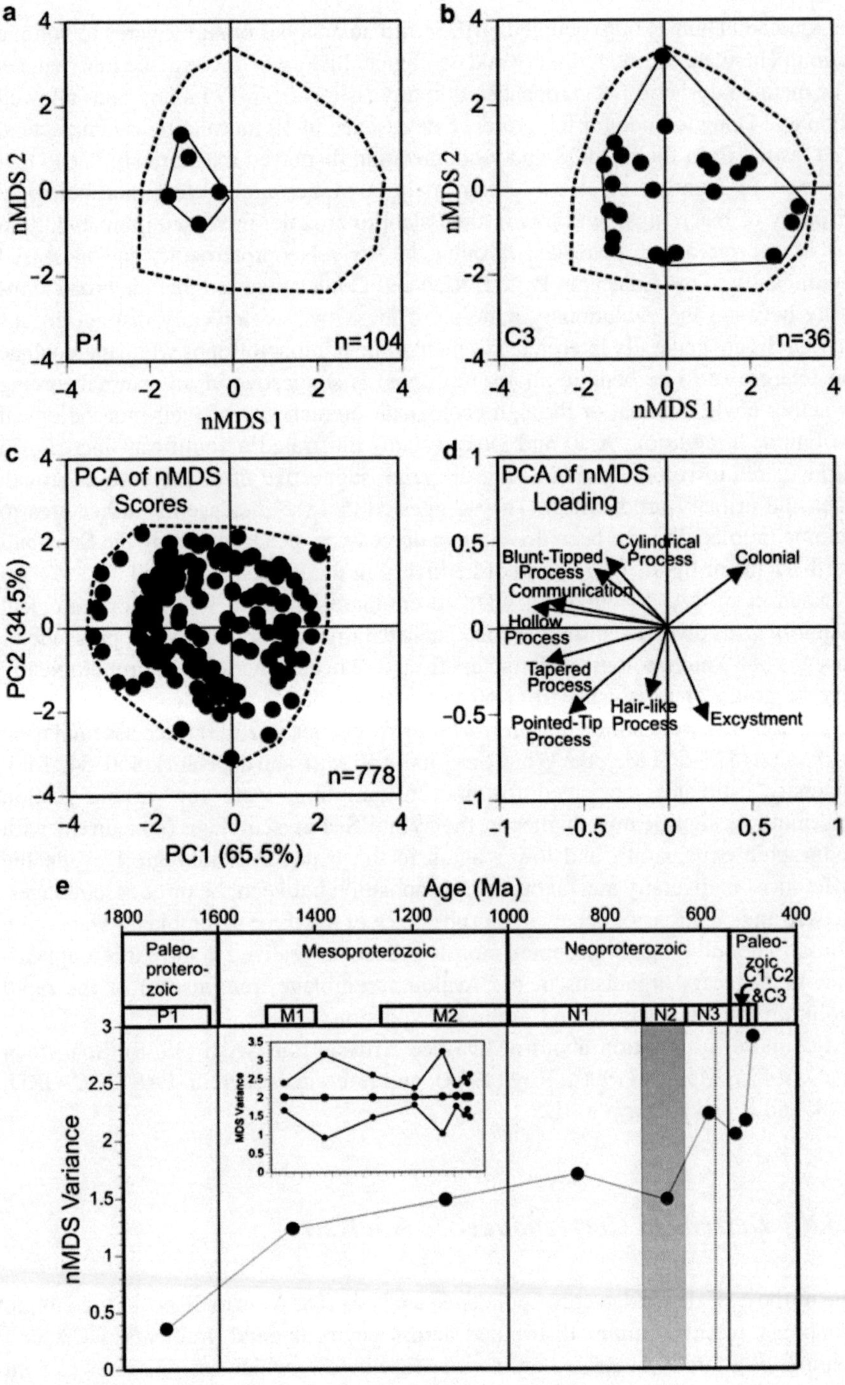
a
nMDS 2
nMDS 1
P1
n=104
b
nMDS 2
nMDS 1
C3
n=36
c
PCA of nMDS Scores
PC2 (34.5%)
PC1 (65.5%)
n=778
d
PCA of nMDS Loading
Cylindrical Process
Blunt-Tipped Process
Colonial
Communication
Hollow Process
Tapered Process
Hair-like Process
Excystment
Pointed-Tip Process
e
Age (Ma)
Paleo-protero-zoic
Mesoproterozoic
Neoproterozoic
Paleo-zoic
C1,C2 &C3
P1
M1
M2
N1
N2
N3
nMDS Variance
MDS Variance

Xiao and Dong (2006) coupled nMDS with an analysis of surface area to volume ratio to elucidate the evolutionary and ecological history of Proterozoic macroalgae. The meta-analysis of 578 carbonaceous compressions from 17 monographs allowed Xiao and Dong to code for the presence or absence of 19 morphological characters. The results from their nMDS and randomization displayed a pattern similar to that reported in Huntley et al.'s (2006a) analysis of acritarchs. The morphological disparity of macroalgae was low in the Paleoproterozoic, increased dramatically in the Mesoproterozoic, plateaued through the early Neoproterozoic, and increased significantly in the Ediacaran Period. Xiao and Dong suggested that the broad similarity between the evolutionary patterns of these two ecologically distinct groups (acritarchs are generally interpreted as planktonic photoautotrophs while macroalgae are interpreted to be benthic photoautotrophs) is suggestive of an external forcing (whether environmental or through ecological interactions) of their morphological evolution. In addition, Xiao and Dong (2006) illustrated a significant increase in surface area to volume ratio in the Ediacaran, suggestive of an increase in growth rates and primary productivity. They suggested that the increase in surface area to volume ratio could have been driven by a decrease in pCO_2 following the Snowball Earth events or by the origination of herbivory in the Ediacaran Period.

Shen et al. (2008) utilized nMDS to compare Ediacara fossil morphological disparity with diversity through time. Their data matrix coded for the presence or absence of 50 morphological characters from 272 occurrences of a morphologically diverse group of organisms from 60 publications and one museum sample. The ordination data were sorted *a posteriori* into Waggoner's (2003) three assemblages: the Avalon (575–565 Ma), the White Sea (560–550 Ma), and the Nama (550–542 Ma). Diversity estimates, corrected for uneven sampling, were low in the Avalon Assemblage, significantly higher in the White Sea Assemblage (concurrent with geographic expansion), and lower again in the Nama Assemblage. Despite the differences in diversity and taxonomic composition between the three assemblages, the size and positions of occupied morphospace of the three assemblages were quite similar. The full range of occupied morphospace was realized in the earliest appearance of Ediacara organisms in the Avalon assemblage, reminiscent of the rapid evolution of metazoans in the Cambrian Explosion.

For more information about nMDS see Kruskal and Wish (1978), Schiffman et al. (1981), Marcus (1990), Roy (1994), and references therein. PAST, PC-ORD, SAS, and R can perform nMDS.

2.3.4 Detrended Correspondence Analysis

The purpose of detrended correspondence analysis (DCA) is to discover taxonomic groupings among samples distributed across environmental gradients. DCA is a member of a family of related ordinations (see below) which assign scores to both species and samples. One can imagine the distributions of many species along an environmental gradient such as temperature or water depth. Some taxa will have narrow distributions and other taxa will have wide distributions according to their environmental tolerances. Many taxa will overlap spatially with one another, but typically the species at either extreme of the gradient will likely not overlap with one another. In a sense, DCA works to reconstruct this distribution. Species that commonly co-occur in samples will have similar ordination scores. Ordination scores are calculated for samples based on the weighted-average ordination scores of their component taxa.

DCA is an iterative analysis similar to Reciprocal Averaging (RA; also called Correspondence Analysis or Analyse Factorielle des Correspondances) which begins by selecting an arbitrary set of scores for the species in the data matrix. These arbitrary scores for species are used to calculate scores for the samples in which they occur (by calculating a weighted average). New scores are then calculated for each species as the average score from each sample in which they occur. These new species scores are used to refine the sample scores which are then used to further refine the species scores. This process continues until the scores stabilize (Hill 1973). DCA improves upon RA by addressing two of its key related flaws: (1) RA tends to warp the first ordination axis into an arch or horseshoe shape and (2) RA does not preserve ecological distances well; such that intermediate scores are widely spaced and peripheral scores are spaced closely together (Hill and Gauch 1980). DCA incorporates detrending to eliminate the arch effect and rescaling in order to preserve ecological distances within the iterative calculation process (Hill and Gauch 1980).

Detrended correspondence analysis requires only a few assumptions about the data. Samples comprise rows and species comprise columns. Species occurrences can be coded as counts, percentages, or presence/absence. DCA will not work well if most or all species co-occur in all samples or, conversely, if few or no species co-occur in samples.

Detrended correspondence analysis and related techniques are commonly used for analyzing gradients in ecological studies and in paleoecological studies. Holland et al. (2001) applied DCA to faunal occurrences (1,337 samples and 46 taxa) in the Ordovician Kope Formation and demonstrated shifting ecological structure along an inferred depth gradient that was not reflected by lithological changes. Holland et al. were able to model preferred depth, depth tolerance, and peak abundance of the most common species in the Kope Formation based upon DCA scores (though not absolute depth). Using the same ordination, Miller et al. (2001) produced a meter scale correlation between five localities based on faunal composition as this was not possible using lithologic characters. Indeed Miller et al.'s DC1 scores were indicative of environmental variables related to water depth. High DC1 scores were associated with species previously interpreted as occurring in shallow water facies and low DC1 scores were associated with the interpreted deeper water species. By ordering sample level DC1 scores according to stratigraphic position Miller et al. (2001) were able to produce a relative water depth curve for each of their five localities. These curves, based solely on taxonomic composition, allowed for high-resolution correlation between locations and revealed a dynamic seafloor where relative water depth changed between locations through time. Scarponi and

Kowalewski (2004), following a similar method, applied DCA to Plio-Pleistocene molluskan associations from the Po River Plain. In this case, Scarponi and Kowalewski were able to interpret absolute (rather than relative) depth estimates by relating DC1 scores to the bathymetric preferences of the modern counterparts living in modern coastal environments of Italy. High resolution sampling coupled with the sensitive environmental proxy provided by DC1 scores allowed the authors to further refine independently-derived lithologically-based sequence stratigraphic interpretations of these Quaternary sequences. Bonelli and Patzkowsky (2008) utilized DCA (in addition to nMDS and ANOSIM) on matrices of faunal occurrences in the Late Mississippian (Chesterian) of the Illinois Basin. From this work they were able to deduce that taxa were more strongly sorted along an environmental gradient prior to the onset of the Late Paleozoic Ice Age (LPIA) than after. Bonelli and Patzkowsky's (2008) conclusion that generalist species fared better during the LPIA at the regional scale rings true with Powell's (2005) conclusion that brachiopods with narrow geographic ranges were selected against during the LPIA at the global scale. However, it should be noted that Bonelli and Patzkowsky did not find evidence of selective extinction of environmentally-restricted species in the Illinois Basin. McFadden et al. (2009) used DCA, in addition to correspondence analysis and canonical variate analysis, to biostratigraphically sub-divide the Ediacaran Doushantuo Formation of South China. The DCA of 1,082 acritarch occurrences from 84 horizons suggested that the taxonomic composition of the samples was not related to environmental gradients or to taphonomic processes, but was dependent upon whether the sample came from the lower or upper Doushantuo Formation, indicating that the acanthomorphic acritarchs are useful index fossils. These results are an important step toward a global biostratigraphic subdivision and correlation of the Ediacaran.

Detrended correspondence analysis can be performed in PAST and PC-ORD. See Hill (1973), Hill and Gauch (1980), Ter Braak (1986), Hammer and Harper (2006), and references therein for further discussion of DCA and related techniques.

2.3.5 *Discriminant Analysis and Canonical Variate Analysis*

Discriminant analysis (DA) and canonical variate analysis (CVA) are related ordinations, which project multivariate data sets to a smaller number of dimensions as well as maximize the differences between pre-defined groups (Albrecht 1980; Blackith and Reyment 1971; Hammer and Harper 2006). DA projects a multivariate data set to one dimension and maximizes the differences between two pre-defined groups. CVA is used with a multivariate data set with three or more groups defined *a priori*. The number of dimensions to which the data are reduced in CVA is equal to the number of pre-defined groups minus one. DA and CVA are useful tools for discovering the combination of variables that account for the differences between groups. DA and CVA report eigenvectors (or loadings; similar to PCA) which relate the DA axis or CVA axes to the original variables. Using the eigenvectors in a linear function allows one to attempt to classify new specimens into a pre-defined group. See Hammer and Harper (2006) for a brief discussion on applying DA to problems of taxonomic classification and testing the reliability of the discriminant function. Though they are useful for questions of taxonomy, one should take care when using DA or CVA because one chooses the groups *a priori* and because these ordinations maximize the differences between the pre-defined groups (for example, as compared to PCA). On these grounds, circular reasoning may come into play when arguing for the veracity of certain groups (e.g., genera, subspecies, sexual dimorphism, etc.). As the two methods are quite similar, the rest of this section will focus only on CVA.

Canonical variate analysis is composed of a series of rotation and rescaling procedures related to within-group and between-group variation (Albrecht 1980). By expanding on examples provided by Albrecht (1980), Fig. 2.5 gives a schematic representation of the CVA procedure. Panels a–d illustrate a successful CVA on data that are multivariate normal and have equal variance–covariance matrices (i.e., the relationships between the variables must be similar among the pre-defined groups). This schematic is of course a simplified two-dimensional illustration of what occurs in a more complex multivariate hyperspace. The first step in CVA can be seen as a rigid rotation of the original coordinates such that the within-group variation is maximized (Fig. 2.5a–b). Within-group variation is standardized in the second step (rescaling; Fig. 2.5c). In the third step between-group variation is maximized or aligned with the primary axis (CV1) (Fig. 2.5d). In this sense CVA can be viewed as two PCA procedures with an intervening rescaling of variation. The primary axis, CV1, explains the majority of the variation in the data and should display the most distinction between the pre-defined groups. Subsequent axes explain the majority of the remaining variation. It is important to stress that CVA works best when the data have a multivariate normal distribution and equal variance–covariance matrices. Figure 2.5e–h illustrate an example of what can occur when the data do not meet these assumptions. Variables 1 and 2 have similar variances in groups A and C but do not covary in the same way. Variables 1 and 2 display a much larger variance in group B than in groups A and C with similar and dissimilar covariances, respectively. The hypothetical end result for the CVA of these groups results in very little distinguishing power along CV1. One can find better distinction among the three groups along CV2 (though still not that great), and of course this should not be the case. Canonical variate analysis works best with continuous variables, but other variable types are allowed. There should be as many (or more) observations as variables. PAST and SAS can perform CVA.

Kowalewski et al. (1997) used CVA to analyze the morphology of seven populations of four *Glottidia* species (lingulid brachiopods). Considered "living fossils", lingulids are morphologically very simple organisms and interpreted to display very slow rates of evolution. By subjecting a data matrix of 162 specimens

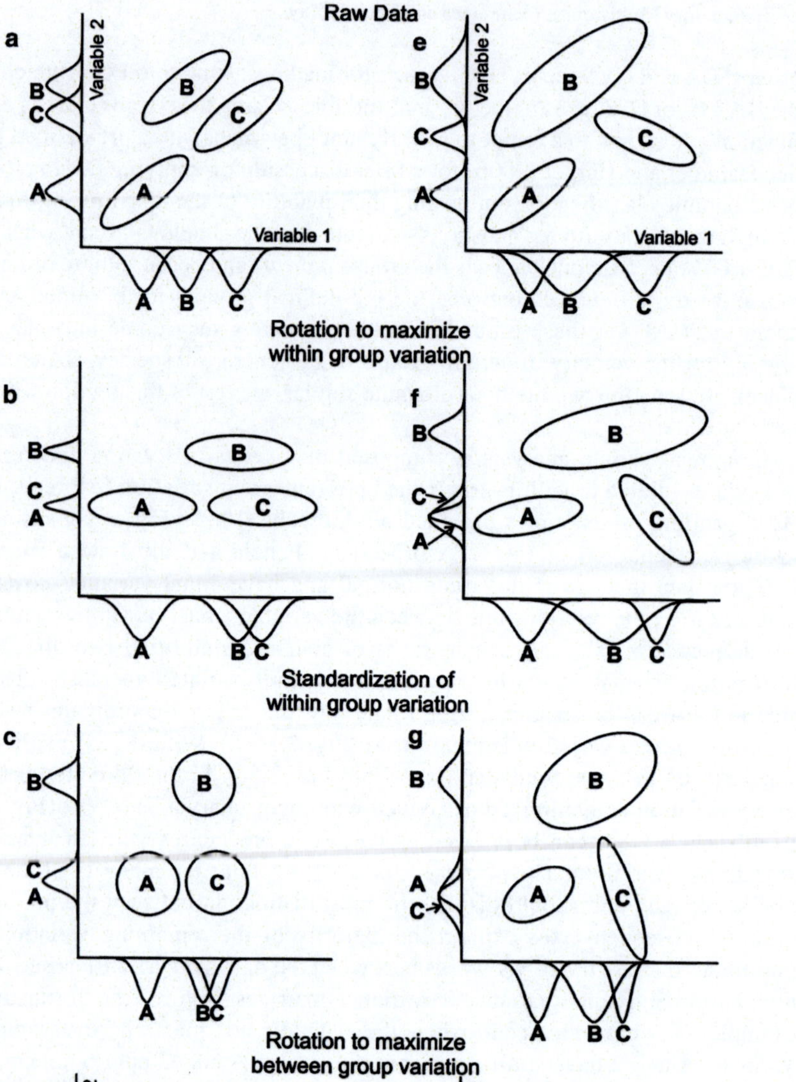

Fig. 2.5 Schematic representation of the stages of Canonical Variate Analysis (CVA). Panels (**a–d**) depict the proper use of CVA on a matrix with three pre-defined groups, whose distribution is multivariate normal, and have equal variance–covariance matrices. Panels (**e–h**) depict the problems which can arise when data with unequal variance–covariance matrices are used with CVA (Modified and expanded after Albrecht (1980))

and six linear measurements to PCA and CVA Kowalewski et al. performed a size free morphological analysis and recognized five distinct morphogroups. Their findings went against the commonly held assertion that lingulid morphology is too simple to preserve useful taxonomic information. McFadden et al. (2009), as previously mentioned (see DCA section) utilized CVA in their biostratigraphic investigation of Ediacaran-aged Doushantuo Formation acanthomorphic acritarchs. So as to take a more conservative approach, McFadden and others performed CVA on the DCA scores (continuous variables) and classified the observations *a priori* according to lithology of the samples. This approach maximized the differences in taxonomic composition between the different lithologies. Even so, the CVA scores grouped according to biozone and not lithology, further strengthening their case for two assemblage biozones.

2.4 Conclusions

I have reviewed a few of the more common types of multivariate ordinations, discussed their utility and limitations, highlighted the data requirements for their proper use, and discussed a few examples of how these ordinations have been utilized in the study of early life. The selection of examples highlighted here, including some Precambrian examples when available and appropriate, is intended to motivate readers to ask new questions about their projects and research interests. Moreover, it should help to demonstrate that there is much work to be done in the study of early life and ecosystems. As the body of published literature on early life grows there will be more and more opportunity for meta-analysis and multivariate ordinations to address both "small-picture" and "big-picture" questions alike. It is my hope that my selection of examples will quickly become dated as this body of work continues its rapid growth.

Acknowledgments I would like to thank Marc Laflamme, Jim Schiffbauer, and Stephen Dornbos for inviting me to contribute to this volume. Michelle Casey, Franz Fürsich, Marc Laflamme, and Arnie Miller provided valuable feedback which greatly improved this contribution. Support for this work was provided by the Alexander von Humboldt Stiftung.

References

Albrecht GH (1980) Multivariate analysis and the study of form, with special reference to canonical variate analysis. Am Zool 20:679–693

Alroy J, Marshall CR, Bambach RK, Bezusko K, Foote M, Fürsich FT, Hansen TA, Holland SM, Ivany LC, Jablonski D, Jacobs DK, Jones DC, Kosnik MA, Lidgard S, Low S, Miller AI, Novack-Gottshall PM, Olszewski TD, Patzkowsky ME, Raup DM, Roy K, Sepkoski JJ Jr, Sommers MG, Wagner PJ, Webber A (2001) Effects of sampling standardization on estimates of Phanerozoic marine diversification. PNAS 98:6261–6266

Alroy J, Aberhan M, Bottjer DJ, Foote M, Fürsich FT, Harries PJ, Hendy AJW, Holland SM, Ivany LC, Kiessling W, Kosnik MA, Marshall CR, McGowan AJ, Miller AI, Olszewski TD, Patzkowsky

ME, Peters SE, Villier L, Wagner PJ, Bonuso N, Borkow PS, Brenneis B, Clapham ME, Fall LM, Ferguson CA, Hanson VL, Krug AZ, Layou KM, Leckey EH, Nürnberg S, Powers CM, Sessa JA, Simpson C, Tomašových A, Visaggi CC (2008) Phanerozoic trends in the global diversity of marine invertebrates. Science 321:97–100

Bambach RK (1983) Ecospace utilization and guilds in marine communities through the Phanerozoic. In: Tevesz MJS, McCall, PL (ed) Biotic interactions in recent and fossil benthic communities. Topics in geobiology, vol 3. Plenum, New York, pp 719–746

Bambach RK, Bush AM, Erwin DH (2007) Autecology and the filling of ecospace: key metazoan radiations. Palaeontology 50:1–22

Blackith RE, Reyment RA (1971) Multivariate morphometrics. Academic, London

Bonelli JR Jr, Patzkowsky ME (2008) How are global patterns of faunal turnover expressed at regional scales? Evidence from the upper Mississippian (Chesterian series), Illinois Basin, USA. Palaios 23:760–772

Boyce CK (2005) Patterns of segregation and convergence in the evolution of fern and seed plant leaf morphologies. Paleobiology 31:117–140

Briggs DEG, Fortey RA, Wills MA (1992) Morphological disparity in the Cambrian. Science 256:1670–1673

da Vinci L (1939) The earth and sea. In: Mather KF, Mason SL (eds) A source book in geology. McGraw-Hill, New York

Efron B (1981) Nonparametric standard errors and confidence intervals. Can J Stat 9:139–172

Foote M (1995) Morphological diversification of Paleozoic crinoids. Paleobiology 21:273–299

Gould SJ (1989) Wonderful life: the Burgess Shale and the nature of history. Norton, New York

Gower JC (1966) Some distance properties of latent root and vector methods used in multivariate analysis. Biometrika 53:325–338

Hammer Ø, Harper DAT, Ryan PD (2001) PAST: paleontological statistics software package for education and data analysis. Palaeontol Electron 4: http://palaeo-electronica.org/2001 1/past/issue1 01.htm

Hammer Ø, Harper DAT (2006) Paleontological data analysis. Blackwell, Oxford

Hill MO (1973) Reciprocal averaging: an eigenvector method of ordination. J Ecol 61: 237–249

Hill MO, Gauch HG Jr (1980) Detrended correspondence analysis: an improved ordination technique. Vegetatio 42:47–58

Holland SM, Miller AI, Meyer DL, Dattilo BF (2001) The detection and importance of subtle biofacies in monotonous lithofacies: the Upper Ordovician Kope Formation of the Cincinnati, Ohio region. Palaios 16:205–217

Huntley JW, Kowalewski M (2007) Strong coupling of predation intensity and diversity in the Phanerozoic fossil record. PNAS 104:15006–15010

Huntley JW, Xiao S, Kowalewski M (2006a) 1.3 billion years of acritarch history: an empirical morphospace approach. Precam Res 144:52–68

Huntley JW, Xiao S, Kowalewski M (2006b) On the morphological history of Proterozoic and Cambrian acritarchs. In: Xiao S, Kaufman AJ (eds) Neoproterozoic geobiology and paleobiology. Topics in geobiology, vol 27. Springer, Berlin/Heidelberg/New York, pp 23–56

Huntley JW, Yanes Y, Kowalewski M, Castillo C, Delgado-Huertas A, Ibáñez M, Alonso MR, Ortiz JE, de Torres T (2008) Testing limiting similarity in quaternary terrestrial gastropods. Paleobiology 34:378–388

Jablonski D (1993) The tropics as a source of evolutionary novelty through geological time. Nature 364:142–144

Jernvall J, Hunter JP, Fortelius M (1996) Molar tooth diversity, disparity, and ecology in Cenozoic ungulate radiations. Science 274:1489–1492

Jolliffe IT (1986) Principal component analysis, 1st edn. Springer, New York

Jolliffe IT (2002) Principal component analysis, 2nd edn. Springer, New York

Kowalewski M, Dyreson E, Marcot JD, Vargas JA, Flessa KW, Hallman DP (1997) Phenetic discrimination of biometric simpletons: paleobiological implications of morphospecies in the lingulide brachiopod *Glottidia*. Paleobiology 23:444–469

Kowalewski M, Goodfriend GA, Flessa KW (1998) High resolution estimates of temporal mixing within shell beds: the evils and virtues of time-averaging. Paleobiology 24:287–304
Kruskal JB, Wish M (1978) Multidimensional scaling sage university paper series on quantitative applications in the social sciences. Sage Publications, Beverly Hills/London
Laflamme M, Narbonne GM, Anderson MM (2004) Morphometric analysis of the Ediacaran frond *Charniodiscus* from the Mistaken Point Formation, Newfoundland. J Paleont 78:827–837
Madin JS, Alroy J, Aberhan M, Fürsich FT, Kiessling W, Kosnik MA, Wagner PJ (2006) Statistical independence of escalatory ecological trends in Phanerozoic marine invertebrates. Science 312:897–900
Marcus L (1990) Traditional morphometrics. In: Rohlf FJ, Bookstein FL (eds) Proceedings of the Michigan Morphometrics workshop. Special Publication Number 2. The University of Michigan Museum of Zoology, Ann Arbor
McCune B, Mefford MJ (2006) PC-ORD, multivariate analysis of ecological data, version 5. MjM Software Design, Gleneden Beach
McCune B, Grace JB, Urban DL (2002) Analysis of ecological communities. MjM Software Design, Gleneden Beach
McFadden KA, Xiao S, Zhou C, Kowalewski M (2009) Quantitative evaluation of the biostratigraphic distribution of acanthomorphic acritarchs in the Ediacaran Doushantuo Formation in the Yangtze Gorges area, South China. Precambrian Res 173:170–190
Miller AI, Holland SM, Meyer DL, Dattilo BF (2001) The use of faunal gradient analysis for intraregional correlation and assessment of changes in sea-floor topography in the type Cincinnatian. J Geol 109:603–613
Payne JL, Boyer AG, Brown JH, Finnegan S, Kowalewski M, Krause RA Jr, Lyons SK, McClain CR, McShea DW, Novack-Gottshall PM, Smith FA, Stempien JA, Wang SC (2009) Two-phase increase in the maximum size of life over 3.5 billion years reflects biological innovation and environmental opportunity. PNAS 106:24–27
Powell M (2005) Climatic basis for sluggish macroevolution during the late Paleozoic ice age. Geology 33:381–384
Raup D (1972) Taxonomic diversity during the Phanerozoic. Science 177:1065–1071
Reyment RA (1991) Multidimensional paleobiology. Pergamon, Oxford
Roy K (1994) Effects of the Mesozoic marine revolution on the taxonomic, morphologic, and biogeographic evolution of a group: apporhaid gastropods during the Mesozoic. Paleobiology 20:274–296
Scarponi D, Kowalewski M (2004) Stratigraphic paleoecology: bathymetric signatures and sequence overprint of mollusk associations from upper quaternary sequences of the Po Plain, Italy. Geology 32:989–992
Schiffman SS, Reynolds ML, Young FW (1981) Introduction to multidimensional scaling: theory, methods, and applications. Academic, New York
Schopf JW (2000) Solutions to Darwin's dilemma: discovery of the missing Precambrian record of life. PNAS 97:6947–6953
Schopf JW (2009) The hunt for Precambrian fossils: an abbreviated genealogy of the science. Precam Res 173:4–9
Sepkoski JJ Jr, Bambach RK, Raup DM, Valentine JW (1981) Phanerozoic marine diversity: a strong signal from the fossil record. Nature 293:435–437
Shen B, Dong L, Xiao S, Kowalewski M (2008) The Avalon explosion: evolution of Ediacara morphospace. Science 319:81–84
Sokal RR, Rohlf FJ (1994) Biometry: the principles and practice of statistics in biological research. Freeman, San Francisco
Stevens SS (1946) On the theory of scales of measurement. Science 103:677–680
Ter Braak CJF (1986) Canonical correspondence analysis: a new eigenvector technique for multivariate direct gradient analysis. Ecology 67:1167–1179
Thomas RDK, Shearman RM, Stewart GW (2000) Evolutionary exploitation of design options by the first animals with hard skeletons. Science 288:1239–1242

Velleman PF, Wilkinson L (1993) Nominal, ordinal, interval, and ratio typologies are misleading. Am Stat 47:65–72

Waggoner B (2003) The Ediacaran biotas in space and time. Integr Comp Biol 43:104–113

Wills MA (1998) Cambrian and recent disparity: the picture from priapulids. Paleobiology 24:177–199

Xiao S, Dong L (2006) On the morphological and ecological history of Proterozoic macroalgae. In: Xiao S, Kaufman AJ (eds) Neoproterozoic geobiology and paleobiology. Topics in geobiology, vol 27. Springer, Berlin/Heidelberg/New York, pp 57–90

Chapter 3
Morphometrics in the Study of Ediacaran Fossil Forms

Marc Laflamme and Michelle M. Casey

Contents

Abstract The evaluation of fossil shapes is the primary goal in the study of paleontology. Unlike biologists who are able to utilize an animal's behavior, soft tissue anatomy, and molecular make-up to study species, paleontologists must concentrate on phenotypical approaches that rely on morphological differences

M. Laflamme
Smithsonian Postdoctoral Fellow, Department of Paleobiology, Smithsonian Institution, National Museum of Natural History, Washington, DC 20013-7012, USA
e-mail: laflammem@si.edu

M.M. Casey (✉)
Department of Geology and Geophysics, Yale University, New Haven, CT 06520-8109, USA
e-mail: michelle.casey@yale.edu

M. Laflamme et al. (eds.), *Quantifying the Evolution of Early Life*, Topics in Geobiology 36,
DOI 10.1007/978-94-007-0680-4_3,

between fossil populations. Morphometric methods which quantify form and shape are extremely powerful tools for assessing questions related to taxonomy, morphospace occupation, morphological disparity or taxon-free estimates of diversity, competition between taxa, ecological character displacement, predatory defense, changes in predatory behavior, anatomical responses to environmental change, and much more. The developing field of morphometrics attempts to mathematically describe shape variations between individuals and can be used to statistically evaluate correct species assignment, isolate shape changes, or identify regions within a phenotype where shape changes are most variable or pronounced. Traditional morphometric approaches focus on the evaluation of several distance measurements taken within a series of specimens, while geometric morphometrics replaces distance-based measurements by evaluating the displacement of corresponding anatomically important locations on specimens which are termed landmarks. Both traditional and geometric morphometric methods allow for the comparison of large samples through a variety of multivariate statistical techniques, many of which will be explored in this chapter. Considering the relative difficulty of characterizing the taxonomic affinities of Precambrian life, paleontologists have adopted morphometric techniques to help elucidate the true diversity of early life.

Keywords Non-Metric Multidimensional Scaling (nMDS) • Principal Components Analysis (PCA) • Principal Coordinates Analysis (PCO) • Ediacara biota • Geometric morphometrics • Landmarks • *Charniodiscus* • *Charnia* • Mistaken point

3.1 Introduction

Morphometrics is the quantitative study of morphological shape, and as such "*examines central tendencies of shape, shape variation, group differences in shape, and associations of shape with extrinsic factors*" (Slice 2007). Stemming from the Greek "morph" for shape, and "metron" meaning measurement, morphometrics is a diverse field of study which, at its core, utilizes statistical and ordinational numerical methods to identify and isolate variations in shapes between sample populations. From a paleontological perspective, such techniques are appealing due to their intrinsic focus on describing biological form and their ability to isolate minor differences in morphology (e.g., Kowalewski et al. 1997).

Body fossils are the source of several types of quantitative data ranging from occurrence data through several types of morphological information. Morphological data pertaining to the size and shape of organisms can be used to evaluate taxonomic affinity, evolutionary history, paleobiology, and paleoecology. Without the luxury of DNA testing, fossil identification of morphospecies will necessarily rely on diagnosable morphological differences between specimens. Furthermore, morphometric assessment of ontogenetic allometry, or changes in shape through ontogeny, can have paleobiological and taxonomic implications (e.g., Houck et al. 1990).

Morphometrics can also provide a taxon-free metric of diversity, e.g., morphological disparity, which is useful when the taxonomic relationships are unknown or problematic (Huntley et al. 2006). Estimates of morphometric similarity in feeding structures have been used to identify and evaluate the effects of competition in the fossil record (Hermoyian et al. 2002). Morphometrics is also well suited to the study of predator-prey interactions. For instance, the anatomical placement of predatory traces on prey organisms can yield information on the behavior of predators (e.g., Casey and Chattopadhyay 2008). Finally, morphometrics can also be used to assess the range of ecophenotypic variation by comparing the morphology of a single taxon across environments or lithologies (e.g., Geary 1992). This is but a small sample of projects which highlight the versatility of morphometric methods as applied to paleobiology.

In this chapter we use two distinct datasets of Ediacara biota from Mistaken Point in Newfoundland, Canada (Fig. 3.1; Laflamme et al. 2004, 2007) to showcase how different morphometric methods can be applied to the study of early life. The Ediacara biota represent an enigmatic grouping of soft-bodied organisms globally distributed, but temporally restricted, to the latest Ediacaran and earliest Cambrian (578–541 Ma; Narbonne 2005; Xiao and Laflamme 2009). These organisms are large (cm-m scale), morphologically complex, and difficult to place within existing higher taxa due to their unique morphological architectures unknown from younger fossils (Narbonne 2004; Narbonne et al. 2009). For these reasons, morphometric studies are apt to shed significant insight into the biology of these organisms and hopefully resolve relationships within the Ediacara biota.

Our objectives with this chapter are to showcase the strengths and limitations of distance-based and landmark-based morphometrics, to outline the types of morphometric data that can be collected to best analyze these fossils, to display the diversity of analytical tools used for data visualization, and to contrast which approaches are better suited for certain types of data.

3.2 Definitions and Terminology

As with all fields of study, morphometrics comes with a detailed set of terminology. Importantly, certain terms which are common in the English vernacular have very specific meanings when used to describe morphometric data or analyses. This can result in confusion and we invite the reader to consult any number of statistical or morphometric textbooks, for example Bookstein (1991), Davis (2002), Dryden and Mardia (1998), Hammer and Harper (2006), Sokal, and Rohlf (1994), and Zelditch et al. (2004). In particular, the reader may wish to consult the SUNY Stony Brook morphometrics website (http://life.bio.sunysb.edu/morph/index.html), which includes an excellent glossary and bibliography to various topics covered in this chapter, as well as Norman MacLeod's morphometrics column Paleo Math 101 published in the Palaeontological Association Newsletters and freely available online at the following website http://www.palass.org/modules.php?name=palaeo_math.

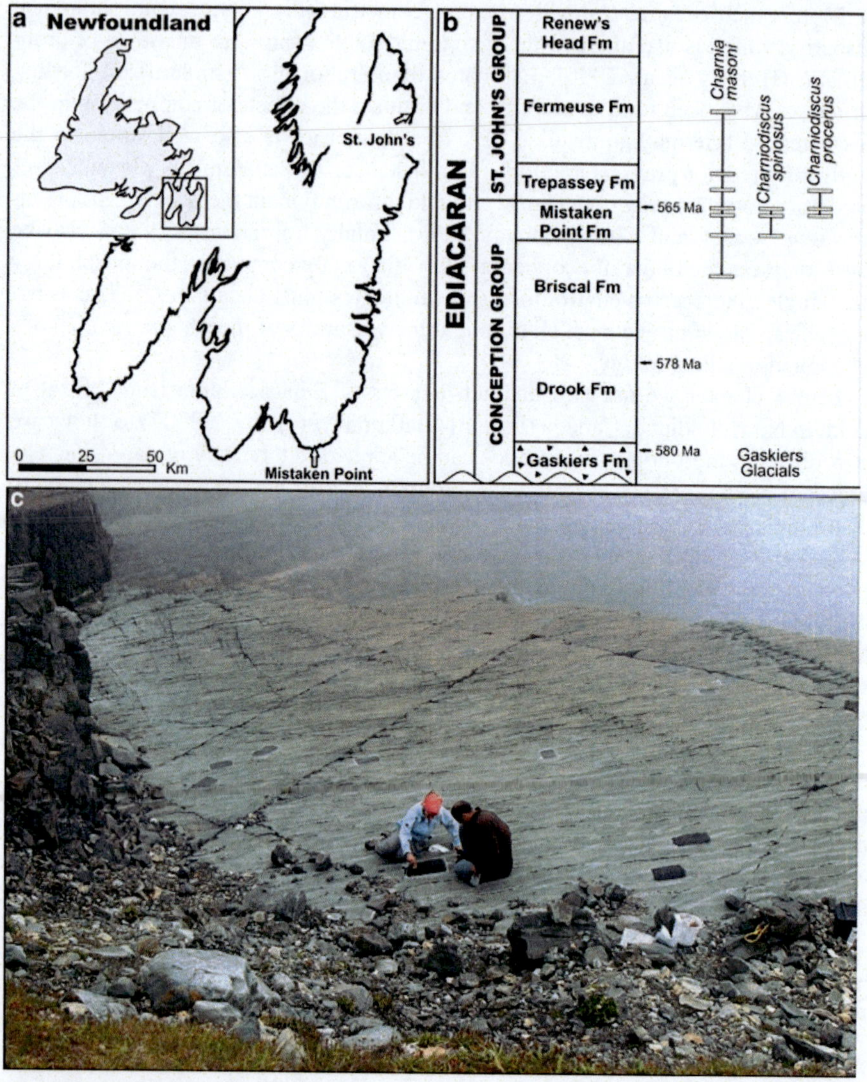

Fig. 3.1 Location map of Mistaken Point, Newfoundland. **(a)** Map of Newfoundland. **(b)** Stratigraphic column with the distribution of *Charnia* and *Charniodiscus*. **(c)** Field photograph of the famous "D" surface within the Mistaken Point Formation

3.2.1 Size, Shape, and Form

The distinction between size, shape, and overall form represent important theoretical questions in morphometric studies. Size is a seemingly straightforward concept however it can sometimes be difficult to define operationally. Size is "*an increase or decrease in the magnitude of linear distances between features (= landmarks) that occurs at the same rate in all regions of the form*" (MacLeod 2009a). A size measure can comprise a single variable (area, volume, distance measurement), or a metric derived from multiple variables which are positively correlated with size (such as centroid size, see Sect. 3.4.2.2). Choosing an appropriate size measure is important as it greatly affects one's ability to evaluate allometry (see below). Shape on the other hand is typically classified as "*all the geometric information remaining in an object (such as a landmark configuration) after differences in location, scale and rotation effects are removed*" (Zelditch et al. 2004). Finally, if you combine the size and shape of an organism, you have defined its form, as form "*includes all the geometric information not removed by rotation and translation*" (Zelditch et al. 2004).

3.2.2 Isometry and Allometry

The way in which organismal form changes through ontogeny can be studied morphometrically. Isometry is the simplest growth vector to understand, as it consists of a "*transformation that leaves distances between points unaltered*" (Zelditch et al. 2004), implying that size is uncorrelated to shape and morphological characters retain the same shape despite an increase in size. Allometry implies that characters show relative shape change with growth and typically follow the power function $Y = bX^a$ where a and b are constants and X and Y are measured variables (Foot and Miller 2007). Allometry can be broken down into several types, including static allometry, ontogenetic allometry and evolutionary allometry (see Cheverud 1982). Allometry can be studied using either bivariate or multivariate morphometric methods, or a combination of the two (e.g., Houck et al. 1990).

Using a single variable as a size measure assumes that a particular variable grows isometrically or nearly isometrically. Since it is difficult to ensure that a chosen size variable is changing isometrically, one must be careful not to conflate the chosen size measure with overall body size. In multivariate analyses it is possible to calculate a theoretical isometric size vector which could be used in place of a single-variable size measure (e.g., PC1 score discussed below). If one were to plot a variable against the chosen size measure, the slope of the resulting line in log-log space is equal to a, from the equation $\mathrm{Y} = \mathrm{bX}^{\mathrm{a}}$ ($\log [\mathrm{Y}] = \mathrm{a} \log [\mathrm{X}] + \log [\mathrm{b}]$). If a is greater than 1.00 the character or skeletal element in question demonstrates positive allometry (becomes relatively larger as the chosen size measure increases); if a is less than 1.00 the character or skeletal element demonstrates negative allometry (becomes relatively smaller as the chosen size measure increases). If $a = 1.00$, the character of skeletal element in question is isometric with respect to the chosen size measure (Foot and Miller 2007).

3.2.3 Discrete, Categorical, and Continuous Variables

The types of variables collected and analyzed as part of a morphometric study will greatly influence the choice of appropriate statistical procedure applied to the data. As such, care must be placed in ensuring that the collected data is appropriate and does not break any fundamental assumptions associated with the statistical procedures. For the purpose of the analyses performed in this study, the following types of variables are most often collected.

Discrete variables, also known as nominal, meristic, or discontinuous variables, include any aspect of morphology which is usually expressed as an integer for which values in between integers are impossible (Sokal and Rohlf 1994; Davis 2002). Examples include aspects of morphology, such as the number of pleural segments in a trilobite, or characteristics which are instead coded, such as color or presence/absence data. A subset within discrete variables is categorical variables which are those used to group specimens a priori. Examples of categorical variables are species, locations, stratigraphic units, geologic ages, etc.

Continuous variables are those which include decimals, such as linear dimensions, angles, weights, and Cartesian coordinates (Sokal and Rohlf 1994). Continuous data can take many forms, including interval scale measurements such as temperature in which the length between each interval is the same, but in which a zero value has no real meaning (Davis 2002). Linear dimensions such as lengths, widths, or heights of structures (e.g., the length of the genal spine in a trilobite), or the distance between such characters (e.g., the distance between the eye sockets in a vertebrate skeleton) represent ratio scales, which are similar to interval scales in that they share an equal increment between each number, but also have a true zero value making negative values impossible (Davis 2002).

3.3 Scale of Morphometric Studies

3.3.1 Assemblage-Level

Typical assemblage-level questions involve studying all organisms (or more likely a predefined subset, such as all microfossils, or all invertebrates, or all mollusks) within a defined paleoecosystem, facies, or time period, and are particularly applicable to studies of disparity and morphospace. Unlike diversity studies which concentrate on how species are distributed, disparity or morphological variety "*refers to the variance in form or the amount of morphological space occupied, irrespective of taxonomic richness*" (Foot 1992). Therefore, disparity studies focus on variations in biological construction irrespective of our ability to identify species (diversity). Studies by (Huntley et al. 2006 and Chap. 2) and Shen et al. (2008) showcase how "taxon-free" approaches can reveal a great deal about how constructional morphologies changed with time and in response to an evolving ecosystem. Briggs et al. (1992) utilized a combination of Principal Components Analysis (PCA) with Phylogenetic Analyses of Parsimony (PAUP) to evaluate the assemblage-level disparity between Cambrian (mostly Burgess Shale) and modern arthropods, demonstrating how taxon-free studies can be used to set constraints on evolutionary processes in deep time.

3.3.2 Taxon-Level

Taxon-level analyses are typically designed to evaluate two or more species based on their forms. The study of *Charniodiscus* serving as our case study (Laflamme et al. 2004; Hofmann et al. 2008) is an example of how morphometrics can be employed to distinguish between species. Studies of evolutionary allometry are another example of a taxon-level analysis (e.g., Gerber et al. 2007). Taxon-level studies can concentrate on differences in mean forms, shapes, or sizes, and can evaluate differences between fossil populations.

3.3.3 Species-Level

Individual-level studies focus on variations within a single species. Individual-based analyses largely focus on questions pertaining to allometry, but can also be useful for isolating sexual dimorphism or phenotypical plasticity associated with environmental and geographical changes.

3.4 Morphometric Datasets

In order to evaluate several different morphometric techniques we have decided to utilize two distinct datasets of Ediacaran fronds from Mistaken Point in Newfoundland. Specimens from Mistaken Point are ideally suited to morphometric analyses because they constitute *in situ* life assemblages of entire Ediacaran communities, which include thousands of complete specimens on several different bedding planes cropping out all along the coast. The paleocommunities were smothered and buried in place by volcanic ash, effectively encasing these organisms like an Ediacaran Pompeii (Narbonne 2005). These life assemblages include presumed juvenile and adult specimens which allow for ontogenetic studies to be performed (Laflamme et al. 2004).

Of the multitude of Ediacaran organisms and body plans described (Xiao and Laflamme 2009), the fronds constitute the most amenable to morphometric studies because their complex morphologies allow for numerous independent measurements to be taken on a single specimen (Fig. 3.2; Laflamme and Narbonne 2008a). Fronds were anchored to the seafloor by a circular basal holdfast from which extended a stem and leaf-like petalodium. The petalodium consists of a series of primary branches which are architecturally distinct and form the basis of recent higher-order taxonomic divisions amongst the fronds (Laflamme and Narbonne 2008a, b). Like most Ediacaran-aged fossils, these fronds were originally soft-bodied and therefore morphometric studies on these fossils requires exquisite preservation, and added care must be taken to account for taphonomic (preservational) biases when collecting morphometric data.

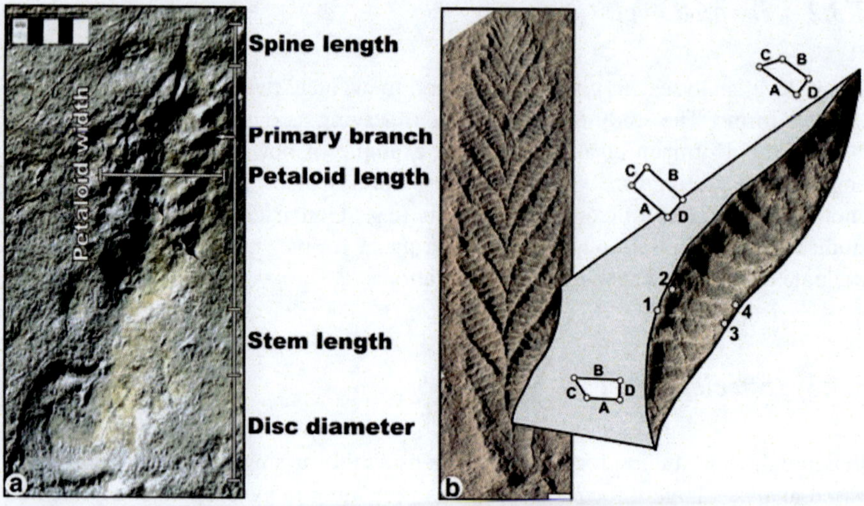

Fig. 3.2 Ediacaran fronds: (**a**) *Charniodiscus spinosus* with labeled measurements taken from each specimen. Scale bar 5 cm. (**b**) *Charnia masoni* with landmarks *1–4* indicated. Scale bar 1 cm

3.4.1 Traditional Morphometric Dataset: Charniodiscus

Charniodiscus is a cosmopolitan frond first described from Charnwood Forest in England but subsequently found at Ediacara in South Australia, the White Sea in Russia, and most recently at Mistaken Point in Newfoundland (see review by Laflamme et al. 2004). Our dataset (Laflamme et al. 2004) was collected with two questions in mind: (1) How many different morphospecies occur within *Charniodiscus* at Mistaken Point and (2) how did *Charniodiscus* grow? Six distinct measurements were collected: Continuous measurements included (1) the petaloid length, (2) petaloid width, (3) stem length, and (4) disc diameter, while discrete variables included (5) number of primary branches, and (6) presence or absence of a spine (Fig. 3.2a).

3.4.1.1 Traditional Morphometric Data Analysis

Traditional morphometric data can be analyzed with univariate (single measurements), bivariate (X-Y cross-plots of two distinct measurements), or more commonly multivariate (more than two distinct measurements) statistical analyses. Examples of univariate studies in Ediacaran paleontology include measuring the diameter of Ediacaran circular fossils such as *Aspidella* (Gehling et al. 2000), in which a highly left-skewed distribution (i.e. high frequency of specimens with small diameters) was used to imply a high rate of juvenile mortality.

Bivariate analyses are common in morphometric studies. Many allometric studies use multiple bivariate plots to compare numerous variables against a size measure (e.g., Houck et al. 1990). Referring to our dataset, comparing the frond length with the stem length (Fig. 3.3a) results in a clear division between two species, with *Charniodiscus spinosus* (white triangles) characterized by a short stem and typically a longer petaloid, while *C. procerus* (black squares) is identified based on the relatively longer stem. In this case, the bivariate analysis clearly differentiates the two species, however if two other variables are compared, such as the petaloid width and the disc diameter, it would be difficult to identify any species differentiation, especially if no *a priori* division was established (Fig. 3.3b).

When it is possible to collect many distinct variables, we recommend evaluating all variables simultaneously through multivariate analyses rather than sequentially as independent pairwise studies. We will evaluate three of the most commonly used analytical methods: Principal Components Analysis (PCA), Principal Coordinates Analysis (PCO), and Non-metric Multidimensional Scaling (nMDS), while paying special attention to the strengths and limitations of all three methods, and differentiating what kinds of data can be evaluated by each technique.

3.4.1.2 Principal Components Analysis (PCA)

Principal Components Analysis (PCA) is an exploratory multivariate analysis that can be used with both traditional and landmark-based morphometric studies. The primary goal of PCA is to allow for the "*projection of a multivariate dataset down to a few dimensions (usually two) in a way that preserves as much variance as possible*" (Hammer and Harper 2006). PCA is responsible for reducing the number of vectors necessary to explain the majority of the variation seen in a dataset, thereby allowing the user to view the bulk of the multivariate variation in a more manageable format as well as analyze it statistically. PCA is a commonly used multivariate analysis, and most multivariate statistical textbooks will have a chapter (or more) dedicated to this method. See Davis (2002), Hammer and Harper (2006), Zuur et al. (2007), and Chaps. 1 and 2 in this volume for instance. PCA is not designed to perform analyses on discrete values (Hammer and Harper 2006).

In our example (Fig. 3.3c–d), we included four continuous measurements (petaloid length, petaloid width, stem length, and disc diameter). Since these variables are in the same units, PCA was run using a covariance matrix. In the resulting analysis PC1 (63.4%) and PC2 (31.6%) account for 95% of the covariance in the dataset, and clearly segregated *Charniodiscus spinosus* (white triangles) from *C. procerus* (black squares). The component loadings (Table 3.1) for PC1 – PCn demonstrate how each distance measurement relates to the PC in question. In cases where most if not all variables have high loading scores on the first PC, it can often be considered as an approximate measure of size (Foot and Miller 2007). Table 3.1 demonstrates that each of the four measurements covaries positively with PC1, suggesting that PC1 is an allometric growth vector, with the smallest specimens located on the left, and the largest specimens on the right (Fig. 3.3c). PC2 on the other hand is dominated almost exclusively by stem length, segregating *C. procerus* (positive PC2) from *C. spinosus* (negative PC2), while PC3 is mostly explained by the petaloid width and

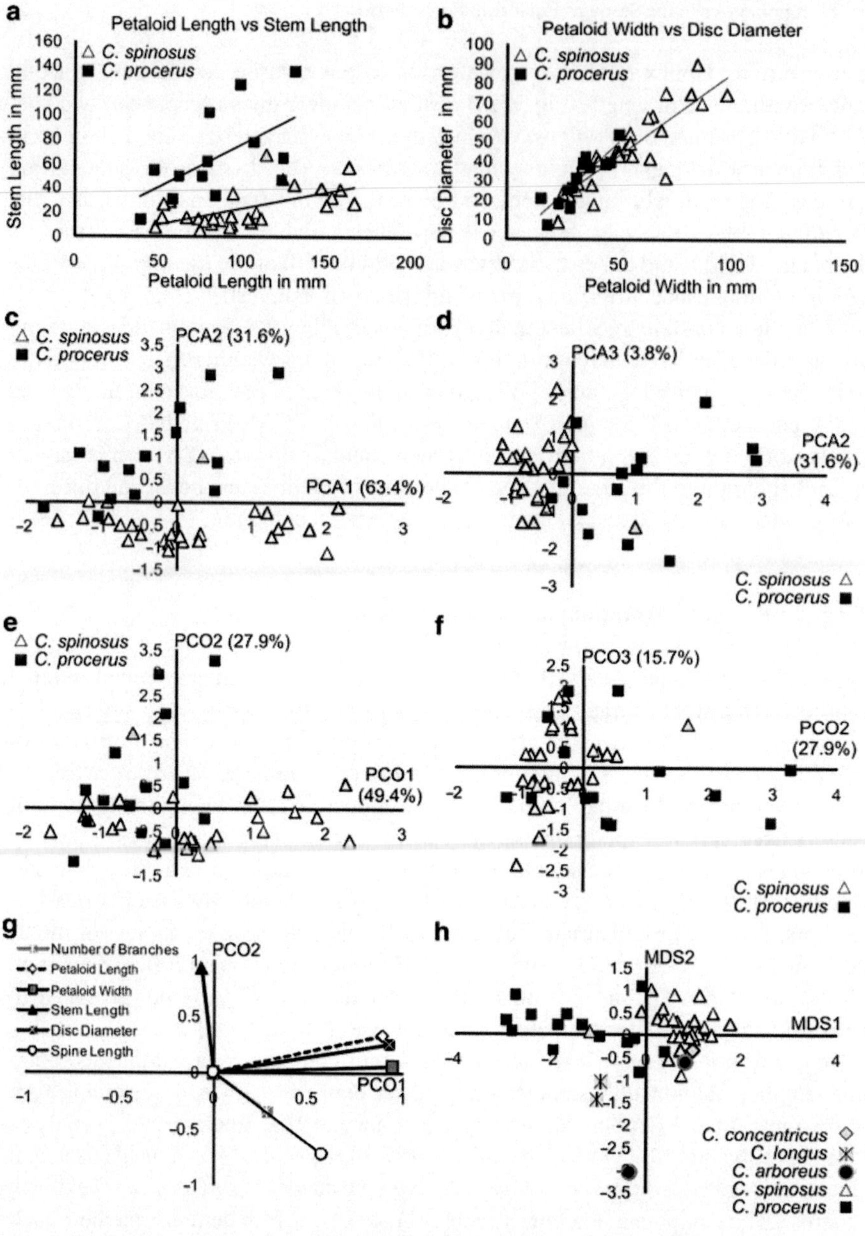

Fig. 3.3 Traditional morphometrics of *Charniodiscus*: **(a)** Bivariate plot comparing the petaloid length with the stem length, demonstrating a clear division between both species of *Charniodiscus* at Mistaken Point. **(b)** Bivariate plot comparing the petaloid width with the disc diameter, which, unlike A, does not display any true differentiation between both species. **(c, d)** Principal Components Analysis (PCA) which includes measurements of the petaloid length, petaloid width, stem length, and disc diameter. **(e, f)** Principal Coordinates Analysis (PCO) which includes the same measurements as C-D in addition to the presence (or absence) of the terminal spine and the number of primary branches. **(g)** Component loadings from the PCO analysis performed in E-F which demonstrates the relevant importance of each character in the construction of PCO 1+2. **(h)** Non-metric Multidimensional Scaling (nMDS) of the same characters as in E-F but which allows for the inclusion of additional *Charniodiscus* species from Charnwood Forest in England and Ediacara in South Australia

Table 3.1 Component loadings and eigenvectors for the PCA analysis of *Charniodiscus* from Mistaken Point, Newfoundland

	PC1	PC2	PC3
Component loadings			
Petaloid length	0.979	−0.089	−0.183
Petaloid width	0.840	−0.396	0.318
Stem length	0.342	0.938	0.057
Disc diameter	0.912	−0.196	0.268
Eigenvectors			
Petaloid length	0.636	−0.116	−1.993
Petaloid width	0.184	−0.174	1.168
Stem length	0.175	0.966	0.488
Disc diameter	0.179	−0.077	0.880

Caption: PC loadings measure the correlation (−1 to 1) of each original variable with the resulting PC axes. All variables are positively correlated with PC1, and as such specimens with higher PC1 scores are necessarily larger. On the other hand, PC2 scores are strongly correlated with the length of the stem. PC loadings are the only way of relating the PCA ordination space to the original distance measurements. Eigenvectors are composed of several eigenvalues that explain the correlation between the measured distances and the PC scores. As such, it is possible to insert new specimens into the analysis without disturbing the original ordination space as follows: PC1 value for unknown = (petaloid length × 0.636) + (petaloid width × 0.184) + (stem length × 0.175) + (disc diameter × 0.179)

disc diameter, which helps little in terms of species differentiation (as demonstrated in 3.3b). A final point of interest is the reoccurring *C. spinosus* specimen (grey triangle) which appears to be nestled within the *C. procerus* morphospace cloud. This specimen was originally misidentified as *C. spinosus* when the analysis was first run due to poor preservation. Following this analysis, we returned to the original source data and concluded that our original identification of this specimen was erroneous. This example goes to show how PCA can also be used to ensure proper identification of morphospecies. Because PCA specifically targets variables with the greatest amount of variance/covariance, it is important to remember that characters with high variance will have a greater influence on the first principle component. Because the ordination PCA produces is based on the original data, each inclusion of additional data/specimens will result in the output of a new ordination.

PCA requires that all data be continuous and present in every sample (i.e. you cannot include any missing data) (Hammer and Harper 2006). Furthermore, PCA is strongly influenced by differences in variance between variables. As such, when all variables were measured in identical units, it is preferred to use a covariance matrix (Fig. 3.5), whereas when units are mixed (e.g., distance measurements in mm vs. temperatures in °C) one should use a correlation matrix. The correlation matrix will standardize all measurements to have a mean of 0.0 and a variance of 1.0 (Davis 2002) in order to account for differences in variance which arise from the mixing of units. However, this standardization can erase meaningful biological differences in variance between characters measured in the same units and is not recommended in those cases (Davis 2002).

As previously mentioned, we also collected data on discrete valuables. In order to include these measurements into our analysis, we must utilize a different statistical approach, and we recommend Principal Coordinates Analysis (PCO).

3.4.1.3 Principal Coordinates Analysis (PCO)

Principal Coordinates Analysis (PCO), also known under the banner of Metric Multidimensional Scaling (MDS), is similar to PCA in using eigenvalues to compare multivariate data, but since PCO can use any measure of association, it can deal with different datasets that PCA could not (Zuur et al. 2007). One major allowance with PCO over PCA is the addition of discrete measurements. Since PCA is based on a correlation or covariance coefficient, it is incorrect to compare continuous with discrete measurements (Zuur et al. 2007). In our case study, PCO allows for the inclusion of two new characters of comparison, the presence/absence of the spine and the number of primary branches.

Figure 3.3e–f demonstrates that the majority of specimens are nicely isolated, but with significant overlap in the third quadrant. This overlap is explained in Fig. 3.3g, which graphically represents the component loadings. Although most measurements are at least moderately correlated with PCO1, which can be considered once again as an approximate growth/size vector, PCO2 is mostly correlated with the stem length and negatively correlated with the presence of a spine and the number of primary branches. Since the presence of a spine is a diagnostic character of *C. spinosus*, and the relatively longer stem length is typical of *C. procerus*, it is not surprising to find these characters opposing each other along PCO2. The number of primary branches is clearly adding to the confusion in the plot, especially since the number of branches is almost identical in both species (between 9 and 13 in well-preserved specimens).

3.4.1.4 Loading Plots

One limitation of multivariate ordinational methods such as PCA and PCO is that changes along the new axes (e.g., PCs) cannot be directly related to changes in the values of specific variables. In order to assess how different quadrants of the new ordination (e.g., PCA or PCO plot) correspond to the original variables, one must consult a loading plot (e.g., Fig. 3.3g). Loading plots break down how much the original variable (i.e. petaloid length) contributes to the displayed component (Hammer and Harper 2006). This plot goes from −1 to 1 on both axes and shows how each of the variables correlates with the axes in question (e.g., PCO1 and PCO2). In our example, stem length is highly positively correlated with PCO2. So specimens with high PCO2 scores will have longer stems than those with low or negative PCO2 scores. It is important to consider and report the amount of variation accounted for by each axis (based on the eignevalues) in both your results and on your graphs.

3.4.1.5 Non-Metric Multidimensional Scaling (nMDS)

Non-metric Multidimensional Scaling (nMDS) is an ordination technique that allows for the inclusion of continuous, discontinuous, and even missing data. As is the case with PCA and PCO, nMDS reduces multidimensional datasets to a few relevant dimensions (two or three) that explain the majority of variation within the samples measured. Unlike PCA and PCO which are based on solving eigenvalue equations, nMDS utilizes numerical optimization methods in order to analyze the variance in a system, although the numerical processing of these models can be computationally overbearing and time consuming (Zuur et al. 2007). NMDS transforms the Euclidean distances between points into ranks, so that although the true Euclidean distances are lost, the relative ranks of these distances between samples are kept (Hammer and Harper 2006). For example, if the distance between point A and B is the largest distance calculated, then it receives the rank of one (1) and as such the distance between these points in the low-dimensional nMDS space will also be the largest. Since nMDS ordination methods are based on ranks rather than Euclidean distances, iterative techniques resample and replot the population over and over again until the best solution is achieved (Hammer and Harper 2006). As with all iterative techniques, the end result may be different with each trial run, and therefore your analysis should be run several times until a stable configuration is reached. It is also important to vary the starting configuration of points in low-dimensional space, since this will greatly influence the final outcome as well.

As nMDS allows for the inclusion of missing data, this technique is ideal for paleontological analyses in which fossils specimens are fragmentary (and may not include all possible distance measurements), disarticulated, taphonomically altered, or in which morphologically disparate taxa are included. In the case of morphologically disparate taxa, missing structures or variables may provide biological information rather than be the result of taphonomic loss or alteration (e.g., Huntley et al. 2006). For instance, if we wished to include a measurement of spine length in our dataset rather than simply a presence/absence metric, all specimens that do not have a spine would have a missing value for this variable. Furthermore, nMDS is the only technique that would allow for a proper evaluation of additional species of *Charniodiscus* from Ediacara, South Australia, and Charnwood Forest, England, because they are missing measurements due to fragmentary/incomplete specimens. Figure 3.3h compares all the same variables utilized in the PCO analysis (petaloid length and width, stem length, disc diameter, number of branches, and presence of a spine) and includes specimens of *C. spinosus*, *C. procerus*, and *C. arboreus* from Mistaken Point, *C. arboreus*, and *C. longus* from South Australia, and *C. concentricus* from England. The nMDS analysis nicely segregates *C. spinosus*, *C. procerus*, and *C. longus*, with some minor overlap of *C. concentricus* and *C. arboreus* with *C. spinosus* likely due to the low number of specimens of *C. concentricus* ($n=1$) and *C. arboreus* ($n=2$). The addition of new specimens would help in isolating the overlapping species, since although the form of their petaloids (describe mostly as a length vs. width ratio) is quite similar, the number of branches, the stem length, and the presence of a spine are all believed to be species-specific traits and would likely isolate these overlapping species into discrete morphospace regions.

3.4.1.6 How to Pick Which Analysis to Use?

How do you decide on which analysis is best suited for your data? It is always up to the research team to decide what is most appropriate, either including more characters and specimens but being restricted in terms of which analyses can be used, or to restrict data collection to complete specimens thus allowing for more powerful statistical tools. Whenever possible, we would recommend comparing both outcomes. Patterns which persist across analyses may be considered robust, while patterns which disappear may need additional consideration. It is certainly acceptable and encouraged to try multiple methods in the exploratory phase of analysis, so long as the initial assumptions of the tests are met. When multiple tests are applied in the confirmatory stage however, it is strongly suggested to report the results of all tests performed, including analyses that do not yield significant results. In many cases, however, you won't have a choice of methods due to constraints imposed by the nature of your data, especially when dealing with specimens that are either damaged, fragmentary, or morphologically distinct enough that not all specimens share the same morphology, making missing data points unavoidable. These concerns are all the more prevalent in paleontological studies due to problems associated with fossil preservation (taphonomy). If one refuses to sacrifice statistical power, missing variables from taphonomically incomplete specimens result in a trade-off between the inclusion of more variables at the expense of decreasing sample size and vice versa (e.g., Casey et al. 2007; Shen et al. 2008).

3.4.2 *Geometric Morphometric Dataset: Charnia*

Charnia (Fig. 3.4) is one of the oldest and longest ranging Ediacaran taxa. First described in the same study that named *Charniodiscus*, *Charnia* is known from Charnwood Forest in England, Ediacara in South Australia, the White Sea in Russia, and Mistaken Point in Newfoundland. The rectangular to pill-shaped secondary modular units in *Charnia* (Fig. 3.2b) are well defined by four distinct pseudo-landmarks (see below) located at each corner of the unit (Laflamme et al. 2007). Our question concerned the nature of the modularity in *Charnia*, and whether secondary modular units were always rectangular in shape and added along the entire length of the branch, or if they were distinct branching units that had clear shape differences and therefore not modular. Geometric morphometrics is ideally suited to testing this hypothesis.

3.4.2.1 Geometric Morphometric Data Analysis

Geometric morphometrics focuses on establishing biologically analogous loci termed landmarks that best explain the shape of the subject in question in 2-dimensional or 3-dimensional space. We will focus exclusively on 2-dimensional geometric morphometrics. The smallest number of landmarks necessary to define a shape is three, and most studies will incorporate substantially more individual landmarks. Traditional (Lele and Richtsmeier 2001) or Type 1 (Bookstein 1991) landmarks are those which correspond to biologically significant structures that represent a point in space at which three structures meet. Type 1 landmarks are ideal for morphometric analyses as they have the least amount of translation or misidentification since they are bracketed on all sides by physical structures. For example, in our dataset on *Charnia*, each corner of an individual secondary modular element was initially identified as a type 1 landmark as it consists of the joining of three different positive ridges on the fossil (Laflamme et al. 2007; Fig. 3.2). However, due to the difficulties associated with the characterization of enigmatic Ediacaran fossils, it is better to consider these points as Type 2 landmarks. Type 2 landmarks (Bookstein 1991) typically represent areas of minimal or maximal curvature in a feature, such as the tip of the spine in *Charniodiscus spinosus*. Type 2 landmarks are troublesome because they are not constrained in all orientations, which means that they are more likely to show operator error in their placement in multiple directions relative to Type 1 landmarks. Type 3 landmarks (Bookstein 1991) are only constrained in one orientation, such as the end of the longest diameter, or the most convex portion of a bivalve shell (umbo). These landmarks are typically the least reliable due to the difficulties in locating the same point on multiple specimens, or even to repeat the same landmark location on the same specimen. Semi-landmarks are often used to define the edge of a curve by placing several equally-spaced points along the curvature of the form (MacLeod 1999). These points are allowed to slide along the curve they define in order to maximize fit among specimens and require specialized statistical treatment which we will not discuss here.

One of the most important aspects of landmark morphometrics is the concern with morphological homology; i.e. that each individual landmark chosen to represent a specific loci on an organism is the same on each specimen analyzed. Homology is an unfortunate term because it has a very specific meaning in biology (shared ancestry) that is not necessarily echoed in landmark selection. For landmark selection, what is necessary is that the points represent the same place on each fossil. As such, landmarks are "*always assumed to represent corresponding parts of locations on the body but they do not always—nor always need to—represent formal homologues in the biological sense of that term*" (MacLeod 2009b). This concern is not always shared by traditional morphometric analyses because distance measurements, such as the widest point in a frond petaloid, can sometimes be taken from slightly different places on each frond specimen. Landmark statistical analyses function with the assumption that your chosen landmarks are morphologically homologous in all of your specimens.

Furthermore, landmarks should be easy to locate on your chosen specimens and be present on all your specimens. Landmark morphometrics is a powerful tool to study minor variations in shape and landmarks that cannot be confidently assigned in all cases will add noise to your data. When a landmark is difficult to isolate in one direction over another, landmark displacement will be biased in that direction and represent an erroneous (untrue) change in morphology. For this reason, it is always recommended to empirically evaluate the repeatability of, or operator error associated with, landmark positioning by repeatedly measuring a single specimen. Landmark morphometric techniques are not always applicable when specimens are composed of discrete articulated parts since the location of the landmarks can sometimes be dependent on the organism's position rather than its actual morphology. This difficulty can be lessened if each skeletal element is analyzed separately as a discrete unit or if all articulated segments are positioned in a standardized fashion. Where individual analysis of every bone is prohibitive (e.g., due to large number of specimens or time constraints) or reposition is impossible because skeletal specimens remain encased in their original sedimentary matrix (e.g., Casey et al. 2007) traditional morphometric techniques may be more appropriate. Landmark methods are suitable for single or fused elements like vertebrate skulls, trilobite cephalons, tests, acritarchs, shells, etc.

Landmark configurations can only provide information contained within the areal coverage of the landmarks, which implies that in order to ensure that all morphological variation is captured by landmark translations, it is essential that an adequate number of landmarks are selected to convey the original shape of the organism under study (Zelditch et al. 2004). Areas of an organism's bodyplan that are found outside of the morphoshape defined by the landmarks will not be evaluated regardless of how much variation is present.

3.4.2.2 Landmark Data Collection and Superposition Methods

Once you have selected a suite of landmarks present on all of your specimens it is necessary to collect the landmark data. We recommend collecting landmarks from digital images of your specimens, making sure that all of your specimens are oriented in the same plane to reduce landmark displacement. This is especially important for specimens that have significant relief. Thankfully, the overwhelming majority of Precambrian fossils, from Acritarchs to carbonaceous filamentous impressions to Ediacaran fossils, are typically preserved as two-dimensional impressions ideal for photography. All photographs must include a scale bar. Landmarks can be collected from digital photographs and processed using TPSDIGW32.EXE (By F. James Rohlf. Version 1.40. http://life.bio.sunysb.edu/morph/) or ImageJ (By Wayne Rasband http://rsbweb.nih.gov/ij/download.html) both available as freeware. The TPSDIGW32.EXE software automatically collects and stores the digitized landmarks into a coordinate matrix and converts these into corresponding vectors of all possible pair-wise distances, which can then be transformed into to shape coordinates. ImageJ automatically records Cartesian coordinates of landmarks in the scale set by the operator. These coordinates can be exported into other statistical programs for superposition.

One of the primary goals of geometric morphometrics is to compare and contrast shapes by removing the influence of size, translation, and rotation from your data (Zelditch et al. 2004). Because the Cartesian coordinates are collected for each landmark using a scale, geometric morphometric methods also retain information related to size or distance between landmarks. These distances can be used to calculate a multi-variable size measure known as centroid size. Centroid size consists of the square root of the summed squared distances from each landmark to the centroid, which represents the average location of all landmarks (Dryden and Mardia 1998). Centroid size can also be used as measure of overall size, since it is uncorrelated with shape variables (Krause 2004; MacLeod 2009a).

Geometric morphometrics uses the raw landmark data collected from scaled digital images to project all the landmarks from all specimens into a single ordination through various different superposition methods. According to Rohlf and Slice (1990), *"Geometrically, this corresponds to superimposing one organism on top of another so that its landmarks align as well as possible (in some sense) with the positions of the corresponding landmarks on the second. Differences in shape are then shown by differences in positions of corresponding landmarks"*. The two most commonly used superposition methods are Procrustes Superposition (more precisely generalized least-squares superposition) and Superposition using Bookstein Coordinates (also sometimes referred to as Bookstein's shape coordinates). Several different incarnations of these methods have been proposed (e.g., Bookstein 1986, 1991; Rohlf 1999; Rohlf and Slice 1990), and we recommend the reader peruse these studies to find the method that best fit their needs. In short, Superposition using Bookstein Coordinates requires two landmarks to be designated as baseline landmarks which are used to center, rotate, and rescale all the other landmarks (Bookstein 1986). Baseline landmarks by definition occupy the Cartesian coordinates (0,0) and (1,0). One of the drawbacks of Bookstein Coordinates is that two landmarks must be sacrificed from the analysis in order to become the baseline. As such, there can be no variation in the placement of the baseline landmarks, which will always occupy points (0,0) and (1,0), regardless of any true morphological variation of those landmarks. Therefore, any variation that would have been present in the baseline landmarks is transferred to all remaining landmarks regardless of their true variation (MacLeod 2009a). Marcus (1990) recommends choosing two landmarks which define the long axis of an object. Superposition using Bookstein Coordinates can also be used to evaluate the anatomical placement of non-homologous traces or structures such as drill-holes or other predation traces (e.g., Casey and Chattopadhyay 2008).

Alternatively, superpositioning of specimens can be performed by overlaying all of your images to reduce the overall distance between all identical landmarks simultaneously. Procrustes Superposition centers, rotates, and scales specimens by simultaneously lining-up all landmarks using either least-square fit or resistant fit optimization techniques (Rohlf and Slice 1990). Procrustes can be performed using the freeware PAleontological STatistics or PAST (By Øyvind Hammer and David Harper http://folk.uio.no/ohammer/past/). In our case, all following steps were performed with the software (CoorDGen6) designed by Davis S. Sheets (http://www3.canisius.edu/~sheets/moremorph.html). Procrustes first calculates the centroid location of each sample, in order to superpose all centroid locations to the centre of a Cartesian graph (0,0), therefore removing the influence of translation on the data. The next step removes size differences between specimens by calculating the centroid size (as above) followed by rescaling all specimens to a centroid size of 1 by dividing all landmarks by their original centroid size. One can think of it as pinning all of your size standardized specimens to a board with a pin occupying the centroid. Finally, in order to eliminate the influence of specimen rotation, all rescaled forms are rotated to allow for an optimal overlap of all landmarks by minimizing the summed squared distances between each corresponding landmark and the reference form. In our example we used the average landmark configuration of all specimens as the reference form. Because Procrustes superimposes specimens using all landmarks, unlike superpositioning using Bookstein coordinates, individual variation in each landmark can be evaluated (e.g., Budd et al. 1994). For a detailed analysis of the mathematics behind Procrustes Superposition and its drawbacks, see Rohlf and Slice (1990) and MacLeod (2009c).

3.4.2.3 *Charnia* Modularity

In our example, four landmarks were placed on every complete secondary modular element from the second to the sixth primary branch of the right side of the holotype of *Charnia masoni* from Charnwood Forest England (Laflamme et al. 2007; Fig. 3.4). Poor preservation of the primary branches on the left side of the specimen, in addition to the incomplete preservation of the furthest branches at the tip of the specimen restricted the analysis to the best six branches on the right side of the holotype. Furthermore, the last few modular units located at the tip of each branch were too poorly preserved to confidently digitize the landmarks (Fig. 3.4).

Once the specimen was digitized, and the coordinate matrix was converted into a procrustes coordinates matrix, it was possible to analyze the samples through standard multivariate statistics such as PCA. The data reconfigured into partial procrustes was performed with PCAGen6N, which performed a PCA analysis on each branch independently in order to demonstrate the variation in form of the secondary modular elements along the primary branches of *C. masoni*.

The first set of interpretations to be drawn from the landmark analysis is the degree of variance within each individual landmark. As seen in Fig. 3.4b, the spread of landmarks 3 and 4 is much greater than the spread surrounding landmarks 1 and 2. Also, the datapoint cloud surrounding each landmark position is greater along the y-axis when compared to the x-axis displacement. Laflamme et al. (2007) argued that the majority of shape variation of the secondary modular elements can be explained by a translation of only two landmarks, and with most of the translation restricted to the y-axis. This hypothesis is further substantiated by the repeatable and predictable shape changes of the secondary modular elements along each individual primary branch (Fig. 3.4c). The modular elements located closest to the midline are characterized by shorter L1-L3 landmark distances (distance A in Fig. 3.2b). Further along the branch, the central secondary modular elements are more rectangular in shape resulting from similar L1-L3 and L2-L4 distances (distances A and B in Fig. 3.2b). The most distal secondary modular elements located near the tip of each individual branch have a distinctively shorter L2-L4

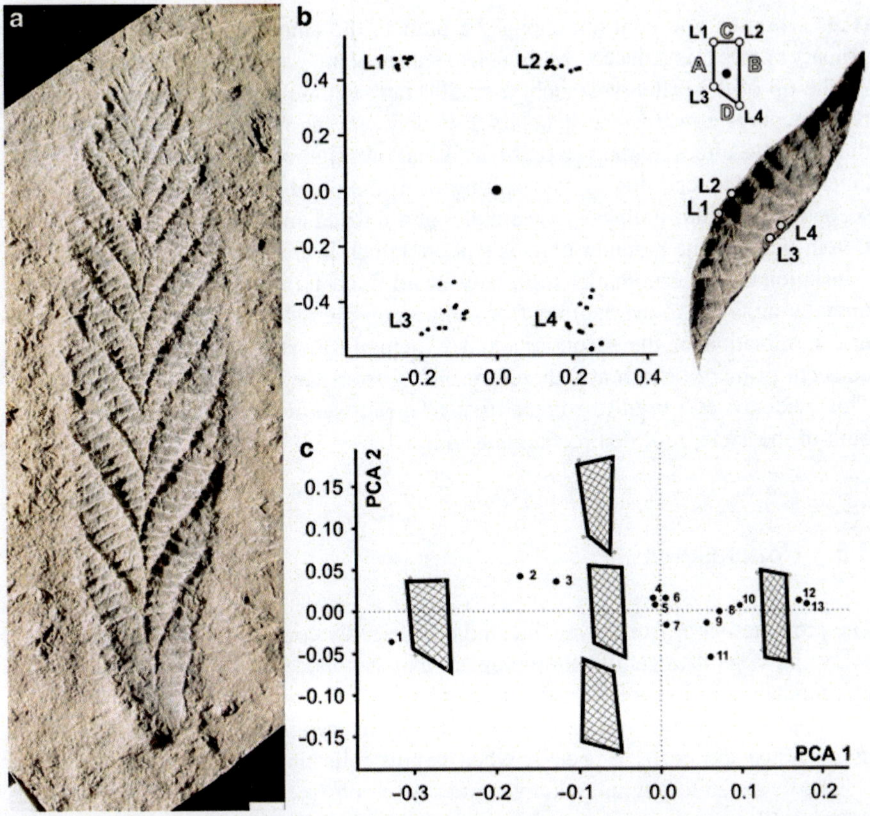

Fig. 3.4 Landmark morphometrics of *Charnia*: (**a**) Holotype of *Charnia masoni* from Charnwood Forest in England. (**b**) Landmark distribution along the second primary branch along the right side of (**a**). (**c**) Landmark migration along the second primary branch on the right hand side, demonstrating the predicable change in the position of the landmarks and the resulting deformation of the box-shaped secondary module. Grey trapezoid indicates average landmark location in that area of the scaterpot. Each individual secondary modular unit is indicated by the associated number (*1* closest to the midline; *13* most distal). Scale bar 1 cm

landmark distance (distance B in Fig. 3.2b). These landmark motion paths are in stark contrast to the L1-L2 (distance C in Fig. 3.2b) or L3-L4 (distance D in Fig. 3.2b) distances, which remain relatively uniform in shape distances throughout the entire length of the primary branches. Taken together, Laflamme et al. (2007) hypothesized that this implied a repeating, semi-rectangular shape for each secondary modular element, and that the trapezoidal or rhombic shape resulting from a shorter L1-L3 or L2-L4 distance represents a taphonomic response to overlying of adjacent primary branches.

The reason that the sections of the branches closest to the midline have a shorter L1-L3 distance is because the branches on the opposite side of the midline cross-over and physically cover the modular unit. As each primary branch alternates and overlaps along the midline, the taphonomic response is a zigzagging central midline

(Fig. 3.4a). As you continue along the branch, the amount of overlap with other primary branches is reduced, resulting in a rectangular (L1–L3=L2–L4) modular unit. At the tip of the primary branch, a similar taphonomic artifact is revealed, as the branch curves upwards and tucks itself underneath the following branch. This has the opposite effect as the proximal landmark configuration, with the L2–L4 distance being reduced due to overlapping from the overlying branch. Although the secondary modular units become smaller at the distal end of the branch, the variation in shape of the modular units is taphonomical. If this hypothesis is correct, the translation of the landmarks represent an artificial termination of the secondary modular units due to overlapping from adjacent primary branches, and therefore the true termination of the secondary unit is actually located beneath the overlying adjacent primary branch and therefore hiding from view (Laflamme et al. 2007). This evidence was used to suggest that *Charnia* was in fact a modular organism built of the same, repeated rectangular unit.

3.5 Conclusions

The proliferation of freely available and user friendly computer software capable of analyzing large datasets of morphological data has made traditional and geometric morphometrics accessible to all. Supplementing qualitative studies with rigorous quantitative morphometrics can also help tease out cryptic trends not readily visible in individual specimens, especially when sample collections are large and daunting.

The first step to any quantitative study is to construct a detailed framework of what questions you are interested in, and to construct a large enough sample size to answer your questions in a statistically significant way. A firm understanding of your hypothesis will significantly reduce the amount of time wasted on repetitive data sampling and the likelihood of erroneous application of numerical methods. The types of questions you hope to answer directly dictate which analyses are best suited to your data and which measurements or counts you must make. Following the construction of a reliable dataset, the nature of your data (continuous or discrete, sample completeness, etc.) will necessarily restrict which analyses will offer the most reliable results (Fig. 3.5). In some instances, it will be beneficial to restrict the amount of data included in the analysis in order to perform more powerful analyses, while other cases will dictate the opposite approach. More importantly, there is no reason why you cannot explore all of these options, as repeated results between analyses might reflect robust trends, while caution should be taken for trends that are restricted to single analyses.

As morphometrics focuses on statistically evaluating geometric relationships measured from large datasets, these analyses are powerful tools to tease out biological questions such as growth strategies, population dynamics, and even speciation events. Most importantly, taxon-free morphometric studies can be applied to groups of organisms with poorly understood phylogenies so as to gain insights on early life studies of stemgroups and problematica. Morphometric studies of Acritarchs and

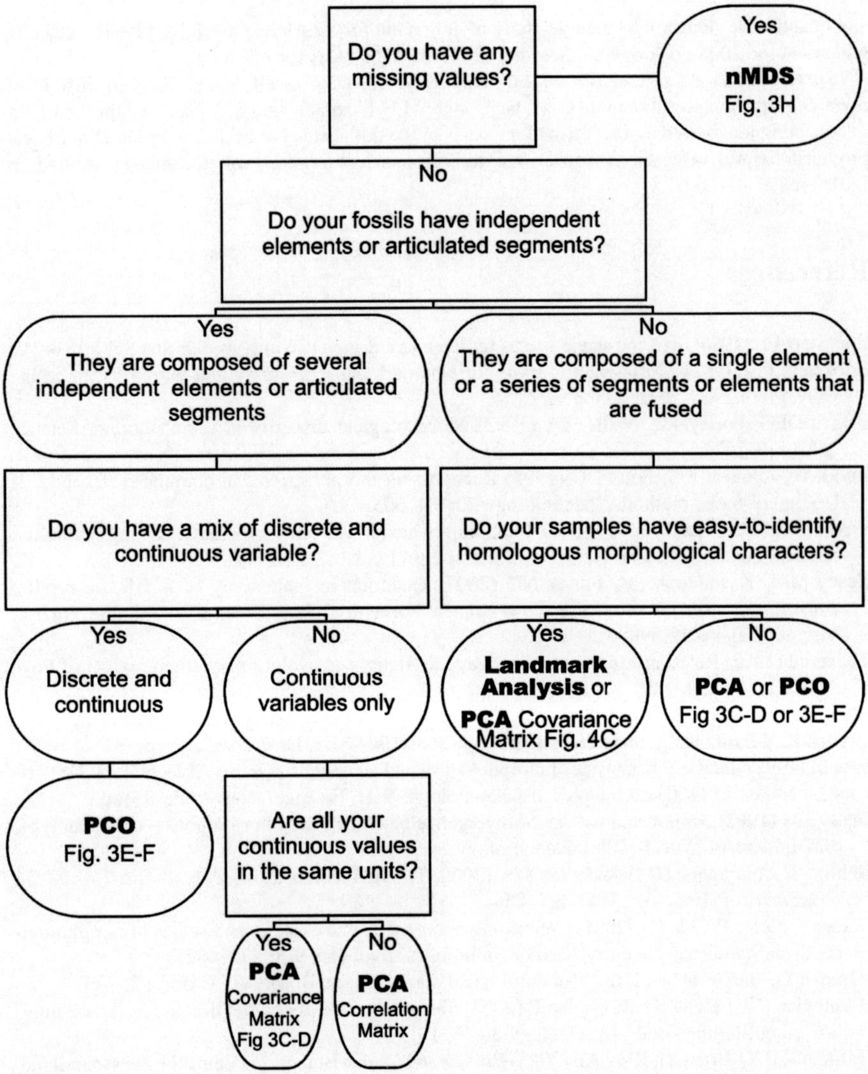

Fig. 3.5 Summary diagram highlighting the major distinctions between PCA, PCO, nMDS and landmark analyses

the Ediacara biota have become increasingly important in understanding Precambrian ecosystems and the strange denizens that constructed them.

Dedication and Acknowledgments We would like to dedicate this chapter to Hans Hofmann, a friend and colleague who passed away while this manuscript was in review. Hans was instrumental in our understanding of early Life, and was a pioneer in applying quantitative data to fossil studies. His 2008 article with S.J. O'Brian and A.F. King, which won the prestigious "best paper award" by the Paleontological society, elegantly displayed the benefits of combining qualitative and quantitative studies into a single study of enigmatic fossils. Hans gracefully gave the keynote address at the 2008 GSA Special Session that formed the basis for this book.

Special thanks are extended to P. Hull, R.A. Krause, M.G. Powell, and C. Tyler for insightful reviews of the manuscript, and to G.M. Narbonne, M.E. Clapham, and S. Villeneuve for their help in collecting the fossil data that formed the basis of the initial study at Mistaken Point. D.S. Sheets provided helpful discussions concerning the geometric morphometric techniques applied in Laflamme et al. (2007).

References

Bookstein FL (1986) Size and shape spaces for landmark data in two dimensions. Stat Sci 1:181–242

Bookstein FL (1991) Morphometric tools for landmark data: geometry and biology. Cambridge Univ. Press, New York, 435pp

Briggs DEG, Fortey RA, Wills WA (1992) Morphological disparity in the Cambrian. Science 256:1670–1673

Budd AF, Johnson KG, Potts DC (1994) Recognizing morphospecies in colonial reef corals. 1. Landmark-based methods. Paleobiology 20:484–505

Casey MM, Chattopadhyay D (2008) Clumping behavior as a strategy against drilling predation: implications for the fossil record. J Exp Mar Biol Ecol 367:174–179

Casey MM, Kowalewski M, Fraser NC (2007) Quantitative taphonomy of a Triassic reptile: *Tanytrachelos ahynis* from the Cow Branch Formation, Dan River Basin, Solite Quarry, Virginia. Palaios 22:598–611

Cheverud (1982) Relationships among ontogenetic, static, and evolutionary allometry. Am J Phys Anthropol 59:139–149

Davis JC (2002) Statistics and data analysis in geology, 3rd edn. Wiley, New York, 638 pp

Dryden IL, Mardia KV (1998) Statistical shape analysis. Wiley, New York, 347pp

Foot M (1992) Rarefaction analysis of morphological and taxonomic diversity. Paleobiology 18:1–16

Foot M, Miller AI (2007) Principles of paleontology. W.H. Freeman, New York, 354pp

Geary DH (1992) An unusual pattern of divergence between 2 fossil gastropods – ecophenotype, dimorphism, or hybridization. Paleobiology 18:93–109

Gehling JG, Narbonne GM, Anderson MA (2000) The first named Ediacara body fossil *Aspidella terranovica*. Palaeontology 43:427–456

Gerber S, Neige P, Eble GJ (2007) Combining ontogenetic and evolutionary scales of morphological disparity: a study of early Jurassic ammonites. Evol Dev 9(5):472–482

Hammer Ø, Harper DAT (2006) Paleontological data analysis. Blackwell, Oxford, 351pp

Hermoyian CS, Leighton LR, Kaplan P (2002) Testing the role of competition in fossil communities using limiting similarity. Geology 30:15–18

Hofmann HJ, O'Brien SJ, King AF (2008) Ediacaran biota on Bonavista Peninsula, Newfoundland, Canada. Journal of Paleontology, 82:1–36

Houck MA, Gauthier JA, Strauss RE (1990) Allometric scaling in the earliest fossil bird, Archaeopteryx-lithographica. Science 247:195–198

Huntley JW, Xiao SH, Kowalewski M (2006) 1.3 billion years of acritarch history: an empirical morphospace approach. Precambrian Res 144:52–68

Kowalewski M et al (1997) Phenetic discrimination of biometric simpletons: paleobiological implications of morphospecies in the lingulide brachiopod Glottidia. Paleobiology 23:444–469

Krause RA (2004) An assessment of morphological fidelity in the sub-fossil record of a terebratulide brachiopod. Palaios 19:460–476

Laflamme M, Narbonne GM (2008a) Ediacaran fronds. Palaeogeogr Palaeoclimatol Palaeoecol 258:162–179

Laflamme M, Narbonne GM (2008b) Competition in a Precambrian world: palaeoecology and functional biology of Ediacaran fronds. Geol Today 24:182–187

Laflamme M, Narbonne GM, Anderson MM (2004) Morphometric analysis of the Ediacaran frond Charniodiscus from the Mistaken Point Formation, Newfoundland. J Paleontol 78:827–837

Laflamme M, Narbonne GM, Greentree C, Anderson MM (2007) Morphology and taphonomy of the Ediacaran frond: Charnia from the Avalon Peninsula of Newfoundland. In: Vickers-Rich P, Komarower P (eds) The rise and fall of the Ediacaran biota, vol 286. Geological Society, London, pp 237–257, Special Publications

Lele SR, Richtsmeier JT (2001) An invariant approach to statistical analysis of shapes. Chapman & Hall/CRC, London, 308pp

MacLeod N (1999) Generalizing and extending the eigenshape method of shape visualization and analysis. Paleobiology 25:107–138

MacLeod N (2009a) PalaeoMath 101: size & shape coordinates. Palaeontol Newsl 69:26–36

MacLeod N (2009b) PalaeoMath 101: Distances, Landmarks & Allometry. Palaeontol Newsl 68:30–39

MacLeod N (2009c) PalaeoMath 101: who is Procrustes, and what has he done with my data? Palaeontol Newsl 70:21–36

Marcus LF (1990) Traditional morphometrics. In: Rohlf J, Bookstein FL (eds), Proceedings of the Michigan Morphometrics Workshop. University of Michigan Museum of Zoology, Ann Arbor, pp 75–122

Narbonne GM (2004) Modular construction of early Ediacaran complex life forms. Science 305:1141–1144

Narbonne GM (2005) The Ediacaran biota: Neoproterozoic origin of animals and their ecosystems. Annu Rev Earth Planet Sci 33:421–442

Narbonne GM, Laflamme M, Greentree C, Trusler P (2009) Reconstructing a lost world: Ediacaran rangeomorphs from Spaniard's Bay, Newfoundland. J Paleontol 83:503–523

Rohlf FJ (1999) Shape statistics: Procrustes superposition and tangent spaces. J Classif 16:197–223

Rohlf FJ, Slice D (1990) Extensions of the Procrustes method for the optimal superposition of landmarks. Syst Zool 39:40–59

Shen B, Dong L, Xiao S, Kowalewski M (2008) The Avalon explosion: evolution of Ediacara morphospace. Science 319:81–84

Slice DE (2007) Geometric morphometrics. Annu Rev Anthropol 36:261–281

Sokal RR, Rohlf FJ (1994) Biometry: the principles and practice of statistics in biological research. Freeman, San Francisco, 887 pp

Xiao S, Laflamme M (2009) On the eve of animal radiation: phylogeny, ecology and evolution of the Ediacara biota. Trends Ecol Evol 24:31–40

Zelditch ML, Swiderski DL, Sheets HD, Fink WL (2004) Geometric morphometrics for biologists: a primer. Elsevier Academic Press, London, 443pp

Zuur AF, Ieno EN, Smith GM (2007) Analysing ecological data. Statistics for biology and health. Springer, New York, 672pp

Websites

CoorDGen6: http://www3.canisius.edu/~sheets/moremorph.html. Designed and last modified: 23 January 2006 by D.S. Sheets

ImageJ: http://rsbweb.nih.gov/ij/download.html. Last modified: 17 November 2004 by W. Rasband

Morphometrics at SUNY Stony Brook: http://life.bio.sunysb.edu/morph/index.html. Last modified on 14 February 2009 by F.J. Rohlf

Paleontological Statistics or (PAST): http://folk.uio.no/ohammer/past/ Designed by P.D. Ryan, J.S. Whalley, Ø. Hammer, and D. Harper

Chapter 4
Analyzing Predation from the Dawn of the Phanerozoic

Lindsey R. Leighton

Contents

Abstract Although there have been numerous reports of predation on Cambrian and older fossils, there have been relatively few quantitative studies conducted on predation during this important interval. Such studies may prove extremely important as predation has been invoked as a primary influence on the Cambrian Explosion. Traces of predatory behavior, such as pursuit traces, crushing and repair scars, and drill-holes, are recommended as the best proxy for predation intensity. This chapter reviews the evidence for predation in the Neoproterozoic and Cambrian and then suggests and explains some techniques for analyzing predation data, with special consideration and examples of analysis of data from this early phase in the history of metazoans.

Keywords Cambrian • Durophagy • Drilling predation • Pursuit trace • Statistics

L.R. Leighton (✉)
Earth and Atmospheric Sciences Department, University of Alberta,
Edmonton, AB, Canada T6G 2E3
e-mail: lleighto@ualberta.ca

M. Laflamme et al. (eds.), *Quantifying the Evolution of Early Life*, Topics in Geobiology 36,
DOI 10.1007/978-94-007-0680-4_4, © Springer Science+Business Media B.V. 2011

4.1 Introduction

4.1.1 Predation and the Cambrian Explosion

Predation is widely-recognized as an important influence structuring communities (Brooks and Dodson 1965; MacArthur 1972; Paine 1974; Menge and Sutherland 1976, 1987; Holt 1977; Navarrete and Menge 1996; Abrams and Matsuda 1997; Finke and Denno 2004) and a major agent of natural selection (Vermeij 1977, 1983, 1987, 2002; Palmer 1979; Vermeij et al. 1980, 1981; Kelley and Hansen 1993; Dietl et al. 2000; Leighton 2003; Baumiller and Gahn 2004). Vermeij (1987) has argued that predation is the single most important such agent – organisms must be able to avoid predation on a daily basis. Failure to respond to one's predators results in death. No other potentially lethal factor is encountered by an organism as frequently as is predation. In Vermeij's hypothesis of escalation, organisms are most likely to respond and adapt to their own enemies, and much of the history of life is driven by adaptive responses to predation. Perhaps the most critical test and possibly the best example of this idea of escalation is the convergent evolution of hardparts in multiple phyla that we call the Cambrian Explosion.

There have been several explanations for the dramatic change in life on Earth at the dawn of the Cambrian but predation has been most frequently invoked as an influence (Evans 1912; Hutchinson 1961; Vermeij 1989, 2002) although both Bengtson (2002) and Marshall (2006) considered predation as a shaper of the events, rather than as the initial trigger. It is difficult to envision another hypothesis that explains the multiple types and materials of skeletons appearing within a few millions of years. McMenamin (1986, 1998) argued in his "Garden of Ediacara" hypothesis that the Neoproterozoic was largely, if not completely, lacking in macropredators. Jensen et al. (1998), examining trace-fossils and sediments associated with the vendobiota concurred that there is no evidence of benthic macropredators. Jensen et al. further argued that the appearance of anomalocarids (Whittington and Briggs 1985) at or immediately prior to the Neoproterozoic-Cambrian boundary radically altered marine ecosystems as organisms were forced to adapt to these new predators, hence, the Cambrian Explosion. So far, this has been the closest we have to evidence supporting a predatory origin for the convergent evolution of hard skeletons.

That said, it is both ironic and surprising that there are relatively few studies of predation during the Cambrian Explosion or even the longer interval of the Cambrian and immediate pre-Cambrian. Those studies on Cambrian predation that do exist are usually isolated examples of inferred predation, sometimes reported almost as an aside to the main focus of the paper. These studies are critical in that they provided the first evidence that predation may have occurred but we clearly need further and more detailed explorations of early predation.

Paleontological research in predation has historically followed this pattern. There have been examples of predation and injuries to Paleozoic brachiopods reported as early as the 1930s (Fenton and Fenton 1931, 1932), but early studies were largely ignored because of a prevailing view that Paleozoic predation was unimportant or didn't even exist. Thirty to 40 years ago, we saw the first few papers

(Brunton 1966; Sheehan and Lesperance 1978; Ausich and Gurrola 1979) that suggested that there were predatory drillholes in the Paleozoic. As both Pete Sheehan and Bill Ausich have told me independently, these new ideas were met with considerable resistance, even though the papers provided solid evidence for drilling predation. Now, after numerous subsequent studies, we have enough data that we even have broad-scale reviews of drilling predation in the Paleozoic (Kowalewski et al. 1998; Huntley and Kowalewski 2007).

I think that we are observing a similar trend in Cambrian studies. After important initial papers describing examples of Cambrian predation (Rudkin 1985; Conway-Morris and Jenkins 1985; Zhu et al. 1994; Jago and Haines 2002), the idea that predation was present and important in the Cambrian is more widely-accepted. Now we are at the point where we need to develop more – and more quantitative – studies of Cambrian predation. The following review of Cambrian predation is not meant to be exhaustive, but rather to provide a general background and to provide examples of quantitative studies or studies amenable to quantitative analysis. For a more thorough review of Cambrian predation, especially of and by trilobites, I recommend Babcock (2003).

The stratigraphic nomenclature of the Cambrian has gone through considerable revision as of late (Babcock and Peng 2007). Lower, Middle, and Upper Cambrian have been provisionally replaced by the Terreneuvian Series, Series 2, Series 3, and the Furongian Series. The Terreneuvian is roughly equivalent to the pre-trilobite portion of the Early Cambrian, whereas Series 2 consists of the remainder of the Early Cambrian, after the first appearance of trilobites. The base of Series 3 is close to the Early-Middle Cambrian boundary in many parts of the world, and Series 3 is an expanded version of the Middle Cambrian, incorporating the lower part of the traditional Upper Cambrian as well. The Furongian Series is a reduced version of the Upper Cambrian. For a more detailed explanation of the new Cambrian stratigraphy, see Babcock and Peng (2007). Throughout this chapter, I will attempt to use the modern terminology when possible, but in most cases, for example when referring to older citations, I will use the traditional stratigraphy when it was not immediately evident in which series the specimens of interest belonged.

4.1.2 A Brief Review of Cambrian Predation

Traces of predation, whether in the form of gut contents, coprolites, traces on the prey skeleton (crushing or drilling scars), or traces in the sediment, are examples of direct evidence of predation. Gut contents are rare but provide compelling evidence for the predator–prey interaction (or at least of one animal consuming another), and are unusual in that the identities of both predator/scavenger and prey can be established. The Cambrian has its share of such fossils. Priapulids from the Middle Cambrian Burgess Shale contain fragments of hyoliths (Conway-Morris 1977). The Middle Cambrian arthropods *Sidneyia* (Burgess Shale, British Columbia), *Utahcaris* (Spence Shale, Utah), and an arthropod similar to *Fuxianhuia* (Kaili Formation, South China) all have been found with trilobite material preserved in the gut

(Bruton 1981; Conway-Morris and Robison 1988; Zhu et al. 1994). Brachiopod fragments have also been preserved in the gut of *Sidneyia* (Conway-Morris and Whittington 1979). The Spence Shale containing *Utahcaris* also contains coprolites containing trilobite, brachiopod, and echinoderm fragments (Conway-Morris and Robison 1988). Vannier and Chen (2005) described two categories of fossil aggregations from the Early Cambrian (Series 2) Maotianshan Shale of Southwestern China that they interpreted respectively as coprolites from anomalocarids and from predatory worms such as priapulids. The former contained a mix of trilobite, other arthropod, and hyolith fragments, while the latter primarily contained hyolith material. These cases are extremely informative but unfortunately, evidence from gut contents and coprolites generally are rare, isolated examples.

The oldest example of a predation trace on the prey skeleton was initially documented by Bengtson and Yue (1992) who examined round borings in the Neoproterozoic genus *Cloudina*. *Cloudina* formed calcareous tubes which have been interpreted to be exoskeletons; the genus is believed to be the oldest organism to do so. It is fascinating that the very first organism with a hard skeleton already has evidence for predation associated with it, although the drilling frequency was only 2.7%. Boring-width and tube-size track each other, suggesting that the holes were made by a predator that exhibited prey selection based on size, although the sample size of borings is quite small, so the significance of the result was in question. Debrenne and Zhuravlev (1997) argued against a predatory origin for the holes, suggesting that the holes were formed by dolomite crystal dissolution but subsequent work by Hua et al. (2003) provided further corroboration for the predatory hypothesis. Hua et al. (2003) documented a much higher drilling frequency (20.0%) in a separate population of *Cloudina*. Moreover, these authors noted that in addition to a failure to find any dolomite crystals present, the tubes, with one exception, never had more than one hole, a finding which would be unlikely if the origin of the hole was chemical. Perhaps the strongest evidence for predation on *Cloudina* was that no holes were found on another tube-forming organism, *Sinotubulites*, found in the same bed, suggesting that the predator had a distinct preference. *Sinotubulites* also has a significantly thicker skeletal-wall, a finding consistent with the argument for prey preference. Even at the dawn of macropredators, we already observe fairly complex predation patterns.

There have also been reports of drilling predation in the Cambrian as well. Bengtson (1968) noted the existence of possibly predatory borings on the enigmatic organism, *Mobergella*. *Mobergella* is an early Cambrian, phosphatic organism that Bengtson interpreted as the operculum of a tubular organism. Conway-Morris and Bengtson (1994) subsequently noted stereotypy of location of these borings towards the apical region of the shell, but also acknowledged a small sample size and a possible taphonomic bias towards preserving this region of the shell.

Predatory borings have been documented on linguliform brachiopods, including both lingulids and acrotretids. Miller and Sundberg (1984) described 13 Late Cambrian acrotretid specimens with borings which were consistent in morphology with predatory borings, but the sample size is too small for statistical analysis, particularly as the specimens were collected from more than one bed. Middle Cambrian acrotretids of the genus *Linnarssonia* from southern Sweden and the

Danish island of Bornholm also bear predatory borings (Conway-Morris and Bengtson 1994). Drilling frequencies among the four samples varied considerably, from roughly 1% to greater than 15%. There is no obvious pattern for stereotypy with respect to location on the valves, but in some samples, there may be a preference for the brachial over the pedicle valve; evidence of stereotypy points towards a predatory origin for the holes (see discussion below). Conway-Morris and Bengtson declined to test for stereotypy statistically because there was a strong imbalance in the preservation of brachial versus pedicle valves. Other examples of drilled Cambrian acrotretids are figured in Lochman (1956) and Grant (1965).

Robson and Pratt (2007) demonstrated predatory drilling on both acrotretids and lingulids from the Middle Cambrian to Early Ordovician Deadwood Formation of South Dakota, USA. Similar to the Conway-Morris and Bengtson study, stereotypy for site was not evident, but stereotypy for valve may have been. As is common with brachiopods, one valve was preferentially preferred within each species, which, as in the previous example, might bias any valve preference. Later in this chapter, I provide an appropriate statistical solution to this sort of problem using data from Conway-Morris and Bengtson as examples. Based on those analyses described below, I summarize the results here (Table 4.1A, B).

Table 4.1 Summary of chi-square tests for stereotypy, using data from two studies discussed in the text

A. Raw data from Conway-Morris and Bengtson (1994) and results for test of stereotypy with respect to valve on two species of *Linnarsonnia*

	S70-2-SB	S70-3-SB	S72-13-SB	S72-22-SB	Totals
Linnarsonnia sp A					
Brachial valves	2,552	1,156	277		3,985
Pedicle valves	2,165	1,174	360		3,699
Drilled Br. V.	18	10	6		34
Drilled Pd. V.	15	14	3		32
Chi-square p-value	0.959	0.436	0.161		0.955
Linnarsonnia sp B					
Brachial valves	738	91	125	169	1,123
Pedicle valves	319	48	80	76	523
Drilled Br. V.	23	8	18	26	75
Drilled Pd. V.	7	3	2	3	15
Chi-square p-value	0.414	0.613	**0.008**	**0.016**	**0.002**

B. Raw data from Robson and Pratt (2007) and results for test of stereotypy with respect to valve for lingulids and acrotretids

	Lingulids	Acrotretids
Brachial valves	216	276
Pedicle valves	421	561
Drilled Br. V.	24	21
Drilled Pd. V.	4	91
Chi-square p-value	**<0.001**	**0.001**

Bold results indicate significant p-values which corroborate the hypothesis that there were significantly more borings on one valve

The Conway-Morris and Bengtson (1994) study observed drilling on two species of *Linnarssonia*, which they referred to as *L. sp. A* and *L. sp. B*. My statistical analysis did not corroborate any stereotypy for valve selection on L. sp. A., but did find a statistically significant preference for the brachial valve of *L. sp. B* in two of the four samples, and for the total of all four samples ($p=0.016$ or lower). In addition, for all of the three samples in which both species were present, there was an extremely significant ($p<0.001$) preference for *L. sp. B*. These stereotypic results corroborate the hypothesis that the holes were predatory in origin, at least those holes on *L. sp. B*.

In the Robson and Pratt (2007) study, the analyses conducted herein indicate that the lingulids were preferentially drilled on the brachial valve ($p<0.001$) whereas the acrotretids were preferentially drilled on the pedicle valve ($p=0.001$). As the lingulids were not present in sufficient numbers at one of the two localities, and the acrotretids were not drilled at all at the other locality, a comparison of predator preference between the two taxa is either unwarranted or unnecessary. See the sections on stereotypy, data organization, and statistical tests for a full description of the methods used in these cases.

The predatory borings described for Cambrian brachiopods in all of these studies are Type-A (sensu Ausich and Gurrola 1979) drill-holes; they are small (typically around 1 mm diameter), circular, and cylindrical. These holes are similar to modern muricid drill-holes (although Paleozoic examples were not made by muricids as these neogastropods did not evolve until the Mesozoic) and have been reported from numerous Paleozoic studies (Sheehan and Lesperance 1978; Ausich and Gurrola 1979; Smith et al. 1985; Leighton 2001, 2003). One interesting feature regarding the Cambrian borings is that they do not usually show site-stereotypy, whereas later such borings routinely do so (Leighton 2001, 2003). Another point of interest is that to my knowledge, there have been no reports of Cambrian predatory drill-holes on members of the subphylum Rhynchonelliformea (which includes all of the former Class Articulata plus some members of the old Class Inarticulata). Similarly, predatory drill-holes on Cambrian molluscs are unknown. Were Cambrian drillers incapable of drilling through thicker calcareous shells?

In the modern, crushing predation, in which the predator attempts to break the prey's skeleton, is a greater source of mortality for most hard-bodied prey than is drilling. Crushing predators are usually significantly larger, relative to their prey, than are drillers, and crushers thus will typically need to consume more food and so have greater attack rates. Examples of crushing predation have also been observed as early as the Cambrian. In the aforementioned study by Robson and Pratt (2007), the authors speculated that given the morphology of the holes observed on acrotretids, some of the holes were more likely to be made by a predator attempting to puncture the shell. Robson and Pratt noted that some predatory nematodes have a stylet suitable for puncturing shells in this manner; the nematode may also use the stylet for injecting digestive enzymes and for consumption.

The vast majority of examples of Cambrian crushing traces are found on trilobites and their relatives. The oldest examples of crushing predation in the fossil record are from Early Cambrian (Series 2) specimens of the trilobitomorph *Naraoia* (Nedin 1999) and from Series 2 specimens of the trilobite *Buenellus* (Babcock and Peel 2007). The injuries on *Naraoia* usually involve an arcuate bite-mark on one side of the animal matched with minor injuries on the opposite side. Nedin (1999) argued that these injuries represented attacks from an anomalocarid. Previous workers (Whittington and Briggs 1985) had noted that anomalocarids had insufficiently sclerotized jaws which could not occlude, suggesting that anomalocarids would have been incapable of taking hard-bodied prey. Nedin (1999) suggested that the anomalocarid would have grasped the trilobite in its jaws from one side while grasping the prey on the opposite side with the anomalocarid raptorial appendages; the trilobite would then have been rapidly flexed, causing stress perpendicular to the direction in which the sclerites articulated. This sort of flexure would have exposed the exoskeleton to alternating torsion and compression sufficient to cause the skeleton to fail. Nedin (1999) noted that the injuries on opposite sides of *Naraoia* were consistent with such an attack. It should be noted, however, that *Naraoia* was only weakly calcified compared to trilobites, so it is possible that it was easier prey for an anomalocarid. Nedin did observe that similar paired injuries also have been documented in studies of predation on Cambrian trilobites (Rudkin, 1979; Conway-Morris and Jenkins 1985; Jell 1989). Grapple and flexure attacks of this type would be consistent with Babcock's (1993) observation that the great majority of attacks on trilobites are on one side of the trilobite, rather than from the front.

The oldest evidence of crushing predation on a trilobite is from six specimens of *Buenellus higginsi* from the Buen Formation (Series 2) of North Greenland (Babcock and Peel 2007). The predation scars are typically arcuate and show evidence of repair, with distortion of the healed area (Fig. 4.1). Many trilobite injuries show repair and were clearly healed (Conway-Morris and Jenkins 1985). Although simple injuries to the pleural tips might have occurred during molting, the damage and repair on these trilobite specimens is extensive enough that the injuries should be attributed to predation. Conway-Morris and Jenkins (1985) noted wrinkling of the pleura in the damaged areas and raised the interesting possibility that such wrinkling was a case of a subsequent molt accommodating the wound. These authors also raised the interesting possibility that many of these attacks took place shortly after molting, when the exoskeleton had not fully sclerotized, and when the trilobite would have been more vulnerable.

Babcock (1993, 2003) performed a detailed analysis of injuries on trilobites through time. To date this is the largest data-set of trilobite injuries; the data were compiled from previous studies and from Babcock's own research, and covered the Cambrian through Pennsylvanian. 75% of sublethal (repaired) trilobite injuries are located on the posterior half of the body, from the posterior thorax to the pygidium. Interestingly, these injuries are also commonly (69%) on the right-side of the trilobite, a phenomenon which Babcock (1993) attributed to left-right behavioral asymmetry on the part of the predators and/or their trilobite prey. The occurrence of these injuries on the posterior half and on the right side of the trilobites is significantly greater than would be expected under a random model.

Data in Babcock (2003) also permit a rough comparison of Cambrian and post-Cambrian predation on trilobites. Post-Cambrian trilobites still showed a significantly greater occurrence of injuries on the right side, but the results were stronger for Cambrian trilobites. Cambrian repaired injuries were also more common than those from all of the post-Cambrian combined (59% vs. 41%), although it is unclear if this pattern is real, or may possibly be an artifact of a greater number of studies on

Fig. 4.1 Specimen of *Buenellis higginsi* from the Sirius Passet Biota, Buen Formation, Cambrian Series 2, North Greenland, showing repaired damage from a bite (*arrow*). Note distortion of repaired region compared to normal morphology of the trilobite (Image courtesy of Loren Babcock)

Cambrian trilobites. If the pattern is real, then the decrease in stereotypy and the decrease in repaired injuries through time suggest that the post-Cambrian predators were different predators – as well as stronger and more lethal. The decrease in asymmetry suggests that post-Cambrian predators did not need to resort to the grapple-and-flex strategy, but may have been able to bite or crush through the trilobite exoskeleton immediately. Post-Cambrian predators, including both "fish" and arthropods, evolved attack structures (jaws, gnathobases etc.) that were strong enough to deliver such bites (Moy-Thomas and Miles 1971; Denison 1978; Signor and Brett 1984); many Paleozoic arthropods had more heavily-calcified attack structures. This would also explain the decrease in repair scars; greater success on the part of the predator would mean more kills and thus fewer repairs. Alternatively, the decrease in repair frequency might be a function of fewer attacks on trilobites.

For the remainder of this chapter, I hope to provide the reader with a basic background on the techniques to use when analyzing predation. The tools involved are applicable to any time period; however, throughout the section, I will endeavor to point out special considerations relating to work on Cambrian predation.

4.2 Measuring Predation

How do we measure predation in the fossil record? More exactly, what are we trying to measure? This depends on the question, but in many paleobiological studies of predation, we are trying to assess predation intensity, the mortality due to predation. Alternatively, we are trying to assess the evolutionary impact of predation – the role of predation as an agent of natural selection. Mortality and evolutionary impact do not necessarily correlate (Vermeij 1982), but for predation to have an evolutionary impact, there must be a real threat of differential mortality due to predation.

I have argued elsewhere (Leighton 2002) that the best way to measure predation intensity is by using direct evidence such as traces of predation, e.g., crushing scars and drill-holes. Indirect, coarse-scale measures, such as predator diversity, proportional predator diversity, and predator abundance may have little to do with predation intensity, and by extension, the evolutionary impact of predation on prey. The problem with these indirect measures is that they do not take into account the abilities of the predator relative to the prey, which obviously influences the attack and success rates of the predator and the mortality of the prey. For example, Madin et al. (2006) used the global-scale proportion of predators (out of all animals) in their study. Although this was not the intent of the study, their data highlight the lack of a relationship (or at least an unexpected one) between proportional predator diversity (or abundance) and predatory ability. There have been two major pulses in marine predator evolution, during which predators evolved to be bigger, stronger, and deadlier; one pulse starting in the Devonian (Signor and Brett 1984), and a second pulse starting roughly in the Cretaceous (Vermeij 1977). Both of these radiations also involved an increase in the diversity of attack types to which marine prey were subjected. Madin et al. (2006) data show that the proportional diversity of predators actually was low during these two intervals. At the opposite end of the spectrum, the proportion of predators in the Cambrian is very high, but Cambrian predators were much weaker and less specialized than their later counterparts and most were probably incapable of taking highly armored or highly mobile prey. Similarly, one would not expect predator abundance to be a reliable indicator of predation intensity. Larger, stronger, predators near the top of the food-web are going to be deadlier than other predators, both in the sense of having greater success rates and also in the sense of having greater consumption rates. However, the abundances of large predators will be low if for no other reason than trophic limitations, i.e. given the low uptake efficiency at any given level of a trophic pyramid, and the increasing size of organisms at higher levels, there is going to be a finite (and small) limit to the number of individuals at the higher levels. Given that predation is a non-equilibrium process (a process that prevents ecosystems from going to population equilibrium, and thereby prevents rival prey species from excluding each other competitively for resources [Paine 1966, 1974; Holt 1977, 1984; Jeffries and Lawton 1984; Tschanz et al. 2007]), the ecological prediction is that as predators become more effective, prey diversity should stabilize or increase (Madin et al. assumed the opposite).

Prey evolution, particularly morphological evolution, also has been used as evidence for predation (Vermeij 1977, 1983; Signor and Brett 1984), but this type of argument can quickly become circular as it assumes that the prey morphology is anti-predatory. However, in conjunction with predation trace data, any of these other measures can be useful and important.

Predation traces may not directly measure the strength of a predator – they may not even identify the predator – but traces provide an assessment of predatory intensity, which is the measure we ultimately are trying to capture. Ideally, predation traces can be used to assess three important metrics – attack frequency, the proportion of individual prey attacked out of the total prey population; success frequency, the proportion of killed prey out of the attacked prey population; and kill frequency, the proportion of killed prey out of the total prey population. Before these values are calculated, it is necessary to confirm that the traces truly are predatory in origin.

There are three primary types of predation traces that are preserved to the fossil record: traces in the sediment indicative of predatory behavior; and traces of crushing and of predatory drilling, both of which are traces that appear on the prey skeleton. One important caution about these three types of traces: although analyzing all three types within the same community would be of great interest as it would provide a more complete picture of community predation, we definitely should not assume *a priori* that the three types of predation will reveal similar patterns or results. The three types of predation traces usually will be made by three entirely different groups of predators. Crushing predators will typically be much larger than their prey, and also much larger than drilling predators. Both crushing and drilling predators will usually be attacking prey with skeletons; in contrast, many of the prey revealed by sedimentary traces will be soft-bodied. Although some predators are capable of taking both hard- and soft-bodied prey, many predators will lack the attack structures necessary to break or penetrate a hard skeleton, and so will be restricted to soft-bodied prey. Given all of the above, not only are the predators making the three different types of traces different, but they also may occupy different parts of the food-web, and different trophic levels. For example, in the Cambrian, crushing predators would include anomalocarids, while the pursuit-trace-maker predators would include trilobites. Trilobites would thus have been predators in one case, but would have been prey to the anomalocarids. There isn't any reason to assume that if anomalocarid attacks on trilobites increased through time or space that this would increase attacks by trilobites on soft-bodied prey (in fact, predation by anomalocarids might even inhibit predation by trilobites). One should definitely not assume that any one type of predation trace can serve as a proxy for any of the other types.

4.2.1 Traces in Sediment of Predatory Behavior

A common type of sedimentary traces indicative of predation is what I will describe as "pursuit traces". Pursuit traces are examples of the pursuit phase of a predator–prey encounter, the phase during which the predator has identified that there is an acceptable prey item within attack range and the predator actively moves towards its prey (the prey may or may not attempt to escape). Pursuit traces will usually involve two distinctly different prey traces in the sediment, one of which is the predator and one of which is the prey. Many readers will no doubt note here that there is an extensive ichnological terminology for the exact morphology and inferred behavior of traces, and that I have deliberately avoided the use of this terminology. The point is that first, we do not *a priori* wish to assume predation, and second, the exact morphology and identity of the two different traces is not the issue.

Perhaps the most easily testable example of potential pursuit traces are cases in which the traces are both vertical (or at least start that way): a prey organism has built a domicile or is mining the sediment, and a predator foraging for food on the surface finds the burrow of the prey and digs down to excavate and capture the prey. There are three criteria supporting the hypothesis that the traces are predatory: (a) one of the traces (the predator's) is distinctly larger in diameter than the other; (b) the supposed predator's trace starts near to, and runs parallel to (if not actually cutting through) the supposed prey trace, and ultimately intercepts the prey trace; and (c) the predator's trace achieves its lowest point in the sediment at the termination of the prey trace – in other words, the predator has no need to burrow any further down than the bottom of the prey's burrow, so if the burrow truly is a pursuit trace, then we should not expect to see the larger burrow penetrate below the depth of the smaller burrow.

Jensen (1990) conducted a detailed study of presumed trilobite traces (*Rugophycus* from the Early Cambrian Mickwitzia Sandstone of Sweden) that appear to be a case of predatory pursuit of some soft-bodied, worm-like prey. Jensen statistically tested whether the co-occurrence (proximity of traces and parallelism of traces) of the two types of burrows was more likely than would be expected by chance. The results indicated that these co-occurring burrows probably were predatory; there was a significant, strong-positive correlation between deviations in the worm burrows and the trilobite burrows. We need more studies of this type on possible pursuit traces.

As will be apparent from much of this discussion of predation, single, isolated examples of possible predation may be of interest, but we cannot measure predation or even convincingly demonstrate that an individual case is predatory in origin without having multiple examples. Put simply, the co-occurrence of two different but parallel burrows may just be a coincidence but finding this pattern repeatedly within the same bed, as Jensen (1990) did, enables quantitative and rigorous testing of a hypothesis of predation.

4.2.2 *Traces of Predation on the Prey Skeleton*

Drilling traces, when demonstrated as predatory, provide direct evidence of predator activity and predation intensity (Vermeij 1987; Leighton 2002). Direct evidence is ideal, but unfortunately, drilling is the only form of predation in which successful predation can be identified consistently on hard-bodied prey in the marine fossil record. Other attack methods, when successful, either leave no evidence that the prey died from predation (e.g. seastars prying open clams) or may destroy the prey and so fail to preserve the predatory act to the fossil record (e.g. crushing).

Yet durophagous (shell-crushing) predators may be the greatest source of mortality for many shelled organisms. Drilling typically is slow; Recent drillers may take 2–3 days attacking, killing, and consuming the prey (Vermeij 1987). In contrast, Recent xanthid crabs are capable of crushing and eating 15 oysters in a single night (Leighton, personal observation). Although there is not always a positive correlation between mortality and selection (Vermeij 1982), the impact of crushing predation in the fossil record needs to be considered when examining predation intensity. In addition, there is no *a priori* reason to assume that drilling predation intensity would be correlated with crushing predation intensity through time. In fact, similar to the example of anomalocarids and trilobites mentioned previously, drilling success might decrease in settings where crushing success increases because most predatory drillers, such as gastropods, are themselves frequent victims of crushing predation (Vermeij et al. 1980). Smaller, or less mobile predators, such as drillers, might be inhibited by the threat of their own predators (Lima 1998; Morton and Chan 1999; Morissette and Himmelman 2000).

Evidence for crushing comes in the form of traces of crushing and repair (Figs. 4.1 and 4.2). Because repair scars are traces of *unsuccessful* predation, interpretation of repair scar frequency across time, space, and taxa can be difficult.

Fig. 4.2 Specimen of *Elrathia kingi* from the Wheeler Formation, Cambrian Series 3, Utah, USA, showing a typical arcuate bite-mark (Image courtesy of Loren Babcock)

For example, strongly biconvex taxa, such as globose rhynchonellide brachiopods, resist point-loading better than more planar shells. However, once a strongly biconvex shell fails, it will be shattered completely (Alexander 1986a, 1990). In contrast, some concavo-convex, strophomenate brachiopods, have features which may limit propagation of fracturing once it occurs (Alexander 1986b, 1989). Given this, concavo-convex brachiopods should have higher repair scar frequencies than do biconvex brachiopods. Alexander (1986a), examining a large data set of brachiopods from multiple localities throughout the Phanerozoic, demonstrated that concavo-convex brachiopods had higher repair frequencies, and most biconvex brachiopods had lower repair frequencies, than expected under a random model. Do concavo-convex brachiopods have higher repair frequencies because they were preferred prey, or do the low repair frequencies on biconvex brachiopods indicate higher kill rates (and shell destruction) for biconvex taxa? The typical concavo-convex brachiopods, the strophomenates, had not yet evolved during the Cambrian, but this sort of problem, in which it is difficult to ascertain exactly what an increase in repair frequency indicates, can potentially be an issue for any group of organisms.

A similar problem can exist when examining the part of the body typically attacked. For example, both Conway-Morris and Jenkins (1985) and Babcock (1993) each examined large collections of predation-scarred trilobites, and both studies found that the posterior of the trilobite (either the pygidium or the posterior lateral lobes) were far more likely to bear scars than more anterior parts of the trilobite. This sort of stereotypy is actually strong evidence for predation (see below), but it is unclear if the increased scarring rate on pygidia is a function of the fact that the posterior end of the trilobite would be most likely to be grasped by a pursuing predator; if pygidial scars are more likely to be preserved because the pygidium itself is the most frequently preserved sclerite; or if attacks on the cephalon would have been more likely to be lethal and so would be more likely to be removed from the fossil record.

Even within a single taxon or lineage, trends in crushing frequency are not interpreted easily. An increase in repair frequency could indicate either an increase in the number of attacks, possibly due to an increase in predators, a switch in prey preference, or a decrease in predator success (Vermeij 1987; Leighton 2002). Conversely, a decrease in repair frequency could be caused by a decrease in attacks or by an increase in predator success. For example, Alexander (1986a) and Vermeij (1987) studied repairs through time for Paleozoic brachiopods and Mesozoic gastropods respectively. Alexander (1986a) noted a decrease in repair frequency, whereas Vermeij (1987) observed an increase. Although the patterns were different, both authors argued that predation intensity increased, and both may be correct. Alexander suggested that the Paleozoic decline in repair frequency was due to an increase in efficiency and power of Late Paleozoic durophagous predators. Vermeij (1987) argued that the Mesozoic increase in repair frequency was a function of increasing attacks. This example illustrates the problems of using repairs as a metric of predation intensity.

Leighton (2002) suggested a means of circumventing these problems. Numerous ecological studies (Werner and Hall 1974; Elner and Hughes 1978; Kitchell et al. 1981; Seed and Hughes 1995) have demonstrated that predators often prefer a particular size of prey. Prey smaller than the preferred size may not provide sufficient energy for the predator; large prey may take too long or be impossible to kill. Intermediate prey are optimal: the attack is more likely to be successful, and the prey thus provides the most net energy per unit time. Of course, different predators might have different optimal size ranges. Nonetheless, an examination of size classes of the prey, in conjunction with repair frequency, should prove enlightening.

Within a time interval, repair scar frequency should be low for the size class that the predator prefers, because the attack is likely to succeed. With increasing prey size within the population, the predator's probability of success declines and consequently, repair frequency will increase, assuming a similar attack frequency. This scenario is analogous to the increase in incomplete drill holes with increasing prey size observed in naticid predator systems (Kitchell et al. 1981; Anderson 1992). There is a prey size at which the predator will not attack the prey. This refuge may be behavioral (the predator could take the prey, but easier prey are available) or functional (the predator is incapable of taking the prey). Regardless of the reason, that size class will have a repair frequency statistically indistinguishable from zero – which identifies the size refuge (Fig. 4.3). In practice, the decrease in repair frequency might not be so abrupt. The predator species varies in size, which may create variation

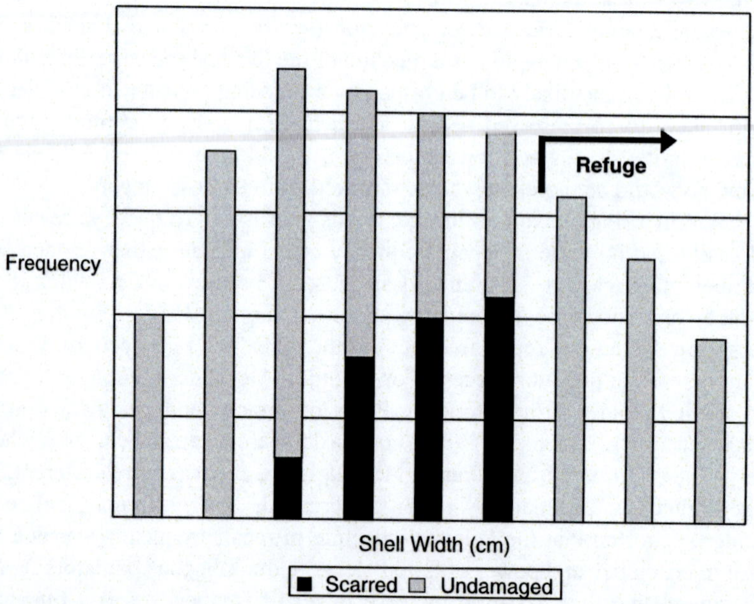

Fig. 4.3 Determination of size refuge for a hypothetical prey population. Prey size increases to the right. At small prey sizes, repair frequency is less because most attacks would be lethal. As prey items get larger, the predator is more likely to fail and the prey more likely to survive and repair the injury. Frequency and proportion of repair scars reaches a peak, reflecting increasing occurrence of failure by the predator as prey size increases. At still larger sizes, the prey achieves a size refuge at which the predator does not attack the prey; repair frequency drops to zero at the size of refuge

in the position of the size refuge. Although the decrease in repair frequency may be gradational rather than instantaneous, the position of the decrease would still be unambiguous so long as the sample size is sufficiently large.

An additional consideration is that the prey might grow subsequent to successful repair; thus, the size of the prey at death might be larger than the size of the prey at the time of the repair. To account for this, each specimen with a crushing scar should be measured for the minimum possible shell size at which the attack occurred; this size would be used in the refuge analysis.

An increase in repair frequency, and a decrease in the size at which the refuge is achieved (this does not imply a change in absolute size), through time or space, suggests that the predator's attack abilities are less effective relative to the prey's defense (which could occur due to improved defenses). Given two prey items that are of the same species, and which differ only in size, then all other things being equal, a predator given a choice is going to eat the larger of the two individuals because that prey item provides more energy. If the predator chooses the smaller prey item, then all other things are not equal – for some reason, the predator is either incapable of killing the larger prey item, or the cost in energy or time of killing the larger prey item is too great for the predator. So if at one time, the predator chooses the larger individuals of a species, but at a subsequent time, the larger individuals are never eaten, then the only reasonable explanation is that the predator's attacking abilities have decreased relative to the prey's defensive abilities (usually a function of the prey evolving better defenses) because otherwise the predator would still attack the larger individuals.

This result, in conjunction with morphological data (e.g., increase in prey ornament or shell thickness), would corroborate a hypothesis that the morphological change was a response to predation. In contrast, an increase in repair frequency, without a related decrease in the position of the size refuge, would argue for an increase in attacks (rather than success), possibly due to an increase in the number of individual predators. As such, this last scenario would not corroborate that a change in prey morphology was an anti-predatory response. This type of analysis is important as a test of whether morphological changes in the prey through space or time really are anti-predatory – for example, we often assume that an increase in ornament must have been an evolutionary response to deter predators but this assumption may be invalid, particularly when we are examining organisms that have no modern descendants or analogs, a common situation in the Cambrian.

Recently, Ahmed and Leighton (2009) used this general approach to demonstrate that spinose brachiopods would achieve a size refuge at smaller shell-sizes than would non-spinose brachiopods that were otherwise morphologically similar. The spines would be a cost-effective means of making the brachiopod effectively larger, without actually increasing the shell size or thickness. Crushing predators, which are often gape-limited, would have had to gape more widely, or take more time, to crush the spinose brachiopod.

This entire approach should work just as well with trilobites as with brachiopods or molluscs. As trilobites have distinct instars during ontogeny, and there is a progressive "release" of sclerites from the pygidium to the thorax, determining the smallest size at which the trilobite could have been attacked would be straightforward. To my knowledge, there have been no studies examining the possibility of size refugia in trilobites, but this would be an interesting avenue for future research.

4.2.3 Crushing Traces

Care must be taken to confirm that a scar is due to predation, and not to other causes (e.g. post-mortem compaction). Crushing traces may appear on shelled organisms (brachiopods or molluscs) or on arthropods. Evidence for predatory scars on shelled organisms includes: (a) The scar is a point-fracture. Predation typically involves inflicting pressure on a small area, not the whole shell (as would compaction) (Elliot and Bounds 1987). (b) Scar shape is non-random, e.g., trapezoidal, triangular, circular, arcuate. Scar shape will reflect the shape of the attack structure of the predator, e.g. teeth, chelae. For example, Alexander (1986a) noted that many scars observed on Paleozoic brachiopods were consistent with the shape of chondricthyian teeth. (c) The scar will not be parallel to growth lines. This criterion is conservative; it assumes that shells break along concentric margins as a result of transport etc. As mentioned previously, some strophomenates had morphologies that may have prevented propagation of fractures beyond a concentric growth band (Alexander 1989). This hypothesis was confirmed by biomechanical experiments in which the anterior margins of shells were crushed by a simulated chela (Alexander 1989). Thus, some predatory scars might be parallel to growth lines. However, scars that are not parallel to growth lines are more likely to have been inflicted by a predator. (d) Scars on opposite sides of the shell match, implying that a predator enclosed and attempted to crush the prey. (e) The damage is repaired. This is the best evidence for predatory attack. Repair to the damaged area, which usually involves distortion of the growth lines near the damaged area (Fig. 4.2), demonstrates that the prey was alive at the time of damage. This metric is also conservative; Alexander (1981), examining Mississippian brachiopods, noted that subadult brachiopods had much higher repair frequencies than adults; older individuals might have more difficulty, or less need, to repair a damaged shell. Evidence for a, b, and either c, d, or e, together would be good evidence for a scar caused by crushing.

Similar criteria can also be applied for attacks on arthropods, especially trilobites. Predatory biting/crushing scars on trilobites will typically be arcuate (Figs. 4.1 and 4.2), or arcuate with a gentle invagination. Scars on easily damaged regions or features that would form a zone of weakness (e.g., facial sutures, spines, minor damage to margins) should not be attributed to predation; predatory scars should cut relatively deeply into the prey body. Two similar scars that would line up if the trilobite had been enrolled would be strong evidence for a biting attack on an enrolled trilobite. As with shelled prey, the best evidence for a predatory attack on a trilobite would be that the damage was repaired. In some cases, evidence for repair, such as pronounced wrinkling of the exoskeleton towards the damage, may even appear on a post-injury molt (Conway-Morris and Jenkins 1985).

4.2.4 Drilling Traces

As with crushing scars, care must be taken to ensure that the drilling traces are predatory in origin; holes also can be created by point solution (Lescinsky and Benninger 1994) and by organisms drilling for domiciles (Richards and Shabica 1969; Wilson and Palmer 2001). Predatory boreholes (a) are circular; (b) perpendicular to the shell surface; (c) penetrate only one valve; (d) are drilled from the outside of the shell; and (e) involve only one successful hole per individual prey (Carriker and Yochelson 1968; Ausich and Gurrola 1979; Leighton 2001). Exceptions to this last criterion are cases of parasitism, which may involve multiple complete holes (Baumiller 1990, 1993). In addition, octopods may drill more than one hole to inject venom (Bishop 1975), and multiple juvenile gastropods have been observed attacking the same individual (Brown and Alexander 1994). There are also exceptions to (a) and (b), but these criteria serve as conservative indicators.

4.2.5 Stereotypy

One of the best types of evidence for predation is that the traces, regardless of whether they are crushing or drilling traces, are stereotypic, and that this stereotypy is biologically meaningful. The idea behind stereotypy is simple – predators tend to attack their prey in a relatively consistent manner, and so the traces, when examined across multiple specimens, will exhibit a recognizable and consistent pattern. In contrast, if the traces are post-mortem and taphonomic in origin, then the damage is more likely to be random (but see below for some important caveats). Some examples of stereotypy:

1. Stereotypy with respect to taxon or morphotaxon. Predators usually favor certain prey items. This prey preference may be a function of optimal foraging on the part of the predator – for a given community, there will be certain prey items that provide a greater benefit (food converted to energy) minus cost (energy spent on the hunt and kill) relative to the time spent foraging. Some prey items will be too difficult or too dangerous for the predator to attack; other prey would be easy kills but would provide too little food to compensate for the time and energy spent by the predator. As a result of this, some potential prey will be attacked and others will not. Without going into all of the specifics of optimal foraging theory (see Charnov 1976, and Stephens and Krebs 1986, for more detailed explanations) it should be noted that the total and relative abundances of the different prey items also influence the selection of prey. For example, if a preferred prey is very rare, then the predator is going to have to attack less optimal prey to ensure that the predator gets enough food. If the preferred prey is very common, then attacking other prey may not be necessary. The point is that these preferences can change very quickly over ecological time-scales as prey populations change, and a preference for one prey observed in a given fossil bed should not

be assumed to hold for a different bed, even if the exact same prey taxa are present in both beds. This may actually help the paleobiologist – a stereotyped signal could be diffused or masked by time-averaging, so the presence of stereotypy, despite time-averaging, should be very strong evidence for predation.

So if one taxon has a significantly greater number of possible predatory traces than other taxa in the community, relative to the prey taxa's proportional abundances, then those traces are more likely to be predatory in origin. This assumes that the preservation potential of the taxa and traces are relatively similar. For example, if you are examining an assemblage of calcareous brachiopods, and one taxon is consistently hit more often than others, then this would be a strong argument that the traces are predatory. In contrast, if you are examining a mixed assemblage of brachiopods and bivalves, and the bivalves are preserved primarily as molds, then obviously the results will be biased towards greater traces on a brachiopod. The test will be more powerful when restricted to taxa that are similar in preservation potential, life-mode etc.

2. Stereotypy with respect to location on the prey. Because predators often attack a prey-type in a consistent manner, predatory traces will often be found in the same location of the prey body. Consistent attack strategies may occur for a variety of reasons: the location may lead to a quicker kill (hitting vital organs of the prey), may make it easier to get to more nutritious food (e.g. some muscle and organ tissues may be more easily digested or provide greater protein), or may be a constraint of the grappling mechanism of the predator. For example, modern naticid gastropods, which are predatory drillers, envelope their bivalve prey in a large foot. The location of the naticid's predatory drill-holes may be constrained by the need to hold the bivalve in a particular orientation to inhibit escape. Whatever the reason, it is common for predators to attack a particular location on the prey; if one location is hit significantly more frequently than other locations, then the traces are probably predatory.

   This sort of stereotypy can be examined on more than one scale, depending on the prey species involved. For example, one could examine stereotypy on the brachial versus pedicle valves of a brachiopod, but one also could look for stereotypy on the body versus mantle cavities of the brachial valves of the same taxon. Such fine-scale tests of stereotypy would be more appropriate for small (relative to the prey) traces, such as borings; crushing scars that are too large and cover too much of the prey would not permit such a test as easily.

3. A related form of stereotypy, applicable to pursuit traces, would be to determine how frequently the supposed predatory trace starts in approximately the same position as the prey burrow. Similarly, the relationship – specifically the convergence – of the two burrows at their lowest points in the sediment could be examined.

Stereotypy is useful evidence and has the advantage that it can be easily quantified and tested statistically, but some caution must be exercised when applying these methods. First, we assume that stereotypy must be biological in origin, but it is important to ascertain whether observed stereotypy could be expected to be the result of taphonomy. As described above, damage due to compaction usually will have a very different appearance from that of predatory damage. Post-mortem damage will also not be repaired, but we cannot always rely on the presence of repairs. In most cases, stereotypy will be biological because taphonomic damage will tend to be random with respect to location, but there are exceptions. If there is a natural plane of weakness in a shell, then that region may fracture under any force, predatory or taphonomic, so if stereotypy is demonstrated, then it is important to evaluate the possibility that the region with frequent damage is not an obvious zone of weakness. Of course, predators may deliberately strike this region exactly because it is weak, and the damage really may be predatory, but excluding such cases is conservative.

Second, just because stereotypy is biological does not automatically mean that it is predatory. For example, epibionts which encrust a host shell may also exhibit strongly stereotypic behavior. The borings of epibionts seeking to stabilize or construct their home might, for example, tend to be located on the highest region of the host shell or near beneficial currents generated by the host (Taylor and Wilson 2003).

Along these lines, it is important to keep in mind that stereotypy should be biologically meaningful. This may be challenging if we know little about the biology of the organism under scrutiny but tests for stereotypy will be more meaningful if there is an underlying biological basis for such stereotypy as opposed to a simple geometric pattern with no obvious biological relevance. These considerations can be particularly important when setting up the test, as described below.

4.3 Data and Statistics

4.3.1 Organizing the Data

One of the best ways to test for stereotypy is to use a "goodness of fit" test, such as a chi-square, binomial, Kolmogorov-Smirnov (a non-parametric test), likelihood-G, or Fisher's exact test. Before discussing the tests, we will discuss the organization of the data. Typically, these tests will be set up with a 2 × 2 matrix. One column will be the observed results and the other column will be the expected results. The two rows will be two separate and clearly distinct categories of results (Table 4.2A). For example, imagine we are interested in testing whether a particular species of trilobite has been attacked more frequently than other trilobites in the assemblage. Our observed column contains data for the species of interest. Our expected column contains data for all other trilobites in the assemblage. The two rows are "number of specimens attacked" and "number of specimens not attacked". The test we choose will determine whether the observed data are significantly different from the expected data.

Another example: we wish to determine if the body cavity region of a particular brachiopod species is attacked more frequently (stereotypically) than the rest of the brachiopod. Our observed column contains the data for the body cavity region (for all

Table 4.2 Examples of setting up a 2 × 2 matrix for goodness-of-fit tests

A. Abstract example; a, b, c, d represent frequencies

	Observed	Expected
Category 1	a	c
Category 2	b	d

B. Hypothetical example, including data

Specimens	Body-cavity	Mantle-cavity
Attacked	15	5
Not-attacked	10	20

C. A hypothetical illustration of the utility of the 2 × 2 matrix

Attacks on	Observed	Expected
Body-cavity	15	12
Mantle-cavity	5	8

The brachiopod species was attacked 20 times. The matrix is organized to determine if the body cavity is preferentially attacked. In this example, the expected frequencies are determined by the fact that the body cavity comprises 60% of the surface area of the brachiopod; therefore the expectation under the random model is that 60% of the total attacks should be over the body cavity. A goodness-of-fit test then determines if the number of the observed attacks is significantly different from this expectation

of the specimens of that species). The expected column contains data for the mantle cavity region. Again, the two rows are "attacked" and "not attacked" (Table 4.2B).

All of these goodness of fit tests for stereotypy start with such a matrix. These 2 × 2 tests are very powerful and very flexible, largely because the user gets to determine the null model – the expected column – and thus can make the null model more biologically relevant and also a more exact test of stereotypy.

For example, in our second example above, testing stereotypy for location on the prey, we can also frame the question and organize the data somewhat differently (Table 4.2C). Note that there are a total of 20 attacks. We independently determine that the body cavity region of the brachial valve constitutes 60% of the valve. Therefore, the expectation is that, if the attacks are completely random with respect to location on the valve (not stereotyped), then 60% of the attacks (12/20) should be over the body cavity. So instead of using "attacked" and "unattacked" as our rows, we will use "body cavity" and "mantle cavity" as our rows. For our observed column, we use the actual observed number of attacks for our observed column (=15). For our expected column, we calculate 60% of the total attacks on the brachial valve (=12), and we use this as our expected number of attacks in the expected column for the body cavity row. Similarly, the remaining 40% (=8) of expected attacks would be on the mantle cavity, and this value fills the last cell (Table 4.2C).

Note that this organization of the data actually takes into consideration the relative proportions of shell surface over the body and mantle cavities. If we had just assumed that there was a 50% chance of attacking either region, then we might get a different result. Because of the flexibility of the matrix in allowing us to determine a more exact null model, we have a better statistical test for stereotypy.

4.3.2 *Statistical Tests*

I don't want to turn this into a lecture on statistics, but some background is necessary. The most frequently used and widely recognized goodness of fit test is the chi-square test (also written as the X^2 test). Aside from the fact that the test is well-known, and so your readers will be more likely to understand it, the chi-square test has several other advantages, most notably that the mathematics involved are so simple that you probably would be able to do it in your head, in the field. However, the chi-square test also is frequently misused, and there are a number of underlying assumptions that are commonly ignored. I'm going to describe the chi-square test in some detail, but if you have a good statistics package, I would recommend that you use the Fisher's Exact Test to test for stereotypy. Technically, the Fisher's Exact Test is not a statistical test – it's an exact test, hence the name. What this means is that the test calculates every possible arrangement of the data and determines where the observed results fall within the distribution of those possible arrangements. You cannot do this in your head, and more complicated examples would take weeks to do by hand. In fact, up until a few years ago, when you tried to perform the Fisher's Exact Test, many statistical packages would warn you that the analysis was going to take long periods of time. I recall once bringing a personal computer to its knees in this way for over a week. With the increasing power and speed of modern computers, these considerations are no longer an issue, and there is no reason not to use Fisher's Exact Test to test for stereotypy, particularly as it does not suffer from some of the limitations of the other tests. That said, Fisher's Exact Test will not be available in every statistical package. For the majority of cases, most goodness of fit tests will give similar results, and the chi-square is still a perfectly valid, well-recognized, and intuitive, test for stereotypy, when applied properly.

The formula for the chi-square test is:

$$X^2 = \sum((o_i - e_i)^2) / e_i$$

where o_i is the frequency observed in category i and e_i is the frequency expected in category i. Using our last example (Table 4.2C), $o_1=15$ and $e_1 = 12$, $o_2=5$ and $e_2=8$, so $((o_1 - e_1)^2)/e_1=9/12=0.75$ and $((o_2 - e_2)^2)/e_2=9/8=1.125$ and the summation $=X^2=1.875$.

This result is then compared to a standard table for significance. For a 2 × 2 matrix (which has one degree of freedom), the threshold for a significant (where $p=0.05$) result $=3.84$; if $X^2>3.84$, then we can refute the null hypothesis, that the observed and expected results are indistinguishable. In our example, $X^2=1.875$, so we cannot refute the null hypothesis; the results do not corroborate stereotypy with respect to location. Note that if we had simply inferred that there was an equal chance of attacking the body cavity or the mantle cavity (50%), then the expected number of attacks on the body cavity = 10 as would the expected number of attacks on the mantle cavity. In that case, $X^2 = 5.0$ (try calculating this for yourself), and we would refute the null hypothesis and corroborate the hypothesis that there was a preference for the body cavity. This example illustrates the utility of a goodness-of-fit test, but it also illustrates how important it is to develop a biologically meaningful and valid null hypothesis. If we develop a poor null hypothesis – for example, if we fail to note that the area of the body cavity is greater than that of the mantle cavity and mistakenly assume that they are equal – then we might produce a completely different, and mistaken, result.

Several cautions and considerations related to the use of the chi-square test and goodness of fit tests in general:

1. The two columns, observed and expected, must be equal in total frequency. Often, you will not be so fortunate that your data happen to produce equal totals. Some statistical packages will make the adjustments for you, but just in case, the standard practice is to adjust the column with the greater total frequency so that it is equal to the other column. We use the totals for the lesser of the two columns because most goodness-of-fit tests, including the chi-square, are more conservative with smaller numbers.

   For example, let us use the data from the Conway-Morris and Bengtson (1994) study described early for predation on *Linnarssonia*. The raw data for *L. sp. B*, Sample S70-2-SB are shown in Table 4.1A and organized into a 2 × 2 matrix in Table 4.3A. We use a and b to indicate the observed values for the brachial and pedicle valves respectively and use c and d to indicate the expected values for the brachial and pedicle valves. We are interested in determining if there are significantly more holes on the brachial valve than on the pedicle valve. It appears that there are, but as Conway-Morris and Bengtson (1994) noted, there is also a taphonomic bias towards preservation of the brachial valves. Thus, we will need to define our expected values – essentially our null hypothesis – to take this bias into account. Our expectation will be based

Table 4.3 Illustrative examples of adjusting the 2 × 2 matrix so that the column totals are equal, using actual data from Sample S70-2-SB of Conway-Morris and Bengtson (1994)

A. Example of organizing data to test for stereotypy by valve on *Linnarsonnia* sp. *B*

Attacks on *L.* sp *B*	Observed	Adjusted exp.	Expected
Drilled Brachial V.	23	70% × 30 = 21	738
Drilled Pedicle V.	7	30% × 30 = 9	319
Totals	30	30	1,057

B. Example of organizing data to test for prey preference for one species

Attacks in S70-2-SB	Observed	Adjusted exp.	Expected
L. sp. A	33	82% × 63 = 51	4,717
L. sp. B	30	18% × 63 = 12	1,057
Totals	63	63	5,774

See Table 4.1A for raw data

on the observed number of valves. However, note that the frequency of valves is, of course, greater than the number of holes. Thus, the total frequency for the observed results $(a+b)=(23+7)=30$ but total frequency for the expected results $(c+d)=(738+319)=1{,}057$. To calculate the new values of the expected results (c′ and d′) and to adjust the columns so that they are equal, we simply calculate the proportion of c out of (c + d) and then apply that percentage to the sum of (a + b). $c/(c+d)=0.698$. $(a+b)=30$. $0.698 \times 30=20.946=c'$. Then $d'=30-c'=9.054$. Note that $(a+b)=(c'+d')$.

We then proceed as before and calculate the chi-square statistic. $o_1=23$ and $e_1=21$, $o_2=7$ and $e_2=9$, so $((o_1-e_1)^2)/e_1=4/21=0.19$ and $((o_2-e_2)^2)/e_2=4/9=0.44$ and the summation $=X^2=0.63$. This X^2 has an associated p-value $=0.414$ (note that the $X^2<3.84$) and so we cannot refute the null hypothesis – even though holes on the brachial valve are three times more common than on the pedicle valve, there is no statistically significant preference for the brachial valve; the observed pattern is at least in part due to the preservational bias towards brachial valves.

2. There will be some cases in which the expected frequencies are based directly on existing data – in essence, the test is comparing two different categories, and arbitrarily using one as the observed and one as the expected. Although using the test in this way is acceptable, it often does matter which category is assigned to the expected column. A quick examination of the chi-square formula should make this evident – the expected frequency appears in the denominator, but the observed frequency does not. In a case such as this, in which the test is simply between two categories and there is not truly an expectation, just a comparison of categories, then we must choose the more conservative (least significant) of the two results as our result.
3. Goodness of fit tests are scale-dependent. In other words, the size of the data set affects significance; the greater the size of the data set, the easier it is to refute the null hypothesis. This is why we always adjust to the smaller of the two columns. We examine one more example from Sample S70-2-SB of the Conway-Morris and Bengtson (1994) study. In this case, we are interested in whether there is a preference for *L. sp. B* over *L. sp. A* (Table 4.1A). There actually are fewer total borings on *L. sp. B*, but there are many more valves of *L. sp. A*, which might bias the result towards *L. sp. A*. Is it possible that this bias masks the predator's preference for *L. sp. B*?

   First, we adjust our raw data so that the two columns are equal, and as noted above, we adjust to the column with the smaller frequency (Table 4.3B). There are 33 borings on 4,717 valves of *L. sp. A* and 30 borings on 1,057 valves of *L. sp. B*. The observed column $=33+30=63$, whereas the expected column $=4{,}717+1{,}057=5{,}774$. We adjust the expected column so that it also equals 63. $4{,}717/5{,}774=0.817$. $0.817 \times 63=51.467$ gives us our adjusted expected value for borings on *L. sp. A*. The adjusted expected value for borings on *L. sp. B* $=63-51=12$. Again, we calculate the chi-square value: $o_1=33$ and $e_1=51$, $o_2=30$ and $e_2=12$, so $((o_1-e_1)^2)/e_1=324/51=6.35$ and $((o_2-e_2)^2)\ /\ e_2=324/12=27.00$ and the summation $=X^2=33.35$. This X^2 has an associated p-value <0.001 (note that the $X^2 >> 3.84$) and so we refute the null hypothesis – there is a very significant preference for *L. sp. B*.

Scale-dependence is very important even in terms of how we set up the data collection. For example, several prior studies overlaid a grid onto the prey and then tabulated the number of attacks within each square of the grid to test for stereotypy. There is nothing inherently wrong with this approach, but the problem is whether there is any justification for the number of cells in the grid. For example, suppose you have a 3 × 3 grid with nine cells. If the decision to have nine cells is arbitrary, then you could just as easily have a 3 × 6 grid with 18 cells, a 6 × 6 grid with 36 cells, or a 3 × 1 grid with 3 cells. It is easier to refute the null hypothesis with 36 cells than with nine, but harder to refute the null hypothesis with three cells than with nine. Without a biological justification for choosing the number of cells, it is possible to bias the statistics towards a favorable result even though the position or total number of the predatory traces on the prey has not changed at all! My recommendation is that rather than specifically using a grid, the worker identifies distinct, biological meaningful regions on the prey – a predator will attack a recognizable region of the prey consistently but probably does not care about a geometrically-defined region identified by a human. For example, one could use the cephalon versus post-cephalon on a trilobite, or the body-cavity (typically indicated by the muscle field) versus mantle cavity of a brachiopod etc.

Another problem with the issue of scale dependence is that many workers assume that the easiest solution to the problem described previously in (1) is to convert the data to percentages. Conversion to percentages is a common data-transformation for a variety of reasons but it is invalid when performing a goodness of fit test. If you actually have less than 100 total observed data points, then converting to percentages has just increased the size of your data set to 100 – and thereby increased your likelihood of refuting the null hypothesis. There are corrections that can be applied to a chi-square when using percentage data, but if you are testing for stereotypy, you will already have the raw data, so it is more conservative, probably more accurate, and no more difficult to perform the standardization described above rather than converting to percentages and applying the correction.

4. With the exception of the binomial test, goodness of fit tests can be applied to matrices larger than a 2 × 2. For example, if there were three distinctive regions on the prey body (e.g. the cephalon, thorax, and pygidium of a trilobite), we could organize the data into a 2 × 3 matrix. At first glance, this would seem ideal for many cases. However, if you use this approach, keep in mind that you get a significant result if any *one* of the rows is different – it does not mean that *all* of the rows are different from each other. The other data could be very similar, but if one row is different, that would be enough to produce a significant result.
5. Some statisticians will argue that chi-square tests are technically parametric because they are based on an existing distribution. However, unlike most other parametric tests, the chi-square test is based on a non-normal distribution. The chi-square distribution is right-skewed. The consequence of this is that although the goodness of fit tests become more conservative with smaller sample sizes, there is a point at which the sample size is too small to get a valid result. There

are two rules of thumb for addressing the sample size. First, if any of the cells = 0, then the assumptions of the chi-square test have been violated. Second, if one-quarter or more of the cells have frequencies < 5, then the assumptions have been violated. As we are generally using a 2 × 2 matrix, this means that if any one of the cells has a value < 5, we should not be using a chi-square. Unfortunately, this situation is very common for cases examining stereotypy – in fact, some of the cases above violate this assumption, although I obtained similar results when using a Fisher's Exact Test.

An additional assumption of the chi-square is that the distribution is unimodal. Again, this assumption may commonly be violated when testing for stereotypy. These assumptions do not apply to the Fisher's Exact Test, which is one of the reasons why I recommend its use.

4.3.3 Predation Traces as Landmarks

As described previously, the use of grids has often been used as a means of quantifying the location of predation traces on prey, but the arrangement of the grid and the choice of grid-size may be problematic. As I mentioned, I recommend using biologically meaningful regions of the prey, rather than a geometrically-defined grid. In addition, Peter Roopnarine (Roopnarine and Buessink 1999) developed a clever solution to comparing the location of predation traces on different specimens, and his method has the additional advantages that it provides a very exact location for the predator trace and is also actually easier to quantify the results. Geometric morphometrics is an often-used technique for describing shape (Marcus et al. 1996; Bookstein 1997). Consistent, preferably homologous, points on the specimens are treated as landmarks and plotted in an x-y Cartesian coordinate space; in this way, the entire organism is treated as a constellation of points and the relationship among points is maintained. It is beyond the scope of this chapter to give a detailed lesson on geometric morphometrics, but as a start, I will provide a very basic introduction to one geometric morphometric method, Bookstein Shape Coordinates, within the context of predation traces. I also recommend appendix A-4 of Bookstein et al. (1985), which gives the relevant formulas for Bookstein Shape Coordinates, as well as providing a nice graphical explanation of the method. There are more recent and sophisticated works on the topic, but the simple "shape triangle" method described therein is sufficient for the worker interested in applying this technique to the analysis of predation traces. Roopnarine's approach is to perform geometric morphometrics on the prey specimens while treating the predation trace as an additional landmark (Fig. 4.4a). In this way, other landmarks act as a baseline for size and orientation of the specimens, and the positions of the predation traces are revealed relative to other landmarks. This technique can be particularly powerful if any of the landmarks are associated with biologically meaningful structures.

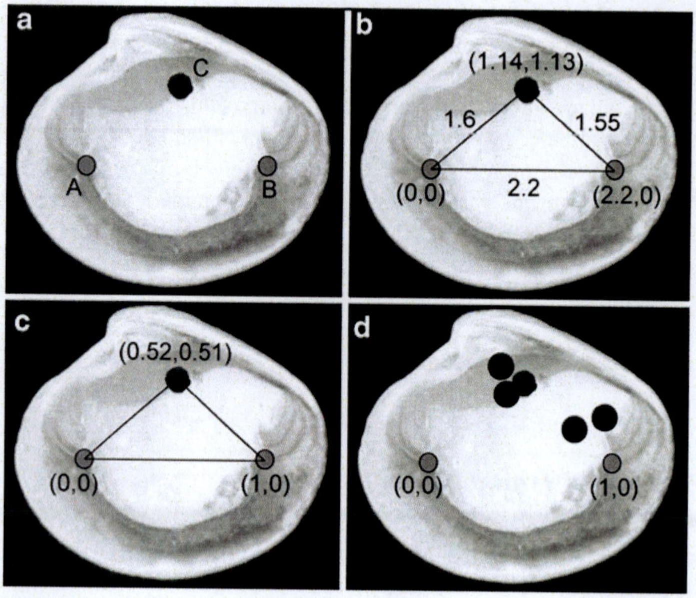

Fig. 4.4 Illustration of using Bookstein Shape Coordinates to compare the location of predation scars (in this case, a drill-hole) among specimens. (**a**) Three landmarks, labeled, *A*, *B*, *C*. *A* and *B* (*grey circles*) are homologous landmarks which can be found on all specimens; *C* (*black circle*) marks a predatory drill-hole. (**b**) Triangle of line-segments between the three landmarks, showing the lengths of the line-segments which can be used to calculate the Cartesian x-y coordinates of the three landmarks trigonometrically, prior to transformation. (**c**) Triangle of points, showing the transformed x-y coordinates. See text for method. (**d**) Graph of shape coordinates for five specimens, each with a single drill-hole, that have been baseline-transformed, so that all specimens now have their *A* and *B* landmarks on the same positions: (0,0) and (1,0) respectively. *Black landmarks* indicate the position of the five drillholes. Graph superimposed on image of specimen to show the position of the drillholes relative to each other and relative to the landmarks on the specimens

As an example, imagine that we are studying predation traces on the valves of several bivalve specimens. We begin by choosing two data points (landmarks) to form a line segment which will be our baseline, which serves as the framework around which other points will be placed. The baseline ideally will be between two homologous points and will be relatively long (which minimizes the effects of imprecision in user measurements). We choose the junctions of the two adductor muscle scars with the pallial line to represent our a (anterior adductor) and b (posterior adductor) landmarks (Fig. 4.4a). Our point of interest, a drill-hole, will be landmark c.

We then capture x-y coordinates for our point(s) of interest. These data can be captured by overlaying a grid over the specimen, either electronically or by hand; there are numerous software packages designed to do this. Alternatively, the line-segments between the three points can be measured directly, and the x-y coordinates then determined by trigonometry. In our example, the lengths of the line-segments are Line ab = 2.2, Line ac = 1.6, Line bc = 1.55. We will set up our Cartesian space so that point a marks the origin (0,0), and Line ab is the x-axis. As such, point b=(2.2,0) by definition. We now find the xy coordinates of the c point, which we shall refer to as xc and yc. To avoid confusion, I use *, rather than x, to indicate multiplication in all of the equations in this section. ab refers to line segment ab etc. From trigonometry,

$$\text{Angle bac (in radians)} = \frac{\text{acos}\ (bc^2 - ab^2 - ac^2)}{(-2 * ac * ab)}$$

$$xc = ac * \cos\ (\text{angle bac})$$

$$yc = ac * \sin\ (\text{angle bac})$$

The results of our calculations are shown in Fig. 4.4b. The three landmarks and their xy coordinates can be calculated in this manner for each specimen in our sample. So far, this is just trigonometry. What we wish to do is to arrange all of our specimens so that all of the a points move to (0,0) and all of the b points move to (1,0). The baseline between a and b would thus define the x-axis, define the basic unit of measure, and most importantly, standardize all of the specimens to the same size (because for all specimens, the post-transformation distance between a and b=1). Ultimately, we want to determine where all of the point c values (the drill-holes) fall in this new baselined morphospace. We will refer to these landmarks in the transformed, baselined morphospace as A, B, and C, and their x and y coordinates as xA, yA etc. Keep in mind that small letters a, b, and c, refer to the original landmark data, and that capital letters A, B, and C refer to the landmarks post-transformation.

Note that we don't actually need to solve for xA, yA, xB, or yB – these landmarks are sent to (0,0) and (1,0) by definition; we only need to solve for xC and yC. We have two long, but simple formulas to solve for xC and yC:

$$xC = \frac{(((xb - xa) * (xc - xa)) + ((yb - ya) * (yc - ya)))}{((xb - xa)^2 + (yb - ya)^2)}$$

$$yC = \frac{(((xb - xa) * (yc - ya)) + ((yb - ya) * (xc - xa)))}{((xb - xa)^2 + (yb - ya)^2)}$$

Note that because we set up the specimens on a grid such that the point a is at (0,0) and point b is at (x,0) to begin with, then the formulas reduce to much shorter versions:

$$xC = \frac{(xb * yc)}{(xb^2)}$$

$$yC = \frac{(xb * yc)}{(xb^2)}$$

We show the results of the baseline transformation in Fig. 4.4c. Our transformation has determined where the C landmark in the new baselined morphospace will be, once the B landmark was transformed from (2.2,0) to (1,0). This entire approach has numerous advantages. The result is intuitive – we essentially are rescaling and overlaying the specimens on top of each other; all of the A points for all specimens are now at (0,0) and all of the B points for all specimens are now at (1,0). One can examine the plot and visualize where the landmarks fall on a specimen. Within this framework, we also can observe exactly where the predation traces, represented by the C point for each specimen, are positioned relative to the other landmarks. Figure. 4.4d illustrates a hypothetical result with five specimens. Remember, the A point for all five specimens is at (0,0) whereas the B point for all five specimens is at (1,0). Thus the five black dots each represent the position of the drillhole on one specimen, relative to the adductor-pallial line junctions of that specimen. Figure. 4.4d shows that we could superimpose the C points on to an image of one specimen so that we could easily observe the relative positions of the drillholes on the different specimens on one plot.

It should be emphasized that (xC, yC) is a shape-coordinate – it is not simply an expression of the position of the C landmark but is actually the third point of a shape-triangle in which A is at (0,0) and B is at (1,0). Thus, C is the position of the predation trace *relative to the rest of the specimen*. In this regard, it can be useful to identify other landmarks that would help identify regions of the prey for stereotypy analysis. One can repeat the entire process for multiple landmarks, always using the same a and b for a baseline and in this way, the position of the trace is always identified relative to other prey landmarks, and more importantly for a stereotypy analysis, the trace is located relative to biologically meaningful regions of the prey defined by those landmarks. Counts of traces within a given region can then be compared to counts outside the region using a goodness of fit test to test for stereotypy.

The use of a predation trace as a landmark is not just for examining whether traces exhibit stereotypy. This approach also permits comparisons of groups of traces. For example, we may wish to compare the location of predation traces across different taxa, or from different environments, different regions, or different times. In our illustrated example, three of the drillholes were located near the umbo, and two of the holes are close to the posterior adductor scar. Imagine that the specimens with holes near the umbo are from a shallow setting while the other specimens are from deeper depths. Thus, in our simple example, the data clouds for predation in the two environments are in very different positions on the shell. This might suggest that there are two different predators involved, one in each environment. For more complicated examples with more data, the data clouds for the two environments (or species, regions, times etc.) can be tested for difference using a Hoteling T-square test, which functions as a multidimensional *t*-test. These methods can be used to examine stereotypy and predation through time and space.

4.3.4 *Predation Frequencies*

Once predatory traces have been identified and tabulated, predation frequency (really a percentage) can be calculated. This can be more complicated than one might think, as the choice of numerator and denominator can have a major influence on the results. There is nothing inherently more correct about any of these decisions; it is important to let the question of interest guide your choice – and essential that you explicitly identify your decision in your work so that there is no confusion on the matter.

The first decision is whether one calculates the percentage of the number of traces relative to the number of prey individuals, or instead calculates the percentage of specimens that have been attacked relative to the number of individuals. The first approach is an attack percentage, which records the frequency of attacks, whereas the second indicates the number of prey within a population that have experienced an attack at some point during their lives. In many studies, these two numbers will not be very different. However, there are some circumstances, notably when examining repair scars over a broad range of prey sizes, where the values might be very different. Larger specimens, which presumably have lived longer, have had more time to experience attacks and accumulate damage. Similarly, a previously-damaged specimen, especially if the damage is comparatively recent, might be a more tempting target for a predator, as such a specimen may already be weakened. On the other hand, older specimens of many species do not grow as quickly as their younger counterparts, and so may be less able to repair (or have less need to repair). In either case, a prey population may have certain individuals that have experienced a greater concentration of repaired attacks, and the results for attack frequency and attacked individual frequency may be very different. In many studies, there is no reason not to report both results if they are very different; this facilitates comparisons with other predation studies and also provides insight into whether attacks may have been more concentrated on certain individuals or whether attacks were extremely common on the prey population.

One consideration in using attack frequency is that one must be reasonably sure that different cases of damage on an individual really originated from different attacks. As noted previously, scars on opposite sides of the prey may be matching scars from a single attacker trying to engulf the prey, or as in the example of anomalocarid attacks, one scar is the actual bite while the opposite scar is where the prey was grappled – in which case, the two scars may not look alike and might be mistaken for two separate attacks. Similarly, a drilling predator that is interrupted by the motion of the prey trying to escape will usually start boring a new hole. Thus, an incomplete hole and a complete hole positioned close together might have been the result of a single encounter. For these reasons, I sometimes prefer to use the attacked individuals frequency, as this is an extremely conservative estimate of the attack rate.

A second decision is related specifically to predation (especially drilling) on bivalved animals, such as brachiopods and bivalves, if the prey taxa are prone to disarticulation. This problem is not an issue if the prey specimens are articulated, but disarticulated valves will often lead to situations in which some valves have no evidence for predation even though the valve came from an animal that was damaged – the damage is on the opposite valve. To illustrate this problem, envision an assemblage of 100 bivalves. Ten percent are killed by a drilling predator. All specimens disarticulate, regardless of cause of death. There are ten valves with complete predator drill holes, and 190 valves without holes. If you calculated the attack frequency by dividing the number of kills by the number of valves, you would get 5%, which is only one-half of the true drilling frequency. This would be a major problem, particularly if you were comparing two taxa with fundamentally different disarticulation rates.

There are two end-member scenarios for taxa that disarticulate, which I will refer to as the "bivalve scenario" and the "brachiopod scenario". The "bivalve scenario" will occur in cases where the prey taxon has two valves that are equal in size and mirror each other, such as heteroconch bivalves. In this case, there is less likely to be a preferred up-down living orientation, no reason for a preservational bias, and a predator probably won't distinguish between the two valves. Consequently, there probably won't be a strong valve preference on the part of the predator. As such, the best approach, as recommended by Kowalewski (2002), is to include all valves in the analysis and divide the number of attacks by one-half the number of total valves. This method solves the problem described above. As noted by Kowalewski (2002), dividing by one-half of the total valves is statistically preferable to multiplying the number of attacks by two – both methods give the same percentage, but the former method is conservative as the latter method artificially increases the sampling size.

The "brachiopod scenario" may occur when the two valves of the prey are different in shape or size; this is typical for brachiopod brachial and pedicle valves but also is true for many bivalves, such as pteriomorphs. Such organisms often have a preferred orientation in which one of the valves is topmost, there may be a preservational bias towards one valve (usually the less convex valve), and predators can distinguish between valves and so will often focus on one of the valves. Consequently, there may be a strong prey preference for one valve. If the preference is very strong, then the best way of assessing drilling frequency would be to analyze and count only that particular valve, regardless of whether a specimen is articulated or disarticulated, i.e., if the pedicle valve were strongly preferentially drilled, you would include and count articulated specimens and disarticulated pedicle valves in the analysis, but not disarticulated brachial valves. This approach also avoids potential problems with the possible preservational bias towards one valve.

Unlike most Cambrian bivalves and brachiopods, trilobites are often large enough that crushing damage is limited to one region, so the problem of disarticulation must be addressed. Trilobites commonly disarticulate, typically involving separation of the cephalon from the rest of the body. As mentioned previously, the post-cephalon region is both much more likely to be preserved and much more likely to be attacked. Thus, the "brachiopod scenario" would also be a "trilobite scenario" for crushing predation, and the most conservative approach would probably be to limit the analysis to only post-cephalons, i.e. obtain the repair frequency by dividing the number of post-cephalic repairs by the number of post-cephalic individuals (including disarticulated post-cephalic regions and articulated trilobites but not disarticulated cephalons).

An additional problem with calculating damage frequencies for trilobites is that trilobites molt. For any individual, there is the potential for multiple instars to be preserved, thus artificially increasing the number of individuals. Moreover, molts may retain scars from damage incurred during earlier instars, so it is even conceivable that a given repair could be counted more than once. Although the use of body-sizes to assess and interpret repair frequency as described above would be of particular interest when applied to trilobites, I would recommend restricting the actual calculation of repair frequency to a particular instar – instars should show distinct modes in a size distribution of the population, so they should be readily recognizable – or to restrict the calculation to specimens that clearly are not molts.

The last consideration when deciding which predation metrics to report is whether to calculate predation frequencies at the taxon or assemblage level. There are four frequently used metrics: (1) predation frequency for a given taxon of interest; (2) predation frequency of the most frequently attacked or damaged taxon; (3) predation frequency of an entire assemblage; and (4) predation frequency of all of the taxa with predation frequencies greater than zero within an assemblage. Kowalewski (2002) gives a thoughtful review of the advantages and disadvantages of all of these measures, and I highly recommend that the reader examine that review. The choice of measure is largely decided by the scale of the question, and of course, by the question itself. While examining an individual taxon or lineage has potentially fewer artifacts (e.g., biases due to differential preservation), it may not be possible to analyze a single taxon or lineage through increasingly greater spatial or temporal scales. On the other hand, assemblage-level measures run into the problem that not all taxa will be attacked – abundant taxa that are never attacked will artificially lower the predation frequency for an assemblage (Leighton 2002). If predators are never interested in particular taxa as prey, then the inclusion of those additional taxa is misleading. Of course, it may not be possible to determine if a predator would consider a particular species as possible prey just by looking at one assemblage in space or time, and this is especially true during the Cambrian. Foraging dynamics are sufficiently complex that the predator's decision to attack a given species may vary with slight changes in environment or in abundance of the potential prey species, so the decision to exclude a species from analysis may not hold in a different time or place. As mentioned earlier, the attack frequency on a prey species may vary as a function of the abundance of other prey, and switching from one prey to another is a common response to changes in prey abundance. Thus, even in a study of predation on one taxon, other prey may influence changes in predation frequency on the taxon of interest. Given these complications, the use of multiple measures is favored. In addition, calculating predation frequencies for each prey species in an assemblage should be done more frequently, as such an approach may permit recognition of prey switching and its influence on predation frequencies among prey.

4.4 Cambrian Considerations

It is particularly frustrating that predation is commonly viewed as a primary influence on the Cambrian Explosion and wide-spread, convergent radiations of skeletonized animals, and yet this hypothesis is virtually impossible to test directly because it would require an examination of predation on soft-bodied fauna in comparison with hard-bodied prey. Even putting this hefty problem aside, there are a number of logistic challenges specific to the Cambrian (or earlier). In order to examine predation traces, the traces have to be preserved, which normally requires prey with a hard skeleton, but this in turn means that predators must have been willing to attack these prey. As noted previously, although Cambrian predators may have been diverse and common, few of them would have been capable of regularly taking hard-shelled prey. Not surprisingly, Cambrian crushing frequencies on brachiopods and molluscs are relatively small, typically <5% (data in Huntley and Kowalewski 2007). Borings of presumed predatory origin are controversial prior to the Devonian (Wilson and Palmer 2001), although Cambrian and earlier borings interpreted as predatory have been reported in several studies described above, and the statistical results here support stereotypy, consistent with predation, for at least two of those studies. Cambrian drilling frequencies have also generally been <10%. These low crushing and drilling frequencies suggest that on average, it would require a minimum sample size of 200 to produce 20 attacks. A sample with less than 20 attacks could still be used as part of a study examining predation across a gradient, but a small sample size would make it difficult (or meaningless) to test for stereotypy, size refugia, or success rates. Workers interested in assessing Cambrian predation on brachiopods or molluscs should be prepared to sample very large populations. To some extent, this is true for the Paleozoic in general, but the Cambrian would appear to be a worst-case example. However, the situation may also be analogous to that facing post-Cambrian Paleozoic predation workers 20 years ago – it is possible that the low predation frequencies of the Cambrian may owe as much to a lack of sampling effort as to biological reality; there simply have not been many studies whose purpose has been to examine predation in the Cambrian.

Cambrian predators could and did take trilobites, as cited above. As the chitinous exoskeleton of arthropods is not as resistant to crushing as would be a calcareous skeleton, early predators, even those that lacked mineralized attack structures, would have been able to break the prey skeleton. Moreover, the hard protective surface of the trilobite exoskeleton was largely restricted to the dorsal side; the ventral surface would have been more vulnerable to predators. Attacks to the cephalon or to the axial lobe would presumably have been fatal, so repairs are more likely to be on the pleural lobes and posterior (Babcock 2003). Attacks also may have been more likely to come from the side; lateral bites potentially would have caused more damage as such attacks would cut across the natural articulation of the sclerites (Babcock 2003). Like other arthropods, trilobites would have had a hemocoel, which would have made them extremely vulnerable if the attack penetrated the body wall. Thus, trilobites would be expected to try to repair damage as quickly as possible to prevent bleeding out.

Trilobites thus are a promising potential field for further predatory studies. That said, although there are many studies reporting trilobite injuries, most such studies are examples of single, isolated cases of possible predation. Few studies (Babcock 1993, 2003 are notable exceptions) have developed data sets on trilobite injuries sufficient to enable quantitative testing. Is this a function of lack of material or of lack of effort in these directions? In some cases, the lack of material is definitely a problem – removal of trilobites from matrix would make finding and examining large numbers of predatory scars logistically challenging. However, I should point out that I have conducted several studies on predatory drilling on Paleozoic brachiopods and in most of these examples, the fauna and localities that I examined had been under study for most of the past century, but no one had previously examined these brachiopods for drill holes. We don't notice what we don't look for. Perhaps this situation may also apply to the early stages of metazoan life on Earth. Given the potential importance of predation as a factor in the early evolution of metazoan life, we need to search for more primary data – traces of predation – from the Cambrian and earlier. I'm looking forward to the results.

Acknowledgments The author would like to thank Greg Dietl and Thomas Hegna for thoughtful reviews, and James Schiffbauer, Marc Laflamme, and Stephen Dornbos for inviting this contribution. The author would also especially like to thank Loren Babcock for providing the two trilobite injury photographs and several relevant citations. The author gratefully acknowledges financial support from NSERC for research on predation.

References

Abrams PA, Matsuda H (1997) Adaptive foraging by predators as a cause of predator-prey cycles. Evol Ecol 6:56–72

Ahmed M, Leighton LR (2009) Anti-predatory spines on brachiopods? In: GSA Abstracts with Programs, Annual Meeting, vol 41, no 7, p 389

Alexander RR (1981) Predation scars preserved in Chesterian brachiopods: probable culprits and evolutionary consequences for the articulates. J Paleontol 55:192–203

Alexander RR (1986a) Frequency of sublethal shell-breakage in articulate brachiopod assemblages through geologic time. In: Racheboeuf PR, Emig CC (eds) Les Brachiopods Fossils et Actuels, First International Brachiopod Congress. Biostratigraphie du Paleozoique 4:159–166

Alexander RR (1986b) Resistance to and repair of shell breakage induced by durophages in Late Ordovician brachiopods. J Paleontol 60:273–285

Alexander RR (1989) Influence of valve geometry, ornamentation, and microstructure on fractures in Late Ordovician brachiopods. Lethaia 22:133–147

Alexander RR (1990) Mechanical strength of selected extant articulate brachiopods: implications for Paleozoic morphologic trends. Hist Biol 3:169–188

Anderson LC (1992) Naticid gastropod predation on corbulid bivalves: effects of physical factors, morphological features, and statistical artifacts. Palaios 7:602–620

Ausich WI, Gurrola RA (1979) Two boring organisms in a Lower Mississippian community of southern Indiana. J Paleontol 53:335–344

Babcock LE (1993) Trilobite malformations and the fossil record of behavioral asymmetry. J Paleontol 67:217–229

Babcock LE (2003) Trilobites in Paleozoic predator-prey systems, and their role in reorganization of Early Paleozoic ecosystems. In: Kelley PH, Kowalewski M, Hansen TA (eds) Predator-prey interactions in the fossil record, vol 20, Topics in geobiology. Kluwer/Plenum, New York, pp 55–92

Babcock LE, Peel JS (2007) Palaeobiology, taphonomy and stratigraphic significance of the trilobite *Buenellus* from the Sirius Passet Biota, Cambrian of North Greenland. Mem Assoc Australas Palaeontologists 34:401–418

Babcock LE, Peng S (2007) Cambrian chronostratigraphy: current state and future plans. Palaeogeogr Palaeoclimatol Palaeoecol 254:62–66
Baumiller TK (1990) Non-predatory drilling of Mississippian crinoids by platyceratid gastropods. Palaeontology 33:743–748
Baumiller TK (1993) Boreholes in Devonian blastoids and their implications for boring by platyceratids. Lethaia 26:41–47
Baumiller TK, Gahn FJ (2004) Testing predator-driven evolution with Paleozoic crinoid arm regeneration. Science 305:1453–1455
Bengtson S (1968) The problematic genus *Mobergella* from the Lower Cambrian of the Baltic area. Lethaia 1:325–351
Bengtson S (2002) Origins and early evolution of predation. In: Kowalewski M, Kelley PH (eds) The fossil record of predation. Paleontol Soc Papers 8:289–317
Bengtson S, Yue Z (1992) Predatorial borings in Late Precambrian mineralized exoskeletons. Science 257:367–369
Bishop GA (1975) Traces of predation. In: Frey RW (ed) The study of trace fossils. Springer, New York, pp 261–281
Bookstein FL (1997) Morphometric tools for landmark data: geometry and biology. Cambridge University Press, Cambridge
Bookstein FL, Chernoff B, Elder R, Humphries J, Smith G, Strauss R (1985) Morphometrics in evolutionary biology, the geometry of size and shape change, with examples from fishes, vol 15. Philadelphia Academy of Sciences, Philadelphia, Special Publication
Brooks JL, Dodson SI (1965) Predation, body size, and composition of plankton. Science 150:28–35
Brown KM, Alexander JE (1994) Group foraging in a marine gastropod predator: benefits and costs to individuals. Mar Ecol Prog Ser 112:97–105
Brunton CHC (1966) Predation and shell damage in a Visean brachiopod fauna. Palaeontology 9:355–359
Bruton DL (1981) The arthropod *Sidneyia inexpectans*, Middle Cambrian, Burgess Shale, British Columbia. Phil T Roy Soc B 295:619–656
Carriker MR, Yochelson EL (1968) Recent gastropod boreholes and Ordovician cylindrical borings. USGS Professional Paper B 593-B, p 26
Charnov EL (1976) Optimal foraging: the marginal value theorem. Theor Popul Biol 9:129–136
Conway-Morris S (1977) Fossil priapulid worms. Palaeontology 20:1–95 (Special Papers)
Conway-Morris S, Bengtson S (1994) Cambrian predators: possible evidence from boreholes. J Paleontol 68:1–23
Conway-Morris S, Jenkins RJF (1985) Healed injuries in Early Cambrian trilobites from South Australia. Alcheringa 9:167–177
Conway-Morris S, Robison RA (1988) More soft-bodied animals and algae from the Middle Cambrian of Utah and British Columbia. Univ Kans Paleontol Contrib 122:1–48
Conway-Morris S, Whittington HB (1979) The animals of the Burgess Shale. Sci Am 241:122–133
Debrenne F, Zhuravlev AY (1997) Cambrian food webs: a brief review. Geobios 20:181–188
Denison R (1978) Placodermi. In: Schultze HP (ed) Handbook of paleoichthyology. G.F. Verlag, Stuttgart, p 128
Dietl GP, Alexander RR, Bien WF (2000) Escalation in Late Cretaceous-early Paleocene oysters (Gryphaeidae) from the Atlantic Coastal Plain. Paleobiology 26:215–237
Elliot DK, Bounds SD (1987) Causes of damage to brachiopods from the Middle Pennsylvanian Naco Formation, central Arizona. Lethaia 20:327–335
Elner RW, Hughes RN (1978) Energy maximization in the diet of the shore crab, *Carcinus maenus* Hermann. J Anim Ecol 47:103–116
Evans JS (1912) The sudden appearance of the Cambrian fauna. In: 11th international geological conference. Compte Rendu 1:543–546
Fenton CL, Fenton MA (1931) Some snail borings of Paleozoic age. Am Midl Nat 12:522–528

Fenton CL, Fenton MA (1932) Orientation and injury in the genus *Atrypa*. Am Midl Nat 13:63–74

Finke DL, Denno RF (2004) Predator diversity dampens trophic cascades. Nature 429:407–410

Grant RE (1965) Faunas and stratigraphy of the Snowy Range Formation (Upper Cambrian) in southwestern Montana and northwestern Wyoming. Geol Soc Am Mem 96:1–171

Holt RD (1977) Predation, apparent competition and the structure of prey communities. Theor Popul Biol 12:197–229

Holt RD (1984) Spatial heterogeneity, indirect interactions, and the coexistence of prey species. Am Nat 124:377–406

Hua H, Pratt BR, Zhang L-Y (2003) Borings in *Cloudina* shells: complex predator-prey dynamics in the terminal Neoproterozoic. Palaios 18:454–459

Huntley JW, Kowalewski M (2007) Strong coupling of predation intensity and diversity in the Phanerozoic fossil record. P Natl Acad Sci USA 104:15006–15010

Hutchinson GE (1961) The biologist poses some problems. In: Sears M (ed) Oceanography. Am Assoc Adv Sci 67:89–94

Jago JB, Haines PW (2002) Repairs to an injured early Middle Cambrian trilobite, *Elkedra raea*, Northern Territory. Alcheringa 26:19–21

Jeffries MJ, Lawton JH (1984) Enemy-free space and the structure of ecological communities. Biol J Linn Soc 23:269–286

Jell PA (1989) Some aberrant exoskeletons from fossil and living arthropods. Qld Mus Mem 27:491–498

Jensen S (1990) Predation by Early Cambrian trilobites on infaunal worms – evidence from the Swedish Mickwitzia Sandstone. Lethaia 23:29–42

Jensen S, Gehling JG, Droser ML (1998) Ediacara-type fossils in Cambrian sediments. Nature 393:567–569

Kelley PH, Hansen TA (1993) Evolution of the naticid gastropod predator-prey system: an evaluation of the hypothesis of escalation. Palaios 8:358–375

Kitchell JA, Boggs CH, Kitchell JF, Rice JA (1981) Prey selection by naticid gastropods: experimental tests and application to the fossil record. Paleobiology 7:533–552

Kowalewski M (2002) The fossil record of predation: an overview of analytical methods. In: Kowalewski M, Kelley PH (eds) The fossil record of predation. Paleontol Soc Paper 8:1–42

Kowalewski M, Dulai A, Fürsich FT (1998) A fossil record full of holes: the Phanerozoic history of drilling predation. Geology 26:1091–1094

Leighton LR (2001) New example of Devonian predatory boreholes and the influence of brachiopod spines on predator success. Palaeogeogr Palaeoclimatol Palaeoecol 165:53–69

Leighton LR (2002) Inferring predation intensity in the marine fossil record. Paleobiology 28:328–342

Leighton LR (2003) Morphological response of prey to drilling predation in the Middle Devonian. Palaeogeogr Palaeoclimatol Palaeoecol 201:221–234

Lescinsky HL, Benninger L (1994) Pseudo-borings and predator traces: artifacts of pressure-dissolution in fossiliferous shales. Palaios 9:599–604

Lima SL (1998) Non-lethal effects in the ecology of predator-prey interactions. Bioscience 48:25–34

Lochman C (1956) Stratigraphy, paleontology, and paleogeography of the *Elliptocephala asaphoides* strata in Cambridge and Hoosick quadrangles, New York. GSA Bull 67:1331–1396

MacArthur RH (1972) Geographical ecology. Princeton University Press, Princeton, 269 p

Madin JS, Alroy J, Aberhan M, Fursich FT, Kiessling W, Kosnik MA, Wagner PJ (2006) Statistical independence of escalatory ecological trends in Phanerozoic marine invertebrates. Science 312:897–900

Marcus LF, Corti M, Loy A, Naylor G, Slice DE (eds) (1996) Advances in morphometrics. NATO ASI Series A: Life Sciences, vol 284

Marshall CR (2006) Explaining the Cambrian "explosion" of animals. Annu Rev Earth Pl Sc 34:355–384

McMenamin MAS (1986) The garden of Ediacara. Palaios 1:178–182
McMenamin MAS (1998) The garden of Ediacara. Columbia University Press, New York
Menge BA, Sutherland JP (1976) Species diversity gradients: synthesis of the roles of predation, competition, and temporal heterogeneity. Am Nat 110:351–369
Menge BA, Sutherland JP (1987) Community regulation: variation in disturbance, competition, and predation in relation to environmental stress and recruitment. Am Nat 130:730–757
Miller RH, Sundberg FA (1984) Boring Late Cambrian organisms. Lethaia 17:185–190
Morissette S, Himmelman JH (2000) Decision of the asteroid *Leptasterias polaris* to abandon its prey when confronted with its predator, the asteroid *Asterias vulgaris*. J Exp Mar Biol Ecol 252:151–157
Morton B, Chan K (1999) Hunger rapidly overrides the risk of predation in the subtidal scavenger *Nassarius siquijorensis* (Gastropoda: Nassariidae): an energy budget and a comparison with the intertidal *Nassarius festivus* in Hong Kong. J Exp Mar Biol Ecol 240:213–228
Moy-Thomas JA, Miles RS (1971) Palaeozoic fishes. W. B. Saunders, Philadelphia
Navarrete SA, Menge BA (1996) Keystone predation and interaction strength: interactive effects of predators on their main prey. Ecol Monogr 66:409–429
Nedin C (1999) *Anomalocaris* predation on nonmineralized and mineralized trilobites. Geology 11:987–990
Paine RT (1966) Food web complexity and species diversity. Am Nat 100:65–75
Paine RT (1974) Intertidal community structure: experimental studies on the relationship between a dominant competitor and its principal predator. Oecologia 15:93–120
Palmer AR (1979) Fish predation and the evolution of gastropod shell sculpture: experimental and geographic evidence. Evolution 33:697–713
Richards RP, Shabica CW (1969) Cylindrical living burrows in Ordovician dalmanellid brachiopod beds. J Paleontol 43:838–841
Robson SP, Pratt BR (2007) Predation of late Marjuman (Cambrian) linguliformean brachiopods from the Deadwood Formation of South Dakota, USA. Lethaia 40:19–32
Roopnarine P, Buessink A (1999) Extinction and naticid predation of the bivalve *Chione* Von Muhlfeld in the Late Neogene of Florida. Palaeontol Electronica 2(1):33
Rudkin DM (1979) Healed injuries in *Ogygopsis klotzi* (Trilobita) from the Middle Cambrian of British Columbia: Royal Ontario Museum Life Sciences Occasional Papers 3:1–8
Rudkin DM (1985) Exoskeleton abnormalities in four trilobites. Can J Earth Sci 22:479–483
Seed R, Hughes RN (1995) Criteria for prey size-selection in molluscivorous crabs with contrasting claw morphologies. J Exp Mar Biol Ecol 193:177–195
Sheehan PM, Lesperance PJ (1978) Effect of predation on population dynamics of a Devonian brachiopod. J Paleontol 52:812–817
Signor PW, Brett CE (1984) The mid-Paleozoic precursor to the Mesozoic marine revolution. Paleobiology 10:229–245
Smith SA, Thayer CW, Brett CE (1985) Predation in the Paleozoic: gastropod-like drillholes in Devonian brachiopods. Science 230:1033–1037
Stephens DW, Krebs JR (1986) Foraging theory. Princeton University Press, Princeton
Taylor PD, Wilson MA (2003) Palaeoecology and evolution of marine hard substrate communities. Earth Sci Rev 62:1–103
Tschanz B, Bersier L-F, Bacher S (2007) Functional responses: a question of alternative prey and predator density. Ecology 88:1300–1308
Vannier J, Chen J (2005) Early Cambrian food chain: new evidence from fossil aggregates in the Maotianshan Shale biota, SW China. Palaios 20:3–26
Vermeij GJ (1977) Patterns in crab claw size: the geography of crushing. Syst Zool 26:138–151
Vermeij GJ (1982) Unsuccessful predation and evolution. Am Nat 120:701–720
Vermeij GJ (1983) Shell-breaking predation through time. In: Tevesz MJS, McCall PL (eds) Biotic interactions in recent and fossil benthic communities. Plenum, New York, pp 649–669
Vermeij GJ (1987) Evolution and escalation, an ecological history of life. Princeton University Press, Princeton, p 527
Vermeij GJ (1989) The origin of skeletons. Palaios 4:585–589

Vermeij GJ (2002) Evolution in the consumer age: predators and the history of life. In: Kowalewski M, Kelley PH (eds) The fossil record of predation. Paleontol Soc Papers 8:375–393

Vermeij GJ, Zipser E, Dudley EC (1980) Predation in time and space: peeling and drilling in terebrid gastropods. Paleobiology 6:352–364

Vermeij GJ, Schindel DE, Zipser E (1981) Predation through geological time: evidence from gastropod shell repair. Science 214:1024–1026

Werner EE, Hall DJ (1974) Optimal foraging and the size selection of prey by the bluegill sunfish (Lepomis macrochirus). Ecology 55:1042–1052

Whittington HB, Briggs DEG (1985) The largest Cambrian animal, *Anomalocaris*, Burgess Shale, British Columbia. Phil T Roy Soc B 306:569–609

Wilson MA, Palmer TJ (2001) Domiciles, not predatory borings: a simpler explanation of the holes in Ordovician shells analyzed by Kaplan and Baumiller, 2000. Palaios 16:524

Zhu M-Y, Vannier J, Van Iten H, Zhao Y-L (1994) Direct evidence for predation on trilobites in the Cambrian. Proc Roy Soc Lond B Suppl. 271:S277–S280

Chapter 5
Ecospace Utilization During the Ediacaran Radiation and the Cambrian Eco-explosion

Andrew M. Bush, Richard K. Bambach, and Douglas H. Erwin

Contents

Abstract A theoretical ecospace is a multi-parameter system for classifying the ecological properties of organisms; because they are viewed in terms of their ecological and functional capabilities, morphologically and phylogenetically disparate organisms can be compared and contrasted. In the ecospace used here, marine animals are classified according to three parameters that can be determined relatively easily

A.M. Bush (✉)
Department of Ecology and Evolutionary Biology and Center for Integrative Geosciences, University of Connecticut,75 North Eagleville Road, Storrs, CT 06269-3043, USA
e-mail: andrew.bush@uconn.edu

R.K. Bambach
Department of Paleobiology, MRC-121, National Museum of Natural History, Smithsonian Institution, P.O. Box 37012, Washington, DC 20013-7012, USA
e-mail: bambachr@si.edu

D.H. Erwin
Department of Paleobiology, MRC-121, National Museum of Natural History, Smithsonian Institution, P.O. Box 37012, Washington, DC 20013-7012, USA
and
Santa Fe Institute, 1399 Hyde Park Road, Santa Fe, NM 87501, USA
e-mail: erwind@si.edu

M. Laflamme et al. (eds.), *Quantifying the Evolution of Early Life*, Topics in Geobiology 36,
DOI 10.1007/978-94-007-0680-4_5, © Springer Science+Business Media B.V. 2011

from fossils: tiering (position relative to the sediment-water interface), motility level, and feeding mechanism. Analyses of faunas from the Ediacaran, Cambrian, and Recent suggest that the ecological richness (number of ecological lifestyles) of the marine fauna rose through time, although the pace of increase slowed after the early Phanerozoic. However, the Ediacaran biota was quite distinct from Phanerozoic faunas in terms of which tiers, motility levels, and feeding mechanisms were employed; thus, the rise to dominance of bilaterians during the Cambrian Explosion caused a fundamental transformation in marine ecology. Changes in marine animal ecology since the Cambrian Explosion were of lesser magnitude.

Keywords Paleoecology • Burgess Shale • Chengjiang fauna • Precambrian • Cambrian explosion • Ediacaran biota

5.1 Introduction

Our burgeoning understanding of the Ediacaran biota (Narbonne 2005; Fedonkin et al. 2007a; Xiao and Laflamme 2009; references therein) allows us to analyze them in systematic ways and compare them to the better-known animals of the Phanerozoic. Theoretical ecospace is a system of analysis that is based on the autecological characteristics of organisms—the ways in which they interact with each other and the environment. Previously, it has been used to track the ecological expansion of the marine fauna through time, to explain why certain ecologic strategies have evolved while others have not, to compare the ecological composition and disparity of different faunas, to test the effects of biases on the fossil record, and to examine the role of competition in structuring fossil assemblages (Bambach et al. 2007; Bush et al. 2007a, b, 2008; Novack-Gottshall 2007; Bush and Daley 2008). Theoretical ecospace is particularly useful for examining the earliest animals because they can be compared with later metazoans using a "common currency" of analysis despite their apparent morphologic oddness and phylogenetic inscrutability; it is one of the few types of analysis that currently provides such breadth of scope.

Ecospace analysis is similar to pictorial reconstructions of fossil faunas (e.g., McKerrow 1978; Algeo and Scheckler 1998; Fedonkin et al. 2007a); in both cases, the fossils and rocks are used to interpret the ecological lifestyles of ancient animals, and the entire assemblage is recreated (it is worth noting that such illustrations almost invariably are time-averaged, inflating the diversity of species present at a single location). Ecospace analysis is simply a method for subjecting these reconstructions to quantitative analysis and hypothesis testing. As with illustrations, a given ecospace analysis may require updating as new ideas and fossils come to light (reconstructions of Ediacaran communities have certainly changed over the years; e.g., Jenkins 1992; Clapham and Narbonne 2002), but this is true of many endeavors in science.

Here, we summarize previous applications of theoretical ecospace to the faunas of the Ediacaran and Cambrian Periods and present a few new analyses. The ecospace approach complements other methods of analyzing the biotic radiations of the late Neoproterozoic and early Phanerozoic, such as analyses of taxonomic diversity (Shen et al. 2008), morphospace (Antcliffe and Brasier 2007; Shen et al. 2008), phylogeny (e.g., Xiao and Laflamme 2009; Brasier and Antcliffe 2009), biogeography (Waggoner 1999), local paleoecology (Clapham et al. 2003; Droser et al. 2006), and development (Knoll and Carroll 1999; Valentine et al. 1999; Erwin and Davidson 2002; Davidson and Erwin 2006; Erwin 2008; Antcliffe and Brasier 2008; Brasier and Antcliffe 2009). A complete understanding of these radiations will integrate all of these approaches; for now, we present initial analyses of ecospace occupation by early animals and compare them to equivalent analyses of diversity and morphology.

5.2 Theoretical Ecospace

A theoretical ecospace is a formal, multidimensional space in which organisms are placed according to their interactions with each other and their environment (Bush et al. 2007a; Novack-Gotshall 2007). Organisms occupying the same position in ecospace should be ecologically similar, at least with respect to the parameters used to define the ecospace, and organisms occupying different positions should be dissimilar. An ecospace can be defined with more or less detail, depending on one's needs and data; more ecological parameters result in a higher dimensional space, and parameters can be divided up with greater or lesser degrees of detail (cf. Bush et al. 2007a; Novack-Gotshall 2007).

The ecospace used here was outlined by Bush et al. (2007a, b), Bambach et al. (2007), and Bush and Daley (2008). It categorizes marine animals using three ecologically important parameters that are relatively easy to determine from fossils (Fig. 5.1): tiering, motility, and feeding mechanism. Tiering refers to the position of an animal (or other organism) relative to the sediment-water interface (Ausich and

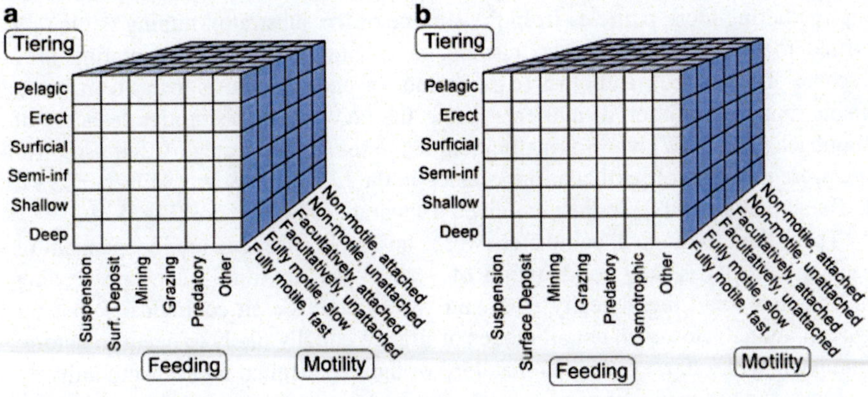

Fig. 5.1 Theoretical ecospace for marine animals, comprised of three ecologically important variables: tiering with respect to the sediment-water interface, motility level, and feeding mechanism. **(a)** Original ecospace of Bush et al. (2007a) and Bambach et al. (2007), modified from those sources. **(b)** Modified ecospace in which osmotrophs are separated from the "other" feeding category

Bottjer 1982, 1985; Bottjer and Ausich 1986), which determines where the animal seeks resources and the types of disturbance it faces. Motility is the ability of an animal to move under its own power (Bambach et al. 2002), and it affects the animal's ability to gather food, cope with disturbances, and otherwise interact with its environment. Attachment is one way to control movement and is included here under motility (Bush et al. 2007a; Bambach et al. 2007). Feeding mechanism is the physical method of gathering food, as opposed to diet (type of food), which can be difficult to determine for many fossil organisms (Bush et al. 2007a; Bambach et al. 2007).

These parameters were chosen because they represent fundamental ecological properties of any animal. Every organism (a) lives in some habitat setting that imposes particular physical demands upon it, (b) must cope with or respond to mechanical stresses and disturbance (including attacks) in some fashion, and (c) requires nutrient input to live. The three ecospace parameters were originally each split into six values, resulting in 216 unique combinations, referred to as "modes of life" (Fig. 5.1a). The tiering categories include pelagic/nektonic (living within the water column), erect epifaunal (on the sediment surface, extending into the water column), surficial (on the sediment surface, not extending appreciably into the water column), semi-infaunal (partly enclosed in the substrate), shallow infaunal (enclosed in the top ~5 cm of the substrate), and deep infaunal (enclosed in the substrate, more than ~5 cm deep).

The motility categories include fully motile (regularly moving as part of the living/feeding routine), which is split into "fast" and "slow" categories. "Fast" organisms are typically not bound strongly to a substrate and move using legs, fins, or other appendages. "Slow" organisms are in constant contact with the substrate and move by creeping over it or pushing through it (see Bush et al. 2007a). The motility categories also include facultatively motile (moving only when disturbed or attacked) and non-motile (not capable of movement). Both of these categories are split into attached and unattached forms (Fig. 5.1). The feeding mechanisms include suspension feeding (capturing small food particles from the water), surface deposit feeding (gathering loose particles from the surface of the substrate), mining (gathering buried food), grazing (scraping, rasping, or nibbling), predation (capturing prey capable of resistance), and other (e.g., chemo- or photosymbiosis, parasitism, etc.). Osmotrophy (absorption of nutrients across the body wall; "absorptive feeders" of Bambach et al. 2007) was originally included in the "other" category, but osmotrophs were probably of particular importance in the Ediacaran biota (Seilacher 1984; Laflamme et al. 2009), so they are given a separate category here (Fig. 5.1b).

These three ecological parameters were chosen because they are independent of each other; that is, any combination of vertical tier, motility level, and feeding mechanism could theoretically exist and would describe an ecologically distinct type of animal. However, not all modes of life are equally likely to occur in nature; Bambach et al. (2007) estimated that 45% of the 216 original modes were unlikely to occur due to functional or energetic constraints.

Novack-Gotshall (2007) independently designed a theoretical ecospace that shares many features with the one used here, but it included additional ecological parameters and subdivided others more finely. Thus, more subtle ecological distinctions could be drawn among organisms, although fossils must be interpreted to a greater extent in order to draw these distinctions.

5.3 Paleoecology of Ediacaran Animals

To some extent, the ecospace approach is independent of phylogeny because distantly related taxa can be ecologically similar and because some ecological assignments will not vary with phylogenetic interpretation (i.e., a fossil's morphology may strongly suggest a particular mode of life, regardless of phylogenetic relationship). However, inferring phylogenetic affinity is critical for interpreting the paleoecology of some fossils, and this problem is greater for the earliest animals, whose phylogenetic positions are most uncertain. Ediacaran fossils were traditionally assigned to various modern phyla, such as the cnidarians, annelids, arthropods, and echinoderms (e.g., Sprigg 1949; Wade 1972; Glaessner 1984; Gehling 1991; Jenkins 1992; Valentine 1992). However, this interpretation has been challenged because many Ediacaran fossils lack synapomorphies that would align them with these groups, or even with the bilateria; Seilacher (1984, 1989, 1992) proposed instead that most Ediacaran animals should be placed outside the metazoa in the extinct kingdom Vendobionta, which had unique morphologies and modes of development. Many paleontologists today regard Ediacaran animals as a phylogenetically diverse menagerie that spanned the base of the metazoan tree (e.g., Fig. 2 of Xiao and Laflamme 2009). They represent a range of stem-group animals between sponges and acoels whose unique morphologies are difficult to place exactly on a phylogeny (Erwin 2009), as well as possible stem-group bilaterians. Although biomarker evidence demonstrates that sponges were present at this time (Love et al. 2009), they are not common in Ediacaran fossil assemblages (see Gehling and Rigby 1996 for an example). This interpretation of the Ediacaran fauna is generally adopted here (for a complete discussion, see Bambach et al. 2007, text and online supplementary material).

Many Ediacaran animals were probably osmotrophs (Xiao and Laflamme 2009; Laflamme et al. 2009). There is no evidence that these species had mouths, guts, anuses, or filtering apparatuses; many, such as the fractally branched rangeomorphs, had morphologies suggestive of maintaining a low ratio between volume and the surface-area of the body wall (Laflamme et al. 2009), and most were sessile. Most of these organisms probably absorbed food directly from the water column and/or across their lower surface. Some forms may have lived more or less infaunally (Grazhdankin and Seilacher 2002, 2005). *Yorgia* and *Dickensonia* have been found in conjunction with series of resting traces that may represent dissolution scars on the underlying microbial mat substrate (Ivantsov and Malakhovskaya 2002; Dzik 2003; Gehling et al. 2005); they may have been intermittently motile, absorbing nutrients through their flat soles. Sponges, such as *Palaeophragmodictya* (Gehling and Rigby 1996), were suspension feeders (trappers of particulate matter from the water column) as well as osmotrophs, and calcified fossils (Germs 1972b;

Grant 1990; Grotzinger et al. 2000; Wood et al. 2002) probably represent the skeletons of suspension feeders. Some other Ediacaran animals also may have been suspension feeders (Bambach et al. 2007).

Kimberella has been interpreted as a bilaterian body fossil, perhaps a stem mollusk, and it occurs in conjunction with traces interpreted as radula-like scratch marks on the substrate (Fedonkin and Waggoner 1997; Fedonkin et al. 2007b). Although its phylogenetic position remains unclear, *Kimberella* was presumably motile, epifaunal, and a grazer. Horizontal burrows provide evidence of additional motile animals interpreted as surficial to shallow infaunal, surficial deposit and mining deposit feeders. Many of these traces have been reinterpreted as tubes (i.e., body fossils; Droser et al. 2005; Jensen et al. 2006), so this mode of life may not be as abundantly occupied as once thought. Potential boreholes in the calcified tubes of *Cloudina* represent the only evidence of predation in the Ediacaran Period (Bengtson and Zhao 1992; Hua et al. 2003); Bambach et al. (2007) conservatively omitted the possible tracemaker from their ecospace analysis, but we include it here as a surficial, slow motile, predator.

As in our previous analyses of ecospace, this analysis is limited to macroscopic animals, so seaweed, phytoplankton, and phosphatized embryos were not considered (e.g., Xiao et al. 1998; Huntley et al. 2006; Xiao and Dong 2006).

5.4 Ecospace Occupation During the Ediacaran

Ediacaran animals have been grouped into three assemblages corresponding to three age ranges (Waggoner 2003; Narbonne 2005): Avalon (579–560 Ma), White Sea (560–550 Ma), and Nama (550–542 Ma). These assemblages differ in their environments of occurrence (Grazhdankin 2004), but are generally thought to reflect changes through time in the fauna, not just different facies associations (e.g., Waggoner 2003; Narbonne 2005; Xiao and Laflamme 2009). The modes of life interpreted for each assemblage are shown in Fig. 5.2a–c. The Avalon assemblage, the earliest of the three, lived in deep-water paleoenvironments and was ecologically the least diverse (Fig. 5.2a). It was dominated by fractally branched rangeomorphs, interpreted as osmotrophs, accompanied by circular holdfasts and a few other forms (e.g., Clapham and Narbonne 2002; Clapham et al. 2003; Narbonne and Gehling 2003; Gehling and Narbonne 2007; Hofmann et al. 2008; Narbonne et al. 2009; Brasier and Antcliffe 2009). Osmotrophs are represented by erect and non-erect epifauna (Fig. 5.2a). It is not clear whether specific macrofossils represent suspension feeders in these assemblages, but sponges must have been present (Love et al. 2009), presumably as erect and/or non-erect epifauna. These two modes are indicated with some degree of uncertainty in Fig. 5.2a. Recently described trace fossils indicate that motile, epifaunal animals were present at this time (Liu et al. 2010). We have not assigned the tracemakers to a specific mode of life because the feeding mechanism is not clear; rather, the surficial, facultatively motile modes of life in Fig. 5.2a are suggested as possible locations for the tracemaker in ecospace. The Avalon assemblage thus had up to five inferred modes of life (cf. Bambach et al. 2007; Xiao and Laflamme 2009).

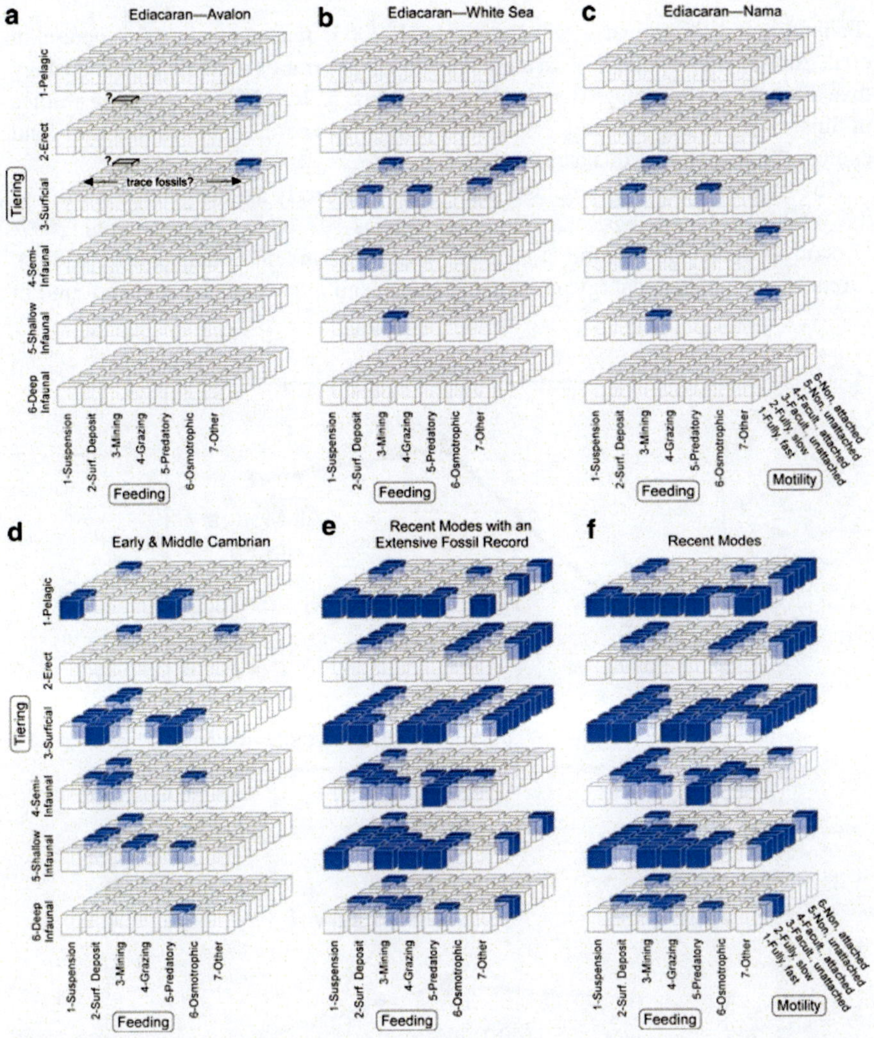

Fig. 5.2 Ecospace utilization in early animal faunas and the modern marine fauna. *Colored boxes* denote modes of life occupied by at least one species in a fauna, and *white boxes* denote unoccupied modes. *Gray boxes* and *question marks* in (**a**) indicate uncertainty; see text (Modified from Bambach et al. 2007; Xiao and Laflamme 2009). (**a**) The Avalon assemblage of the Ediacaran Period, (**b**) the White Sea assemblage of the Ediacaran Period, (**c**) the Nama assemblage of the Ediacaran Period, (**d**) the early-middle Cambrian Period, (**e**) Recent animals that have an extensive fossil record (see Bambach et al. 2007), and (**f**) all Recent animals

The White Sea assemblage (Fig. 5.2b) lived in shallower water than the Avalon assemblage and contained rangeomorphs, erniettomorphs, bilateral forms, and a great variety of disks and radial forms (Jenkins et al. 1983; Fedonkin 1992; Gehling and Rigby 1996; Fedonkin and Waggoner 1997; Gehling 2000; Droser et al. 2006). A motile fauna is represented by *Kimberella* and horizontal trace fossils

(Fedonkin and Waggoner 1997; Droser et al. 2005; Jensen et al. 2006; Fedonkin et al. 2007b). The White Sea assemblage was considerably more diverse in ecology than the Avalon assemblage (ten modes of life; Figs. 5.2b and 5.3a). Ongoing studies of this fauna in Russia and Australia promise to increase the taxonomic (and ecologic?) diversity of this assemblage.

The final Ediacaran assemblage, the Nama, had nearly as many modes of life as the White Sea (nine; Figs. 5.2c and 5.3a) (Germs 1972a, b; Narbonne et al. 1997; Grotzinger et al. 2000; Wood et al. 2002; Grazhdankin and Seilacher 2002, 2005). It represents a higher-energy, nearshore setting, and metazoans have been found in

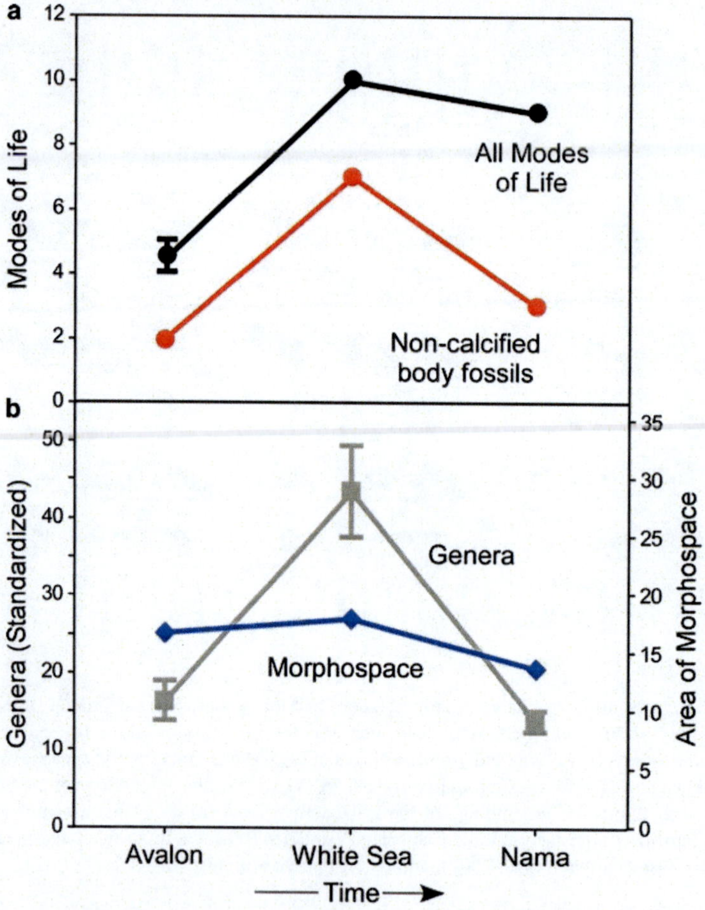

Fig. 5.3 Ecological, taxonomic, and morphologic diversity during the Ediacaran. (**a**) Number of modes of life in the three Ediacaran assemblages. As in Fig. 5.2, we indicate some uncertainty in the number of modes in the Avalon assemblage (in reality, of course, all values are uncertain and subject to change pending further study). (**b**) Number of genera, standardized by rarefaction, and morphospace occupation, calculated as area of the convex hull occupied in morphospace (Data in (**b**) from Shen et al. 2008)

two distinct habitats. In sandy environments, erniettomorphs were common, along with a diverse assemblage of poorly studied tubes, some other soft forms, and some horizontal trace fossils. On microbial reefs, calcified metazoans like *Cloudina* and *Namacalathas* first appeared in Nama assemblage times (Germs 1972b; Grant 1990; Grotzinger et al. 2000; Wood et al. 2002). It is worth noting that the Nama is the least well-studied of the three assemblages.

5.5 Comparing Diversification in Ecology, Taxonomy, and Morphology During the Ediacaran Period

Shen et al. (2008) compared the taxonomic diversification of the Ediacaran fauna to its occupation of morphospace and found that the two were decoupled: taxonomic richness (number of genera) increased from the Avalon to the White Sea assemblage, but declined again in the Nama assemblage, whereas morphospace occupation remained relatively constant (Fig. 5.3b) (see also Freeman 2009; Xiao et al. 2009). They were agnostic as to whether developmental or ecological constraints prevented morphological expansion of the fauna, but likened their documented patterns to trends in diversity and morphospace occupation during the Cambrian Explosion and the rest of the Phanerozoic (e.g., Gould 1991; Briggs et al. 1992; Foote et al. 1992; Foote 1997).

Ecological richness (number of modes of life) increased markedly from the Avalon assemblage to the White Sea assemblage, perhaps doubling, but remained fairly constant from the White Sea to the Nama assemblage, declining by only one mode (Fig. 5.3a). At first blush, therefore, the temporal occupation of ecospace by the Ediacaran fauna appears to be intermediate in pattern between its occupation of morphospace and its taxonomic diversification: the fauna did expand ecologically at first (similar to taxonomic richness), but did not decline markedly thereafter (similar to morphospace occupation). However, Shen et al. (2008) examined only body fossils of soft-bodied animals, which included most Ediacaran taxa but excluded trace fossils (which would be difficult to include in a body-fossil morphospace) and calcified taxa such as *Cloudina*. Traces and calcified taxa made up a substantial portion of the ecological diversity of the Nama assemblage; when these fossils (and inferred suspension feeders in the Avalon assemblage) are removed from the ecological data set, ecological richness parallels taxonomic richness (Fig. 5.3). It is not currently possible to do the opposite—add in the taxonomic diversity represented by trace fossils and calcareous fossils—although it would probably not compensate for the decline in the diversity of soft-bodied forms from the White Sea assemblage to the Nama assemblage.

Bambach et al. (2007) suggested that taxonomic and ecologic richness might be coupled during the Phanerozoic, at least on some time scales. For non-calcified body fossils, these results are consistent with their suggestion: when commensurate data are compared, taxonomic and ecologic diversity appear to be related in the Ediacaran. If this coupling is supported by further data, the causal relationships need to be determined. Do higher-diversity faunas have more ecological diversity simply because there are more taxa, and thus more chances to explore ecospace? Or is something more interesting going on: did higher diversity allow (or even compel) the exploration of new modes of life through the effects of ecological interactions such as competition or disturbance?

The disconnect between morphospace occupation and taxonomic/ecologic richness is intriguing. Apparently, the morphologic changes that determine both ecology and genus- to species-level taxonomy are not the same as the body-plan-level parameters often used to delineate morphospaces, and these sets of characters have different temporal dynamics (e.g., Foote 1997).

5.6 Trends in Ecospace Utilization from the Ediacaran to the Phanerozoic

The comparison of faunas from different times and places motivates theoretical ecospace analysis—ecospace provides a common system for understanding how faunas differ, as well as how they are similar, even when taxonomic composition does not overlap. Thus, it is well suited for comparing Ediacaran and Phanerozoic animals that are difficult to compare in terms of morphology and phylogeny. In this section, we compare the Ediacaran fauna (all assemblages combined) to those of the Early-Middle Cambrian and the Recent. Thus, we are comparing the ecologic forms produced by the initial radiation of animals in the Ediacaran, those generated during the Cambrian Explosion, and those resulting from additional evolution between the Cambrian and the Recent. Preservation of soft-boded animals in the Ediacaran and Cambrian (e.g., the Burgess Shale and Chengjiang biotas) (Briggs et al. 1994; Gehling 1999; Hou et al. 2004; Gaines et al. 2008) allows fairly direct comparisons of these faunas to the modern fauna, and the modern fauna can also be degraded to reflect only those modes of life likely to be preserved in the fossil record (Bambach et al. 2007).

The occupation of ecospace could have changed through time in many ways, including the number, identity, relative abundance, geographic distribution, and environmental distribution of modes of life. However, we are only beginning to combine information on the ecospace occupation, abundance, and distribution of fossils (e.g., Bush et al. 2007a, b; Bush and Daley 2008); these integrated approaches will be interesting for early animal assemblages, but for now, we take a more general approach, focusing on the number and identity of modes of life from the Ediacaran to the Phanerozoic.

The obvious parameter to consider first is ecological richness, the number of modes of life in each fauna. Considering all Ediacaran assemblages combined, the first burst of animal diversification produced only 13 recognized ecological lifestyles (Figs. 5.2a–c and 5.4). Ecospace use then increased greatly during the Cambrian Explosion as metazoans (and particularly bilaterians) adopted many new ecological lifestyles—the known ecological richness of the Early-Middle Cambrian

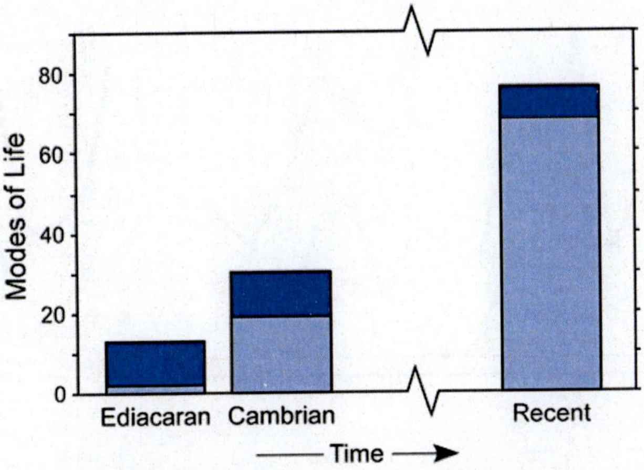

Fig. 5.4 Number of modes of life in the Ediacaran, Cambrian, and Recent faunas. *Lower bars*: skeletal taxa. *Upper bars*: all taxa. See Bambach et al. (2007) for details

was 30 modes of life (Bambach et al. 2007), more than twice that of the Ediacaran (Figs. 5.2d and 5.4). Excluding "other" feeding mechanisms such as parasitism and symbiosis, which may be difficult to recognize from fossils, the modern world has 76 modes of life, over two and a half times that of the Cambrian (Figs. 5.2f and 5.4). Some of this additional ecological diversification occurred during the Ordovician Radiation, and some of it came after (Bambach et al. 2007). The effects of sampling biases on these figures are not entirely clear (Hendy et al. 2009); we obviously have a better ability to sample the modern world than the fossil record, but the Ediacaran and Cambrian data sets include soft-bodied organisms and are extensively time-averaged, which must offset this bias to some degree. Also, if one only counts modern taxa with a fossil record, thus eliminating modern soft-bodied taxa, modern ecospace use still exceeds that of the Cambrian (counting soft-bodied taxa) by more than a factor of 2 (68 modes of life; Figs. 5.2e and 5.4). Thus, in a multiplicative sense, the Cambrian Explosion accomplished about half of the ecological expansion that occurred between the Ediacaran and the Recent, regardless of how one considers biases; arithmetically, the number of modes of life in the Cambrian is closer to that in the Ediacaran (Fig. 5.4).

In terms of ecological richness, the Cambrian represents perhaps the halfway position between the Ediacaran and the Recent, but what about the *identities* of the modes of life in the faunas—are Cambrian modes of life most similar to Ediacaran ones, Recent ones, or is the Cambrian fauna a mixture of Ediacaran and Recent lifestyles? It is difficult to visually assess this question from Fig. 5.2, so this information is reorganized in Fig. 5.5. For each fauna, Fig. 5.5a shows the number of modes of life in each tiering, motility, and feeding subcategory. The "other" feeding category was again deleted because these feeding types may be undercounted in fossil faunas. In these raw data, the most obvious pattern is the already-demonstrated

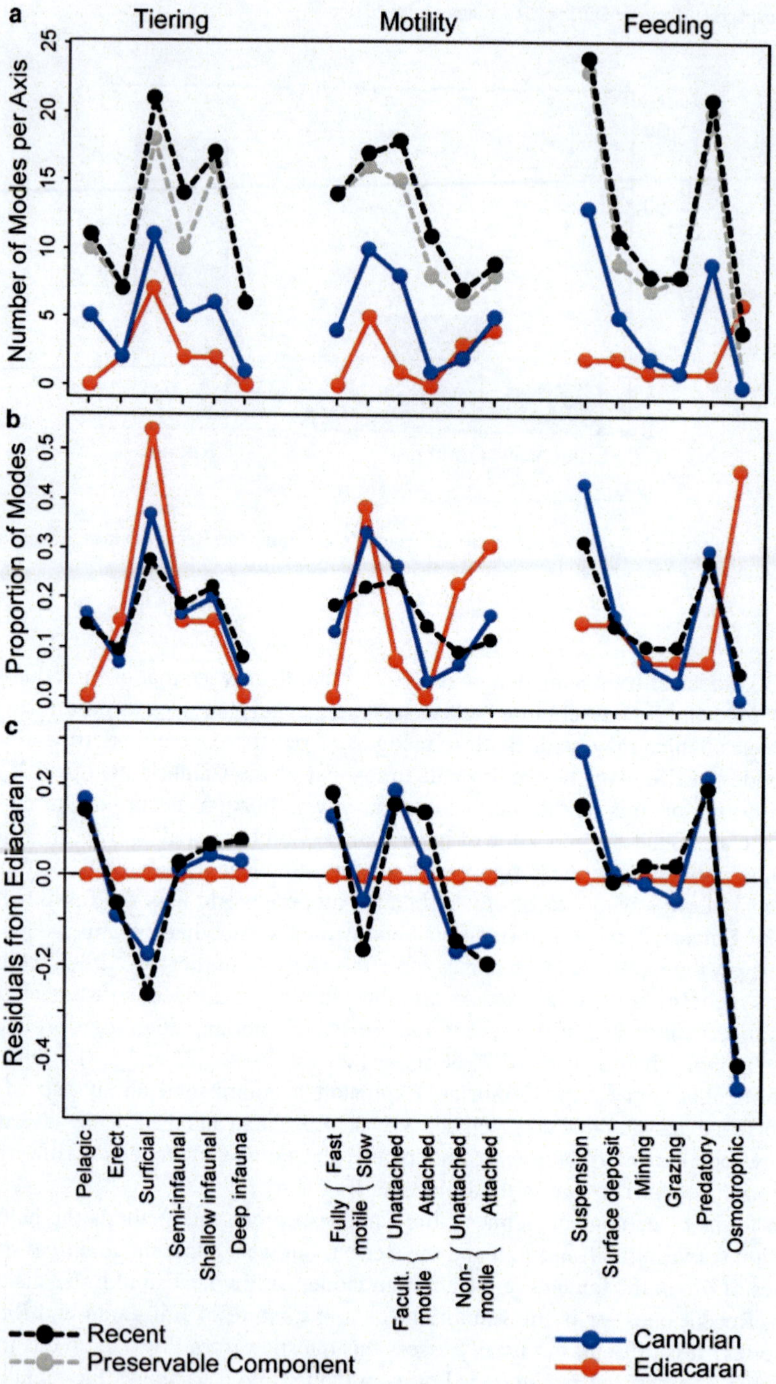

Fig. 5.5 Ecological composition of the Ediacaran, Early-Middle Cambrian, and Recent faunas. **(a)** Total number of modes of life per ecological category. The portion of the Recent fauna likely to be preserved in the fossil record is also shown, but because it is similar to the entire Recent fauna,

increase in ecological richness through time (i.e., the Recent values generally plot above the Cambrian values, which plot above the Ediacaran ones).

To normalize for differences in total ecological richness so that the relative importance of the ecological categories can be compared, the raw data were converted to proportions by dividing the number of modes in each ecological category by the total number of modes (Fig. 5.5b). These proportions do not necessarily represent the proportions of species or individuals in each category; more refined data would be necessary to calculate those values. Rather, they represent the proportion of modes of life in a fauna in each category, and modes of life can vary in species richness/abundance. However, these values probably bear at least some relation to abundance (e.g., see Bush et al. 2007a). All three data sets differ in proportions to some extent, but the Cambrian and Recent are generally similar, with the same general "shapes" to the curves (Fig. 5.5b). The changes from the Ediacaran to the Phanerozoic (as well as the similarities between the Cambrian and Recent) are more easily seen in the residuals of the Cambrian and Recent proportions on the Ediacaran proportions, calculated by subtracting the proportion of modes in each ecological category in the Ediacaran from the proportion in the Cambrian and Recent (Fig. 5.5c).

Based on these data, the Ediacaran was distinctly different from the Phanerozoic; some common Phanerozoic ecologic types were missing entirely (e.g., pelagic animals, deep infauna, rapidly motile animals), and the proportions of others were dissimilar compared to the two Phanerozoic faunas (Fig. 5.5b). Non-motile animals were proportionally more common in Ediacaran times, whereas facultatively motile animals became common in the Phanerozoic. Exclusively osmotrophic feeders played a large role in the Ediacaran and have mostly disappeared, whereas suspension feeders (some of which, like sponges, may also feed osmotrophically) rose to greater prominence in the Phanerozoic. Predators are, of course, a hallmark of the Phanerozoic, and their rise from the Ediacaran to the Cambrian plays a role in many ecological explanations of the Cambrian explosion (Evans 1912; Hutchinson 1961; Stanley 1973; Vermeij 1990; Bengtson and Zhao 1992; Bengtson 2002; Marshall 2006).

Although the Cambrian and Recent are surprisingly similar in terms of the proportions of ecologic types that were present, there are, of course, differences in ecology between them (Fig. 5.5b, c). Since the Cambrian, burrowing (especially deep burrowing), facultative motility, and grazing have increased in importance (e.g., Vermeij 1977, 1987; Stanley 1977; Thayer 1979, 1983; Ausich and Bottjer 1982, 1985; Bambach 1983; Steneck 1983; Bottjer and Ausich 1986; Bellwood 2003; Bush et al. 2007a, b). Predators show no increase from the Cambrian to the Recent in these data, which is perhaps surprising given the abundant literature on the topic (Vermeij 1977, 1987; Signor and Brett 1984; Kowalewski et al. 1998;

Fig. 5.5 (continued) it is not shown in subsequent plots. (**b**) Proportion of modes in each ecologic category, calculated separately for each of the three parameters (tiering, motility, and feeding). (**c**) Proportion of modes in each ecologic category, calculated as residuals on the Ediacaran biota

Bambach et al. 2002; Kowalewski and Kelley 2002; Kelley et al. 2003; Huntley and Kowalewski 2007). However, much of this literature concerns an increase in durophagous predation, and many of the Cambrian predators, such as *Opabinia*, apparently victimized non-shelled prey (Whittington 1975). Also, Novack-Gottshall (2004) found that Cambrian assemblages were enriched in predators relative to assemblages from later in the Paleozoic.

5.6.1 Quantitative Tests

These differences between the Ediacaran, Cambrian, and Recent faunas should be unsurprising to anyone familiar with these biotas, although this analysis brings them into sharp relief, and the similarities between the Cambrian and Recent were greater than we expected (although see Dunne et al. 2008). Using these quantitative data, we can take this analysis a step further: the degree of similarity or difference among faunas can be calculated explicitly, and the pace of change in ecological composition can be compared directly to the pace of change in ecological richness (Fig. 5.4).

The dissimilarity in ecological composition between each pair of faunas was calculated as Manhattan distance based on the proportions of the 18 ecological categories shown in Fig. 5.5b. To do this, the proportion of modes in each ecological category was treated as a variable, and the difference between the two samples was calculated for each of the 18 variables. To calculate Manhattan distance, the 18 difference values were simply summed to give the total dissimilarity. The minimum value was 0.0 (all proportions equal) and the maximum value was 3.0, because the maximum value was 1.0 for the tiering variables combined, the motility variables combined, and the feeding variables combined. Note that this approach does not require that two faunas possessed the exact same modes of life (i.e., boxes in Fig. 5.2) to be similar, as long as they had similar distributions of proportions on the tiering, motility, and feeding axes. For example, a sample with a shallow infaunal predator and an erect suspension feeder would register as identical to a sample with a shallow infaunal suspension feeder and an erect predator.

The Ediacaran fauna was virtually equidistant from the two Phanerozoic faunas: 2.26 units from the Cambrian fauna and 2.46 from the Recent fauna (a similar value was obtained if only the fossilizable component of the Recent was used). In contrast, the dissimilarity between the Cambrian and Recent faunas was only 0.93. These differences are shown graphically in Fig. 5.6a. These values suggest the following. (1) A major ecological change in animal ecosystems occurred during the Cambrian Explosion when the "basal animal ecosystem" of the Ediacaran was replaced by the "bilaterian-dominated ecosystem" of the Phanerozoic. (2) Ecological changes in the marine fauna since the Cambrian Explosion have been relatively minor compared to those that occurred during the Cambrian Explosion, especially considering that half a billion years have elapsed. The changes that have occurred—increased durophagous predation, infaunalization, motility—have merely been

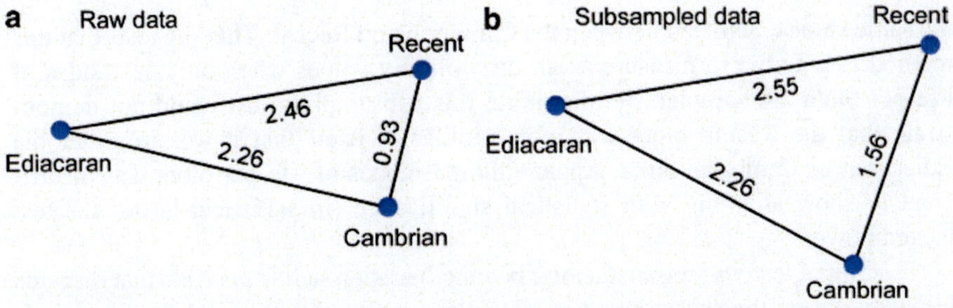

Fig. 5.6 Ecological dissimilarity between the Ediacaran, Cambrian, and Recent biotas. Calculated as Manhattan distance based on the proportion of modes of life in the tiering, motility, and feeding categories shown in Fig. 5.5b. (**a**) Dissimilarity based on the raw data. Values are indicated on the sides of the triangle. (**b**) Dissimilarity based on the subsampled data

variations on the bilaterian-dominated ecosystem blueprint established during the Cambrian (also see Dunne et al. 2008). (3) In terms of ecological composition, the modern marine fauna has not become appreciably more dissimilar from the Ediacaran fauna than the Cambrian fauna was (Fig. 5.6)—it has just become richer (Fig. 5.4).

There is a potential problem with this analysis. The Ediacaran fauna is measurably different in ecological composition from the Cambrian and Recent faunas, but it is also contains many fewer modes of life, and it is possible that measured dissimilarity is affected by number of modes of life. Without further testing, it is possible, for example, that any low-number set of modes of life would appear as dissimilar from the Cambrian and Recent faunas as the Ediacaran does. We examined this problem with subsampling routines in which modes of life were randomly removed from the Cambrian and Recent data sets until the number of modes was equivalent to that in the Ediacaran data set. Thus, dissimilarity values could be compared more directly. Each subsampling routine was run 10,000 times to get a stable value for average dissimilarity, as well as 95% confidence intervals.

The most obvious way to subsample was to reduce both the Cambrian and Recent data sets to 13 modes of life, the number in the Ediacaran. This approach demonstrated that reducing the number of modes of life in the data sets did in fact increase the measured dissimilarity between them (i.e., the dissimilarity metric is biased). The average subsampled dissimilarity values rose to 2.45 for the Ediacaran-Cambrian comparison, 2.70 for the Ediacaran-Recent comparison, and 1.82 for the Cambrian-Recent comparison. We tested whether or not the Cambrian data were significantly more similar to the Recent data than to the Ediacaran data by taking the difference between the Cambrian-Ediacaran and Cambrian-Recent dissimilarity values for each iteration of the subsampling routine. The 95% confidence interval around this difference included zero, so this test failed to show that the Cambrian was more similar to the Recent than the Ediacaran, in contrast to our interpretation of the raw data. However, several additional analyses indicate that this test was not terribly useful. When two independent subsamples of the Recent data set were compared using this approach, the average dissimilarity was 1.80—almost exactly the same value calculated between the Cambrian and Recent. This suggests that this method is not able to measure small dissimilarity values when only 13 modes of life per biota are sampled. Furthermore, this subsampling test could not demonstrate that the Recent biota was more similar to itself that it was to either the Ediacaran or Cambrian biota. Apparently, 13 modes of life per biota was insufficient to show anything with statistical significance (in statistical terms, the test lacked power).

This initial test was unsatisfactory, but another approach is possible that discards less data during the subsampling process, thus raising the power of the test. One's initial instinct is to reduce every sample to the same number of modes of life, thus making them equivalent for comparison. However, every *sample* does not need to contain exactly the same number of modes of life for the *comparisons* to be equivalent, because each comparison involves two samples. As long as the same *two* sample sizes are used in every comparison, then there should be no bias. Out of the three data sets, the two smallest are the Ediacaran and Cambrian, with 13 and 30 modes of life, respectively. As long as each dissimilarity computation involves a sample with 13 modes and a sample with 30 modes, this bias should be controlled.

The Cambrian-Ediacaran comparison did not require subsampling because the sample sizes were 13 and 30 already; the dissimilarity between these samples remained 2.26 units. For the Ediacaran-Recent comparison, the Recent data set was subsampled to 30 modes of life; the resulting dissimilarity was 2.55, a slight increase from the raw data (Fig. 5.6b). The Cambrian-Recent comparison was performed two ways: by leaving 30 modes in the Cambrian data set and subsampling the Recent to 13 modes, and by subsampling the Cambrian to 13 modes and the Recent to 30 modes. The results were virtually identical (1.56 and 1.55), as one would hope. The average dissimilarity between subsamples of the Recent data set of size 13 and 30 was also quite similar (1.51), suggesting that the Cambrian-Recent dissimilarity is not necessarily substantial. After subsampling, the Cambrian-Recent dissimilarity fell within the 95% confidence interval 1.03–2.15, which is less than the Ediacaran-Cambrian dissimilarity of 2.26. Thus, the Cambrian is more similar to the Recent than to the Edicaran at the 0.05 significance level. Unsurprisingly, the Recent was also more similar to the Cambrian than it was to the Ediacaran at the 0.05 significance level. Thus, the randomizations support the original observation that the Ediacaran fauna was distinct in ecological composition from the Cambrian and Recent faunas.

5.6.2 *Ecological Change in the Cambrian Explosion: Tiering, Motility, Feeding*

Animal ecosystems changed dramatically during the Cambrian Explosion, but did one ecospace parameter (tiering, motility, or feeding mechanism) change more than others? When each ecospace axis was considered individually, the distance values between the Ediacaran and Cambrian faunas were 0.52 for tiering, 0.71 for motility, and 1.03 for feeding. Thus, feeding mechanisms changed the most in the raw data, reflecting the shift from the osmotrophy-dominated communities of the Ediacaran to the suspension-feeding and predation-dominated communities of the Phanerozoic (Fig. 5.5b, c). However, statistical significance was difficult to evaluate with these data, and it would be interesting to pursue this line of inquiry further with an assemblage-based data set in which rigorous statistical analyses were possible.

5.6.3 A Comment on Ediacaran-Phanerozoic Trends

In highlighting the great ecological changes that occurred during the Cambrian Explosion, we mean to emphasize that no changes of similar substance have occurred in the marine realm ever since, despite devastating mass extinctions (McGhee et al. 2004; Erwin 2006) and other notable ecological events (Vermeij 1977, 1987). It is not our intent, however, to downplay the ecological changes that probably occurred during the Ediacaran Period itself. The initial rise of macroscopic animal ecosystems was obviously as important as anything that came after, and the great increase and shift in ecological lifestyles from the Avalon assemblage to the White Sea assemblage may equal the Cambrian explosion in ecological importance—in fact, the measured dissimilarity between these two Ediacaran assemblages (2.93) is slightly greater than that between the Ediacaran fauna and the Phanerozoic faunas. However, small numbers prevent us from emphasizing this figure, and future work on assemblage-level data will better address this issue.

5.7 Future Work

> There are known knowns; there are things we know we know. We also know there are known unknowns; that is to say we know there are some things we do not know. But there are also unknown unknowns—the ones we don't know we don't know. (Rumsfeld 2002)

Much remains to be discovered about early animals—the depictions of ecospace occupation in Fig. 5.2a–c will assuredly require updating as new fossils are discovered, well-known fossils are reinterpreted, and (perhaps???) entire paradigms are overturned. The pace of new discoveries in this arena has been phenomenal, and the number of unknown unknowns may still be considerable. The ecological summaries presented herein are a formalized description of how many paleontologists currently reconstruct the Ediacaran fauna and are expected to change as refinements become available. In addition to expanding our knowledge of the fauna and refining our interpretations of the fossils, future studies of the Ediacaran fauna will likely do much to document the distribution of ecological types among individual assemblages (e.g., Clapham et al. 2003; Droser et al. 2006), geographic regions (Waggoner 1999), and habitats, as well as documenting the changing nature of the marine biota in considerably more temporal detail than has been attempted here. Hopefully, the general analyses presented here have pointed the way towards some interesting questions.

5.8 Conclusions

Theoretical ecospace provides a system for classifying animals based on autecology and for comparing the ecological diversity and composition of faunas of different ages. Ecospace utilization increased from the Avalon assemblage (up to five modes of life) to the White Sea assemblage (ten modes), but decreased slightly in the Nama assemblage (nine modes). This pattern is not entirely similar to previous descriptions of either taxonomic diversity, which rises and falls symmetrically from the Avalon to the White Sea to the Nama, or morphospace occupation, which stays relatively constant. However, the ecospace analysis included trace fossils and calcified fossils; when these were excluded, trends in ecospace use mirrored trends in taxonomic richness. It is possible that ecological richness and taxonomic richness are linked on some scales of observation, although the reasons for this linkage are not known.

Ecospace analysis highlights the ways in which the Ediacaran fauna was similar and different from Phanerozoic faunas. It was similar (or "modern") only to the extent that it was not ecologically uniform; multiple vertical tiers, motility levels, and feeding mechanisms were present. Thus, as in the Phanerozoic, animals used varied ecological strategies to live and eat, but otherwise the Ediacaran fauna was distinct ecologically. It encompassed limited ecological diversity, with only 13 ecological lifestyles known from the whole period, and only 10 known from any one assemblage. The Early-Middle Cambrian boasted more than twice as many modes of life, and ecological richness increased again by a similar factor between the Cambrian and the Recent. The Ediacaran is maximally dissimilar to the Phanerozoic, however, when one examines the ecological composition of the fauna—that is, which vertical tiers, motility levels, and feeding mechanisms were occupied. The Ediacaran fauna was distinct from both the Cambrian and Recent (>2.2 units), whereas the Cambrian and Recent were relatively similar (0.93 units). In the raw data, feeding mechanisms changed more than tiering and motility during the Cambrian Explosion, but this pattern needs to be evaluated more rigorously.

The extinction of most Ediacaran animals and the rise to dominance of the bilaterians during the Cambrian Explosion completely reorganized the use of ecospace during several tens of millions of years, and represents a fundamental ecological discontinuity in the history of marine animal ecology. The subsequent half billion years have witnessed only minor modifications of this Cambrian ecological blueprint as ecospace progressively filled. Interestingly, Dunne et al. (2008) have noted that Cambrian food webs are remarkably similar to modern marine food webs, also suggesting that the general ecologic structure of the Phanerozoic was set during the Cambrian Explosion. It is often noted that all major animal body plans evolved during the Cambrian Explosion and that evolution has merely tinkered with them ever since; marine ecology appears to have followed a similar tempo of change. Although some cast members have come and gone in the years since, the stage of the modern ecological theater was set in the Cambrian.

Acknowledgments Thanks to P. Novack-Gottshall, R. Krause, and M. Laflamme for helpful reviews and comments.

References

Algeo TJ, Scheckler SE (1998) Terrestrial-marine teleconnections in the Devonian: links between the evolution of land plants, weathering processes, and marine anoxic events. Philos Trans R Soc Lond B 353:113–130

Antcliffe JB, Brasier MD (2007) Towards a morphospace for the Ediacara biota. In: Vickers-Rich P, Komarower P (eds) The rise and fall of the Ediacaran biota. Geological Society, London, pp 377–386

Antcliffe JB, Brasier MD (2008) *Charnia* at 50: developmental models for Ediacaran fronds. Palaeontology 51:11–26

Ausich WI, Bottjer DJ (1982) Tiering in suspension-feeding communities on soft substrata throughout the Phanerozoic. Science 216:173–174

Ausich WI, Bottjer DJ (1985) Phanerozoic tiering in suspension-feeding communities on soft substrata: implications for diversity. In: Valentine JW (ed) Phanerozoic diversity patterns. Princeton University Press, Princeton, pp 255–274

Bambach RK (1983) Ecospace utilization and guilds in marine communities through the Phanerozoic. In: Tevesz MJS, McCall PL (eds) Biotic interactions in recent and fossil benthic communities. Plenum, New York, pp 719–746

Bambach RK, Knoll AH, Sepkoski JJ Jr (2002) Anatomical and ecological constraints on Phanerozoic animal diversity in the marine realm. Proc Natl Acad Sci USA 99:6854–6959

Bambach RK, Bush AM, Erwin DH (2007) Autecology and the filling of ecospace: key metazoan radiations. Palaeontology 50:1–22

Bellwood DR (2003) Origins and escalation of herbivory in fishes: a functional perspective. Paleobiology 29:71–83

Bengtson S (2002) Origins and early evolution of predation. In: Kowalewski M, Kelley PH (eds) The fossil record of predation, Paleontological Society Papers 8. Yale University Reprographics and Imaging Services, New Haven, pp 289–317

Bengtson S, Zhao Y (1992) Predatorial borings in late Precambrian mineralized exoskeletons. Science 257:367–369

Bottjer DJ, Ausich WI (1986) Phanerozoic development of tiering in soft substrata suspension-feeding communities. Paleobiology 12:400–420

Brasier MD, Antcliffe JB (2009) Evolutionary relationships within the Avalonian Ediacara biota: new insights from laser analysis. J Geol Soc Lond 166:363–384

Briggs DEG, Fortey RA, Wills MA (1992) Morphological disparity in the Cambrian. Science 256:1670–1673

Briggs DEK, Erwin DH, Collier FJ (1994) The fossils of the Burgess Shale. Smithsonian Institution Press, Washington

Bush AM, Daley GM (2008) Comparative paleoecology of fossils and fossil assemblages. In: Kelley PH, Bambach RK (eds) From evolution to geobiology: research questions driving paleontology at the start of a new century. Paleontol Soc Paper 14:289–317

Bush AM, Bambach RK, Daley GM (2007a) Changes in theoretical ecospace utilization in marine fossil assemblages between the mid-Paleozoic and late Cenozoic. Paleobiology 33:76–97

Bush AM, Kowalewski M, Hoffmeister A, Bambach RK, Daley GM (2007b) Potential paleoecologic biases from size-filtering of fossils. Palaios 22:612–622

Bush AM, Bambach RK, Daley GM (2008) Were local ecological interactions linked to alpha diversity trends in level-bottom marine communities? Geol Soc Am Abs Prog 40:232–236

Clapham ME, Narbonne GM (2002) Ediacaran epifaunal tiering. Geology 30:627–630

Clapham ME, Narbonne GM, Gehling JG (2003) Paleoecology of the oldest known animal communities: Ediacaran assemblages at Mistaken Point, Newfoundland. Paleobiology 29:527–544

Davidson EH, Erwin DH (2006) Gene regulatory networks and the evolution of animal body plans. Science 311:796–800

Droser ML, Gehling JG, Jensen SR (2005) Ediacaran trace fossils: true or false? In: Briggs DEG (ed) Evolving form and function: fossils and development. Yale Peabody Museum Publications, New Haven, pp 125–138

Droser ML, Gehling JG, Jensen S (2006) Assemblage paleobiology of the Ediacara biota: the unabridged edition? Palaeogeogr Palaeocl Palaeoec 232:131–147
Dunne JA, Williams RJ, Martinez ND et al (2008) Compilation and network analysis of Cambrian food webs. PLoS Biol 6:e102:0693–0708
Dzik J (2003) Anatomical information content in the Ediacaran fossils and their possible zoological affinities. Integr Comp Biol 43:114–126
Erwin DH (2006) Extinction: how life on Earth nearly ended 250 million years ago. Princeton University Press, Princeton
Erwin DH (2008) Wonderful Ediacarans, wonderful cnidarians? Evol Dev 10:263–264
Erwin DH (2009) Early origin of the bilaterian developmental toolkit. Philos Trans R Soc Lond B 364:2253–2261
Erwin DH, Davidson EH (2002) The last common bilaterian ancestor. Development 129:3021–3032
Evans JS (1912) The sudden appearance of the Cambrian fauna. 11th Int Geol Cong, Stockholm. Compte Rendu 1:543–546
Fedonkin MA (1992) Vendian faunas and the early evolution of metazoa. In: Lipps JH, Signor PW (eds) Origin and early evolution of the Metazoa. Plenum, New York, pp 87–129
Fedonkin MA, Waggoner BM (1997) The late Precambrian fossil *Kimberella* is a mollusc-like bilaterian organism. Nature 388:868–871
Fedonkin MA, Gehling JG, Grey K et al (eds) (2007a) The rise of animals: evolution and diversification of the kingdom Animalia. Johns Hopkins University Press, Baltimore
Fedonkin MA, Simonetta A, Ivantsov AY (2007b) New data on *Kimberella*, the Vendian mollusc-like organism (White Sea region, Russia): paleontological and evolutionary implications. In: Vickers-Rich P, Komarower P (eds) The rise and fall of the Ediacaran biota. Geological Society, London, pp 157–179
Foote M (1997) Evolution of morphological diversity. Annu Rev Ecol Syst 28:129–152
Foote M, Gould SJ, Lee MSY et al (1992) Cambrian and Recent morphological disparity: discussions and reply. Science 258:1816–1818
Freeman G (2009) The rise of bilaterians. Hist Biol 21:99–114
Gaines RR, Briggs DEG, Yuanlong Z (2008) Cambrian Burgess Shale-type deposits share a common mode of fossilization. Geology 36:755–758
Gehling JG (1991) The case for Ediacaran fossil roots to the metazoan tree. Mem Geol Soc India 20:181–224
Gehling JG (1999) Microbial mats in terminal Proterozoic siliciclastics: Ediacaran death masks. Palaios 14:40–57
Gehling JG (2000) Environmental interpretation and a sequence stratigraphic framework for the terminal Proterozoic Ediacara Member within the Rawnsley Quartzite, South Australia. Precambrian Res 100:65–95
Gehling JG, Rigby JK (1996) Long expected sponges from the Neoproterozoic Ediacara fauna of South Australia. J Paleontol 70:185–195
Gehling JG, Narbonne GM (2007) Spindle-shaped Ediacara fossils from the Mistaken Point assemblage, Avalon Zone, Newfoundland. Can J Earth Sci 44:367–387
Gehling JG, Droser ML, Jensen SR et al (2005) Ediacara organisms: relating form to function. In: Briggs DEG (ed) Evolving form and function: fossils and development. Yale Peabody Museum Publications, New Haven, pp 43–66
Germs GJB (1972a) The stratigraphy and paleontology of the lower Nama Group, south west Africa. Univ Cape Town Dept Geol Precambrian Res Unit Bull 12:1–250
Germs GJB (1972b) New shelly fossils from the Nama Group, south west Africa. Am J Sci 272:752–761
Glaessner MF (1984) The dawn of animal life: a biohistorical study. Cambridge University Press, Cambridge
Gould SJ (1991) The disparity of the Burgess Shale arthropod fauna and the limits of cladistic analysis: why we must strive to quantify morphospace. Paleobiology 17:411–423
Grant SWF (1990) Shell structure and distribution of *Cloudina*, a potential index fossil for the terminal Proterozoic. Am J Sci 290-A:261–294

Grazhdankin D (2004) Patterns of distribution in the Ediacaran biotas: facies versus biogeography and evolution. Paleobiology 30:203–221

Grazhdankin D, Seilacher A (2002) Underground Vendobionta from Namibia. Palaeontology 45:57–78

Grazhdankin D, Seilacher A (2005) A re-examination of the Nama-type Vendian organism *Rangea schneiderhoehni*. Geol Mag 142:571–582

Grotzinger JP, Watters WA, Knoll AH (2000) Calcified metazoans in thrombolite-stromatolite reefs of the terminal Proterozoic Nama Group, Namibia. Paleobiology 26:334–359

Hendy AJW, Aberhan M, Alroy J et al (2009) A 600 million year record of ecological diversification. Geol Soc Am Abs Prog 41:508

Hofmann HJ, O'Brien SJ, King AF (2008) Ediacaran biota on Bonavista Peninsula, Newfoundland, Canada. J Paleontol 82:1–36

Hou X-G, Adridge RJ, Bergström J et al (2004) The Cambrian fossils of Chengjiang, China: the flowering of early animal life. Blackwell, Malden

Hua H, Pratt BR, Zhang L (2003) Borings in *Cloudina* shells: complex predator-prey dynamics in the terminal Neoproterozoic. Palaios 18:454–459

Huntley JW, Kowalewski M (2007) Strong coupling of predation intensity and diversity in the Phanerozoic fossil record. Proc Natl Acad Sci USA 104:15006–15010

Huntley JW, Xiao S, Kowalewski M (2006) On the morphological history of Proterozoic and Cambrian acritarchs. In: Xiao S, Kaufman AJ (eds) Neoproterozoic geobiology and paleobiology. Springer, Berlin/Heidelberg/New York, pp 23–56

Hutchinson GE (1961) The biologist poses some problems. In: Sears M (ed) Oceanography. Am Assoc Adv Sci Publ 67, Washington, DC, pp 85–94

Ivantsov AY, Malakhovskaya YE (2002) Giant traces of Vendian animals. Dokl Earth Sci 385A:618–622

Jenkins RJF (1992) Functional and ecological aspects of Ediacaran assemblages. In: Lipps JH, Signor PW (eds) Origin and early evolution of the Metazoa. Plenum, New York, pp 152–171

Jenkins RJF, Ford CH, Gehling JG (1983) The Ediacara Member of the Rawnsley Quartzite: the context of the Ediacara assemblage (late Precambrian, Flinders Range). J Geol Soc Aust 30:101–119

Jensen S, Droser ML, Gehling JG (2006) A critical look at the Ediacaran trace fossil record. In: Xiao S, Kaufman AJ (eds) Neoproterozoic geobiology and paleobiology. Springer, Berlin/Heidelberg/New York, pp 115–157

Kelley PH, Kowalewski M, Hansen TA (eds) (2003) Predator-prey interactions in the fossil record, vol 20, Topics in geobiology. Plenum/Kluwer, New York

Knoll AH, Carroll SB (1999) Early animal evolution: emerging views from comparative biology and geology. Science 284:2129–2137

Kowalewski M, Kelley PH (eds) (2002) The fossil record of predation. Paleontological Society Special Papers 8. Yale University Reprographics and Imaging Services, New Haven

Kowalewski M, Dulai A, Fürsich FT (1998) A fossil record full of holes: the Phanerozoic history of drilling predation. Geology 26:1091–1094

Laflamme M, Xiao S, Kowalewski M (2009) Osmotrophy in modular Ediacara organisms. Proc Natl Acad Sci USA 106:14438–14443

Liu AG, McIlroy D, Brasier MD (2010) First evidence for locomotion in the Ediacara biota from the 565 Ma mistaken point formation, Newfoundland. Geology 38:123–126

Love GD, Grosjean E, Stalvies C et al (2009) Fossil steroids record the appearance of Demospongiae during the Cryogenian period. Nature 457:718–721

Marshall CR (2006) Explaining the Cambrian "Explosion" of animals. Annu Rev Earth Pl Sc 34:355–384

McGhee GR Jr, Sheehan PM, Bottjer DJ, Droser ML (2004) Ecological ranking of Phanerozoic biodiversity crises: ecological and taxonomic severities are decoupled. Palaeogeogr Palaeocl Palaeoec 211:289–297

McKerrow WS (ed) (1978) The ecology of fossils: an illustrated guide. MIT Press, Cambridge

Narbonne GM (2005) The Ediacara biota: Neoproterozoic origin of animals and their ecosystems. Annu Rev Earth Pl Sc 33:421–442
Narbonne GM, Gehling JG (2003) Life after snowball: the oldest complex Ediacaran fossils. Geology 31:27–30
Narbonne GM, Saylor BZ, Grotzinger JP (1997) The youngest Ediacaran fossils from southern Africa. J Paleontol 71:953–967
Narbonne GM, Laflamme M, Greentree C et al (2009) Reconstructing a lost world: Ediacaran rangeomorphs from Spaniard's Bay, Newfoundland. J Paleontol 83:503–523
Novack-Gottshall PM (2004) Ecological disparity of deep-subtidal, soft-substrate assemblages during the Cambrian through Devonian. Geol Soc Am Abs Prog 36:457
Novack-Gottshall PM (2007) Using a theoretical ecospace to quantify the ecological diversity of Paleozoic and modern marine biotas. Paleobiology 33:273–294
Rumsfeld, DH (2002) United States Department of Defense news briefing, 12 February 2002. http://www.defenselink.mil/transcripts/transcript.aspx?transcriptid=2636
Seilacher A (1984) Late Precambrian and Early Cambrian Metazoa: preservational or real extinctions? In: Holland HD, Trendall AF (eds) Patterns of change in Earth evolution. Springer, Berlin/Heidelberg/New York, pp 159–168
Seilacher A (1989) Vendozoa: organismic construction in the Proterozoic biosphere. Lethaia 22:229–239
Seilacher A (1992) Vendobionta and Psammocorallia: lost constructions of Precambrian evolution. J Geol Soc London 149:607–613
Shen B, Dong L, Xiao S et al (2008) The Avalon explosion: evolution of Ediacara morphospace. Science 319:81–84
Signor PW III, Brett CE (1984) The mid-Paleozoic precursor to the Mesozoic marine revolution. Paleobiology 10:229–245
Sprigg RC (1949) Early Cambrian "jellyfishes" of Ediacara, South Australia and Mount John, Kimberley District, Western Australia. T Roy Soc South Aust 73:72–99
Stanley S (1973) An ecological theory for the sudden origin of multicellular life in the Late Precambrian. Proc Natl Acad Sci USA 70:1486–1489
Stanley SM (1977) Trends, rates, and patterns of evolution in the Bivalvia. In: Hallam A (ed) Patterns of evolution, as illustrated by the fossil record. Elsevier, Amsterdam, pp 209–250
Steneck RS (1983) Escalating herbivory and resulting adaptive trends in calcareous algal crusts. Paleobiology 9:44–61
Thayer CW (1979) Biological bulldozers and the evolution of marine benthic communities. Science 203:458–461
Thayer CW (1983) Sediment-mediated biological disturbance and the evolution of the marine benthos. In: Tevesz MJS, McCall PL (eds) Biotic interactions in recent and fossil benthic communities. Plenum, New York, pp 479–625
Valentine JW (1992) *Dickinsonia* as a polypoid organism. Paleobiology 18:378–382
Valentine JW, Jablonski D, Erwin DH (1999) Fossils, molecules and embryos: new perspectives on the Cambrian explosion. Development 126:851–859
Vermeij GJ (1977) The Mesozoic marine revolution: evidence from snails, predators, and grazers. Paleobiology 3:245–258
Vermeij GJ (1987) Evolution and escalation: an ecological history of life. Princeton University Press, Princeton
Vermeij GJ (1990) The origin of skeletons. Palaios 4:585–589
Wade M (1972) *Dickinsonia*: polychaete worms from the late Precambrian Ediacara fauna, South Australia. Mem Queensland Museum 16:171–190
Waggoner BM (1999) Biogeographic analyses of the Ediacara biota: a conflict with paleotectonic reconstructions. Paleobiology 25:440–458
Waggoner B (2003) The Ediacaran biotas in space and time. Integr Comp Biol 43:104–113
Whittington HB (1975) The enigmatic animal *Opabinia regalis*, Middle Cambrian, Burgess Shale, British Columbia. Philos Trans R Soc Lond B 271:1–43

Wood RA, Grotzinger JP, Dickson JAD (2002) Proterozoic modular biomineralized metazoan from the Nama Group, Namibia. Science 296:2383–2386

Xiao S, Dong L (2006) On the morphological and ecological history of Proterozoic macroalgae. In: Xiao S, Kaufman AJ (eds) Neoproterozoic geobiology and paleobiology. Springer, Berlin/Heidelberg/New York, pp 57–90

Xiao S, Laflamme M (2009) On the eve of animal radiation: phylogeny, ecology and evolution of the Ediacara biota. Trends Ecol Evol 24:31–40

Xiao S, Zhang Y, Knoll AH (1998) Three-dimensional preservation of algae and animal embryos in a Neoproterozoic phosphorite. Nature 391:553–558

Xiao S, Kowalewski M, Shen B et al (2009) The rise of bilaterians: a reply. Hist Biol 21:239–246

Chapter 6
Quantifying Bioturbation in Ediacaran and Cambrian Rocks

Katherine N. Marenco and David J. Bottjer

Contents

K.N. Marenco (✉)
Department of Geology, Bryn Mawr College, Bryn Mawr, PA 19010, USA
e-mail: kmarenco@brynmawr.edu

D.J. Bottjer
Department of Earth Sciences, University of Southern California,
Los Angeles, CA 90089, USA
e-mail: dbottjer@usc.edu

M. Laflamme et al. (eds.), *Quantifying the Evolution of Early Life*, Topics in Geobiology 36,
DOI 10.1007/978-94-007-0680-4_6, © Springer Science+Business Media B.V. 2011

Abstract Trace fossils, which record the activities of benthic metazoans, typically are bedding-parallel and are preserved along bedding planes in Ediacaran-Cambrian siliciclastic rocks. The quantity of bioturbation present within a bed can reflect a variety of ecological and environmental factors, including oxygenation, salinity, sediment accumulation rate, substrate consistency, and benthic community structure. Thus, quantitative data can augment qualitative descriptions of early bioturbation, in turn facilitating the detailed characterization of facies and the meaningful comparison of Ediacaran-Cambrian sedimentary data. In this chapter, we review and describe a number of new and established methods for estimating quantities of bioturbation, both from bedding plane surfaces and from vertical outcrops or core sections. These methods are designed to complement descriptive paleontological, ichnological, and sedimentological data-collection procedures, and many could be incorporated into a variety of early-life studies.

Keywords Bioturbation • Grid analysis • Image analysis • Ichnology • Semiquantitative methods

6.1 Introduction

Trace fossils have been integral to the study of the early evolution of life since the 1950s (Seilacher 1955, 1956). Ediacaran trace fossils provide convincing evidence of bilaterians (e.g., Valentine 1995; Budd and Jensen 2000) as early as 555.3 Ma (Martin et al. 2000) and are a key source of information about benthic behavior prior to the advent of mineralized skeletons. Cambrian trace fossils record the behavioral evolution of the radiating benthic fauna and the early colonization of new habitats by metazoans.

The character of preserved bioturbation changed markedly from the Ediacaran to the Cambrian. An increase in the complexity and diversity of trace fossils during the Ediacaran-Cambrian boundary interval, roughly coincident with the radiation of nearly all major metazoan groups, was first reported by Seilacher (1956). A Cambrian increase in maximum trace fossil size was subsequently recognized (e.g., McIlroy and Logan 1999; Jensen 2003). In addition, the average depth and intensity of bioturbation increased during the Ediacaran-Cambrian transition (Ausich and Bottjer 1982; Bottjer and Ausich 1986; Droser 1987; Droser and Bottjer 1988, 1993; McIlroy and Logan 1999), which represented a pronounced change in benthic ecology (Seilacher and Pflüger 1994; McIlroy and Logan 1999; Seilacher 1999; Bottjer et al. 2000) and likely led to changes in seawater chemistry (e.g., Brasier and McIlroy 1998; McIlroy and Logan 1999; Canfield and Farquhar 2009). Through critical re-examination of the Ediacaran trace fossil record, numerous recent studies (e.g., Jensen 2003; Seilacher et al. 2003; Droser et al. 2005; Jensen et al. 2005; Seilacher et al. 2005; Jensen et al. 2006) have reinforced the notion that trace fossils were morphologically simple, small, and shallow until the latest Ediacaran, mostly consisting of non-branching horizontal excavations produced close to the sediment-water interface (e.g., Jensen 2003; Jensen et al. 2006).

Variations in the quantity of bioturbation preserved in a rock unit have been interpreted to reflect, for example, differences among depositional environments (e.g., Bottjer and Droser 1991), and fluctuations in sediment accumulation rate (e.g., Howard 1975; Frey 1978) and bottom-water oxygenation (e.g., Savrda et al. 1984; Ekdale and Mason 1988; Savrda and Ozalas 1993). In Ediacaran and Cambrian sedimentary rocks, degree of bioturbation may additionally reflect the evolutionary development of the benthic fauna and the initial expansion of early benthic communities. Therefore, the need for methods to assess quantities of bioturbation, recognized already by numerous workers (e.g., Taylor and Goldring 1993; Taylor et al. 2003; McIlroy 2004), is particularly great for studies of the Ediacaran-Cambrian interval.

In this chapter, we describe a number of new and established methods for estimating quantities of bioturbation, both from bedding plane surfaces and from vertical outcrops or cores, and evaluate their potential utility for Ediacaran-Cambrian research. We have grouped these methods by degree of subjectivity: semi-quantitative (Sect. 6.2) or quantitative (Sect. 6.3). All of the methods discussed are designed to complement descriptive paleontological, ichnological, and sedimentological data-collection procedures. Quantities of bioturbation are not meaningful apart from their sedimentological and paleontological context (e.g., Taylor and Goldring 1993; McIlroy 2004; MacEachern and Bann 2008), and we recommend that all quantitative analyses be accompanied by detailed qualitative (descriptive) evaluation of the studied material.

6.2 Semi-Quantitative Methods

Semi-quantitative methods are the type most commonly used to estimate bioturbation quantity. These methods typically employ visual estimation of the percentage of the sediment disrupted by bioturbation and are designed for efficient application in the field. Visual estimates of total bioturbation are ranked using grades or scores that represent different ranges of percentage (see examples discussed below). Most schemes for ranking bioturbation estimates consist of an ordinal scale: a series of ranked, but unequal, categories (e.g., Stevens 1946; Jager and Looman 1995). It is reasonable to find the median and mode of data that are based on an ordinal scale, but the mean and standard deviation of such a dataset are widely considered to be statistically meaningless (e.g., Stevens 1946; but see also Velleman and Wilkinson 1993, and references therein). Bioturbation estimates generated using semi-quantitative methods have limited mathematical utility and, therefore, are not fully quantitative. At the same time, semi-quantitative methods are efficient and offer greater precision than description alone because the data-collection procedures are standardized and the ranking schemes help to constrain results. Whether a semi-quantitative or quantitative method is the best choice for a given study depends on the research questions being posed.

Based on the examination of modern bioturbated sediments, a number of workers developed similar schemes, beginning in the 1950s, to visually assess the degree of sediment disturbance by bioturbation (e.g., Moore and Scrutton 1957; Reineck 1963, 1967; Howard and Reineck 1972; Howard and Frey 1975). These sets of bioturbation grades were difficult to implement, however, because none was accompanied by an explicitly-defined methodology. To fulfill the need for a more practical and reproducible approach, Droser and Bottjer (1986) and Taylor and Goldring (1993) modified and updated aspects of these earlier methods.

6.2.1 Ichnofabric Indices (Droser and Bottjer 1986)

Droser and Bottjer (1986) developed a flashcard-based method for the efficient classification of bioturbation intensity according to six ichnofabric indices (ii's): one (no bioturbation), two (up to 10% of bedding disrupted), three (10–40%), four (40–60%), five (60–100%), and six (sediment homogenized) (Fig. 6.1). The accompanying ichnofabric flashcards, designed to facilitate consistency through pattern recognition, depict a series of hypothetical vertical cross-sections representing ii's one through five. Following the initial publication of the ichnofabric index method with flashcards for carbonate-dominated shelf environments (Droser and Bottjer 1986), additional flashcards were produced for the nearshore (Droser and Bottjer 1989b) and deep sea (Droser and Bottjer 1991).

The ichnofabric index method can be applied equally well to vertical outcrop exposures and split cores of sedimentary material. In the field, ichnofabric indices should be logged from outcrop in a continuous fashion (weathering and diagenesis permitting), either during or subsequent to the measurement and description of a stratigraphic section. Flashcards appropriate to the studied lithofacies may be used as visual aids to improve the precision of ichnofabric scoring. The width and vertical thickness of each scored interval must be recorded as field logging proceeds. Scored intervals may vary in vertical thickness (above a one centimeter minimum), but a consistent horizontal field of view should be maintained throughout the logging process. Using a horizontal field of view of 50 cm for ichnofabric evaluation is recommended (Bottjer and Droser 1991). Because results may vary greatly depending on the chosen horizontal field of view, this width value should always accompany ichnofabric index data when the latter are reported.

When a set of ichnofabric index data have been collected from the same sedimentary lithofacies, these data can be converted into percentages and presented together in a specialized histogram, or "ichnogram" (Bottjer and Droser 1991). The total thickness of strata scored with a particular ichnofabric index can be represented as a percentage of the total measured outcrop thickness. An ichnogram provides a means of visually comparing the proportions of total outcrop thickness that were scored using different ichnofabric indices.

Once ichnofabric index data have been converted to percentages, an average ichnofabric index can be calculated for the outcrop or locality in question. Because ichnofabric indices comprise an ordinal scale, averages of these data are not statistically valid but merely provide reference points for use in broad-scale comparisons (Droser and Bottjer 1989a). The average ichnofabric index can be computed from a set of data as follows (Droser and Bottjer 1989a):

$$\text{average ii} = (\%\ \text{of strata scored as ii}\,1 * 1) + (\%\ \text{ii}\,2 * 2) + \ldots + (\%\ \text{ii}\,6 * 6) \quad (6.1)$$

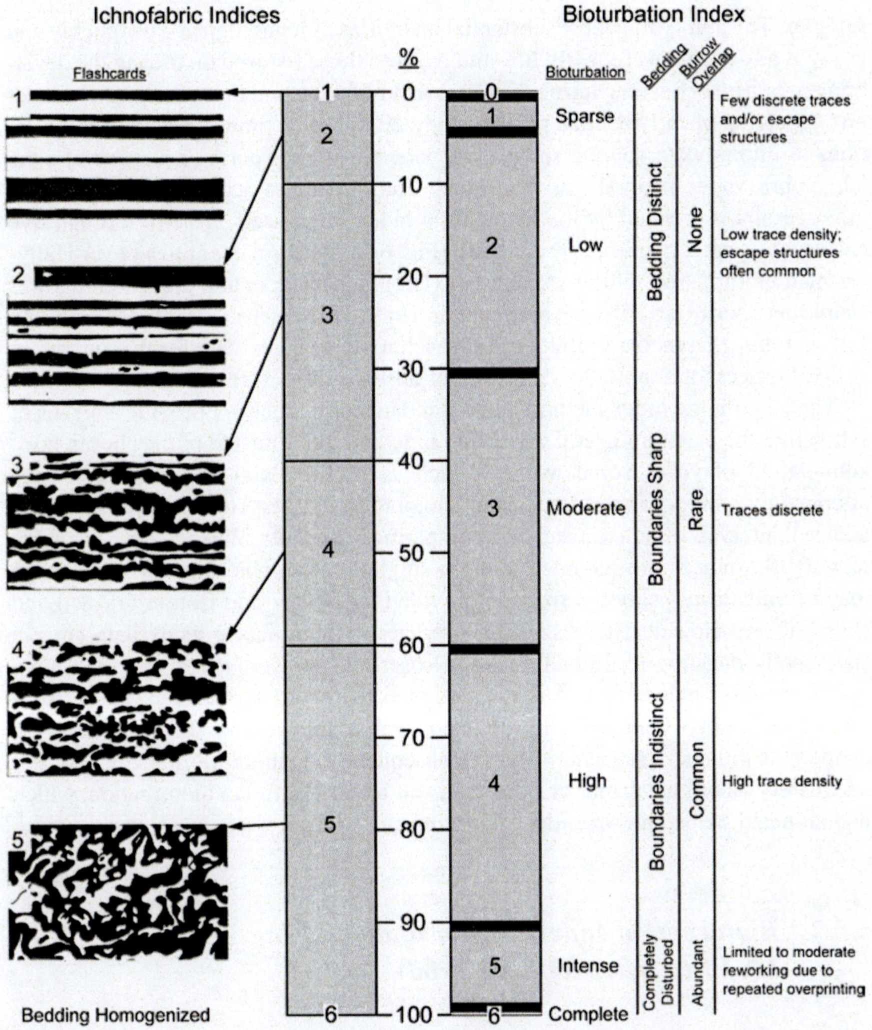

Fig. 6.1 Comparison of ichnofabric indices (*left*) and bioturbation index grades (*right*). The range of percentages encompassed by each ichnofabric index (ii) and bioturbation index grade is shown, in *gray*, in the middle portion of the figure. *Heavy black lines* in the bioturbation index column denote the additional demarcation (of 1%) between these grades compared to the ii's. Flashcards for ii's one through five, modified with permission from Droser and Bottjer (1986, Fig. 1, p. 558), are shown to the left of the ii scale. To the right of the bioturbation index scale, the defining characteristics of each bioturbation index grade are summarized in four columns (Flashcard images: copyright (1986) by SEPM (Society for Sedimentary Geology))

The ichnofabric index method is a simple, efficient means of estimating quantities of bioturbation in vertical outcrop exposures and core material. No specialized equipment is required to carry out the method, and large volumes of ichnofabric index data can be collected in the field without the necessity of obtaining physical samples. The ability to collect substantial quantities of ichnological data quickly and easily is advantageous for early life studies, both those focused on tracing the development of new behaviors during the Cambrian radiation and those seeking a broader understanding of early benthic paleoecology. At the same time, the statistical limitations of ordinal data and the subjectivity inherent in the scoring process make the ichnofabric index method less than ideal. The maximum accuracy of ichnofabric index results is dictated by the ichnofabric index scale itself; ii's two through five each encompass a range of 10% or more (Fig. 6.1). Because quantities of vertically-oriented bioturbation seldom exceed 10% (ii2) in Ediacaran and pre-trilobite lower Cambrian rocks (e.g., Droser and Bottjer 1988; Droser et al. 1999; McIlroy and Logan 1999; Marenco and Bottjer 2008), ichnofabric index data for this important interval reflect little more than presence or absence of bioturbation.

The ichnofabric index method, in its standard form, does not provide a means of estimating the extent of lateral variability in ichnofabric outside of the chosen horizontal field of view. Some workers, such as McIlroy and Logan (1999), have attempted to compensate by recording ichnofabric indices from multiple laterally-adjacent intervals of outcrop and then comparing the results. When widely-disparate ichnofabric indices are recorded along a single stratigraphic interval, an average may be calculated using the formula provided by Droser and Bottjer (1989a) and stated above. However, as discussed, averages of ichnofabric index data are not statistically meaningful. In addition, as Droser and Bottjer (1989b) acknowledge, the ichnofabric index method is not designed to record sediment mixing that is more horizontally- than vertically-oriented, such as the work of shallow-burrowing "undermat miners" (Seilacher 1999). Consequently, ichnofabric indices may not adequately record the extent of Ediacaran and lower Cambrian bioturbation, which is dominated by bedding-parallel excavations.

6.2.2 *Bioturbation Index (Taylor and Goldring 1993; Modified After Reineck 1963)*

Taylor and Goldring (1993) sought to combine the attributes of descriptive and semi-quantitative schemes for representing the character and intensity of bioturbation in cores and outcrops. These authors used the 'bioturbation index' of Reineck (1963), which is comprised of seven grades, each defined by a range of percentages and degrees of burrow abundance, burrow overlap, and sediment-fabric sharpness (Fig. 6.1). Unlike the ichnofabric index scale, the bioturbation index begins at zero (Reineck 1963). The ranges of percentage represented by grades one through six are intended to be used as guides only (Taylor and Goldring 1993). The bioturbation index grades are, briefly, as follows: zero (no bioturbation), one (1–4% 'sparse' bioturbation), two (5–30% 'low' bioturbation), three (31–60% 'moderate' bioturbation), four (61–90% 'high' bioturbation), five (91–99% 'intense' bioturbation), and six (100% 'complete' bioturbation) (Reineck 1963; Taylor and Goldring 1993). Taylor and Goldring (1993) developed detailed descriptions to accompany the numerical bioturbation index grades. The following are brief summaries of these descriptions. Grades one and two represent situations in which physical sedimentary structures are more prominent than biogenic structures. Grades three and four represent the transition toward more intense bioturbation, in which physical sedimentary structures are disrupted and burrows overlap one another. Grades five and six represent the dominance of biogenic structures, eventual complete disruption of bedding, and reworking of the sediment.

Taylor and Goldring (1993) recommend that the bioturbation index be applied on a bed-by-bed basis, recording small-scale within-bed variations in ichnofabric. It is also acceptable to assign single grades to larger packages of material in which the ichnofabrics are similar, as long as the analytical field of view is kept consistent throughout. The bioturbation index is intended to be used in conjunction with ichnofabric constituent diagrams, which are graphical representations of ichnofabrics based on the relative proportions of their various components. These diagrams are capable of conveying a substantial amount of information about ichnofabrics in a sophisticated manner. However, because ichnofabric constituent diagrams neither incorporate bioturbation index data directly nor facilitate the quantitative assessment of bioturbation, we will not discuss them further in this chapter.

In practice, the bioturbation index and the ichnofabric index scale differ only in the total number of grades used to classify bioturbation and the ranges of percentage represented by those grades (Fig. 6.1). Consequently, these two methods share similar advantages and disadvantages. Both methods provide simple, efficient means of scoring vertically-oriented bioturbation in the field, and using one or the other is a matter of personal preference.

6.2.3 Bedding Plane Bioturbation Indices (Miller and Smail 1997)

To complement the numerous published methods designed for quantifying bioturbation from vertical exposures, Miller and Smail (1997) developed the bedding plane bioturbation index method, a semiquantitative technique for evaluating the amount of bioturbation on bedding plane surfaces. Designed to be used in conjunction with ichnofabric indices, the five bedding plane bioturbation indices (BPBI's) represent the same ranges of percentage as ii's one through five. Miller and Smail also emulated the ichnofabric index flashcards, producing four sets of diagrams depicting commonly-encountered scenarios of trace fossil morphology and distribution on bedding planes: same size and shape, even distribution; same size, different shape, and even distribution; different size and shape, even distribution; and different size and shape, uneven distribution.

A consistent two-dimensional field of view should be used when applying the bedding plane bioturbation index method to field exposures. Miller and Smail recommend a field of view of 625 cm^2, which can be delineated on a studied bedding plane using a 25-by-25 cm (internal dimensions) frame constructed of PVC pipe, cardboard, or another sturdy material. A portion of a bedding plane can be randomly selected for study by using a random number generator or by tossing the frame onto the bedding plane surface. Multiple portions of larger individual bedding planes may be evaluated to assess potential heterogeneity in the spatial distribution of bioturbation. Once a field of view has been selected for study, a bedding plane bioturbation index can be assigned to the bedding plane area through visual comparison with BPBI flashcards. Additionally, Miller and Smail recommend that studied bedding planes be photographed so that field estimates can later be compared with more precise image analysis results.

Like the ichnofabric index method, the bedding plane bioturbation index method is highly efficient for field scoring and does not require specialized equipment or the collection of physical samples. Because bedding-parallel bioturbation is often more common than vertically-oriented disruption in Cambrian and earlier rocks, the bedding plane bioturbation index method is particularly well-suited for use in early-life studies. However, because the five BPBI's are identical to ii's one through five, data collected using the bedding plane bioturbation index method has the same limitations as those discussed for ichnofabric indices. In addition, the scale and intensity of bioturbation on a bedding plane exposure can be difficult to interpret, as Miller and Smail (1997) discussed at length. For example, one or two larger burrows may cover the same total area as many widely-dispersed smaller burrows, but the latter will likely appear to achieve greater coverage of the bedding surface. We recommend (as do Miller and Smail) that the bedding plane bioturbation index method be used in conjunction with a quantitative laboratory technique (see Sect. 6.3) to check for consistency and to correct any pattern-recognition errors.

6.3 Quantitative Methods

In general, quantitative methods employ direct measurement as the primary mode of data collection. Quantitative data can be subjected to more extensive statistical manipulation than semi-quantitative data. At the same time, quantitative methods typically are more time-consuming than semi-quantitative approaches because greater care is required to produce more accurate results. Relatively few quantitative methods have been developed for assessing quantities of bioturbation in rocks and sediment cores. One of the most prominent obstacles to bioturbation analysis is the difficulty of distinguishing between biogenic and abiogenic sedimentary structures, especially when bioturbation is preserved indistinctly. Increasing visual contrast through digital image enhancement can be helpful, but this is not possible in all cases.

6.3.1 *Grid-Based Estimation of Percentage Area*

Although grid-based methods are common in vertebrate ichnology for the study of trackways (e.g., Mezga and Bajraktarević 1999; Swanson and Carlson 2002;

Getty 2005), only a few invertebrate ichnological studies have made use of such techniques (Pickett 1972; Miller 1977; Pemberton and Frey 1984; Heard and Pickering 2008; Heard et al. 2008; Marenco and Bottjer 2010). Three of these latter studies have examined the spatial relationships among vertical burrows, such as *Skolithos* and *Diplocraterion*, through nearest-neighbor analysis of bedding plane surfaces (Pickett 1972; Miller 1977; Pemberton and Frey 1984). Nearest-neighbor grid analysis can provide insight into the dispersion of bioturbating organisms, but this method does not generate an estimate of the total bioturbation present.

One other type of grid-based method has been employed in invertebrate ichnological studies: percentage cover estimation. Originally developed for plant ecology, this type of method has been used widely in modern ecological studies to determine the relative importance of different species in an ecosystem (Floyd and Anderson 1982). Percentage cover of bioturbation, or percentage area bioturbated, can be estimated using grids from both bedding planes and vertical cross-sections (but see discussion below).

6.3.1.1 Bedding Plane Analysis: 10 × 10 cm Grids (Heard and Pickering 2008)

In a study of Middle Eocene deep-marine siliciclastic strata, Heard and Pickering (2008) used 10 × 10 cm grids to estimate the average quantity of bioturbation on bedding plane exposures. Each grid was comprised of 100 one-by-one cm cells. Following random placement of a grid on a bedding plane surface, the authors counted each grid cell that contained bioturbation. The total number of bioturbated grid cells was then used as an estimate of the percentage area bioturbated within the grid. Heard and Pickering (2008) used these estimates of percentage area bioturbated to help characterize distinct deep-marine environments and recognize trends in trace fossil density and diversity within the studied strata.

The grid method used by these authors resembles a plant ecology technique, termed the 'rectangle method' by Korva (1996), that is used to estimate the percentage ground cover of green foliage. Using the rectangle method, percentage cover is found by counting each grid cell that is more than half-filled by foliage and then dividing this number by the total number of cells in the grid (Burstall and Harris 1983; Korva 1996). Korva (1996) found that the rectangle method tends to over- or underestimate percentage cover depending on the size and distribution of foliage patches relative to the size of the grid cells. For example, if many grid cells are only slightly more than half-filled by foliage, the result will likely be an overestimate of percentage cover. The potential for inaccurate estimates is even greater with the method used by Heard and Pickering (2008), in which grid cells are counted on a presence absence basis. Underestimation is not possible with this form of presence-absence evaluation, but the total amount of bioturbation present may be overestimated considerably if trace fossils are scattered and are much smaller than the individual grid cells. We discourage the use of this type of method due to the likelihood of obtaining exaggerated percentage cover estimates.

6.3.1.2 Analysis of Cores: 10 × 6 cm Grids (Heard et al. 2008)

Also in 2008, Heard and colleagues published an analysis of the bioturbation in a 230 m-long, 6 cm-wide well core from the same Middle Eocene succession as the previously-discussed study. The authors applied a 10 × 6 cm grid to successive 10 cm-long intervals of the cut surface of the core. Each time the grid was applied, the authors counted the number of one-square-centimeter grid cells that contained bioturbation. They divided this number by the total number of grid cells (60) and multiplied the result by 100 to obtain a percentage area bioturbated. In addition, the authors recorded the abundance of individual ichnotaxa by counting the number of grid cells occupied by each ichnotaxon of interest.

Applying a grid-based percentage cover method to vertical cross-sections of strata is more complicated than applying such a method to bedding planes because vertical exposures reflect a succession of sedimentary events over time. When placed on a vertical exposure or core, a two-dimensional grid necessarily lumps together the information from a particular vertical interval of strata. Like bulk-sampling for geochemical data, this technique may obscure details that are expressed at a finer scale than the sampling interval. A relatively large sampling interval may be practical, however, if the thickness of strata to be analyzed is substantial, as in the study presented by Heard et al. (2008). Because this method is nearly identical to Heard and Pickering's (2008) use of grids for bedding plane analysis, the probability of overestimating bioturbation quantities is also high. Because of these drawbacks, we do not recommend the use of this or other grid-based methods for estimating quantities of bioturbation from vertical exposures.

6.3.1.3 Intersection Grid Method for Bedding Planes (Marenco and Bottjer 2010)

We developed a grid-based approach, the intersection grid method, for estimating the percentage of bioturbation on bedding plane surfaces (Marenco and Bottjer 2010). The intersection grid method is based on the point intercept method (Goodall 1952; Floyd and Anderson 1982; Greig-Smith 1983; Floyd and Anderson 1987), a plant ecology technique for estimating percentage cover. In the point intercept method, a pin is placed vertically at every point where two lines intersect within a square or rectangular grid. The grid dimensions and number of line intersections are chosen based on the size of the test area and the study objectives. After the grid is positioned over a vegetated study area, percentage cover is determined by counting the number of pins that touch a species of interest and dividing by the total number of pins in the grid.

We adapted the point intercept method for the analysis of bioturbation by reducing the average sample area to one square meter or smaller and emphasizing a digital grid-application process. Taking digital photographs of bedding planes for later analysis is logistically simpler than constructing a physical grid in the field, and greater analytical precision is possible if grids are digitally applied to these photographs.

However, if large-scale analyses are attempted, it may be more practical to construct a grid on the bedding plane than to later piece together and analyze a large composite image file.

The intersection grid method involves a combination of field and laboratory procedures. In the field, after documenting the important attributes of a studied bedding plane, a square or rectangular frame (we recommend a frame no larger than one square meter in size) should be used to designate a portion of the surface for analysis using the intersection grid method (Fig. 6.2a, d). On larger bedding planes, one or more random samples can be obtained by tossing the frame onto the surface. A digital photograph should be taken of each framed portion of the bedding plane. If possible, position the camera directly above the center of the frame, and select a high image-quality (large file size). Once in the lab, use drawing software such as Adobe Illustrator to construct a grid and superimpose it onto the framed area of each photograph (Fig. 6.2b, e). Ensure that the grid lines are evenly distributed and that the outer dimensions of the grid match the inner perimeter of the frame as closely as possible. For the most accurate estimates, reduce the width of the grid lines as much as the software will permit. Use the software to mark, with a colored dot or other symbol, each grid-line intersection point that touches a bioturbated portion of the gridded photograph (Fig. 6.2c, f). These marks should be made on a separate layer of the drawing document from those containing the grid and bedding plane photograph. Points where grid lines touch the outer frame should be excluded from analysis. Magnify the gridded image if necessary to make accurate determinations. Once the entire grid has been evaluated, obtain the total number of marked intersection points by selecting all of the marks using the drawing software. (Adobe Illustrator, for example, contains a function that will display the total number of objects in a selection.) Divide the preceding result by the total number of grid-line intersection points to generate an estimate of the percentage of the framed bedding plane area that is bioturbated. Finally, save the modified image file in the drawing program as a permanent record of the intersection grid analysis.

Figure 6.2 illustrates, with two bedding plane examples, the process of applying a grid to a studied bedding plane and analyzing it using the intersection grid method. Both bedding planes shown in the figure were studied from exposures of the lower Cambrian portion of the Wood Canyon Formation in eastern California, USA. A 600 cm^2 (24 × 25 cm^2) frame was used to designate a portion of each bedding plane for analysis. The bedding plane pictured in Fig. 6.2a–c, from the upper member of the Wood Canyon Formation, consists of medium-grained sandstone and contains the large bedding-parallel trace fossil *Psammichnites*. A square grid containing 961 grid-line intersections was applied to this bedding plane surface. During intersection grid analysis, 545 intersection points were marked, resulting in an estimate of 60% bioturbation for the studied bedding plane area. Pictured in Fig. 6.2d–f is a varnished quartzite bedding plane from the middle member of the Wood Canyon Formation that contains abundant *Skolithos* burrow entrances. A square grid containing 4,900 grid-line intersections was applied to the bedding plane surface. A total of 1,944 intersection points were marked during analysis, indicating that approximately 41% of the surface was bioturbated.

Fig. 6.2 Steps for analyzing a bioturbated bedding plane using the intersection grid method, as applied to two lower Cambrian bedding planes from the Wood Canyon Formation, eastern California, USA. (**a**) Bedding plane, containing *Psammichnites* trace fossils, with a 600 cm^2 (24 × 25 cm) frame denoting the specific area to be analyzed. (**b**) A square grid (961 grid-line intersections) applied digitally to the framed area of the photograph. (**c**) Black dots are placed on grid-line intersections that touch bioturbated portions of photograph. 545 grid-line intersections were marked in this study image. (**d**) Bedding plane exposure with abundant *Skolithos* burrow entrances. 600 cm^2 (24 × 25 cm^2) frame used. (**e**) Square grid (4,900 grid-line intersections) applied digitally. (**f**) 1,944 grid-line intersections were marked during analysis

The ratio of average trace fossil size to total sample area must be considered when undertaking analyses using the intersection grid method. When most trace fossils are small relative to the sample area, a finely-subdivided grid will be required to capture as much detail as possible. Conversely, when trace fossils are large

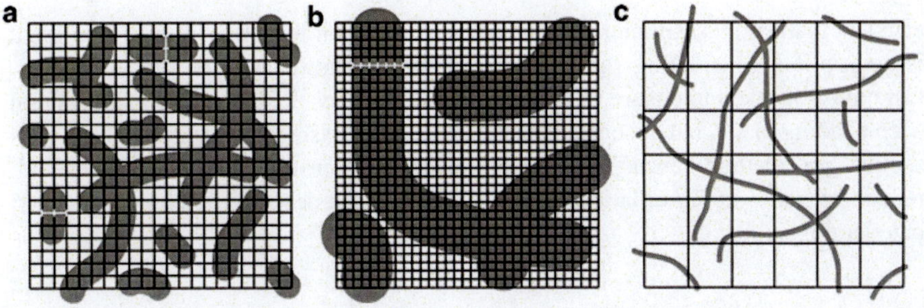

Fig. 6.3 Drawings of hypothetical bedding planes illustrating the importance of proper grid scaling for intersection grid method analysis. (**a**) Grid is scaled appropriately for bioturbation present: the width of an average trace fossil encompasses three adjacent grid-line intersections (*white brackets*). (**b**) Grid is too finely-subdivided for the trace fossils present. Seven adjacent grid-line intersections are encompassed by the width of an average trace fossil. (**c**) Grid is too coarsely-subdivided. None of the trace fossils within the grid are touching grid-line intersections

relative to the sample area, fewer grid-line intersections are needed for obtaining an accurate estimate. If a grid is too finely subdivided, time and effort will be wasted to complete the analysis. We recommend the following rule-of-thumb for scaling grids to optimize accuracy and efficiency: the average width/diameter of the trace fossils present should be two to five times greater than the width of one grid cell (or the distance between two adjacent grid lines). In other words, the width of a horizontal trace fossil oriented parallel to one set of grid lines should encompass at least three adjacent grid-line intersections (Fig. 6.3a). Depending on the quantity of bioturbation present, an analysis using a grid scaled within this range requires between 30 min and 4 h. Based on our tests of the method, using one of the recommended grid scales should reliably produce results within 1% of the actual percentage bioturbated (Marenco and Bottjer 2010). Scaling a grid outside the recommended range will reduce the accuracy and/or efficiency of the analysis (Fig. 6.3b, c).

The intersection grid method and the bedding plane analysis method employed by Heard and Pickering (2008) are comparably precise because both are based on presence/absence data. However, the intersection grid method is more accurate because each unit of data is a point rather than a two-dimensional area. Theoretically, each grid-line intersection point is dimensionless and, therefore, can record only the presence or absence of bioturbation (Greig-Smith 1983). In contrast, each grid cell consists of an area in which bioturbation coverage is 0–100%. When the bioturbation within grid cells is evaluated on a presence-absence basis, as in the study by Heard and Pickering (2008), this potential range of variation is not taken into account.

We feel that the intersection grid method provides a satisfactory means of estimating the percentage of bioturbation on bedding planes and is well-suited for evaluating the bedding-parallel bioturbation that is typical of Ediacaran and lower Cambrian strata. Analyses using the method are readily reproducible when grids are scaled according to the recommended range, and a record of each analysis can be saved for publication and archiving. The intersection grid method is somewhat less efficient than similar, simpler methods, but we feel that the extra time commitment is appropriate considering the accuracy of the results that the method produces. Care must be taken to ensure that the grid scale matches the bioturbation present on each bedding planc, but accomplishing this becomes easier with practice. The intersection grid method can be used effectively to refine estimates made in the field, recognize trends in bioturbation patterns, and augment descriptive ichnological and paleontological data.

6.3.2 *Image Analysis*

Computer-aided image analysis can help to reduce or eliminate inconsistencies in ichnological datasets that commonly result from the use of more subjective data collection procedures (e.g., Francus 2001). A number of ichnological studies have employed image analysis techniques, either to enhance the contrast of digital images so that bioturbated areas can be distinguished more easily (Magwood and Ekdale 1994) or to assess the quantity of bioturbation in sedimentary rocks (e.g., Francus 2001; Löwemark 2003). Images that contain at least some initial contrast between bioturbation structures and surrounding material generally produce more robust image analysis results (e.g., Magwood and Ekdale 1994).

Magwood and Ekdale (1994) used digital image enhancement techniques to improve the visibility of complex ichnofabrics in photographs of deep-sea box cores. Edge detection and enhancement, for example, can be useful for recognizing individual biogenic structures and cross-cutting relationships that might not otherwise be visible due to low contrast. Other techniques, such as spatial autocorrelation, allow minor variations in the distribution of trace fossils to be discerned. Histograms can be generated from the manipulated image data to summarize visual information, such as the dominant orientation(s) of biogenic structures. Magwood and Ekdale (1994) did not, however, employ these types of techniques to estimate quantities of bioturbation.

6.3.2.1 Thin Section Image Analysis or the H Method (Francus 2001)

Based on the dominant characteristics of hemipelagic sediments, Francus (2001) developed a method for estimating quantities of bioturbation in thin sections of these fine-grained sediments. Hemipelagic sediments accumulate in deep marine settings and lakes through the gradual settling of very fine-grained particles from suspension. Francus studied numerous hemipelagic fabrics and concluded that elongated sediment particles typically accumulate in a bedding-parallel fashion unless disrupted, most commonly by bioturbation. Thus, quantities of bioturbation in these sediments can be estimated by determining the degree to which the original bedding-parallel fabric has been disturbed from a dominantly-horizontal orientation.

Francus prepared thin sections from freeze-dried soft-sediment lake cores for analysis. After the sedimentary fabric of each thin section had been described using conventional petrography, the thin sections were carbon-coated and examined under a scanning electron microscope with a back-scattered election detector. Grayscale digital photomicrographs, 512 × 512 pixels in size, were generated using the microscope. These images were then processed using the public-domain NIH Image program to produce black-and-white images in which white pixels represented the clay-rich matrix and black pixels represented larger sediment grains. The types of image processing applied to each image included filtering for noise reduction, thresholding, and binary image editing.

After image processing was complete, each of the larger grains was analyzed individually. The lengths of each grain's major (L) and minor (l) axes were measured, and the angular difference between horizontal and the grain's major axis (α) was determined by measuring in a counter-clockwise direction. Area (A) was found by counting the number of pixels that comprised the image of the grain. These measurements were then compiled in a spreadsheet. Francus calculated an approximate grain size (D_0) for each grain, in microns, with the following equation:

$$D_0\left(\mu m\right) = 2\sqrt{\frac{A\left(\mu m^2\right)}{\pi}} \tag{6.2}$$

in which A represents the area of each grain. Using the various grain measurements, plus the approximate grain size value (D_0), the morphology and position relative to horizontal of each grain (H_i) were found using the following equation:

$$H_i = \left(\frac{L}{l} - 1\right)\left(\left(2\cos\left(2\alpha\right)\right)\left|\cos\left(2\alpha\right)\right|\right)\left(\frac{D_0 - MD_0}{sD_0}\right) \tag{6.3}$$

in which MD_0 and sD_0 are the mean and standard deviation, respectively, of all grains in a single thin section image. The first term of Eq. 6.3 in parentheses, adjusts the H_i according to the shape of the grain such that elongated grains result in high H_i values, whereas spherical grains result in an H_i of zero. The second term of Eq. 6.3 assigns values based on the orientation of the grain, as follows: horizontal grains receive high positive values, vertical grains receive negative values, and intermediate grains (at 45° and 135°) receive zero. The last term of Eq. 6.3 provides a value proportional to the grain's size. The end member results of this analysis are high H_i values for large, elongated, horizontal grains and low H_i values for small spherical grains.

Francus (2001) defines the index H as the horizontality of the grain fabric, or the mean of the H_i values for grains that have a D_0 greater than MD_0. Eliminating smaller grains from the calculation of H prioritizes the more prominent grains in the fabric and prevents the H_i values of the more numerous smaller grains, which may not be oriented as strongly because of their size, from masking the dominant orientation of the fabric.

The *H* method is effective at classifying recent hemipelagic sediments according to varying degrees of sediment disturbance (Francus 2001). The data collection procedures are standardized and objective, provided that the worker can consistently distinguish abiogenic sediment disruption from bioturbation in order to select appropriate samples for analysis. Because the *H* method can detect microbioturbation, this technique can be used to determine whether bioturbation is present within massive deposits. At the same time, the *H* method is less sensitive to variations in the degree of bioturbation within intensely-bioturbated samples, which are represented by a limited range of *H* index values. The thin-section-sized field of view necessitated by the *H* method is limiting, especially when a significant amount of study material, such as a lengthy sediment core, must be evaluated. In addition, the method has very limited applicability to paleoecological studies because samples must consist of uniformly fine-grained sediment that has not been diagenetically altered. The *H* method cannot, for example, compensate for changes in grain orientation that result from current activity in shallow-water settings or for compaction during burial. Thus, this method is not useful for studies of Ediacaran and Cambrian rocks, which have undergone significant burial alteration.

6.3.2.2 Automated Image Analysis of X-Radiographs (Löwemark 2003)

Löwemark (2003) recognized an abundance of pyritized microburrows in x-radiographs of five Late Quaternary sediment gravity cores collected from the Portuguese continental slope. The microburrows, which stand out distinctly in x-radiographs due to their high pyrite content, consist of two types: *Trichichnus*, thin (usually less than one 1 mm in diameter) cylindrical burrows that can reach several tens of centimeters in length; and "*Mycellia*," randomly-oriented filamentous burrows that are tenths of a millimeter in diameter and up to several centimeters long. To estimate the abundance and distribution of these burrows in x-radiographs, Löwemark used the image analysis program DIAna, which was designed to automatically recognize linear features, such as roads, in aerial photographs and satellite images. Linear features that are expressed within a set gray-value threshold are detected by DIAna, and the program records the length, width, and curvature of these features.

To produce images suitable for analysis, Löwemark divided each x-radiograph into 10 × 5 cm sections and scanned each at a resolution of 300 dpi. These smaller images were analyzed using the DIAna program. To avoid including data on elongated non-burrow objects, Löwemark disregarded all linear features less than 2 mm in length that were detected by DIAna. Using DIAna, all of the remaining detected linear features were subdivided into very small segments, and the orientation of each segment was measured. A cumulative frequency distribution was calculated automatically using the measured orientations of the linear features in each image. The results of this calculation were plotted in a rose diagram.

The main advantages of using this method over manual analysis of biogenic structures are objectivity and efficiency. As Löwemark notes, collecting the same volume of data manually would be prohibitively time-consuming and considerably more subjective. Subjectivity is unavoidable during the image-preparation stage, however. The gray values in x-radiographs can vary considerably due to differences in sediment composition, water content, and thickness of the x-rayed slab. In order to avoid the inclusion of non-burrow objects in the automated analysis, the gray values of each image must be adjusted manually. Consequently, comparisons between sample results are only meaningful if the samples were taken from the same core or from similar lithological units.

Like the *H* method, this automated technique is designed for a specialized application, and samples must meet certain criteria to be suitable for analysis. Most importantly, the contrast between burrows and the surrounding sediment must be uniformly high to generate an accurate estimate of burrow abundance. Pyritization produces the highest-possible contrast in x-radiographs; the fill of most burrows, even if distinctly different from the surrounding sediment, may not be detected by the DIAna program. In addition, burrows must be well-defined and linear in overall morphology in order to be recognized by DIAna. This method estimates the quantity of discrete burrows in a sample rather than the total bioturbation present. We see the potential for two specific types of applications in early life studies: analysis of Cambrian *Skolithos* piperock in instances where the burrow-sediment contrast is sufficiently high, and quantification of diagenetically-pyritized burrows in Ediacaran and Cambrian rocks.

6.3.3 Imaging Techniques (X-Radiography, CT, MRI)

Imaging techniques such as x-radiography, computed tomography, and magnetic resonance imaging are minimally destructive and can provide useful three-dimensional information about trace fossils and ichnofabrics. Numerous studies have employed x-radiography to examine bioturbation in sediments (e.g., Howard 1969; Rhoads and Boyer 1982; Grimm et al. 1996) and in sedimentary rocks (e.g., O'Brien 1987; Marenco and Bottjer 2008). X-radiography is relatively inexpensive and can be useful for identifying vertically-oriented trace fossils. However, relatively thin, uniform slabs of material are necessary for proper exposures (Bouma 1964; Howard 1969; Hamblin 1971; Dufour et al. 2005), and bedding-parallel trace fossils often are not well-expressed in x-radiographs (Fu et al. 1994).

Recently, a number of studies have explored the use of magnetic resonance imaging (e.g., Gingras et al. 2002a; Gingras et al. 2002b) and computed tomography (e.g., Fu et al. 1994; Dufour et al. 2005, 2007) for visualizing trace fossils in three dimensions. Both techniques scan samples repeatedly and generate a series of cross-sectional image 'slices' showing the density contrast of the sample within the plane of each scan. Magnetic resonance imaging (MRI) is useful for studying the bioturbation-enhanced pore space of sedimentary rocks because it emphasizes differences in density between fluid- or air-filled pore spaces and non-porous rock (Gingras et al. 2002b). Relatively few sedimentary studies have employed MRI technology, however, because MR waves attenuate strongly in many materials, resulting in low-quality images (Gingras et al. 2002b). Gingras et al. (2002b) used MRI to study an 84 cm^3

block of sandstone dominated by *Macaronichnus* burrows. Despite the relatively low quality of the MRI images, the authors identified at least four different burrow morphologies and located porous zones within the sample (Gingras et al. 2002b).

Computed tomography (CT) has been used extensively to study the distribution of biogenic structures in unconsolidated sediments (e.g., Fu et al. 1994; Capowiez et al. 1998; Mermillod-Blondin et al. 2003; Dufour et al. 2005). CT scanning produces higher-quality images than MRI, and the thickness and frequency of scanned slices can be adjusted to suit the study parameters (Fu et al. 1994). In an early application of CT to the study of bioturbation, Fu et al. (1994) scanned two sediment gravity cores from the North Atlantic. The cores were scanned continuously in 3 mm slices, generating more than 300 CT images per meter of core. Using these bedding-parallel images, Fu and colleagues identified details of *Scolicia* and *Zoophycos* burrows and reconstructed cross-cutting relationships that were unclear or invisible in vertical x-radiographs of the same core material.

The first three-dimensional reconstructions of biogenic structures using CT images were produced by workers studying earthworm burrows (e.g., Joschko et al. 1991; Joschko et al. 1993; Capowiez et al. 1998). These reconstructions, generated by transforming CT data into simpler binary images, showed the layout and approximate dimensions of earthworm burrow systems but did not accurately reflect burrow volume (Dufour et al. 2005). Recently, techniques have been developed for calculating accurate burrow volumes and producing realistic three-dimensional reconstructions from CT data (e.g., Pierret et al. 2002; Bastardie et al. 2003; Capowiez et al. 2003; Dufour et al. 2005, 2007). Below we discuss the first of these studies to generate three-dimensional, volumetrically-accurate reconstructions of bioturbation in marine sediments.

6.3.3.1 Calculating Burrow Volume Using CT (Dufour et al. 2005)

Dufour et al. (2005) collected three sediment cores from a high intertidal zone in Québec and seven additional cores from marshes along the Gaspé Peninsula, Canada. All cores were 10 cm in diameter, with lengths ranging from 8 to 25 cm. After most of the water had drained out, the cores were CT-scanned continuously at 140 kV in successive 1-mm-thick slices. Three-dimensional reconstructions of the biogenic structures within each core were produced using OsiriX freeware and the CT scanner software. To specify which CT data reflected biogenic structures, the authors selected ranges of tomographic intensity values that correspond to soft organic matter and air. Tomographic intensity (TI) reflects the degree to which x-rays attenuate according to the density of the scanned material, relative to the density of water (Dufour et al. 2005). The software then constructed three dimensional maps using the distributions, within the CT scan slices, of these particular ranges of TI values.

In order to calculate the percentage of core volume comprised by biogenic structures and organic matter, Dufour and colleagues used two automated calculations programmed in MATLAB. In the first, all of the TI values for a single CT slice are plotted against the number of pixels in the slice. The result is two or more curves,

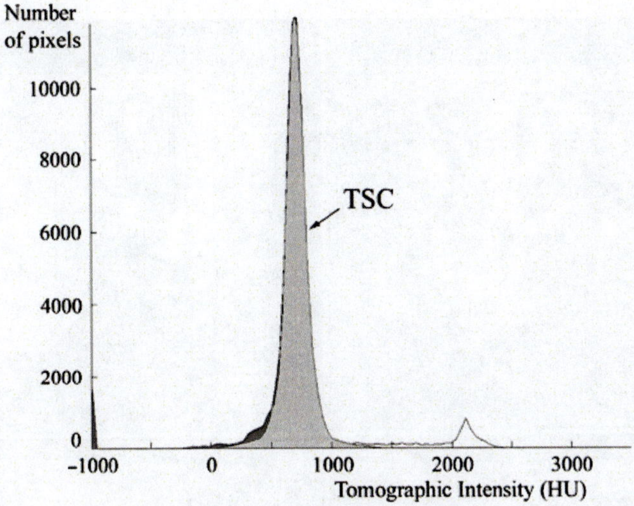

Fig. 6.4 Graph of tomographic intensity (TI) versus number of pixels for a 0.5 mm-thick CT slice from a sediment core collected in Québec, Canada, by Dufour et al. (2005). From left to right, successive peaks on the TI curve correspond to air, organic matter, sand, and gravel, respectively. The area under the portion of the TI curve that corresponds to air and organic matter is shaded in *dark gray*, and area under the theoretical sediment curve (TSC) is shaded in *light gray* (Graph is reproduced, with permission, from Dufour et al. (2005, Fig. 1, p. 375). Copyright (2005) by the American Society of Limnology and Oceanography, Inc.)

with approximately normal distributions, representing the different types of materials present in the slice (Fig. 6.4). A theoretical sediment curve (TSC), or the curve expected if sediment alone were present in the plane of the CT slice, is then superimposed on the data plot. The two peaks to the left of the TSC (with lower TI values and, therefore, densities) represent air-filled space and organic matter. The area under the portion of the TI curve that is located to the left of the TSC is, thus, proportional to the area of the two-dimensional slice occupied by materials less dense than sediment, which are most likely biological in origin (Fig. 6.4). By subtracting the area under the TSC from the area representing air and organic matter, and then dividing the result by the total area under the TI curve, an approximate percentage area occupied by organisms can be obtained. After this calculation is performed (automatically) for each CT slice, it is possible to determine the percentage of the total core volume that is made up of organisms and burrow structures. For comparison, a second calculation was performed to determine the distribution of air-filled space within the core. The proportion of TI values in the range representing air space was found for each CT slice and then plotted to show changes over the length of the core. A similar calculation was performed for organic matter in some core slices.

The three-dimensional reconstructions produced by Dufour and colleagues reveal the surfaces of burrows (most likely the sediment-burrow interface), including possible meiofaunal burrows, as well as the outlines of bivalves, eelgrass roots, and clumps of organic matter (Fig. 6.5). Upon dissecting the cores, the

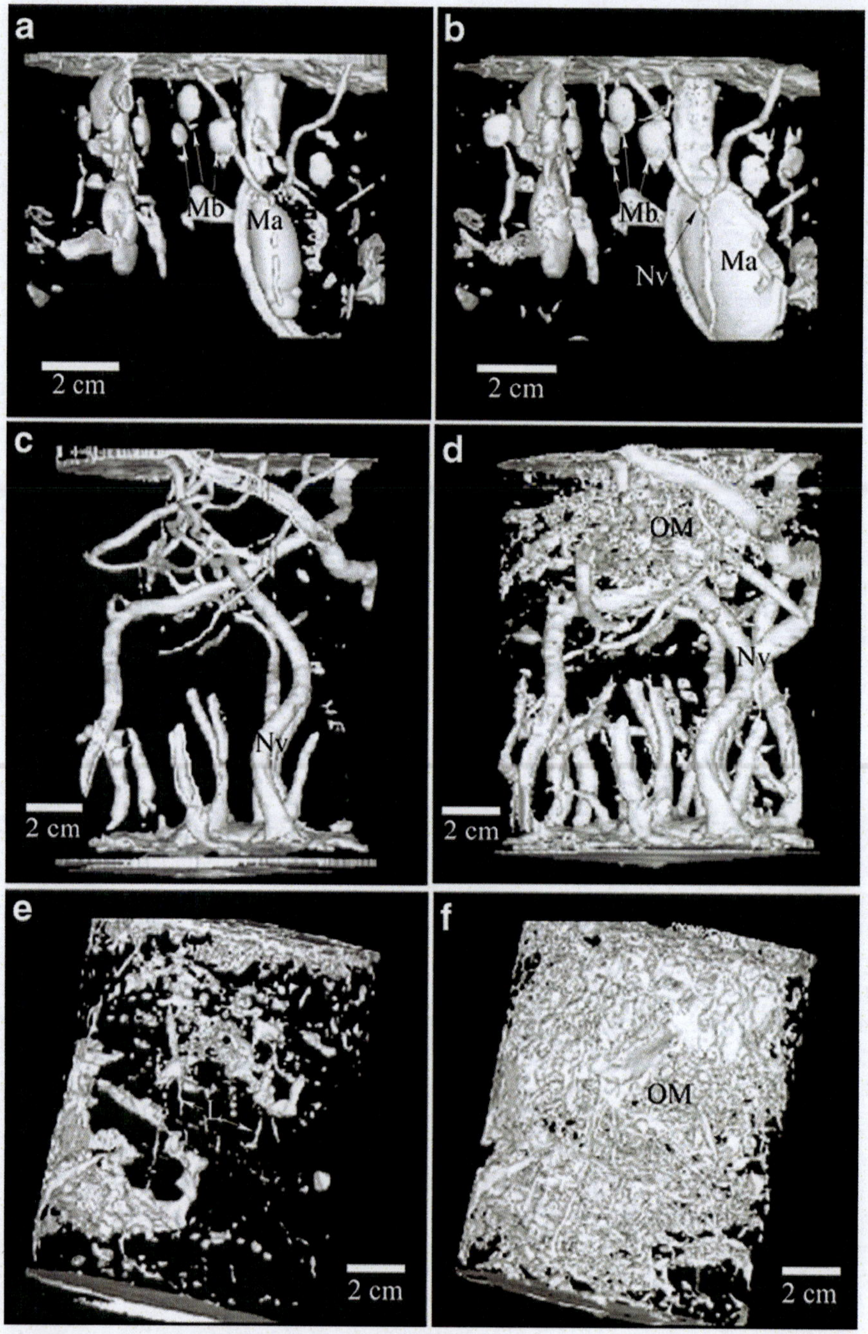

Fig. 6.5 Three-dimensional reconstructions of the space occupied by air (*left column*) and organic matter (*right column*) in three sediment cores collected by Dufour et al. (2005). **(a)** and **(b)** The soft parts of the bivalves *Mya arenaria* (Ma) and *Macoma balthica* (Mb) are visible, along with

authors identified the two bivalve species seen in outline (Fig. 6.5a, b) and discovered five nereid worms within burrows that are visible in the reconstructions (Fig. 6.5c, d). Using the volumetric data and reconstructions, Dufour and colleagues compared the intensities and types of bioturbation, as well as the organic matter content, among the studied cores.

In 2007, Dufour and colleagues revised their method for calculating burrow volume after determining that the TSC approach can lead to under- or overestimation depending on the composition of core sediment. The authors re-analyzed three of the original cores using a more complicated and time-intensive technique, which involves fitting a Gaussian curve to the TI peak generated by each material that is present in the sediment core. The results of these follow-up analyses, which display a narrower range of variation in burrow volume, appear to be more accurate than the original estimates (Dufour et al. 2007).

The ability to calculate the volume of a sample occupied by biogenic structures, without having to destroy the sample in the process, is ideal for studying and comparing ichnofabrics. In a recent study, Gagnoud et al. (2009) successfully applied the techniques developed by Dufour et al. (2005) to three sediment cores comprised of modern sediment and material deposited during the postglacial marine transgression of eastern Quebec (<12,500 years BP). Based on publications to date, these postglacial cores are the oldest material to be successfully analyzed using Dufour and colleagues' advanced CT scanning techniques. Calculating the volume of bioturbation in ancient rock samples may be possible using CT technology (Dufour et al. 2005; Gagnoud et al. 2009). A likely complicating factor for such analyses would be the greater range of possible sample densities resulting from varying degrees of lithification and diagenesis. In addition, burrow fill density would vary significantly from one sample to another, necessitating additional compensatory calculations for determining volumes of bioturbation. In very old rocks, the densities of the matrix and the burrow fill would be equal, or nearly so, precluding application of the technique altogether. These challenges aside, CT scanning is a promising avenue for future quantitative studies of ancient bioturbation, including Ediacaran and Cambrian ichnofabrics.

6.4 Summary

Trace fossils preserved in Ediacaran and lower Cambrian rocks are commonly simple, bedding-parallel, and shallow. At the same time, the Ediacaran-Cambrian transition is marked by a noticeable increase in the average depth, intensity, and complexity of

Fig. 6.5 (continued) a mucus-lined burrow constructed by *Nereis virens* (Nv) (**b**). (**c**) and (**d**) Burrows of *Nereis virens* (Nv) and a patch of organic matter (OM) (**d**). (**e**) and (**f**) Eelgrass (*Zostera marina*) roots constitute most of the organic matter (OM) in (**f**) and possibly some of the structures in (**e**) (Figure is reproduced, with permission, from Dufour et al. (2005, Fig. 2, p. 376). Copyright (2005) by the American Society of Limnology and Oceanography, Inc.)

bioturbation, leading to significant changes in the structure of benthic communities. Being able to quantify the bioturbation preserved in Ediacaran and Cambrian rocks is essential for studying the evolution of behavior and the impact of new behaviors on the substrate. A number of semi-quantitative and quantitative techniques have been developed for estimating the amount of bioturbation preserved on vertical outcrop exposures, rock and sediment core slabs, and bedding plane surfaces. These techniques are designed to be used in conjunction with descriptive paleontological, sedimentological, and ichnological methods rather than as stand-alone analyses.

Semi-quantitative methods for estimating quantities of bioturbation include ichnofabric indices (Droser and Bottjer 1986, 1989a, b; Bottjer and Droser 1991), the bioturbation index (Reineck 1963; Taylor and Goldring 1993), and bedding plane bioturbation indices (Miller and Smail 1997). These methods involve visual estimation of the percentage bioturbation within a portion of a vertical outcrop or bedding plane exposure and the assignment of a numerical score based on that estimate. The use of ordinal scales for assigning scores is a drawback of these methods because ordinal data have limited statistical utility. At the same time, semi-quantitative analysis is efficient and practical for field use.

Quantitative estimates of bioturbation intensity can be made using grid-based techniques (e.g., Heard and Pickering 2008; Heard et al. 2008; Marenco and Bottjer 2010), computer-aided image analysis (e.g., Francus 2001; Löwemark 2003), and advanced imaging technology (e.g., Dufour et al. 2005; Gagnoud et al. 2009). Grid-based techniques, such as the intersection grid method (Marenco and Bottjer 2010), require little specialized equipment and are well-suited for evaluating the bedding-parallel bioturbation that is common in Ediacaran and lower Cambrian rocks. The computer-aided image analysis methods developed by Francus (2001) and Löwemark (2003) are designed for specialized applications. However, Löwemark's method may be useful for the examination of high-contrast ichnofabrics in Cambrian rocks. Using CT scanning to visualize ichnofabrics in three dimensions and calculate the total volume of biogenic structures is an exciting new technique that has not yet been applied to lithified material (Dufour et al. 2005; Gagnoud et al. 2009). We are optimistic about the potential for using this technology to study ancient ichnofabrics in greater detail.

Acknowledgments The development of the intersection grid method was supported by grants from the Geological Society of America, the International Association of Sedimentologists, and the University of Southern California. We are grateful to Murray Gingras and Marc Laflamme for their insightful comments on an earlier version of this manuscript. Thanks to Pedro Marenco, the USC Geobiology Group, and the attendees of the Ninth International Ichnofabric Workshop for helpful discussions.

References

Ausich WI, Bottjer DJ (1982) Tiering in suspension-feeding communities on soft substrata throughout the Phanerozoic. Science 216(4542):173–174

Bastardie F, Capowiez Y, de Dreuzy J-R, Cluzeau D (2003) X-ray tomographic and hydraulic characterization of burrowing by three earthworm species in repacked soil cores. Appl Soil Ecol 24:3–16

Bottjer DJ, Ausich WI (1986) Phanerozoic development of tiering in soft substrata suspension-feeding communities. Paleobiology 12(4):400–420

Bottjer DJ, Droser ML (1991) Ichnofabric and basin analysis. Palaios 6(3):199–205

Bottjer DJ, Hagadorn JW, Dornbos SQ (2000) The Cambrian substrate revolution. GSA Today 10(9):1–7

Bouma AH (1964) Notes on X-ray interpretation of marine sediments. Mar Geol 2(4):278–309

Brasier MD, McIlroy D (1998) *Neonereites uniserialis* from *c*. 600 Ma year old rocks in western Scotland and the emergence of animals. J Geol Soc Lond 155:5–12

Budd GE, Jensen S (2000) A critical reappraisal of the fossil record of the bilaterian phyla. Biol Rev 75:253–295

Burstall L, Harris PM (1983) The estimation of percentage light interception from leaf area index and percentage ground cover in potatoes. J Agric Sci Camb 100:241–244

Canfield DE, Farquhar J (2009) Animal evolution, bioturbation, and the sulfate concentration of the oceans. Proc Natl Acad Sci 106(20):8123–8127

Capowiez Y, Pierret A, Daniel O, Monestiez P, Kretzchmar A (1998) 3D skeleton reconstructions of natural earthworm burrow systems using CAT scan images of soil cores. Biol Fertil Soils 27:51–59

Capowiez Y, Pierret A, Moran CJ (2003) Characterisation of the three-dimensional structure of earthworm burrow systems using image analysis and mathematical morphology. Biol Fertil Soils 38:301–310

Droser ML (1987) Trends in extent and depth of bioturbation in Great Basin Precambrian-Ordovician strata Nevada and Utah, Unpublished dissertation. University of Southern California, Los Angeles

Droser ML, Bottjer DJ (1986) A semiquantitative field classification of ichnofabric. J Sed Petrol 56:558–559

Droser ML, Bottjer DJ (1988) Trends in extent and depth of bioturbation in Cambrian carbonate marine environments, western United States. Geology 16:233–236

Droser ML, Bottjer DJ (1989a) Ordovician increase in extent and depth of bioturbation: implications for understanding early Paleozoic ecospace utilization. Geology 17:850–852

Droser ML, Bottjer DJ (1989b) Ichnofabric of sandstones deposited in high-energy nearshore environments: measurement and utilization. Palaios 4:598–604

Droser ML, Bottjer DJ (1991) Trace fossils and ichnofabric in Leg 119 cores. Proc ODP Sci Results 119:635–641

Droser ML, Bottjer DJ (1993) Trends and patterns of Phanerozoic ichnofabrics. Annu Rev Earth Planet Sci 21:205–225

Droser ML, Gehling JG, Jensen S (1999) When the worm turned: concordance of Early Cambrian ichnofabric and trace-fossil record in siliciclastic rocks of South Australia. Geology 27:625–628

Droser ML, Gehling JG, Jensen S (2005) Ediacaran trace fossils: true and false. In: Briggs DEG (ed) Evolving form and function: fossils and development. Peabody Museum of Natural History, Yale University, New Haven, pp 125–138

Dufour SC, Desrosiers G, Long B, Lajeunesse P, Gagnoud M, Labrie J, Archambault P, Stora G (2005) A new method for three-dimensional visualization and quantification of biogenic structures in aquatic sediments using axial tomodensitometry. Limnol Oceanogr Meth 3:372–380

Dufour SC, Desrosiers G, Long B, Lajeunesse P, Gagnoud M, Labrie J, Archambault P, Stora G (2007) Update: A new method for three-dimensional visualization and quantification of biogenic structures in aquatic sediments using axial tomodensitometry. Limnol Oceanogr Methods. http://www.aslo.org/lomethods/free/2007/2005_0372u.pdf. Cited 29 Jan 2010

Ekdale AA, Mason TR (1988) Characteristic trace-fossil associations in oxygen-poor sedimentary environments. Geology 16:720–733

Floyd DA, Anderson JE (1982) A new point interception frame for estimating cover of vegetation. Vegetatio 50(3):185–186

Floyd DA, Anderson JE (1987) A comparison of three methods for estimating plant cover. J Ecol 75:221–228

Francus P (2001) Quantification of bioturbation in hemipelagic sediments via thin-section image analysis. J Sed Res 71(3):501–507

Frey RW (1978) Behavioral and ecological implications of trace fossils. In: Basan PB (ed) Trace fossil concepts, SEPM short course no. 5. Society of Economic Paleontologists and Mineralogists, Tulsa, Oklahoma, pp 43–66

Fu S, Werner F, Brossmann J (1994) Computed tomography: application in studying biogenic structures in sediment cores. Palaios 9:116–119

Gagnoud M, Lajeunesse P, Desrosiers G, Long B, Dufour S, Labrie J, Mermillod-Blondin F, Stora G (2009) Litho- and biofacies analysis of postglacial marine mud using CT-scanning. Eng Geol 103:106–111

Getty P (2005) Excavated and in situ dinosaur footprints from the Murray Quarry (Early Jurassic East Berlin Formation), Holyoke, Massachusetts, USA. Ichnos 12(3):163–178

Gingras MK, MacMillan B, Balcom BJ (2002a) Visualizing the internal physical characteristics of carbonate sediments with magnetic resonance imaging and petrography. Bull Can Petrol Geol 50(3):363–369

Gingras MK, MacMillan B, Balcom BJ, Saunders T, Pemberton SG (2002b) Using magnetic resonance imaging and petrographic techniques to understand the textural attributes and porosity distribution in *Macaronichnus*-burrowed sandstone. J Sed Res 72(4):552–558

Goodall DW (1952) Some considerations in the use of point quadrats for the analysis of vegetation. Aust J Sci Res 5(1):1–41

Greig-Smith P (1983) Quantitative Plant Ecology, 3rd edn. University of California Press, Berkeley

Grimm KA, Lange CB, Gill AS (1996) Biological forcing of hemipelagic sedimentary laminae: evidence from ODP site 893, Santa Barbara Basin, California. J Sed Res 66(3):613–624

Hamblin WK (1971) X-ray photography. In: Carver RE (ed) Procedures in sedimentary petrology. Wiley and Sons, New York, pp 251–284

Heard TG, Pickering KT (2008) Trace fossils as diagnostic indicators of deep-marine environments Middle Eocene Ainsa-Jaca basin, Spanish Pyrenees. Sedimentology 55:809–844

Heard TG, Pickering KT, Robinson SA (2008) Milankovitch forcing of bioturbation intensity in deep-marine thin-bedded siliciclastic turbidites. Earth Planet Sci Lett 272:130–138

Howard JD (1969) X-ray radiography for examination of burrowing in sediments by marine invertebrate organisms. Sedimentology 11:249–258

Howard JD (1975) The sedimentological significance of trace fossils. In: Frey RW (ed) The study of trace fossils. Springer-Verlag, New York, pp 131–146

Howard JD, Frey RW (1975) Regional animal-sediment characteristics of Georgia estuaries. Senckenb Marit 4:81–123

Howard JD, Reineck H (1972) Physical and biogenic sedimentary structures of the nearshore shelf. Senckenb Marit 4:47–79

Jager JC, Looman CWN (1995) Data collection. In: Jongman RHG, ter Braak CJF, van Tongeren OFR (eds) Data analysis in community and landscape ecology. Cambridge University Press, Cambridge, pp 10–27

Jensen S (2003) The Proterozoic and Earliest Cambrian trace fossil record; patterns, problems and perspectives. Integr Comp Biol 43:219–228

Jensen S, Droser ML, Gehling JG (2005) Trace fossil preservation and the early evolution of animals. Palaeogeogr Palaeoclimatol Palaeoecol 220:19–29

Jensen S, Droser ML, Gehling JG (2006) A critical look at the Ediacaran trace fossil record. In: Xiao S, Kaufman AJ (eds) Neoproterozoic geobiology and paleobiology. Springer, Dordrecht, pp 115–157

Joschko M, Graff O, Muller PC, Kotzke K, Lindner P, Pretschner DP, Larink O (1991) A nondestructive method for the morphological assessment of earthworm burrow system in three dimensions by X-ray computed tomography. Biol Fertil Soils 11:88–92

Joschko M, Muller PC, Kotzke K, Dohring W, Larink O (1993) Earthworm burrow system development assessed by means of X-ray computed tomography. Geoderma 56:209–221

Korva JT (1996) Grids in ground cover measurements. Potato Res 39(4):533–540

Löwemark L (2003) Automatic image analysis of x-ray radiographs: a new method for ichnofabric evaluation. Deep Sea Res I 50:815–827

MacEachern JA, Bann KL (2008) The role of ichnology in refining shallow marine facies models. In: Hampson GJ, Steel RJ, Burgess PM, Dalrymple RW (eds) Recent advances in models of siliciclastic shallow-marine stratigraphy, SEPM special publication no. 90. Society for Sedimentary Geology, Tulsa, pp 73–116

Magwood JPA, Ekdale AA (1994) Computer-aided analysis of visually complex ichnofabrics in deep-sea sediments. Palaios 9:102–115

Marenco KN, Bottjer DJ (2008) The importance of *Planolites* in the Cambrian substrate revolution. Palaeogeogr Palaeoclimatol Palaeoecol 258(3):189–199

Marenco KN, Bottjer DJ (2010) The intersection grid technique for quantifying the extent of bioturbation on bedding planes. Palaios 25:457–462

Martin MW, Grazhdankin DV, Bowring SA, Evans DAD, Fedonkin MA, Kirschvink JL (2000) Age of Neoproterozoic bilaterian body and trace fossils White Sea, Russia: implications for metazoan evolution. Science 288:841–845

McIlroy D (2004) Some ichnological concepts, methodologies, applications and frontiers. In: McIlroy D (ed) The application of ichnology to palaeoenvironmental and stratigraphic analysis, special publication 228. The Geological Society of London, Bath, p 490

McIlroy D, Logan GA (1999) The impact of bioturbation on infaunal ecology and evolution during the Proterozoic-Cambrian transition. Palaios 14(1):58–72

Mermillod-Blondin F, Marie S, Desrosiers G, Long B, de Montety L, Michaud E, Stora G (2003) Assessment of the spatial variability of intertidal benthic communities by axial tomodensitometry: importance of fine-scale heterogeneity. J Exp Mar Biol Ecol 287(2):193–208

Mezga A, Bajraktarević Z (1999) Cenomanian dinosaur tracks on the islet of Fenoliga in southern Istria, Croatia. Cretaceous Res 20(6):735–746

Miller MF (1977) Middle and Upper Ordovician biogenic structures and paleoenvironments, southern Nevada. J Sed Petrol 47:1328–1338

Miller MF, Smail SE (1997) A semiquantitative field method for evaluating bioturbation on bedding planes. Palaios 12:391–396

Moore HB, Scrutton PC (1957) Minor intertidal structures of some recent unconsolidated sediments. AAPG Bull 41:2723–2751

O'Brien NR (1987) The effects of bioturbation on the fabric of shale. J Sed Petrol 57(3):449–455

Pemberton SG, Frey RW (1984) Quantitative methods in ichnology: spatial distribution among populations. Lethaia 17:33–49

Pickett JW (1972) The ecology of worm populations in the Erins Vale Formation (Late Permian) southern Sydney Basin. J Geol Soc Aust 19:313–320

Pierret A, Capowiez Y, Belzunces L, Moran CJ (2002) 3D reconstruction and quantification of macropores using X-ray computed tomography and image analysis. Geoderma 106:247–271

Reineck H-E (1963) Sedimentgefüge im Bereich der südlichen Nordsee. Abh Senckenbergischen Naturforschenden Ges 505:1–138

Reineck H-E (1967) Parameter von Schichtung und bioturbation. Geol Rundsch 56:420–438

Rhoads DC, Boyer LF (1982) The effects of marine benthos on physical properties of sediments: a successional perspective. In: Tevesz JS, McCall PL (eds) Animal-sediment relations. Plenum, New York, pp 3–52

Savrda CE, Ozalas K (1993) Preservation of mixed-layer ichnofabrics in oxygenation-event beds. Palaios 8:609–613

Savrda CE, Bottjer DJ, Gorsline DS (1984) Development of a comprehensive oxygen-deficient marine biofacies model: evidence from Santa Monica San Pedro and Santa Barbara Basins, California continental borderland. AAPG Bull 68:1179–1192

Seilacher A (1955) Ecological significance of fossil tracks and trails. GSA Bull 66(12):1663

Seilacher A (1956) Der Beginn des Kambriums als biologische Wende. Neues Jb Geol Paläontol Abh 103:155–180

Seilacher A (1999) Biomat-related lifestyles in the Precambrian. Palaios 14(1):86–93

Seilacher A, Pflüger F (1994) From biomats to benthic agriculture: A biohistoric revolution. In: Krumbein WE, Paterson DM, Stal LJ (eds) Biostabilization of Sediments. Bibliotheks und Informationssystem der Carl von Ossietzky Universität, Oldenburg, pp 97–105

Seilacher A, Grazhdankin D, Legouta A (2003) Ediacaran biota: The dawn of animal life in the shadow of giant protists. Paleontol Res 7(1):43–54

Seilacher A, Buatois LA, Mángano MG (2005) Trace fossils in the Ediacaran–Cambrian transition: Behavioral diversification, ecological turnover and environmental shift. Palaeogeogr Palaeoclimatol Palaeoecol 227:323–356

Stevens SS (1946) On the theory of scales of measurement. Science 103:677–680

Swanson BA, Carlson KJ (2002) Walk, wade, or swim? Vertebrate traces on an Early Permian lakeshore. Palaios 17(2):123–133

Taylor A, Goldring R (1993) Description and analysis of bioturbation and ichnofabric. J Geol Soc Lond 150(1):141–148

Taylor A, Goldring R, Gowland S (2003) Analysis and application of ichnofabrics. Earth Sci Rev 60:227–259

Valentine JW (1995) Late Precambrian bilaterians: grades and clades. In: Fitch WM, Ayala FJ (eds) Tempo and mode in evolution: genetics and paleontology 50 years after Simpson. National Academy Press, Washington, DC, pp 87–107

Velleman PF, Wilkinson L (1993) Nominal, ordinal, interval, and ratio typologies are misleading. Am Stat 47(1):65–72

Chapter 7
Assessing the Role of Skeletons in Early Paleozoic Carbonate Production: Insights from Cambro-Ordovician Strata, Western Newfoundland

Sara B. Pruss and Hannah Clemente

Contents

Abstract The trend in the abundance of skeletal organisms does not exhibit a monotonic rise beginning with the Cambrian Explosion, but rather has oscillated through major biodiversification events and mass extinctions. With the appearance of several skeletonized clades in the early Cambrian, organisms began exerting some control on carbonate production, but it is not until later in the Paleozoic that those organisms account for a significant proportion of total carbonate production. Here, it is proposed that changing abundances of skeletal organisms can be quantified by examining their relative contributions to sedimentary rock fabrics. This method provides a quantitative way to determine changes in skeletal contribution to carbonates through time, for example, to track the expansion of skeletal organisms from the Cambrian through the Ordovician radiation. Analysis of carbonates from

S.B. Pruss and H. Clemente (✉)
Department of Geosciences, Clark Science Center, Smith College, Northampton, MA 01063, USA
e-mail: spruss@smith.edu; hclement@smith.edu

M. Laflamme et al. (eds.), *Quantifying the Evolution of Early Life*, Topics in Geobiology 36, DOI 10.1007/978-94-007-0680-4_7, © Springer Science+Business Media B.V. 2011

Cambro-Ordovician strata of western Newfoundland demonstrates that the skeletal contribution to Cambrian carbonates is low compared to their Middle Ordovician counterparts. The point count data reported here thus reveal that important ecological changes, such as the increase in carbonate skeletal abundance through the Early Paleozoic, can be gleaned by this approach.

Keywords Port au Port Group • St. George Group • Table Head Group • Point-count • Ordovician radiation

7.1 Introduction

Traditional methods used to quantify fossil abundances include collecting and/or analyzing existing fossil abundance data (e.g., Novack-Gottshall and Miller 2003; Clapham et al. 2006; many others), as well as counting shell bed thicknesses and abundances through stratigraphic intervals (e.g., Kidwell 1991; Kidwell and Brenchley 1994; Li and Droser 1997, 1999). Point counting, a method that involves counting and identifying components in thin section within a grid, involves examining not only the abundance of fossils but also the relative contribution of all carbonate types, including skeletal carbonate, to carbonate production. During intervals such as the Cambrian, when shelly taxa were generally smaller and less robust than those of the later Paleozoic (e.g., Li and Droser 1997; Knoll 2003), point counting thin sections provides a particularly useful way to examine and quantify fossil material that may not be apparent in outcrop.

Like any paleobiological method, point counting analyses have both advantages and limitations that must be considered when employed. Whereas many studies examine relative abundances using existing data, point counting has the potential to measure absolute abundances of fossil groups within a particular stratigraphic interval or environment (e.g., Flügel 1982, 2004; Carozzi 1961). Furthermore, point counting quantifies all material that comprises a carbonate rock, so other patterns, such as changes in the contribution of microbial grains and ooids to total carbonate, can also be gleaned. Finally, organisms that disarticulate and/or are very small may be missed with studies that focus on bedding plane analyses; this small material is less likely to be overlooked in thin section (Jaanusson 1972). As in most paleobiological studies, samples that have been altered by diagenesis, including fabric-destructive dolomitization must, in general, be excluded from analysis. Also, many, although not all, fossils preserved in thin section cannot be identified beyond the class-level, so diversity patterns are often only coarsely culled relative to paleoecological analyses based on counting specimens. Consequently, it is important to be clear about what patterns are being evaluated by the point counting methodology. While there are limitations to the point-counting method, it has been demonstrated that when other methodologies are used in concert with point counting, such as an equal-area approach, there is a general agreement of the data (Jacobsen et al. in press).

Previous work has demonstrated the utility of point counting thin sections in studies ranging from characterization of microfacies (e.g., Carozzi 1961; Walpole and Carozzi 1961; Flügel 1982, 2004; and many others) to understanding ecological changes in reefs (e.g., Rowland et al. 2008). Since point counting thin sections provides a way to quantify the changing contributions of various components to carbonate rocks, it has also been used to assess changing skeletal contribution through major ecological changes, such as during the Permo-Triassic extinction and recovery (Payne et al. 2006).

The goal of this chapter is to review existing work (Pruss et al. 2010) and contribute new data on the changing contribution of mineralized skeletons to carbonate deposits from the Series 3 Cambrian strata to the Ordovician diversification of heavily skeletonized marine organisms in western Newfoundland. Organisms with thin carbonate skeletons first appeared in late Neoproterozoic time (Grotzinger et al. 2000), but the major initial diversification of skeletonized taxa occurred in the Terreneuvian (early Cambrian). This radiation of skeletonized animals may have affected the global carbon cycle (e.g., Zeebe and Westbroek 2003; Maloof et al. 2005; Ridgwell and Zeebe 2005; Higgins et al. 2009), perhaps by shifting the locus of carbonate production from abiotic or microbially-mediated precipitation from the water column (Knoll and Swett 1990) to biologically-mediated skeletal production on the shelf; however, Series 3 Cambrian through the Lower Ordovician intervals are often recognized as a time of low skeletal abundance (e.g., Cowan and James 1993; Rowland and Shapiro 2002; Pruss et al. 2010) following the early Cambrian extinction of heavily calcifying archaeocyathids (Zhuravlev and Wood 1996). Previous observations made on Cambro-Ordovician shell beds in the western United States (Li and Droser 1997, 1999) suggest that the diversity and abundance of skeletal organisms as well as the thickness of shell beds was low relative to later Ordovician shell accumulations. The increase in skeletal abundance that occurs during the Middle Ordovician appears to be rapid and involves several clades (e.g., Droser and Finnegan 2003). The coeval diversification of skeletal taxa in distantly related groups (e.g., calcareous algae, brachiopods, echinoderms, etc.) suggests the influence of extrinsic drivers, either ecological or environmental (Knoll 2003).

This study involved the examination of facies through point-counts of representative thin sections from western Newfoundland and using these data to provide estimates of proportional volumes of skeletal carbonate in measured stratigraphic sections. These results complement those generated by Li and Droser (1997, 1999) for the Basin and Range Province; however, these data were collected from a very different region and tectonic setting and encompass multiple volumetrically significant facies including, but not limited to, shell beds. These results support the notion that the skeletal contribution to shallow subtidal carbonate sediments did increase between the Furongian and Middle Ordovician. Ultimately, quantifying the changing skeletal abundance during Cambro–Ordovician time in thin section may help to illuminate previously identified changes in the nature of the carbon cycle of the Early Paleozoic while demonstrating the utility of the point counting methodology.

7.2 Skeletons and the Early Radiation of Life

While it has been suggested by many workers familiar with the early Paleozoic that Cambrian carbonates are less fossiliferous than those that postdate the Ordovician radiation (e.g., Cowan and James 1993; Rowland and Shapiro 2002), few studies have quantitatively assessed these trends. In previous work, Li and Droser (1997, 1999) examined the nature of shell beds from the Cambrian through the Middle Ordovician in the western United States by quantifying the changes in the thickness, composition and abundance of shell beds. In the Cambrian, shell beds are generally thin and lenticular (1–20 cm), less abundant per total section examined and dominated by small shelly taxa and trilobites, organisms that generally do not secrete robust calcareous skeletons (Li and Droser 1997). Middle Ordovician shell beds show a markedly different pattern. Shell beds are typically polytoxic, thicker, and more abundant per section examined (Li and Droser 1999). These studies demonstrated that the dynamics of the Ordovician radiation not only involved diversification of marine organisms but directly impacted the formation and distribution of shells beds; furthermore, the biomass of skeletal organisms dramatically changed through this event (Li and Droser 1997, 1999).

Analyses of reefs reveal a comparable pattern from the Terreneuvian through the Lower Ordovician. The first appearance of animal reefs occurs within the first 20 million years of the Cambrian. Within these reef ecosystems, sponges act as framework builders and can be quantitatively important carbonate producers (e.g., Rowland and Shapiro 2002; Rowland et al. 2008). After the decimation of archaeocyathid reefs during a Cambrian extinction event (e.g., Zhuravlev and Wood 1996; Rowland and Shapiro 2002), microbial build-ups become prevalent in reef environments through the Early Ordovician. Fossil material can be incorporated into these microbial build-ups, but organisms, such as eocrinoids, lived in these settings as bafflers rather than framework builders (e.g., Bambach 1983). It is not until after the Middle Ordovician radiation that animals again become ecologically important in reef settings.

7.3 Methods

We analyzed Cambro–Ordovician carbonate strata of western Newfoundland (Fig. 7.1) by examining the contribution of skeletal carbonate to total carbonate production in thin section and applying these data to the volume of lithofacies preserved in outcrop. For a variety of units in Newfoundland, thin sections of representative peritidal-to-shallow subtidal carbonate facies were examined to determine how much skeletal material contributed to each carbonate lithology. As in most paleobiological and geochemical analyses, heavily dolomitized samples were excluded (see additional discussion of this in the following section). Mudstones, wackestones, packstones, skeletal grainstones, oolites, and microbialites

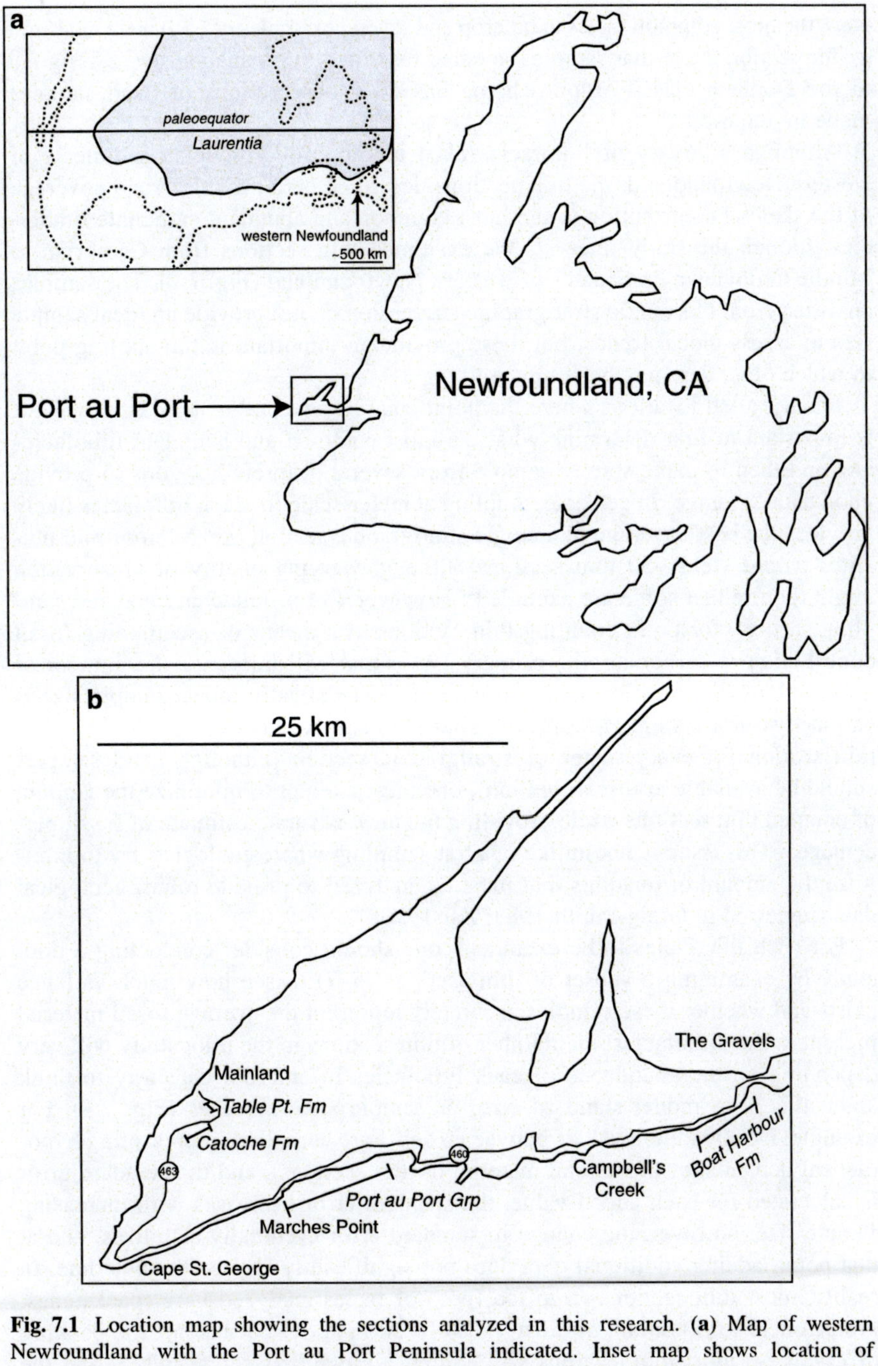

Fig. 7.1 Location map showing the sections analyzed in this research. **(a)** Map of western Newfoundland with the Port au Port Peninsula indicated. Inset map shows location of western Newfoundland on Laurentia during Cambro-Ordovician time (Modified from Pruss et al. 2010). **(b)** Locations on Port au Port Peninsula where sections were sampled. Formation names show were these units were sampled

were the most common facies in outcrop and several examples of each were analyzed in thin section. Note that samples selected for thin section analysis were assigned to the Dunham classification scheme based on observations of hand samples made in outcrop.

Although a variety of lithofacies exist in Cambro-Ordovician sediments of western Newfoundland, the thin sections described here provide broad coverage of the skeletal contribution to the most common and abundant carbonate lithofacies through the study interval. We examined thin sections from Cambrian to Middle Ordovician carbonates of western Newfoundland (Fig. 7.2). The samples analyzed from this single stratigraphic succession do not provide an ideal sample size to assess global trends, but these provide an important initial starting point to which other day may be compared.

Ideally, at all locations where the point counting method is to be employed, it is important to first determine what the most common and abundant lithofacies are, and then to make sure to sample from several different horizons to provide good data coverage. In general, sampling at meter-scale for each lithofacies likely provides the best coverage of a single stratigraphic section (i.e. 50 m of microbialites would yield ~50 thin sections, although varying quality of preservation might require that some are excluded). However, to formulate an idealized sampling strategy for point counting thin sections as a means of ascertaining fossil contribution to carbonate, the primary factor that will influence the number of thin sections that must be analyzed to attain a statistically robust sample size is variance within a single lithofacies. That is to say, in order to avoid counting a thin section for every meter of stratigraphic section (although hand samples should be available at this resolution), one must attempt to minimize the amount of counted thin sections while providing the most accurate estimate of fossil percentages. This issue is not unlike quadrat sampling where ecologists try to minimize the amount of quadrats that must be analyzed to provide robust ecological data (reviewed in Greig-Smith 1983; Zar 1996).

For each lithofacies to be examined, one should consider conducting a pilot study by examining a subset of thin sections (5–7) to see how much variance exists and whether these samples accurately represent the average fossil material present in this lithofacies; the number of thin sections in the pilot study will vary depending on the abundance of each lithofacies in outcrop. One way to think about this is to model standard error as sample size increases (Fig. 7.3). For example, if 500 random values between 0 and 1 are compiled (representing hypothetical percentages of skeletal material of thin sections), and the standard error is calculated for each added value, the standard error decreases with increasing sample size; however, the change in standard error eventually stabilizes, and at that point, adding additional data does not significantly change the outcome. In reality, most thin sections from the types of facies analyzed here (packstones, grainstones, etc.) contain at most ~50% fossil material because of the packing of grains, so most thin sections will exhibit a range between 0 and .5, and the variance of these samples will be smaller than those in Fig. 7.3. Nonetheless, this provides a potentially useful framework for developing a sampling scheme.

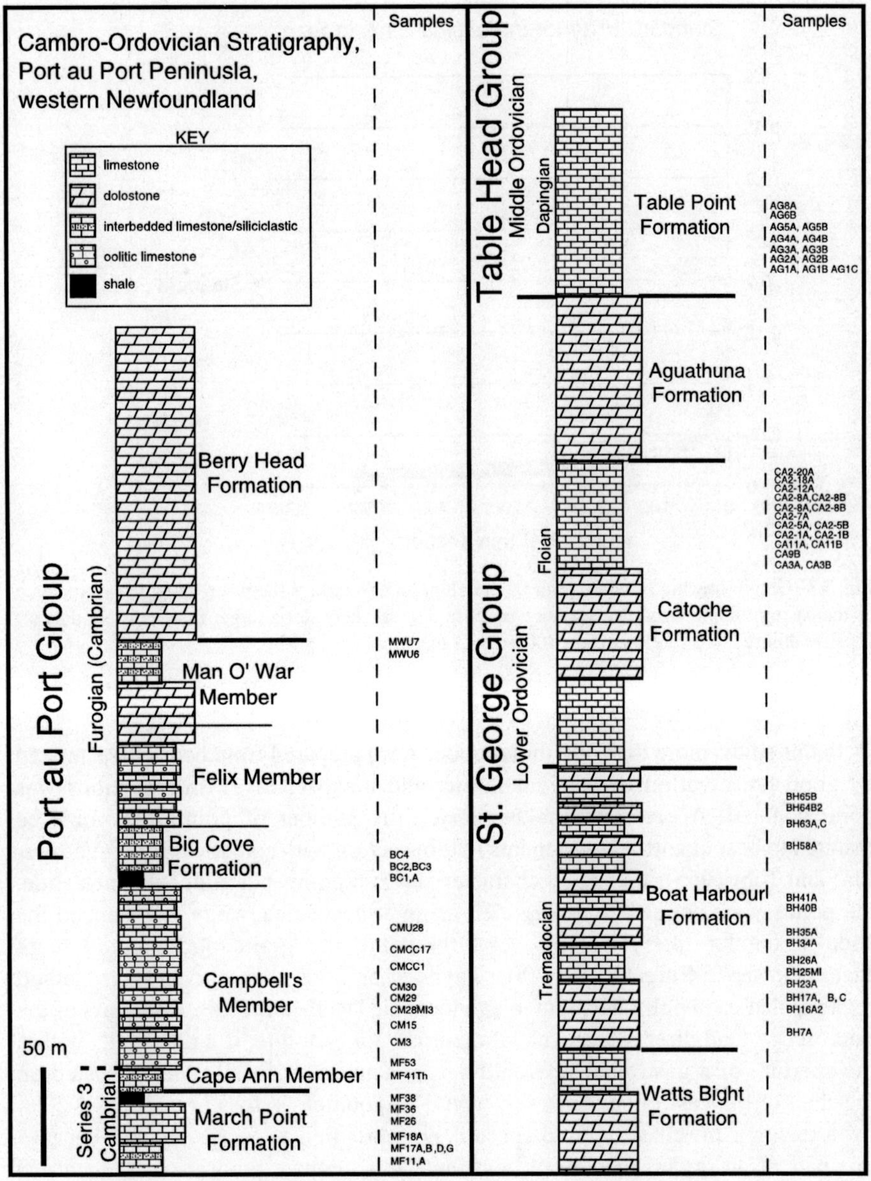

Fig. 7.2 *Stratigraphic column* showing Cambrian and Ordovician stratigraphy of western New foundland. Note all samples are located on the column and correspond to samples detailed in Table 7.1

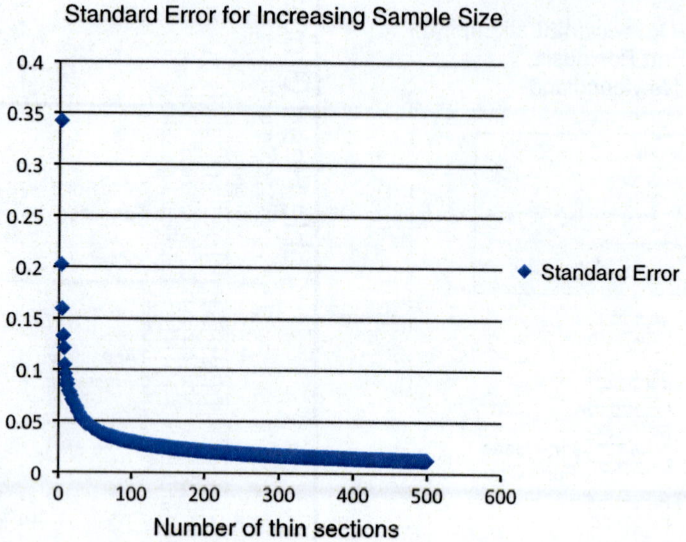

Fig. 7.3 Graph showing standard error of a random set of numbers (between 0 and 1) as a function of increasing sample size. The variance in these data is likely to be larger than any real data set because thin sections rarely contain 100% fossil material

In this study, more than 500 thin sections were prepared from beds characterized by good preservation of original fabric, and a subset of 72 thin sections was point counted. A previous analysis tested the number of points that must be counted to best quantify components in thin section with statistical rigor (Van Der Plas and Tobi 1965); this study characterized 200 points per slide. On each slide, the points were counted and classified using a mechanical stage that stepped the slide at regular intervals (Fig. 7.4), following the grain-solid method (e.g., Jaanusson 1972; Flügel 1982, 2004; Payne et al. 2006). The grain-solid method requires that each point be counted and identified as fossil if the cross-hairs of the microscope land directly on skeletal material; for example, if a point falls within the aperture of a gastropod, it would not be counted as fossil. Points located on the shell of the gastropod, however, would be counted as fossil (see Flügel 2004 for a review). In this way, void space is not classified as fossil material because this part of an organism did not contribute to carbonate production. Cement- or mud-filled void spaces and syntaxial overgrowths extending beyond the fossilized material are considered separately (Flügel 1982, 2004) (Fig. 7.4). Each point is identified until 200 points have been counted (see Fig. 7.4 for examples of indentified points). Fossilized material that cannot be classified, either because it is too small, broken, or recrystallized, is placed in the "unidentified fossil" category to quantify the total amount of fossil material even in the absence of more specific identifications. The benefits of point counting are that it provides a quantitative assessment of all carbonate phases, including but not limited to fossil material and generates data that allow one to compare thin sections to each other within facies,

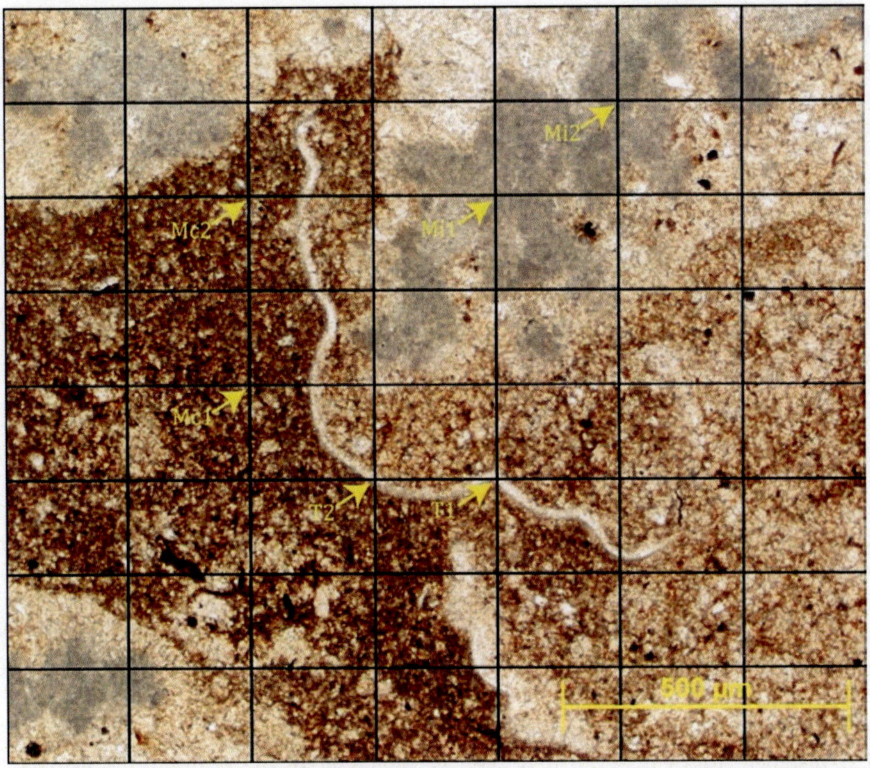

Fig. 7.4 Photomicrograph showing *grid* on a thin section that represents the mechanical stage that is stepped at regular intervals during point counting analysis. Mc1 and Mc2 are points that would be characterized as micrite, T1 and T2 as trilobite fossil material, and Mi1 and Mi2 as microbial textures

between facies, and through time. The major goal of this research was to assess the contribution of fossil carbonate to total carbonate production of individual facies through the Cambro-Ordovician interval. Other valuable data that can be gleaned from this type of work, although not the focus of this study, are ecological changes reflected in the types of fossils preserved as well as assessing changes in the total carbonate components of rocks, a feature not easily quantified in outcrop alone.

Although the point counting method provides information about the importance of skeletons within individual carbonate facies, it is also valuable to consider the relative contributions of different facies to overall carbonate depositional packages as they changed through time. To estimate changes in skeletal contributions at the outcrop scale, we measured the thickness of each facies type in the Series 3 (Cambrian) to Middle Ordovician succession in western Newfoundland. Average proportions of skeletal material, generated by point counting thin sections (discussed above), were multiplied by the thickness of each facies in outcrop and these products were summed and divided by the total thickness to obtain a thickness-normalized estimate of skeletal carbonate contribution at the stratigraphic section scale (e.g., Payne et al. 2006; Pruss et al. 2010). For example, in a 20 m section consisting of 10 m of wackestone with a mean skeletal content of 5% and 10 m of grainstone with mean skeletal content of 15%, the estimate of mean skeletal content for the entire section would be:

$$\left[(10\times.05)+(10\times.15)\right]\div 20=.1$$

7.4 Biases in Thin Section Analysis

Thin sections offer insights into the fine-scale make up of facies, and if preservation is good, they provide details of carbonate rocks not necessarily visible at the outcrop scale. For example, thin or diminutive fossils that may go unnoticed in outcrop are easily viewed in thin section. Also, poorly exposed outcrop (or outcrop that is partially covered by lichen or other plants) can still be analyzed in thin section, which is another advantage. However, point counting data are influenced by certain biases that must be addressed in the final analysis. Firstly, thin sections generally capture only small portions of a given facies and therefore a single thin section may provide a biased assessment of the contribution of carbonate components to a single bed, particularly if significant lateral variability exists. Prevalence of fossil material in certain areas of the unit may be expected. For instance, fossil material may well be abundant in the flanking grainstones surrounding microbial mounds but relatively absent from mounds themselves. To combat this issue, multiple thin sections can be taken from a single bed, particularly one that appears to vary laterally. This may ultimately be the most valuable way to assess local heterogeneity. Nonetheless, thin sections still provide only limited coverage and must be used in concert with good field observations. Secondly, variation exists between similar facies: not all grainstones are identical to each other, so one must generate enough data to capture the variability of such facies, as described above.

Encountering diagenetically altered samples is inevitable, especially in carbonate strata. Some samples will be diagenetically altered to the point where they cannot be meaningfully analyzed, as in the case of heavily dolomitized samples. In other cases, dolomite rhombs are scattered throughout a slide but much of the original fabric is retained; in these examples, the dolomite rhombs can be counted the same way that spar and other cements are quantified. In the case of heavily dolomitized samples, either unaltered portions of the samples should be point counted or the entire sample should be excluded from analyses. This is not a trivial problem, as dolomitization can be quite widespread and require that an entire stratigraphic interval be excluded from analysis (e.g., samples taken from the Watts Bight Formation). As in most paleobiological studies, working in areas where diagenetic alteration is limited is best for quantitative thin section analyses.

It is not clear from the analysis of thin sections from western Newfoundland whether dolomitization selectively affects fossiliferous material or facies devoid of skeletons; however, pertidal facies such as tidally-influenced microbial laminite in the Boat Harbour and Agauthuna formations appear to be more heavily dolomitized than associated subtidal facies. Future work is required to determine if dolomitization provides a selective bias in western Newfoundland or in other areas of interest.

A more general issue that must be taken into account is the degree to which a single geographic locality can represent broader trends. This is not so much a limitation of the point counting method as a result of limited geographic sampling. As described above, the data presented here represent changes in skeletal abundance from the Cambrian to the Ordovician of western Newfoundland, but clearly there are environmental changes occurring through the stratigraphic section, as well. Obviously, this issue cannot be resolved without comparing the newly amassed data to that from other localities to see if the trends here are geographically widespread. Duplicating work at other localities may lend support to the observed trends in skeletal abundance (e.g., Pruss et al. 2010; Ibex area of Utah to western Newfoundland) or provide contrasts that must be considered.

7.5 Geological Setting of Strata in Newfoundland

Laurentia lay over the equator during Cambro-Ordovician time (Fig. 7.1) (Smith 2001; Hodych et al. 2004), largely covered by epeiric seas, and the Series 3–Furongian (Cambrian) Port au Port and Lower–Middle Ordovician St. George groups (Fig. 7.2), exposed in western Newfoundland, were deposited on its southeastern margin. On the Port au Port Peninsula, Cambro–Ordovician strata are well exposed and have been extensively studied along two arms of a broad, gently-dipping anticline that makes up the south-facing shore (e.g., Pratt and James 1982, 1986; Chow and James 1987; Knight and James 1987; James et al. 1989; Cowan and James 1993) (Fig. 7.1). Biostratigraphic and chemostratigraphic constraints on Cambrian strata include the Series 3 *Bolaspidella* through Furongian *Elvinia* trilobite zones (e.g., Westrop 1992; Saltzman et al. 2004) and the geographically widespread SPICE excursion (later Cambrian positive carbon isotope excursion) (Saltzman et al. 2004; Hurtgen et al. 2009). Lower Ordovician strata of the St. George Group preserve biostratigraphically well-constrained assemblages of conodonts (Ji and Barnes 1994) and gastropods (Rohr et al. 2001). The Table Head Formation is also biostratigraphically constrained to the Middle Ordovician by several fossil groups, including trilobites and graptolites (e.g., Whittington and Kindle 1963).

The Series 3 March Point Formation and Series 3–Furongian Petit Jardin Formation make up the Port au Port Group (~190 m; Fig. 7.2). Cambrian carbonates of western Newfoundland contain very few skeletal carbonates, with a few thin lenticular beds of trilobite hash representing the only significant fossil accumulations exposed in the field. In contrast, oolites, microbial buildups, flat-pebble conglomerates, and thinly-bedded carbonate mudstones are common, deposited in environments ranging from deep subtidal to peritidal.

The Lower Ordovician St. George Group consists (from bottom to top) of the Watts Bight, Boat Harbour, Catoche, and Aguathuna formations (Knight and James 1987) and is overlain by the Middle Ordovician Table Head Formation (Fig. 7.2). The majority of the St. George Group was deposited in a shallow carbonate-dominated subtidal to peritidal setting. The Aguathuna Formation is heavily dolomitized in most areas where it is exposed and was therefore not sampled in this research. Along the south-facing shore of the Port au Port Peninsula, complete stratigraphic sections of the Watts Bight (~70 m) and Boat Harbour (~150 m) formations are exposed. We examined these units in outcrop, but only the Boat Harbour samples were analyzed in thin section (the Watts Bight Formation is extensively dolomitized). Incomplete sections of the Floian (Lower Ordovician) Catoche Formation and Dapingian (Middle Ordovician) Table Point Formation are exposed along a roadcut north of Cape St. George (Boyce et al. 2000; see Fig. 7.1), and these outcrops provided preliminary data on the nature of changing skeletal abundances in upper Lower and lower Middle Ordovician strata.

It is important to note that while the depositional environments preserved in the Cambro-Ordovician Port au Port and St. George groups do change through time, the overall sedimentary setting is characterized by tropical carbonate-rich facies deposited in shallow subtidal and peritidal environments (e.g., Knight and James 1987; Cowan and James 1993). No significant latitudinal change occurs during this time period in western Newfoundland. Facies do become more carbonate-rich up-section, and there is a greater abundance of potentially restricted peritidal settings in the Lower Ordovician strata examined in this study (particularly in the Boat Harbour Formation); however, the Lower Ordovician Catoche Formation and Middle Ordovician Table Head Formation were deposited under mostly subtidal conditions. The greatest variation in depositional settings is preserved in Cambrian strata of the Port au Port Group (subtidal, below wave base to peritidal environments), yet it has been observed in outcrop that no facies is particularly fossiliferous (Cowan and James 1993; Pruss et al. 2010).

7.6 Analysis of Point Counting Abundance

7.6.1 Series 3 to Furongian (Cambrian) and Tremadocian (Lower Ordovician) Results

The percentage of skeletal material identified in point-counts of various Cambrian and Ordovician carbonate facies were analyzed from western Newfoundland (Fig. 7.5, modified from Pruss et al. 2010), and all points were identified to the most specific category possible (Table 7.1, modified from Pruss et al. 2010). In Cambrian samples from the Port au Port Group, skeletal material, on average,

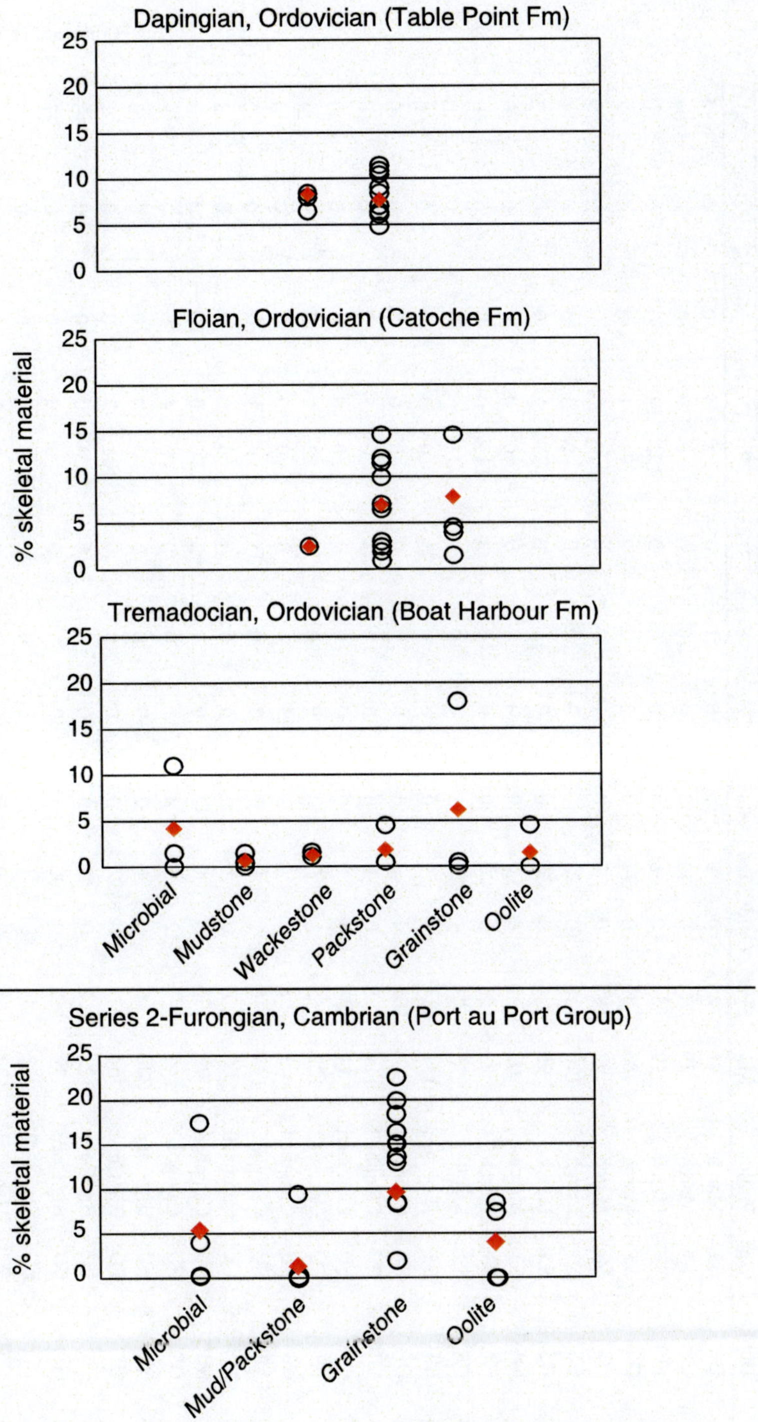

Fig. 7.5 Diagram showing percent of skeletal material in each analyzed thin section. Each *circle* represents a thin section and *diamonds* show averages for each lithofacies analyzed

Table 7.1 Table showing all points counted for each slide analyzed (Modified from Pruss et al. 2010)

Sam.	Dunham	Age	Fmtion	Micr	Spar	Ooid	Trilo	Oncoid	Echino	Brach	Renalc	FP	Unident	Mollusk	Peloids	Other
MF11	GS	S3	March	41	85	0	9	5	16	1	0	5	4	0	0	34
MF11A	GS	S3	March	39	24	0	33	0	0	0	0	0	0	0	0	104
MF17	FP	S3	March	19	54	0	6	0	0	0	0	122	14	0	0	0
MF17B	GS	S3	March	83	36	0	14	0	2	2	0	74	1	0	0	8
MF17D	GS	S3	March	92	95	0	3	0	1	0	0	0	0	0	0	9
MF17G	GS	S3	March	106	19	0	23	47	2	0	0	0	3	0	0	1
MF18A	GS	S3	March	95	6	0	38	0	0	1	0	57	1	0	0	2
MF26	MS/WS	S3	March	200	0	0	0	0	0	0	0	0	0	0	0	0
MF36	FP	S3	March	53	23	8	0	0	0	0	3	113	0	0	0	0
MF38	GS	S3	March	75	71	1	21	5	4	0	0	11	6	6	0	0
MF41Th	MIC	S3	Cape Ann	151	14	0	0	0	0	0	35	0	0	0	0	0
MF53	O	S3	Cape Ann	4	102	94	0	0	0	0	0	0	0	0	0	0
CMCC1	MIC/GS	FU	Campbell's	114	8	0	0	0	30	0	43	0	5	0	0	0
CMCC17	O	FU	Campbell's	11	53	134	0	0	0	0	0	0	0	0	0	0
CM3	FP	FU	Campbell's	122	55	1	0	0	0	0	0	12	0	0	0	0
CM15	GS	FU	Campbell's	3	98	82	0	2	15	0	0	0	0	0	0	0
CM28MI3	MIC	FU	Campbell's	148	8	1	0	0	7	0	35	0	1	0	0	0
CM29	MS/WS	FU	Campbell's	200	0	0	0	0	0	0	0	0	0	0	0	0
CM30	MIC	FU	Campbell's	78	30	0	0	0	0	0	92	0	0	0	0	0
CMU28	GS	FU	Campbell's	140	3	32	2	0	11	0	0	0	4	0	0	8
BC1	MS/WS	FU	Big Cove	200	0	0	0	0	0	0	0	0	0	0	0	0
BC1A	GS	FU	Big Cove	64	1	0	24	0	0	0	0	108	2	0	0	1
BC2	GS	FU	Big Cove	123	2	0	45	0	0	0	0	20	0	0	0	10
BC3	GS	FU	Big Cove	79	10	0	30	55	0	1	0	0	0	0	0	27
MWU6	O	FU	Man o' War	62	34	87	1	0	8	0	0	0	8	0	0	0
MWU7	MS/WS	FU	Man o' War	192	8	0	0	0	0	0	0	0	0	0	0	0
BH7A	FP	TR	B. Harbour	155	20	0	0	0	1	0	0	10	0	0	7	7
BH16A2	PS	TR	B. Harbour	125	30	0	0	0	0	0	0	15	0	1	20	9

BH17A	PS	TR	B. Harbour	19	91	0	4	2	24	0	0	2	5	3	32	18
BHMI7B	MS	TR	B. Harbour	17	35	0	0	0	0	0	137	0	0	0	11	0
BH17C	MS	TR	B. Harbour	93	31	0	1	0	0	0	0	4	0	0	19	52
BHMI12B	MIC	TR	B. Harbour	89	63	0	0	0	0	0	19	0	1	9	19	0
BH12F	GS	TR	B. Harbour	119	68	0	3	0	0	0	0	0	2	0	8	0
BH23A	GS	TR	B. Harbour	97	7	0	1	0	1	2	0	6	3	2	14	67
BH25MI	MIC	TR	B. Harbour	152	3	0	0	0	0	0	0	1	0	22	17	5
BH26A	MS	TR	B. Harbour	132	4	0	0	0	0	0	0	0	0	0	0	64
BH34A	WS	TR	B. Harbour	142	13	0	0	0	2	0	0	0	1	0	39	3
BH35A	FP	TR	B. Harbour	23	56	33	0	0	0	0	0	38	0	0	48	2
BH40B	WS	TR	B. Harbour	34	63	0	0	0	1	0	0	8	1	0	93	0
BH41A	MIC	TR	B. Harbour	105	5	0	0	0	0	0	0	0	3	0	0	87
BH58A	WS	TR	B. Harbour	116	21	0	0	0	1	0	0	18	2	0	29	13
BH63A	O	TR	B. Harbour	27	49	107	0	0	0	0	0	1	0	0	11	5
BH63C	O	TR	B. Harbour	42	56	42	0	0	0	0	6	0	0	0	54	0
BH64B2	MIC	TR	B. Harbour	6	30	0	0	0	0	0	150	0	0	0	0	14
BH65B	O	TR	B. Harbour	38	59	73	1	1	5	0	0	10	2	1	10	0
CA3A	GS	FL	Catoche	78	31	0	0	0	0	0	0	0	3	0	0	88
CA3B	PS	FL	Catoche	95	23	2	0	28	0	0	44	0	2	0	6	0
CA9B	PS	FL	Catoche	0	41	2	1	33	4	0	90	0	0	0	2	27
CA11A	GS	FL	Catoche	0	28	0	1	6	6	0	78	0	1	0	3	77
CA11B	PS	FL	Catoche	0	48	0	0	0	6	0	130	5	0	0	10	1
CA21A	PS	FL	Catoche	0	34	11	1	2	6	1	103	0	21	0	21	0
CA21B	PS	FL	Catoche	0	32	11	2	15	5	0	82	0	15	1	37	0
CA25A	GS	FL	Catoche	0	22	12	1	0	14	0	94	0	14	0	0	43
CA25B	PS	FL	Catoche	1	10	1	0	37	4	0	125	0	0	0	5	17
CA27A	PS	FL	Catoche	81	0	5	2	0	9	0	0	0	3	0	0	100

(continued)

Table 7.1 (continued)

Sam.	Dunham	Age	Fmtion	Micr	Spar	Ooid	Trilo	Oncoid	Echino	Brach	Renalc	FP	Unident	Mollusk	Peloids	Other
CA28A	PS	FL	Catoche	0	49	0	0	0	4	0	103	0	15	0	25	4
CA28B	PS	FL	Catoche	0	40	3	0	0	17	1	98	0	6	0	35	0
CA212A	GS	FL	Catoche	0	16	0	1	0	5	1	105	0	2	0	0	70
CA218A	PS	FL	Catoche	0	47	0	0	0	0	0	104	0	13	0	35	1
CA220A	WS	FL	Catoche	0	64	3	0	93	0	0	0	0	4	1	0	35
AG081A	PS	DAP	Table Pt	0	48	2	0	0	15	0	101	0	6	0	28	0
AG081B	PS	DAP	Table Pt	0	53	0	0	5	4	0	97	0	14	4	0	23
AG081C	PS	DAP	Table Pt	0	49	0	1	0	5	0	127	0	6	0	11	1
AG082A	WS	DAP	Table Pt	0	40	0	1	0	3	0	97	0	11	1	2	45
AG082B	WS	DAP	Table Pt	0	45	0	0	0	5	0	119	0	8	3	4	16
AG083A	PS	DAP	Table Pt	0	50	0	1	0	1	0	107	0	8	1	19	13
AG083B	PS	DAP	Table Pt	0	58	0	0	1	3	0	118	0	7	0	6	7
AG084B	PS	DAP	Table Pt	0	47	0	2	3	2	0	115	0	7	1	14	9
AG084C	PS	DAP	Table Pt	0	56	0	0	0	7	0	95	0	14	2	4	22
AG085B	WS	DAP	Table Pt	0	55	0	0	1	1	0	109	0	7	3	4	20
AG086B	WS	DAP	Table Pt	0	20	0	1	0	6	2	159	0	7	0	4	1
AG088A	WS	DAP	Table Pt	0	63	0	1	2	2	0	119	0	6	3	0	4

Sam. is sample identification, *Dunham* is Dunham classification scheme (*GS* is non-oolitic grainstone, *FP* is flat-pebble, *MS* is mudstone, *WS* is wackestone, *PS* is packstone, *MIC* is microbialite, and *O* is oolite), *Age* is defined as *S3* (Series 3 Cambrian), *FU* (Furongian), *TR* (Tremadocian), *FL* (Floian), *DAP* (Dapingian), and in the classification scheme, *Micr* is micrite, *Renal* is renalcid-like microbial structure and *Other* includes microfacies like dolomite rhombs and organic material accounted for ~8% of points counted from all examined thin sections (n=26). Fossil content ranges up to 23%, but ten thin sections analyzed from various facies contain no fossil material (see Fig. 7.5; Pruss et al. 2010). In total, 12 grainstone thin sections (excluding oolites) were analyzed and these contained the highest average skeletal content (~10%). Oolites contain ~5% average fossil material. Thrombolitic microbialites occasionally preserve relatively high abundances of fossil material (CMCC1 in Table 7.1), particularly in the interstices of mounds (18%), but stromatolites generally contained little to no skeletal material (CM30 in Table 7.1).

Using the thin section point count data, we apply these averages to the outcrop scale to determine the contributions of skeletal material to the stratigraphic section as a whole (Fig. 7.5). The volume of Cambrian carbonate that consists of skeletal material is ~2.4% (Fig. 7.5). This reflects the low abundance of skeletal grainstones in outcrop (Cowan and James 1993; Pruss et al. 2010) and the prevalence of mudstone/packstones that, on average, contain only ~1% fossil material in thin section (see Fig. 7.5). In Cambrian strata, mudstone/packstones and oolites make up much of the stratigraphy (Pruss et al. 2010). Microbialites are also prevalent but skeletal grainstones are rare.

The Tremadocian Boat Harbour Formation was examined, and the results are similar to the Cambrian data discussed above. For example, non-oolitic grainstones preserve a relatively large range of total skeletal material. One of the grainstone thin sections contains the highest amount of skeletal material preserved in all Boat Harbour thin sections (18%) but many analyzed grainstones consist of peloids or broken flat-pebbles and thus contain much less fossil material (0–0.5%) (Fig. 7.5). Like Cambrian examples, thrombolites can contain locally abundant skeletal material preserved in the interstices of mounds (11%) (Fig. 7.5).

The volume of carbonate produced by skeletal material is similarly low in the Boat Harbour Formation. Only ~1.6% of the total section consists of fossil material (Fig. 7.6). In spite of a noticeable increase in shell pavements and the increased preservation of skeletal macrofossils, skeletal organisms continue to make up a quantitatively small amount of total carbonate production while microbial mounds, mudstones, and flat-pebble conglomerates continue to dominate the stratigraphy.

7.6.2 Floian (Lower Ordovician) to Dapingian (Middle Ordovician) Results

In western Newfoundland, a preliminary analysis of Floian and Dapingian carbonates was conducted to provide a first-order point of comparison with the Series 3 Cambrian and Tremadocian samples previously analyzed (Pruss et al. 2010). An oncolite-rich portion of the Floian Catoche Formation was examined in addition to ~25 m of wackestones and packstones of the Dapingian Table Point Formation (Fig. 7.2). In the Catoche Formation, oncolitic packstones represent the most common facies, and these contain on average ~6% fossil material, with as much as 14.5% in a single thin section (Fig. 7.5). Fossil grainstones show a similar range (1.5–14.5%) and contain on average 7% skeletal material (Fig. 7.5). Only one wackestone was analyzed

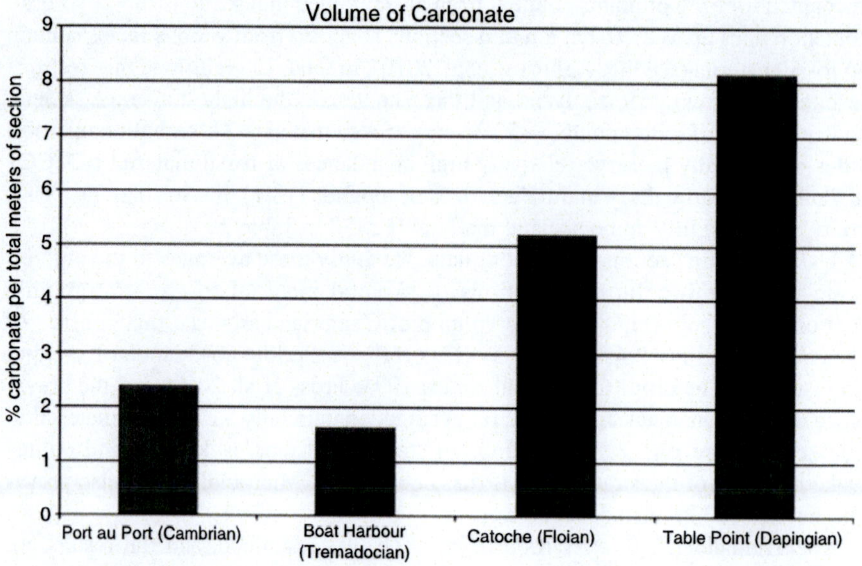

Fig. 7.6 *Bar graph* showing the volume of carbonate as a percent of the total measured section for each of the major stratigraphic units analyzed

(2.5% skeletal). Although facies are limited for this portion of the Catoche Formation, these data show an increase in average skeletal material in two of the three examined facies. In the Dapingian Table Point Formation, packstones and wackestones were analyzed from a small outcrop north of Cape St. George. Both facies contain, on average, ~8% skeletal material. All thin sections analyzed contained some fossil material; the range is 5–11% skeletal material in packstones and 6.5–9% in wackestones (Fig. 7.5).

When examining changes in volume of carbonate, about 5% of the Floian Catoche Formation and 8% of the Table Point Formation consist of skeletal material (Fig. 7.6). The oncolitic packstones that make up much of the stratigraphy all contain at least some fossil material, and this is apparent in outcrop and in thin section. In the Dapingian Table Point Formation, diverse macrofossils are present in outcrop and quantitatively make up a larger proportion of carbonate than in any unit examined stratigraphically below. The change in volume of skeletal carbonate from the Cambrian to the Middle Ordovician represents a fourfold increase (Fig. 7.6).

7.7 Secular Trends in Cambro-Ordovician Skeletal Abundance, Western Newfoundland

Examining Cambro-Ordovician carbonate strata of western Newfoundland at various scales reveals the relative importance of skeletal production of carbonate during the Early Paleozoic. Cambrian strata of the Port au Port Group represent a variety of depositional environments, and in all of them, skeletal material makes up only a small amount of total carbonate production. In thin section, skeletal material can be locally high (up to 23%), particularly in grainstones, but many thin sections contain no fossil material (Fig. 7.5) (Pruss et al. 2010). Interestingly, fossiliferous grainstones make up a small proportion of total Cambrian stratigraphy, so the skeletal volume of carbonate is quite low (~2%). Lowermost Ordovician strata show a similar trend; the Boat Harbour Formation contains little skeletal material in outcrop or in thin section. Importantly, the Boat Harbour Formation was deposited in a subtidal to peritidal setting, so environmental restriction may play a role in the overall trends observed in this section; nonetheless, even in subtidal facies, the amount of total skeletal material in the most fossiliferous thin section was 18%.

A change in amount of fossil material occurs in the Floian Catoche and Dapingian Table Point Formations. Although only limited facies were sampled, these data provide a preliminary means of comparison to the Series 3 Cambrian and Tremadocian units analyzed (Pruss et al. 2010). Oncolite-rich packstones and wackestones are common in the Catoche Formation north of Cape St. George, and all of the thin sections analyzed from this unit contain some fossil material. The same point can be made for thin sections of the Dapingian Table Point Formation; in spite of limited facies analyzed, all of the thin sections contain fossil material; skeletal volume of carbonate for the Catoche and Table Point Formations are 5% and 8%, respectively. This study reveals a fourfold increase in the volume of skeletal carbonate in Cambro-Ordovician formations of western Newfoundland. These data must be interpreted with a firm recognition that facies and environments do not remain unchanged through time, and greater vertical and lateral analysis of facies, particularly in the Ordovician, will further refine the trends reported here. These data do align with observations made from Ordovician strata analyzed in the Ibex area of Utah (Pruss et al. 2010).

Quantifying skeletal material in thin section has proved to be an important tool in assessing the relative contribution of carbonate types to total carbonate production (e.g., Payne et al. 2006; Pruss et al. 2010). For example, not only can the proportion of fossil material be analyzed, but all components of carbonate rocks are also quantified. The proportion of fossil material changed from the Cambrian to the Ordovician, but so did the types of fossil material observed (Pruss et al. 2010). One observation is that more fossils are assigned to the "unidentified fossil" group in the Catoche and Table Point formations; this does not reflect preservation but rather represents an increase in small, broken fossil pieces that cannot be confidently assigned to existing groups (the utility of the unidentified category is that it enables us to quantify total fossil contribution to carbonate). Other common carbonate components in all thin sections from this time interval include micrite, calcitic cement (spar), intraclasts, peloids, ooids, oncoids, and other microbial textures (Table 7.1).

7.8 Future Work

Abundance data from point counting analysis of thin sections provide an important way to determine the relative contribution of skeletal material to total carbonate production. The results here show a promising way to quantify the changing abundances of skeletal material from Cambrian carbonates, which are dominated by non-skeletal microbialites, flat-pebble conglomerates, and mudstones, to Ordovician packstones and wackestones that contain more abundant fossil material. Volumetrically, point-count analyses suggest that fossil material makes up four times more carbonate in the post-Tremadocian Ordovician than in Cambrian units; of course, further work at other localities will refine these trends.

Many areas for future work exist. Firstly, it will benefit this study to analyze a greater variety of facies from Lower and Middle Ordovician strata of western Newfoundland. Our limited sampling of the Floian Catoche and Dapingian Table Point Formations permits a first order comparison to Cambrian and Tremadocian strata (Pruss et al. 2010), but an analysis of more diverse facies would provide important constraints on the nature of skeletal expansion during the Ordovician radiation in western Newfoundland. It would also benefit our analysis of trends to duplicate this effort in other areas of Laurentia to compare to the data generated from western Newfoundland. In this way, we would glean a much more complete picture of the nature of changing carbonates across Laurentia during this critical interval of time.

This type of analysis also holds potential for other intervals of Earth history. Point counting analyses have demonstrated the nature of changing skeletal abundances during the end-Permian extinction and aftermath (Payne et al. 2006). Using this type of analysis for other extinction events could prove quite useful in determining how the record of skeletal abundance is affected by changes in diversity. Another area where greater abundance data would be beneficial is in the study of Terreneuvian Cambrian reefs. Here, it is possible to assess the relative contribution of archaeocyathid, and other skeletal elements, to total carbonate production in early reefs. In this way, it could be determined if the skeletal carbonate factory of the early Cambrian operated in a fundamentally different way than that of later Cambrian oceans. This methodology holds much promise to address the changing abundances of skeletal carbonate during other critical intervals in Earth history.

7.9 Conclusions

A handful of studies have demonstrated that important ecological changes in the Early Paleozoic are not only characterized by an increase in total diversity, but by an increase in animal abundance (e.g., Li and Droser 1997, 1999; Finnegan and Droser 2005; Pruss et al. 2010). Here, these data demonstrate that in western Newfoundland, a four-fold increase in skeletal contribution to the volume of carbonate production occurs from the Cambrian to the Middle Ordovician. Point counting analysis of thin sections documents an increase in skeletal contribution that is generally corroborated by outcrop observations. Furthermore, these observations are in good agreement with similar studies in the Ibex Area of Utah (Pruss et al. 2010). Like previous studies that have focused on the increase in thickness and abundance of shell beds (Li and Droser 1997, 1999), the Ordovician radiation is reflected not only as an increase in diversity but in sedimentological changes to carbonate production. It has previously been noted that quantitative production of carbonates by eukaryotes began during the early Cambrian (e.g., Zeebe and Westbroek 2003) but it has been subsequently demonstrated that carbonate production is not more quantitatively dominated by metazoans until the Ordovician (Pruss et al. 2010). In spite of the limitations of examining facies in one geographic location and with the understanding that the changing environments must affect these results, the nature of carbonate production is shown to change during this time in western Newfoundland.

Acknowledgments We thank the Agouron Institute (postdoctoral fellowship to SP), Harvard University and Smith College for partial funding of this research. We would like to acknowledge A. Breus, A. Mushegian, and D. Schmandt, for thin section preparation and analysis and K. B. Hender, S. Finnegan, M. Laflamme, and J. Payne for helpful discussions and comments that improved this manuscript.

References

Bambach RK (1983) Ecospace utilization and guilds in marine communities through the Phanerozoic. In: Tevesz MJS, McCall PL (eds) Biotic interactions in recent and fossil benthic communities. Plenum, New York, pp 719–746

Boyce WD, Knight I, Rohr DM, Williams SH, Measures EA (2000) The Upper St. George Group, Western Port au Port Peninsula: lithostratigraphy, biostratigraphy, and depositional environments and regional implications. Current Research Newfoundland Department of Mines and Energy, Geological Survey Report 2000-1, pp 101–125

Carozzi AV (1961) Reef petrography in the Beaverhill Lake Formation, Upper Devonian, Swan Hills area, Alberta, Canada. J Sediment Petrol 31:497–513

Chow N, James NP (1987) Cambrian grand cycles: a northern Appalachian perspective. Geol Soc Am Bull 98:418–429

Clapham ME, Bottjer DJ, Jamet CM, Bonuso N, Fraiser ML, Marenco PJ, Dornbos SQ, Pruss SB (2006) Assessing the ecological dominance of Phanerozoic marine invertebrates. Palaios 21:431–444

Cowan CA, James NP (1993) The interactions of sea-level change, terrigenous-sediment influx, and carbonate productivity as controls on Upper Cambrian Grand Cycles of western Newfoundland. Geol Soc Am Bull 105:1576–1590

Droser ML, Finnegan S (2003) The Ordovician radiation: a follow-up to the Cambrian explosion? Integr Comp Biol 43:178–184

Finnegan S, Droser ML (2005) Relative and absolute abundance of trilobites and rhynchonelliform brachiopods across the Lower/Middle Ordovician boundary, eastern Basin and Range. Paleobiology 31:480–502

Flügel E (1982) Microfacies analysis of limestones. Springer, Berlin

Flügel E (2004) Microfacies of carbonate rocks: analysis, interpretation and application. Springer, New York

Greig-Smith P (1983) Quantitative plant ecology, 3rd edn. Blackwell Scientific, Oxford

Grotzinger JP, Watters WA, Knoll AH (2000) Calcified metazoans in thrombolite-stromatolite reefs of the terminal Proterozoic Nama Group, Namibia. Paleobiology 26:334–359

Higgins JA, Fischer WW, Schrag DP (2009) Oxygenation of the oceans and sediments: consequences for the seafloor carbonate factory. Earth Planet Sci Lett 284:25–33

Hodych JP, Cox RA, Kosler J (2004) An equatorial Laurentia at 550 Ma confirmed by Grenvillian inherited zircons dated by LAM ICP-MS in the Skinner Cove volcanics of western Newfoundland: implications for inertial interchange true polar wander. Precambrian Res 129:93–113

Hurtgen MT, Pruss SB, Knoll AH (2009) Evaluating the relationship between the carbon and sulfur cycles in the later Cambrian ocean: an example from the Port au Port Group, western Newfoundland, Canada. Earth Planet Sci Lett 281:288–297

Jaanusson V (1972) Constituent analysis of an Ordovician limestone from Sweden. Lethaia 5:217–237

Jacobsen N, Twitchett RJ, Krystyn L (in press) Palaeoecological methods for assessing marine ecosystem recovery following the Late Permian mass extinction event. Palaeogeogr Palaeoclimatol Palaeoecol, p. 3

James NP, Stevens RK, Barnes CR, Knight I (1989) Evolution of a Lower Paleozoic continental-margin carbonate platform, northern Canadian Appalachians. In: Crevello P, Wilson JL, Sarg JF, Read JF (eds) Controls on carbonate platform and basin development, Society of Sedimentary Geology Special Publication, vol. 44. SEPM, Tulsa, pp. 123–146

Ji Z, Barnes CR (1994) Lower Ordovician conodonts of the St. George Group, Port au Port Peninsula, western Newfoundland, Canada. Palaentologr Canad 11:1–149

Kidwell SM (1991) The stratigraphy of shell concentrations. In: Allison PA, Briggs DEG (eds) Taphonomy, releasing the data locked in the fossil record. Plenum, New York

Kidwell SM, Brenchley PJ (1994) Patterns of bioclastic accumulation through the Phanerozoic: changes in input or in destruction? Geology 22:1139–1143

Knight I, James NP (1987) The Stratigraphy of the Lower Ordovician St. George Group, western Newfoundland – the interaction between eustasy and tectonics. Can J Earth Sci 24:1927–1951

Knoll AH (2003) Biomineralization and evolutionary history. In: Dove PA, Yoreo JD, Weiner S (eds) Biomineralization. Mineralogical Society of America, Washington, DC, pp 329–356

Knoll AH, Swett K (1990) Carbonate deposition during the late Proterozoic era: an example from Spitsbergen. Am J Sci 290-A:104–132

Li X, Droser ML (1997) Nature and distribution of Cambrian shell concentrations: evidence from the Basin and Range Province of the Western United States (California, Nevada, and Utah). Palaios 12:111–126

Li X, Droser ML (1999) Lower and Middle Ordovician shell beds from the Basin and Range province of the western United States (California, Nevada, and Utah). Palaios 14:215–233

Maloof AC, Schrag DP, Crowley JL, Bowring SA (2005) An expanded record of early Cambrian carbon cycling from the Anti-Atlas Margin, Morocco. Can J Earth Sci 42:2195–2216

Novack-Gottshall PM, Miller AI (2003) Comparative taxonomic richness and abundance of Late Ordovician gastropods and bivalves in mollusc-rich strata of the Cincinnati Arch. Palaios 18:559–571

Payne JL, Lehrmann DJ, Wei J, Knoll AH (2006) The pattern and timing of biotic recovery from the end-Permian mass extinction on the Great Bank of Guizhou, Guizhou Province, south China. Palaios 20:63–85

Pratt BR, James NP (1982) Cryptalgal-metazoan bioherms of early Ordovician age in the St-George Group, Western Newfoundland. Sedimentology 29:543–569

Pratt BR, James NP (1986) The St. George Group (Lower Ordovician) in western Newfoundland: tidal flat island model for carbonate sedimentation in shallow epeiric seas. Sedimentology 33:313–343

Pruss SB, Finnegan S, Fischer WW, Knoll AH (2010) Carbonates in skeleton-poor seas: new insights from Cambrian and Ordovician strata of Laurentia. Palaios 25:73–84

Ridgwell A, Zeebe RE (2005) The role of the global carbonate cycle in the regulation and evolution of the Earth system. Earth Planet Sci Lett 234:299–315

Rohr DM, Measures EA, Boyce WD, Knight I (2001) Early Ordovician gastropods of the Barbace Cove Member (Boat Harbor Formation) and Catoche Formation, western Newfoundland. In: Pereira CPG, Walsh DG (eds) Current research, geological survey of Newfoundland and Labrador Geological Survey. Department of Mines and Energy, St. John's, pp 113–126

Rowland SM, Shapiro RS (2002) Reef patterns and environmental influences in the Cambrian and earliest Ordovician. In: Kiessling W, Flügel E, Golonka J (eds) Phanerozoic reef patterns: Society of Sedimentary Geology Special Publication, vol 72. SEPM, Tulsa, pp 95–128

Rowland SM, Oliver LK, Hicks M (2008) Ediacaran and early Cambrian reefs of Esmeralda County, Nevada: non-congruent communities within congruent ecosystems across the Neoproterozoic–Paleozoic boundary. Geological Society or America Field Guides 11, pp 3–100

Saltzman MR, Cowan CA, Runkel AC, Runnegar B, Stewart MC, Palmer AR (2004) The Late Cambrian SPICE event and the Sauk II-Sauk III regression: new evidence from Laurentian basins in Utah, Iowa, and Newfoundland. J Sediment Res 74:366–377

Smith AG (2001) Paleomagnetically and tectonically based global maps. In: Zhuravlev AY, Riding R (eds) The ecology of the Cambrian radiation. Columbia University Press, New York, pp 11–46

Van Der Plas L, Tobi AC (1965) A chart for judging the reliability of point counting results. Am J Sci 263:87–90

Walpole RL, Carozzi AV (1961) Microfacies study of Rundle Group (Mississippian) of Front Ranges, central Alberta. AAPG Bull 45:1810–1846

Westrop SR (1992) Upper Cambrian (Marjuman-Steptoean) trilobites from the Port Au Port Group, western Newfoundland. J Paleontol 66:228–255

Whittington HB, Kindle CH (1963) Middle Ordovician Table Head Formation, Western Newfoundland. Geol Soc Am Bull 74:745–758

Zar JH (1996) Biostatistical analysis, 3rd edn. Prentice-Hall, Upper Saddle River

Zeebe RE, Westbroek P (2003) A simple model for the CaCO3 saturation state of the ocean: the "Strangelove", the "Neritan", and the "Cretan" ocean. Geochem Geophys Geosys 4:1–26

Zhuravlev AY, Wood RA (1996) Anoxia as the cause of the mid-early Cambrian (Botomian) extinction event. Geology 24:311–314

Chapter 8
Exploring the Ecological Dynamics of Extinction

Amelinda E. Webb and Lindsey R. Leighton

Contents

A.E. Webb (✉)
Department of Geology and Geophysics, Yale University, New Haven, CT 06511, USA
e-mail: amelinda.webb@yale.edu

L.R. Leighton
Earth and Atmospheric Sciences Department, University of Alberta,
Edmonton, AB, Canada T6G 2E3
e-mail: lleighto@ualberta.ca

M. Laflamme et al. (eds.), *Quantifying the Evolution of Early Life*, Topics in Geobiology 36,
DOI 10.1007/978-94-007-0680-4_8, © Springer Science+Business Media B.V. 2011

Abstract Extinction events are defined by taxonomic impact, however the ecological impact of extinction has been difficult to quantify and therefore less understood. Measuring taxonomic membership and turnover in communities is relatively straightforward; however quantifying changes to community structure using traditional methods is often hampered by taxonomic turnover. The purpose of this chapter is to introduce the concept of ecological impact and to provide techniques for exploring the ecological dynamics of extinction events. Using an example from the Early Cambrian, this chapter will review traditional techniques used in extinction studies, as well as a new technique that can be used to infer the ecological impact of extinction events independent of taxonomic impact. Although the study of early metazoan life often focuses on major radiations, there were also major extinction events. Pre-cambrian and Cambrian fauna had elevated rates of both origination and extinction. Within a few million years of the Cambrian Explosion there was a mass extinction with a rate comparable to the five major mass extinctions. The low absolute diversity of these early communities contributed to low absolute taxonomic loss, and as a result these extinction events have not received the attention they deserve. The analysis presented here represents the first attempt to examine community structure during the end Early Cambrian mass extinction and reveals elevated stress levels in communities during the extinction interval. The results of this analysis underscore the conclusion that the extinction dynamics of early communities provide an ideal setting for exploring the ecological impacts of extinction.

Keywords Community structure • Ecological impact • Methods • Paleoecology • Rank–abundance curves

8.1 Introduction

Paleontological studies of early life often focus on the development of metazoan life on Earth and the evolution of new phyla (Briggs et al. 1992; Tucker 1992; Babcock et al. 2001; Rowland and Hicks 2004; Li et al. 2007; Maheshwari et al. 2007). However, understanding the dynamics of extinction is crucial for a complete picture of the patterns and processes of life through time. The restructuring of ecosystems by extinction changes the biotic environment affecting subsequent evolution. The ecological impact of extinction has been explored rarely, as it is difficult to quantify ecological loss compared to the simplicity of quantifying taxonomic loss. As both extinction and origination rates were relatively high during the Pre-Cambrian and Cambrian (Bambach et al. 2004), the rapid changes in ecosystems potentially had profound impacts on the course of evolution, making Cambrian faunas particularly interesting candidates for examination of ecological dynamics during extinction events (not to imply earlier faunas are less interesting, rather that logistical constraints preclude robust studies). The purpose of this chapter is to introduce techniques useful for examining the ecological dynamics of extinction events and suggest guidelines for interpreting the results. Examples will be given via new analyses of published data, including a taxonomic range dataset (from Zhuravlev and Wood 1996) and an extinction community dataset (from Landing 1988).

Over the past decade, several paleoecological studies have shown that ecological and taxonomic impact are equally important in understanding extinction, as the magnitude of ecological impact is not necessarily directly related to the magnitude of taxonomic loss of an event (Droser et al. 2000; McGhee et al. 2004; Webb 2008; Webb and Leighton 2008; Webb et al. 2009). For example, although the Cretaceous-Paleogene mass extinction was ranked fifth out of the big five mass extinctions, the ecological impact of the event was ranked as second out of the big five (Droser et al. 2000). In addition, Webb and Leighton (Webb 2008; Webb and Leighton 2008; Webb et al. 2009) have shown there are ecological signals that serve as warning signs before an extinction event, as well as indicating ecological recovery millions of years before taxonomic recovery. Ecological impact can be described as transient or permanent changes in ecosystems such as the number of different ecosystem types or the structure within the ecosystems. Until recently, description of ecological impact was qualitative, as traditional ecological measures could not be used on extinction communities successfully due to the rapid taxonomic turnover that characterizes extinctions. Webb and Leighton (Webb 2008; Webb and Leighton 2008; Webb et al. 2009) developed a way to quantify changes in community structure that reflect the relative stress on a community, providing new opportunities to explore the ecological impact of extinction events.

Extinction is the complete disappearance of a taxon, and as such, extinction studies traditionally focus on taxonomic diversity or the number of taxa present at any given time. Even as evolution is an ongoing process, with periods of explosive radiation alternating with normal background rates of origination, so is extinction. Bambach et al. (2004) showed that extinctions can be statistically divided into two categories: background versus mass extinction. There are long periods with relatively stable rates of extinction (background) interspersed with shorter intervals of high rates of extinction. When the extinction rates of all time bins included in Bambach and others' study (2004) are arranged in increasing order, there is a continuous gradient. However, this pattern is driven by the high rates of extinction (and origination) in the Cambrian and Ordovician, the two periods that dominate the top quartile of intervals with highest extinction rates. Removing Cambrian and Ordovician intervals results in a break around 30% extinction (genera) that distinguishes between the majority of intervals (background extinction) and those with higher extinction rates which comprise the major mass extinctions (Bambach et al. 2004; Bambach 2006). This pattern suggests that although there is a distinction between mass extinctions and background extinction after the Ordovician, extinction in early metazoan faunas is different from that observed after the Ordovician, inviting closer examination of extinction in the Cambrian.

8.2 Taxonomic Signals – How to Identify Extinction Intervals

Traditionally, extinction events are recognized quantitatively by comparing changes in extinction rate across intervals of rock/time. The extinction rate is calculated by counting the number of taxa that went extinct within an interval of time compared to the total number of taxa that were observed in that interval. Extinction rates can be calculated for any taxonomic level (species, genus, order, family, etc.). The use of genera is common because this level provides a balance between high taxonomic resolution and potential systemic biases. The use of species provides the highest possible taxonomic resolution, however the definition of species is problematic. When a taxonomist begins naming species within a time interval, the diversity of that interval will be directly related to whether the taxonomist prefers to lump or split taxa into species. Pseudoextinction, or the perceived extinction of a species based on a changed taxonomic name, is also linked to the division of taxonomic work by time intervals. For example, brachiopod workers generally focus on either the Paleozoic or Mesozoic, such that no single genus of terebratulide brachiopod appears to survive the Permo-Triassic extinction even though the group has many extant genera (Leighton and Schneider 2008). In general, the problem of pseudoextinction is greater at lower taxonomic levels.

Different styles of preservation can affect the recognition of species, as the characteristics of higher taxonomic levels are less difficult to recognize when specimens are preserved poorly. For example, a badly abraded bi-valved shell is still identifiable as either a brachiopod or a bivalve, but further identification may be impossible. The use of higher taxonomic levels (order, family, etc.) reduces such biases, however it becomes more difficult to detect extinction events at higher levels. Specifically, the extinction of a taxon at a given taxonomic level requires the extinction of all individuals and taxa below that level (e.g., the extinction of a species requires the death of every individual of that species, the extinction of a genus requires the extinction of all species in the genus, the extinction of an order requires the extinction of all genera and species within that order, etc.). Therefore, if an extinction event did not include the extinction of all genera in an order, then examination of order level extinctions would not detect that particular extinction event. The choice of which taxonomic level to use in a study should be based on whether it is more important to detect fluctuations in diversity at low taxonomic levels or whether it is more important to avoid possible biases from preservation and taxonomic definitions. When the purpose of the study is to explore ecological dynamics, it is more important to detect the extinctions than to explore high-level, broad patterns in diversity. The loss of even a single species can have profound effects on an ecosystem (Paine 1966); therefore the lowest practical taxonomic level should be utilized to pinpoint extinction intervals for examination of ecologic signals.

Another factor to consider is whether to use published taxonomic ranges, or whether to collect occurrence data from the specific stratigraphic levels at which a taxon is observed. Occurrences can be converted into range data, which assumes that a taxon "occurs" in each time interval between the first and last occurrence of that taxon even if it is not observed in an intervening interval (ghost range). It is advisable to include the ghost ranges of taxa that do not commonly occur in the strata between their first and last occurrence, as a taxon will not originate and go extinct twice unless two taxa have been lumped mistakenly into one taxon. From these datasets it is possible to calculate a diversity curve for the area and time period of interest. Two common diversity measures are standing and peak diversity. Standing diversity represents the number of taxa that occur in both the interval of interest and the subsequent interval, thereby excluding taxa that only occur in one interval (singletons) and taxa that become extinct in the interval of interest. Peak diversity is the raw diversity for an interval or the total number of taxa that were present in an interval (includes singletons and extinction taxa).

Beyond choosing the taxonomic level for an extinction study, a useful time interval must be selected. Again, this choice is a balance between resolution and availability of data. High-resolution data are commonly only available in specific localities where the stratigraphy and preservation are conducive, making it difficult to include multiple localities or broad geographic ranges. If global data are used, then the stratigraphic resolution will be poor. For example, when Webb et al. (2009) investigated microfossil community response to the Paleocene-Eocene thermal maximum, few sedimentary cores were available with continuous sedimentation, good preservation, and sufficient numbers of two different microfossil groups (ostracods and foraminifera), so the geographic scale was sacrificed for temporal resolution and taxonomic diversity. The choice of scale always requires a balance between the breadth and the resolution of the data; the Early Cambrian datasets used in this chapter are prime examples. The taxonomic range dataset (from Zhuravlev and Wood 1996) include global occurrences but sacrifice temporal resolution and community membership information, whereas the extinction community dataset (from Landing 1988) sacrifices geographic range for temporal resolution and community composition information.

8.2.1 Extinction Rates

The simplest method for calculating an extinction rate is to count the number of taxa that became extinct in a single interval and divide it by the number of taxa that were observed during that interval (Sepkoski et al. 1981). Often, taxa that only occur in a single time interval (singletons) are excluded from calculations to avoid inflating the extinction rate or violating statistical assumptions, as the occurrence of singletons in an interval is positively correlated to sampling intensity; however it is possible and potentially advisable to include singletons in diversity analyses (Fitzgerald and Carlson 2006). Related to this simple calculation of extinction rate (the ratio of extinct taxa to total diversity in an interval) is survivorship analysis, which compares the duration of taxa survival over a set of time intervals (i.e. how many taxa from the first interval are either still alive or have gone extinct at the second, third, n-th interval) (Raup 1978). More complicated calculations of extinction rates have been developed to remove potential sampling biases that affect diversity estimates (and the subsidiary extinction rates) (Alroy 2000; Foote 2000, 2001; Alroy et al. 2008). For example, some intervals are sampled more frequently or thoroughly, because more rock is available (either global volume or amount exposed) or because more workers are sampling that region or preservation is better (lagerstätten or siliceous faunas). Such intervals are more likely to have high numbers of rarer taxa that were encountered due to sampling rather than due to an actual high diversity fauna from that interval (Raup 1976; Miller 2000; Peters and Foote 2001). The more complex calculations of diversity and extinction rates assume sampling biases (such as sampling intensity, preservation type [soft body versus original hard parts versus siliceous replacement, etc.], rock volume/exposure, "pull of the recent" [artificial lengthening of taxonomic ranges of recent taxa that have extant members but a poor fossil record] and the Signor-Lipps effect) are at least partly responsible for the patterns observed in diversity curves and use subsampling and probabilities to reduce these biases (Alroy 2000; Foote 2000, 2001; Peters and Foote 2001; Alroy et al. 2008). It is important to address these potential biases when calculating diversity curves. However, re-sampling techniques can be over-used, such that real biological signals are removed along with potential sampling biases. For instance, a re-sampled diversity curve can be generated from sub-samples of the original data that are too small to accurately represent diversity, resulting in a diversity curve that shows no periods of radiation or extinction events. After the iconic Sepkoski diversity curve (Sepkoski et al. 1981) was published as a consensus diversity curve (Miller 2000), subsequent attempts to remove sampling biases from diversity curves have not drastically altered the overall trend of Phanerozoic diversity; the major mass extinctions are still apparent in the various sampling-adjusted diversity curves although the timing of events may have shifted slightly (Alroy et al. 2001, 2008).

Although it is useful to calculate diversity and extinction rates, the numbers alone cannot be used to understand the underlying patterns and processes. The variety of methods employed to calculate diversity have not yielded a single "best" method, compounding the difficulty of employing and interpreting traditional measures of taxonomic impact. The above discussion is intended to provide a starting point for paleoecological studies of extinction.

8.3 Ecological Signals

8.3.1 Exploring Patterns of Extinction Among Taxa

Once an extinction interval has been identified, the next step is to explore potential patterns within the taxonomic loss. Separating different guilds or lineages for calculation of extinction rates can be useful for detecting selection pressures. A guild is a grouping of taxa based on life history, where all members of the guild share particular characteristics. For example, brachiopods are a guild as all members are sessile (primarily) epifaunal suspension feeders, whereas molluscs contain multiple guilds (mobile predatory gastropods, mobile grazing gastropods, epifaunal suspension feeding bivalves, infaunal suspension feeding bivalves, etc.). Often, specific groups of taxa (guilds or clades) will display greater rates of extinction, such as the extinction of non-avian dinosaurs at the Cretaceous-Paleogene extinction (Twitchett 2006). However, divisions for groupings may not be obvious, such that it is useful to employ exploratory methods such as ordinations to reveal groupings within the data (see more information in Chaps. 1 and 2 in this volume or a seminal review such as Chap. 4 of McCune and Grace 2002). When used in paleoecology, ordination is a multivariate technique that uses the abundances (or simply presences/absences) of taxa within samples to plot those samples in a multidimensional space. The distances between samples in the multidimensional ordination space are indicative of differences in taxonomic composition or abundance such that the most dissimilar samples are the farthest apart in ordination space. The cloud of points (where each point is a sample, and each sample is an assemblage or a community) in multidimensional space is then translated into a two- or three-dimensional space that best captures the most variation (most distance) between samples. It is also possible to perform an ordination that places taxa, rather than samples, in ordination space, as a means of assessing how similar taxa are to each other in distribution. Ordinations are useful for discovering what is most likely to be causing variation between samples (or the variables within samples). For ecological-extinction data, samples are time intervals, localities, horizons, or individual collections within any of the previous levels (e.g., multiple collections can be taken from one horizon) and the variables within those samples are the occurrences (presence/absence, abundance, biovolume, etc.) of different taxa. The use of ordinations in intervals with high rates of taxonomic turnover can be difficult, because if there is complete turnover, the ordination will be unable to detect other gradients or groupings within the dataset beyond the gradient imposed by extinction (which is time). This problem can be partly avoided by using higher taxonomic levels, as the turnover of orders and families will be much less than that of species and genera. An ordination of intervals should reinforce the identification of the extinction interval (as the ordination will most likely separate the intervals based on whether the intervals are before or after the extinction event which has already been identified by analysis of extinction rates). An ordination of the taxa across intervals may aid in determining which taxa, clades, or guilds were the most affected by the extinction.

8.3.2 Exploring Paleoecological Data

Once the initial exploration of extinction rates and taxonomic turnover is completed, then analyses can be undertaken to further reveal the patterns of the extinction and more importantly begin to reveal the processes driving the extinction. All mass extinctions are characterized by major environmental changes, although the specific type of change and the rate of that change are unique for each extinction (Twitchett 2006). The stress resulting from environmental change can impact all levels of the biological hierarchy (from cells to the biosphere), and that stress is expressed in a hierarchical manner related to the biological hierarchy (Table 8.1). For example, stress leading to the death of an organism requires the death of all cells within that organism, but may not have a significant impact on the population. Similarly, the extinction of a community requires the extinction of all populations of taxa in that community, which requires the death of all individuals within those populations. The highest levels of stress impact the biosphere, resulting from the extinction of many taxa, which can be traced back down each lower level of the hierarchy (Table 8.1). The lowest levels of stress that can be statistically detected in the fossil record affect communities (although it is possible to find examples of stressed populations, individuals, or even cells, the preservation of such signals is rare). Therefore, to study the ecological signals of extinction events, the highest useful resolution is at the level of communities. Such analyses require more specific, higher resolution data (assemblages) than the previous analyses (ranges/occurrences of taxa). It is the most useful to have multiple assemblages within each time interval during the extinction (before, during, and after). Abundance data are useful for ordinations and are required for metrics describing community structure.

8.3.2.1 Collecting Paleoecological Data

Paleontologists often remark on the incompleteness of the fossil record, with Darwin's apologetic discussion from *Origin of the Species* (1904) on the inadequacies of the rock record setting the tone for subsequent paleontological discussions and eventual arguments for utility of the fossil record (Gould and Eldredge 1993). The wealth of information available from the fossil record far outweighs the difficulties inherent in seeking that information. However, these challenges cannot be dismissed; rather it is necessary to tailor the study design such that sampling, statistical analyses, and interpretation all account for potential biases.

The different preservation potentials of various taxa within an assemblage is commonly one of the first objections raised to attempts to interpret fossil community data (excluding fossil lagerstätten which are too rare for studies of dynamics, extinction or otherwise) (Kidwell and Bosence 1991; Kidwell and Flessa 1995). The loss of soft-bodied organisms from the fossil record complicates paleoecological interpretation, however recent studies support the fidelity of the remaining hard parts to original community composition (Kidwell and Flessa 1995; Zuschin et al. 1999, 2000; Kidwell 2007, 2008). Trace fossils have been used as a proxy for soft-bodied organisms, with changes in diversity, abundance, and depth of bioturbation reflecting changes in environmental conditions (see Droser and Bottjer 1986 for an ichnofabric index that has been used to reveal interesting patterns during mass extinctions by Twitchett and Barras 2004 and Twitchett 2006). However, as trace fossil taxonomy is problematic, trace fossils cannot be directly combined with community data but can be used as a paleoecological proxy or for environmental data.

Table 8.1 Hierarchy of biological stress

Biological level	Stress level	Time scale	Cause	Effect (smallest – most extreme)	Present in fossil record
Biosphere (global diversity)	High	Centuries-millions of years	Disruption of many ecosystems	Loss of diversity – mass extinction	Yes
Ecosystem		Decades-centuries	Disruption of many communities	Ecosystem restructuring – regional extinction	Yes
Community		Months-decades	Disruption of populations	Community restructuring – disappearance of community	Yes
Population		Weeks-years	Death of organisms	Reduced size – disappearance	Rare
Organism		Days-years	Extreme conditions	Reduced fitness – death	Rare (requires death)
Cell/tissue	Low	Hours-weeks	Daily-seasonal fluctuations	Reduced function – cell/tissue death	Rare (requires death)

Each biological level is affected by different relative amounts of stress, with the frequency occurring at different temporal scales. Each higher level can only be impacted when all lower levels are also impacted. Not all levels of biological stress are likely to be preserved in the fossil record

Another common objection to paleocommunity data is the prevalence of time-averaging in fossil assemblages. Hiatuses and condensed intervals are challenges when interpreting paleontological data, however the amount of time that is present in the rock record provides a unique opportunity to observe large-scale, long-term trends. Time-averaging can be a positive factor by averaging out the normal background variations present in high-temporal-resolution data sets. It has even been argued that time-averaged assemblages actually represent a better sample of the average community composition than snap-shot samples from a single hour or day (Walker and Bambach 1971; Olszewski 1999; Kidwell 2008).

When designing a paleoecological study, it is important to keep in mind the question that is being asked as well as potential problems that might prevent the inference of a reasonable answer to that question. For example, if the study seeks to understand global patterns of ecosystem change, it would be insufficient to focus on trilobite communities from one locality. However, if the study seeks to understand a high-resolution record of community response in a particular basin, then the above trilobite section would be appropriate. Designing an appropriate sampling protocol can avoid some biases altogether or at least reduce the prevalence of the bias in a dataset. The following list details the different factors that should go into the design of the study.

1. Selection of Taxa.

    (a) Ecology. In part, selection of taxa is determined by the research question, although in some cases it might be necessary to restrict the purview of a question for practicality. For example, while it might be interesting to review community response to extinctions throughout the entire Cambrian, the practicality of sampling similar communities at a useful temporal resolution would be prohibitive. It would be more useful to examine a single event during the Cambrian, or to choose a single taxonomic group to follow across multiple extinction events. When considering which taxa to include in a study, it is important to either census all taxa within the community or to choose a guild as a proxy. It is acceptable to focus on a guild as a proxy for the entire community, as it is commonly impractical or impossible to sample the entire community (Hubbell 2001, 2005). In addition, the complexity of the entire community can make it difficult to identify the ecological signals that are being analyzed (Brown 1995; Hubbell 2001, 2005). When selecting a guild or guilds to include in the study, remember to consider the different roles each guild is likely to play within a community. If the study question regards the community response in general, then a guild that is abundant, diverse, and ecologically dominant within the community should be selected. If the question deals with the response of a particular taxonomic group or life history, then the guild should be selected accordingly. However, the response of a single guild may not be representative of the community as a whole (Morris 1986; Forcino et al. 2010). Therefore, interpretations of single-guild studies need to address the suitability of the guild and the implications if the guild is not a good proxy. It is advisable to collect as much information about the entire community as possible, as later analyses can be restricted to individual guilds within a dataset but cannot include data that have not been collected.

(b) Taphonomy. Differential preservation can be addressed initially in the selection of taxa to be included (shelly fauna, reef builders, complete assemblage) and/or the quality of preservation at available localities. Comparison of biomineralized organisms with different skeletal compositions requires localities that preserve both mineralogies (such as calcite and aragonite for a comparison of brachiopods and molluscs). Once the taxa and localities have been selected, individual samples must be examined for evidence of biased preservation, and samples with obvious bias should be excluded.

2. Selection of Time Intervals. Often the division of a time period into intervals will be subject to practical constraints such as the current state of stratigraphic divisions, correlations, and absolute dating between localities. These factors will determine the minimum size of temporal intervals within a study. Uncertainty related to the stratigraphy or dating may make it more conservative to increase the duration of intervals used within the study. However, an ecological signal could exist within the middle of a long time interval, but because those samples with a signal are combined with others the signal will be diluted. If there are too few intervals within the time period being studied, statistical analysis will be difficult (it may not be possible to detect significant ecological signals).

3. Avoiding Taphonomic Biases.

   (a) Time-averaging. Extensive time-averaging within a sample can obscure the original taxonomic composition. If fossils display variable or high levels of breakage and abrasion, this is evidence of biological or physical reworking (with high energy involved) and possibly of time-averaging. Differential preservation within a single taxon also can indicate time-averaging or changing taphonomic conditions, both of which can bias the composition of an assemblage. If environmental preferences are known for taxa within an assemblage, the mixing of taxa from different environments can also be an indicator of reworking and greater than average time-averaging. Although it is wiser to avoid evidence of extensive time-averaging, it is possible that an ecological signal will still be present even though the community has suffered substantial taphonomic alteration, which would suggest that the original signal was strong enough to survive the dilution of time-averaging. For example, a microfossil core from an interval of global warming had evidence of reworking that smeared the environmental and taxonomic signals up and down the core; however, it was still possible to detect an ecological response to the warming event despite taphonomic overprinting (Webb et al. 2009).

   (b) Transport. Peri- or post-mortem transport can bias a fossil assemblage against the original (living) composition by selective transport (related to size or shape) or the mixing of assemblages from different environments. Evidence for transport includes size-sorting, imbrication, linear alignment of elements, or turbidite facies. Samples displaying any of these indicators should be excluded from paleoecological analyses.

4. Sample Size.
   (a) Minimum and Maximum Bounds. Generally, between 20 and 30 individuals per community is the bare minimum necessary to begin capturing a community signal, and the likelihood of an accurate representation of the actual population increases with sample size (Hayek and Buzas 1997). Sample sizes of 75–100 individuals have been suggested as a minimum for abundance-based calculations (Chang 1967; Fatela and Taborda 2002). Larger samples are more likely to include rare taxa, but eventually the likelihood of discovering a new taxon decreases to near zero. Many microfossil studies sample until the number of individuals is approximately 300 as this number has been shown to consistently be beyond the point at which it is likely to add new taxa with further sampling, although there is still debate about whether 300 is too many or too few (Fatela and Taborda 2002). Although larger sample sizes are desirable, there will be a point of diminishing returns, where more time is spent gathering data than information or statistical power is gained by increasing sample size. Studies investigating community signals require smaller sample sizes to capture community structure (e.g., with ordinations) than do studies seeking to accurately report total diversity (microfossils: Fatela and Taborda 2002; macrofossils: Forcino and Leighton 2010). For example, Forcino and Leighton (2010) showed that a 2 liter subsample of a 13 liter bulk sample exhibits the same community signal as the original bulk sample based on ordinations and evenness metrics (for original dataset and community patterns see Forcino et al. 2010).
   (b) Sampling Effort. Sample size is also constrained by sampling effort, which should be similar for each sample. If a set number of individuals is used as the sample procedure, then it is advisable to record the amount of effort required to reach that quota (i.e., how much sediment was processed or how large of an area was surveyed). For example, the abundance (density) of fossils or particular taxa can be an important indicator of ecology, environment or sedimentation rate. A shell pavement is very different from a semi-barren mudstone. Similarly, if sampling is conducted for specific amounts of time, it is also important to record how much sediment/area is surveyed (was the entire time spent sitting in one spot, or was it spent moving over the entire outcrop?). Sampling a given amount of sediment or surface area is the most consistent method for quantifying sampling effort, although it often results in variable sample sizes (numbers of individuals). However, if the amount of sediment or size of the area is known, it is possible to standardize differences between samples by sub-sampling to the smallest sample size, as well as to test for statistical biases due to different abundance in samples using Multivariate Analysis of Variance (MANOVA) or linear regression models. Absolute abundance reflects the density of fossils within each sample that can be related to biological factors (species blooms) as well as environmental factors (sedimentation rate, etc.), both of which should be considered when interpreting the results. When possible,

it is useful to collect replicate samples as this allows for a quantification of the lateral variation and as multiple small replicate samples provide a better estimate of the actual population than one large sample (Bennington and Rutherford 1999; Bennington 2003). If variation is greater in sample replicates (lateral variation or patchiness) than between time intervals, then the temporal signal is less significant than the heterogeneity in contemporary communities. Changes in lateral variation through time can also provide an important ecological signal reflecting environmental heterogeneity.

(c) Counting Methods. The taxa included in a study should guide the choice of how to measure for abundance. Abundance can be based on relative abundance (rare, uncommon, common, abundant, etc.), raw abundance (the number of individuals), biovolume, biomass, or percent cover (how much of a given area is covered with a colony or individual, commonly used for plants, encrusting organisms, and colonial organisms). Relative abundance reduces the power of statistical analyses but can be used if necessary. Raw abundance is the most useful for solitary organisms with single element skeletons (such as foraminifera). However, if the taxon easily disarticulates (e.g., echinoderms), molts skeletal elements (e.g., arthropods), or is colonial (e.g., corals), then raw abundance can bias results against solitary organisms. The bias can be partly avoided by estimating the number of skeletal elements that represent an individual (e.g., five crinoid calyx plates equal one individual), only counting adult instars or counting a colony as an individual. When counting individuals that disarticulate into few pieces (such as bivalves), it is possible to treat each skeletal element as an individual without introducing a statistically significant bias (Gilinksy and Bennington 1994). Biovolume (measured directly or estimated using point count grids, linear point intercept, average body size for each taxon, or from biomass and density) (Ausich 1981; Watkins 1996), biomass (measured directly or estimated from biovolume and density) (Forcino et al. 2010), and percent cover avoid most of these biases, except for taxa that molt. Generally, only one abundance measure is used (for examples of paleoecological extinction studies using techniques described in this section see Payne et al. 2006; Layou 2009), however comparisons between biovolume and raw abundance can reveal important differences (e.g., compare the biovolume of a mouse to an elephant, or the number of krill that would be required to equal the mass of a baleen whale) (Forcino et al. 2010; Webb 2008 Webb and Leighton 2008).

5. Environmental Variables. The impact of environmental gradients on species composition has been well documented (Connell 1961; Whittaker 1965; Lomolino and Channell 1995; Peters and Bork 1999; Thomas et al. 2004; Pigliucci 2008). Whenever possible, environmental variables, such as relative water depth, energy, and sediment type, should be collected for each sample. As extinction events are commonly accompanied by environmental change (Twitchett 2006),

it can be difficult to show whether the ecological signal is environmental or from the extinction. For example, Peters (2008) demonstrated that extinction selectivity during Mesozoic and Paleozoic mass extinctions was related to sediment type (siliciclastic or carbonate, respectively). When possible, it is advisable to sample from the same environments or suite of environments within and across time intervals. Environmental factors affect not only species composition, but also community structure. Statistical analyses that are useful for exploring environmental signals within a community dataset include ordinations (mapping environmental variables onto sample ordinations), Multivariate Analysis of Variance (MANOVA), and linear regression models.

Multivariate data sets, especially with multiple data types (species counts, environmental variables, time, etc.), are more difficult to analyze as there are more possible biases that can affect the analysis and interpretation. The guidelines presented above can be used to minimize some biases. The following sections will address how to analyze and interpret paleoecological data and how to avoid overlooking potential biases.

8.3.2.2 Ordinations

Beyond revealing groupings for analysis of guild or lineage extinction rates, ordinations can also be used to explore the nuances of taxonomic turnover before or after an extinction horizon. As stated above, ordinations cannot detect gradients within data with high rates of turnover; in order to use ordinations for exploration of taxonomic turnover, the dataset must be subdivided by comparing taxa from intervals before or after an event. The ordination can be run using an interval-taxon matrix (as discussed in Sect. 8.3.1), or by using a sample-taxon matrix where the samples are communities to be compared within an interval (i.e., before the extinction) or across intervals (i.e., before and during). Once an ordination has been plotted, it is possible to map on independent information about the samples. Each point in the ordination is a specific sample, and information such as time, location, environment, extinction rate, origination rate, most abundant taxon, etc., can be mapped onto each sample. Once the sample information is mapped onto the ordination, it becomes easier to spot if specific samples are grouping together based on time or environment, or some combination. If clear groupings or gradients are observed in a mapped variable, it is possible to use statistical analyses on the ordination scores and the variables that were mapped onto the ordination (classification analyses [e.g., cluster analysis, discriminate function analysis] or tests of similarity/difference [e.g., t-tests, regressions, correlation]). Variables for an ordination of taxa (as opposed to the previous example of a sample ordination) include relative size, mobility, feeding strategy, higher taxonomic affinity, environment, time of extinction, etc. The same matrix is used for sample and taxon ordinations, however the columns and rows are reversed to calculate the new distance measures.

8.3.2.3 Evenness

Simple diversity (sometimes termed "richness"), or the number of species present, is a traditional descriptor of a community (Hayek and Buzas 1997; Magurran and Henderson 2003). However, the simple diversity of a community does not reveal information about the distribution of individuals among the taxa in that community (Magurran and Henderson 2003). For example, two communities could each have 10 individuals that can be divided into three species. In community α, eight individuals are species A, with only one each of species B and C. In community β, there are four individuals of species A and three each of species B and C. In both communities, species A is the most abundant, but the abundances of B and C are very different. To quantify this difference, diversity indices measuring evenness are employed to describe community structure. Evenness describes how evenly distributed the individuals are among taxa within a community (Hayek and Buzas 1997; Magurran and Henderson 2003; Bulinski 2007). A community with perfect evenness would have the same number of individuals in each taxon present in the community (i.e., a community sample of 30 individuals comprised of three taxa each with 10 individuals). If one taxon is much more abundant than all others, then evenness will be very low. In the above example with communities α and β, community α has low evenness and community β has high evenness. Similar to calculations of diversity and extinction rates, there are many different equations for calculating evenness. Some metrics are more sensitive to small sample abundances and become unstable. At larger sample sizes, the different evenness metrics behave similarly (Bulinski 2007). For a discussion of the merits of different evenness metrics, see Bulinski (2007). A more in depth discussion of the derivation and application of various diversity indices is presented in Chap. 13 of Hayek and Buzas' (1997) *Surveying Natural Populations*. For the purposes of this chapter, Buzas and Gibson's evenness (Hayek and Buzas 1997; Buzas and Hayek 2005) will be used, as it is simple to calculate and the results are intuitive. Using Shannon's H and the richness of the community, Buzas and Gibson's evenness is the proportion of equally dominant taxa in the community ($E = e^H/S$, where E is evenness, H is Shannon's H, and S is the number of species in the community; $H = -\sum p_i \ln(p_i)$, where p_i is the proportion of the i-th species) (Buzas and Hayek 2005).

8.3.2.4 Rank–Abundance Curves (RACs)

While evenness can reveal changes in community structure (shifts in the proportion of ecologically dominant species), it does not have the sensitivity to detect changes in abundance among the non-dominant (rare) species. To address this lack of sensitivity, Webb and Leighton (Webb 2008 Webb and Leighton 2008; Webb et al. 2009) developed a novel application of rank–abundance curves (RACs), also known as species-abundance distributions in modern ecological literature (Halloy and Barratt 2007; McGill et al. 2007). A RAC is a visualization of the abundance distribution of taxa within a community (Fig. 8.1). The relative abundance of taxa

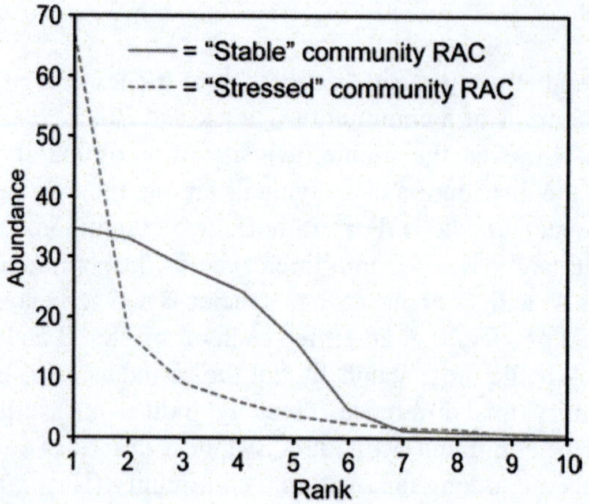

Fig. 8.1 Hypothetical rank–abundance curve. The two curves presented in this plot were created from hypothetical communities with the same number of individuals and taxa, only with different distributions of individuals among those taxa. The *dashed line* represents a community that is dominated by one taxon, and the curve shape is very peaked and convex-down. The kurtosis value for this curve is relatively high (>>3) and would be interpreted as representing a stressed community. The *solid line* represents a community with a more even distribution, and the curve shape is relatively flat and convex-up. The kurtosis value for this curve is low (<3) and would be interpreted as representing a stable or mature community with low stress. RACs are sometimes plotted on log axes, however we prefer to use the un-transformed curve as the log curve visually overemphasizes the tail of the curve

within a community is related to resource partitioning; taxa that are more successful at securing the necessary resources will be more abundant than other taxa (Whittaker 1965; Parrish and Bazzaz 1982). The taxa within a community are ranked according to their abundance in the sample, with the most abundant taxon receiving a rank of one and the least abundant taxon receiving the highest rank (which is the same number as the number of taxa in the sample) (Bazzaz 1975). Taxonomic rank is plotted on the x-axis, and the abundance of each taxon is plotted on the y-axis (Fig. 8.1). RACs are used in modern ecological studies to assess ecosystem health (Gray 1981; Belaoussoff and Kevan 1998; Halloy and Barratt 2007) after Bazzaz (1975) showed that the shape of RACs changes predictably through an ecological succession from a recently disturbed community to a climax community. Recently disturbed communities display a relatively convex-down or peaked curve.

RAC shape can be quantified using kurtosis (Webb 2008; Webb and Leighton 2008; Webb et al. 2009), a statistical measure of the peakedness and the tail weight of a distribution (the amount of the distribution that is shifted to/from the tails compared to a normal distribution) (Balanda and MacGillivray 1988; Joanes and Gill 1998). Kurtosis is used as a measure of the direction in which a distribution differs from a normal or Gaussian curve (which has a kurtosis value of 3); kurtosis can be applied to any uni-modal distribution, not being limited to bell-shaped curves (Joanes and Gill 1998). Traditionally kurtosis is defined as the fourth moment of a distribution (m_4/m_2^2, where $m_r = 1/n \sum (x_i - \overline{x})^r$). A peaked (leptokurtic) curve has kurtosis >3, while a flattened (platykurtic) curve has kurtosis <3.

Although kurtosis is scale-independent, the maximum value is constrained by the number of observations (sample size) because the peak of a distribution of 9:1 (10 individuals, where 9 are one taxon and one is another taxon) is smaller proportionally than 99:1 or 999:1. A community with perfect evenness will have similar kurtosis values, whether there are 10, 100, or 1,000 individuals. However, the maximum possible kurtosis for a community with 1,000 individuals is much higher than for a community with 10 individuals. The result is that for any given sample size, there are fewer possible kurtosis values for platykurtic distributions than there are for leptokurtic distributions. As kurtosis is being used as a proxy for stress, the constraint of sample size on maximum possible kurtosis does present a potential bias, however, because smaller sample sizes have lower maximum possible kurtosis values, the direction of the bias is conservative (indicating lower relative stress levels, rather than artificially inflating stress). Webb et al. (2009) used a foraminifera dataset to examine the effect of sample size on kurtosis by subsampling 50 individuals from the original samples of ~300. The average kurtosis for a sample size of 50 was significantly correlated to the kurtosis of the original samples (Pearson's correlation, $r=0.98$, $p<0.05$), indicating that sample size does not bias kurtosis.

For the purpose of quantifying community structure, the most important feature of a RAC is the relative shape of the curve, rather than the number of taxa (richness) or the absolute abundances of those taxa (evenness). For this reason, RAC kurtosis is calculated for the frequency distribution of the RAC's taxon ranks. This frequency distribution is generated by transforming the original abundances into a distribution of the different taxonomic ranks. Using the hypothetical community β from the evenness example in Sect. 8.2.3, there were 4 individuals of taxon A and 3 each of taxa B and C, which yields a distribution of (4, 3, 3). For the RAC of community β, taxon A would be given the rank of 1 (most abundant), and taxa B and C would be ranked 2 and 3, interchangeably. The RAC frequency distribution for community β is (1, 1, 1, 1, 2, 2, 2, 3, 3, 3). In order to quantify the shape of the RAC, it is important to transform the abundances into rank frequencies, as the distribution of abundances alone is not informative about the curve shape. The frequency distribution assigns each individual a taxon rank, thereby directly representing the frequency (abundance) of that taxon in the community. The frequency of different taxonomic ranks is a better representation of the distribution of individuals in a community than is the distribution of abundances. The transformation of the original abundances into a taxon-rank-frequency distribution allows a more direct measure of the shape of the RAC, rather than a measure of the distribution of abundance which is strongly biased by total abundance and species richness in a sample. Although both total abundance (sample size) and species richness are important components of diversity/evenness, metrics measuring community structure should not be only a reflection of sample size and richness. The kurtosis of a RAC provides a new way to quantify community structure, revealing nuances not available in richness or evenness.

As a community matures (without disturbances), its RAC becomes more convex-up and flat (lower kurtosis). These changes can be scaled up to geological time scales on the logic that fossil assemblages represent an average community structure which can then be quantified by the kurtosis of the RAC generated from the assemblage (Webb 2008; Webb and Leighton 2008; Webb et al. 2009). If a fossil assemblage has experienced disturbances (stress), then the RAC will reflect the recency or prevalence of disturbances. Therefore, an extinction event, which is a period of major stresses leading to taxonomic loss, will leave a distinct signature on communities during the extinction interval (Table 8.1). For example, foraminifera and ostracod communities display changes in community structure before the extinction horizon during the Paleocene-Eocene thermal maximum, with increasing stress levels providing a warning sign before taxonomic loss (Webb et al. 2009). Following the interval of global warming, community structure recovered before diversity returned to pre-event levels. Brachiopod communities during the Ordovician-Silurian mass extinction also display increasing stress before the extinction horizons, followed by a relatively rapid ecological recovery millions of years before the taxonomic recovery Webb 2008; Webb and Leighton (2008). These examples of rapid ecological recovery support Harries et al.'s (1996) models for recovery from mass extinction. Examining modern and fossil community dynamics provide a medium for demonstrating broad ecological paradigms that are largely theoretical. For example, identifying disaster taxa before an extinction is difficult, even in modern systems (Harries et al. 1996), as the logic is often circular (there is an extinction, which is a disaster, and this taxon is abundant, so it must be a disaster taxon). RACs provide a way to identify early warning signs before the extinction event as well as identifying when communities return to less-stressed conditions, regardless of taxonomic composition (Webb 2008; Webb and Leighton 2008; Webb et al. 2009).

Care must be taken when utilizing RACs to compare community structure, as different environments (e.g., forest, subtidal, deep ocean) and different types of communities (e.g., brachiopod, foraminifera, plant) have different baseline kurtosis values (Bazzaz 1975; Peters and Bork 1999; Webb 2008; Webb and Leighton 2008; Webb et al. 2009). In addition, Wagner et al. (2006) used RACs (or species-abundance distributions) to argue that community complexity has increased through the Phanerozoic based on their observation that the average shape of community RACs has changed through time gradually. Therefore, RACs require the establishment of a baseline kurtosis value well before the onset of the extinction event. Another constraint is the requirement of more than two taxa in a community, as it is impossible to generate a RAC from a community with only two taxa (two points create a line, not a curve). Such communities can be considered as being monopolized by one or two taxa. Rather than discarding monopolized communities from the analysis of community stress, the proportion of monopolized communities to the total number of communities in an interval can be calculated. The proportion of monopolized communities should be positively correlated to kurtosis, as a monopolized community is analogous to an extremely peaked curve and by inference a stressed community (where one or two taxa are overwhelmingly more abundant than other taxa within a community). Webb and Leighton demonstrated a positive relationship between the proportion monopolized communities and the average kurtosis of those communities for Ordovician-Silurian brachiopods, even with the removal of kurtosis values >40 (these communities were instead included in the proportion monopolized), providing a conservative indication of relative stress (Webb 2008; Webb and Leighton 2008). As with any statistical method, certain assumptions and conditions must be met, and once the conditions of the RACs and kurtosis are met (quality of data, baseline, etc.), RACs provide a powerful tool for analyzing the relative stress levels of fossil communities.

8.4 Examples

Background extinction rates were much higher before and during the Cambrian than the rates observed through the rest of the Phanerozoic (Bambach et al. 2004). The high extinction rates of the Cambrian are commonly overlooked as researchers focus on the Cambrian as the dawn of metazoan evolution (in which we have a vested interest, being metazoans) rather than a period with remarkable rates of both origination and extinction. Rapid taxonomic turnover characterizes most of the Cambrian, resulting from high rates of origination and extinction. Such elevated rates result in large fluctuations of global diversity, as high origination rates can lead to explosive radiation even as the high rates of extinction can lead to catastrophic extinctions, much like gambling. A high risk bet has a high payoff if the gambler wins; however, if the gambler loses the bet, the financial loss can be ruinous (Kozek 1995). If the gambler suffers a series of losses, he will quickly lose any capital he has gained, as well as his initial bet, resulting in the gambler having to leave the game. Now substitute species for dollars and consider the impact of several high payoffs (originations leading to an explosive radiation) or several losses in a row (extinctions that could have led to the complete extinction of biomineralized/metazoan life on Earth). Obviously macroevolution did not lose the gamble, however the pattern of diversity during the history of early metazoan life records the precarious state of evolution. The Cambrian radiation records the net "winnings," as originations remain more common than extinctions. However, there were significant "losses" during the same interval. For example the Cambrian and Early Ordovician are well known for the distinct biomeres used for biostratigraphy (Stitt 1975; Palmer 1984; Holland 1995; Saltzman et al. 1995; Thomas 1995; Westrop and Cuggy 1999). The biomeres represent rapid pulses of taxonomic turnover, requiring periods of elevated extinction and origination rates interspersed with periods of much lower rates (Thomas 1995).

At the transition from the Early to Middle Cambrian, an extinction event comparable in terms of percentage taxa lost to the Permo-Triassic extinction occurred (Zhuravlev and Wood 1996; Zhuravlev 1996; Erwin 1998, 2001; Bambach et al. 2004). However, as the background extinction rates during the Cambrian were already elevated, the loss of >50% of genera twice within a few million years is not as out-of-ordinary as the drastic increase in extinction rate that marks the Permo-Triassic extinction. Nonetheless the end-Early Cambrian (Botomian) extinction was a unique extinction event in terms of the patterns and processes revealed about extinction in early metazoan life. The Botomian extinction was different from other Paleozoic mass extinctions in that the majority of extinctions occurred among a single group, the diverse archaeocyathans, while extinction rates in other taxonomic groups were much lower (although still elevated from background rates) (Zhuravlev and Wood 1996). Numerous explanations for the extinction cause have been proposed, from ocean anoxia to volcanism to global warming (Zhuravlev and Wood 1996; Hough et al. 2006). Rapid fluctuations in $\partial^{13}C$ occur throughout the extinction interval and have been interpreted as linked to the environmental and biological changes associated with the extinction (Brasier et al. 1994). Regardless of the cause, the extinction was a global phenomenon (Li et al. 2007; Zhuravlev and Wood 1996). The following examples will use data from this interval to explore the ecological dynamics of the Botomian extinction(s).

8.4.1 Extinction Rates

8.4.1.1 Results

Zhuravlev and Wood (1996) compiled a taxonomic dataset for the Early Cambrian (Appendix 8.1). Using their diversity and extinction counts for marine invertebrate taxa (excluding the problematic "small shelly fauna"), the extinction and origination rates were calculated by dividing the number of extinctions or originations by the total diversity for each time interval. The highest rates of extinction occur during the Botomian (intervals 11, 13, and 14), which was also a period of low origination rates (Fig. 8.2). Figure 8.2 clearly shows the large fluctuations in diversity during the Cambrian Radiation that resulted from high rates of both origination and extinction. Similar patterns are observed in all taxonomic groups included in the calculations (Appendix 8.1).

8.4.1.2 Interpretation

The first step in interpreting diversity data is to check for sharp changes in diversity. A visual examination of the diversity curve allows unique intervals to be targeted for further examination. For example, in Fig. 8.2 there are two sharp drops in diversity during the Botomian. Adding in the extinction and origination rates allows further interpretation of the diversity loss. A diversity loss could be caused by a number of scenarios (all of which have been observed during the Phanerzoic as shown by Bambach et al. 2004): (1) no change in origination rates but an increase in extinction rates, (2) a decrease in origination rates and no change in extinction rates, (3) a slight decrease in origination rates and a slight increase in extinction rates, or (4) an increase in origination rates but an even larger increase in extinction rates. Origination rates reflect the relative pace of evolution, as high rates are likely caused by environmental heterogeneity, community complexity, or unoccupied niche space (such as that opened by the evolution of skeletons) (Schluter 2000). Extinction rates are more likely to reflect the relative amount of stress on a system, as each taxon that becomes extinct requires the death of many individuals. Diversity

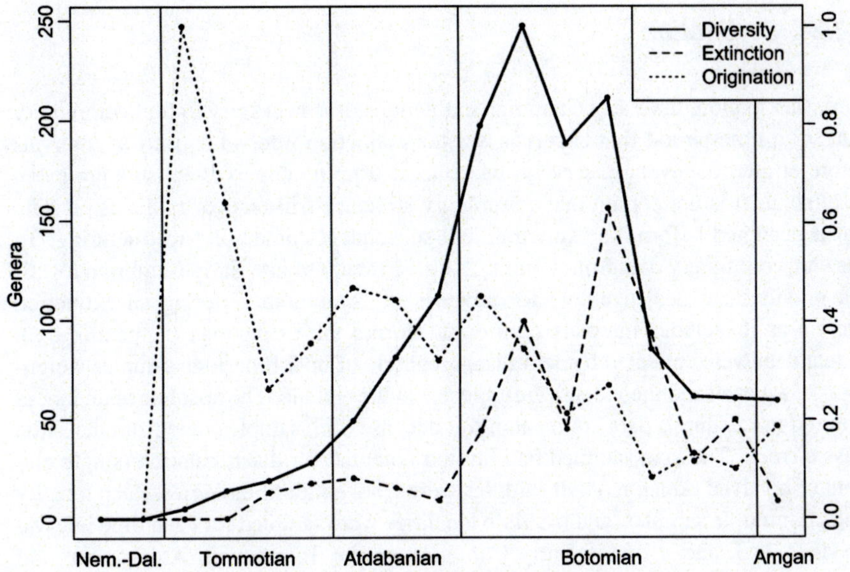

Fig. 8.2 Diversity and evolutionary rates during the Early Cambrian. The data used to generate these curves are presented in Appendix 8.1, which is taken from Zhuravlev and Wood (1996). Nem.-Dal. is an abbreviation of the Nemakit-Daldynian Stage. The Toyonian is combined into the Botomian as per initial divisions of the International Commission on Stratigraphy Time Scale (Gradstein et al. 2008)

declines in the Early Cambrian were caused by high rates of extinction with little change in origination rates. The two spikes of extinction are in accordance with Zhuravlev and Wood's (1996) interpretation of the global diversity data, placing two major extinction events during the Botomian. The extinction rates for the various taxonomic groups included in the analysis reflect a similar pattern, suggesting that the causes of the extinctions were widespread stressors affecting all environments. If different taxonomic groups were affected to various degrees, then it would be possible to infer the type of stresses leading to differential extinction. For example, the Cretaceous-Paleogene extinction rate was 47% for marine genera, but the extinctions were restricted to specific groups, such as ammonites, rudist clams, coccolithophores without resting stages, etc. (Twitchett 2006). This pattern suggests distinct selection pressures, rather than widespread selection pressures such as the almost complete extinction of all life forms at the Permo-Triassic extinction (Twitchett 2006). The pattern in Zhuravlev and Wood's (1996) data does not indicate that there are specific taxa that should be targeted for examination of the ecological dynamics of the extinction, although archaeocyathans are the most diverse group with the highest rates of extinction. However, in the case of Cretaceous-Paleogene communities, it would be important to select taxa that were affected in the extinction, and if desired, a group not affected by the extinction could serve as a control group for comparison of ecological response to the extinction.

8.4.2 Ordination

To further explore the Early Cambrian extinctions, it is necessary to find community data for a time period that encompasses the extinction interval as well as intervals before or after the event to serve as baseline. It is preferable to establish a pre-event baseline, as it is not certain that community structure will recover to the same state that it occupied before the extinction and subsequent community restructuring. To this end, community data from Landing's (1988) study of late Early Cambrian localities provides an ideal example for analysis of communities during an extinction interval as it includes intervals before and during the extinctions (Appendix 8.2). Faunal data were collected by acid-disaggregation of bulk limestone samples weighing ~4.5 kg each and then identifying species in the residue. The absolute abundances reported in Landing's data are meaningful because each sample came from the same mass of rock. The taxa sampled had limited potential for disarticulation (single element or bivalved skeletons). All samples were from similar lithologies. Each locality yielded multiple samples, and multiple localities were sampled for each time interval (~Stages 2, 3, and 4 of the Early Cambrian or the Tommotian, Atdabanian, and Botomian). Three samples were excluded from analysis for having <20 individuals.

8.4.2.1 Results

Using the R statistical packages, 'vegan', 'asbio', and 'timeDate' (R Development Core Team 2009), ordinations of the samples were calculated. Both polar ordination and principle coordinates analysis were explored using the distance measures Sorenson, Jaccard, and Whittaker. Although distance between samples shifted, the relative position of samples in ordination space was not altered. Figure 8.3 is a polar ordination using Sorenson distance. Samples separated into two groups, pre-Botomian and Botomian. Samples did not group by locality (which is equivalent to lithology). An ordination of the phyla occurrences (the total number of individuals of all species within each phylum that were observed in each sample) displayed the same relative pattern as the ordination of the species occurrences.

8.4.2.2 Interpretation

During extinction intervals, ordinations cannot resolve community gradients because the main gradient present in the data is that of taxonomic turnover related to the extinction event. The extinction bias in the ordination is obvious in the Landing dataset as the Botomian samples clustered together, and the other intervals (Atdabanian and Tommotian) clustered together. The division between the extinction and pre-extinction intervals in the ordination is a visual confirmation of the pattern predicted by the diversity curve and extinction rates. Due to the size of the dataset (only 17 samples), and the lack of difference between samples (similar environments, water

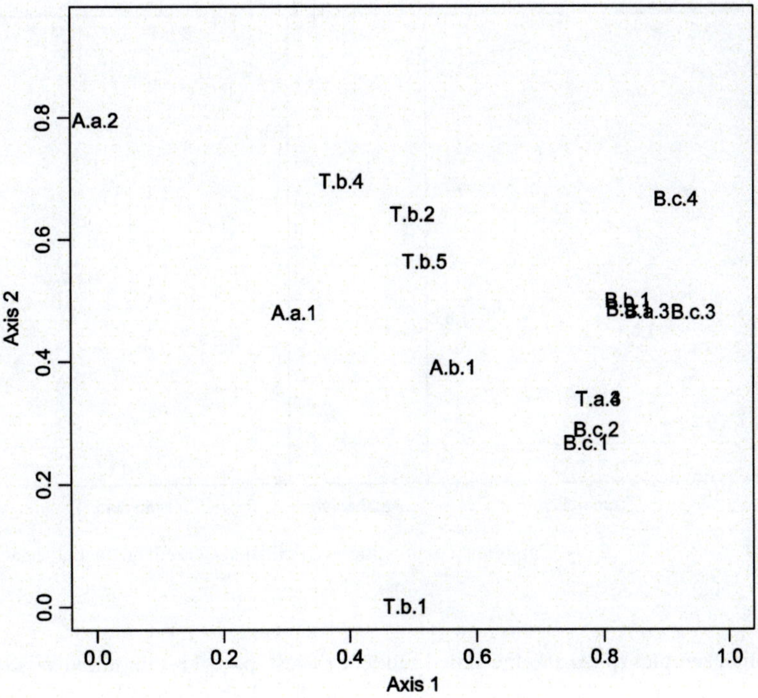

Fig. 8.3 Polar ordination of Early Cambrian communities. The data used to calculate this ordination are presented in Appendix 8.2, which is taken from Landing (1988). The *first letter* of each point refers to the time interval (Tommotian, Atdabanian, Botomian), and the *second letter* refers to the locality with the *numbers* indicating stratigraphic level at each locality

depth, and lithology), there is little to be gained by further ordination analyses. If there were more samples and a greater variety of environmental variables, it would be useful to divide the original dataset into pre-extinction and extinction intervals. Dividing the dataset at extinction boundaries removes the extinction gradient present in the undivided dataset and allows the ordination techniques to reveal other ecological gradients. Comparisons can then be made between the pre-extinction and extinction ordinations as to which variables appear to be structuring the different communities, and if different factors are active at different times.

8.4.3 Community Structure

8.4.3.1 Results

Further analysis of Landing's dataset reveals changes in community structure within Botomian communities. Evenness was slightly, but not significantly ($p>0.10$, one-tailed Student's t-test), lower in Botomian samples. Kurtosis was slightly higher in

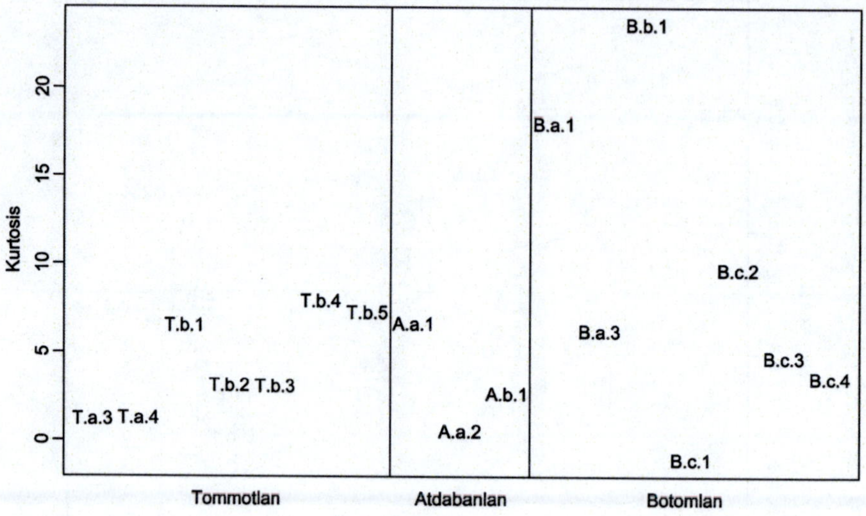

Fig. 8.4 Kurtosis for Early Cambrian communities (Calculations based on faunal data from Landing (1988))

Botomian samples ($p<0.10$, one-tailed Student's t-test; $p=0.11$, one-tailed Wilcoxon rank sum test) as can be seen in Fig. 8.4. The highest kurtosis and lowest evenness values were observed in Botomian communities, and kurtosis values were significantly more variable in Botomian communities (F-test, $p<0.05$). Although there was not a significant difference in richness ($p>0.05$, one-tailed Student's t-test), Botomian samples had slightly lower diversity, including the samples with the lowest diversity observed in the entire study interval. Sample size was not significantly different in Botomian samples. Kurtosis was significantly correlated (Pearson's correlation coefficient, $r>0.48$, $p<0.05$; $r>0.61$, $p<0.01$) with evenness ($r=-0.82$) and sample size ($r=0.63$); however, kurtosis was not significantly correlated to richness. Sample size was significantly correlated with both richness ($r=0.49$) and evenness ($r=-0.69$).

8.4.3.2 Interpretation

Based on the analysis of extinction rates and the separation of Botomian communities in ordination space, the prediction for community structure was that stress levels would increase during the extinction interval. Higher levels of stress would be reflected in the community structure as communities would not have the chance to mature before the next disturbance impacted the area, leading to lower evenness and more peaked RACs. As there is natural variation in community structure, related to community type and small scale stresses, some variation is expected in RAC shape and evenness even during intervals with no extinction. However, variation during intervals with high predicted stress should be well beyond that observed in intervals with low predicted stress (such that the two distributions are statistically distinguishable).

Although it is possible to predict stress based on extinction rates and ordinations, the timing and magnitude of the stress can only be observed and quantified through analyzing RACs. In the Landing dataset, there is some variation in community structure within the Tommotian and Atdabanian samples, which is used as a baseline for normal variations in community structure. When the pre-extinction communities are compared to the Botomian communities, the difference in community structure is visible. The use of a one-tailed Student's t-test (parametric) or Wilcoxon rank sum test (non-parametric) determines if the two distributions of kurtosis values (as a measure of community structure) are not only different, but different in that one distribution has significantly higher or lower values than the other.

This technique can either be used on multiple samples in each of a series of intervals (such as the Landing dataset), or on a time-series of samples such as can be collected at high sampling resolution from a large vertical section of sediment or microfossil samples from a sediment core. Kurtosis and evenness values can be compared to time or environmental variables (such as isotope or sea-level curves), using correlations or t-tests. Although evenness and RACs measure similar aspects of community structure, a comparison of the values for the same samples reveals the differences between these measures. For the Landing dataset, a scatterplot of kurtosis versus evenness reveals the difference (Fig. 8.5). For relatively mature, stable communities kurtosis and evenness change in tandem. However, in communities

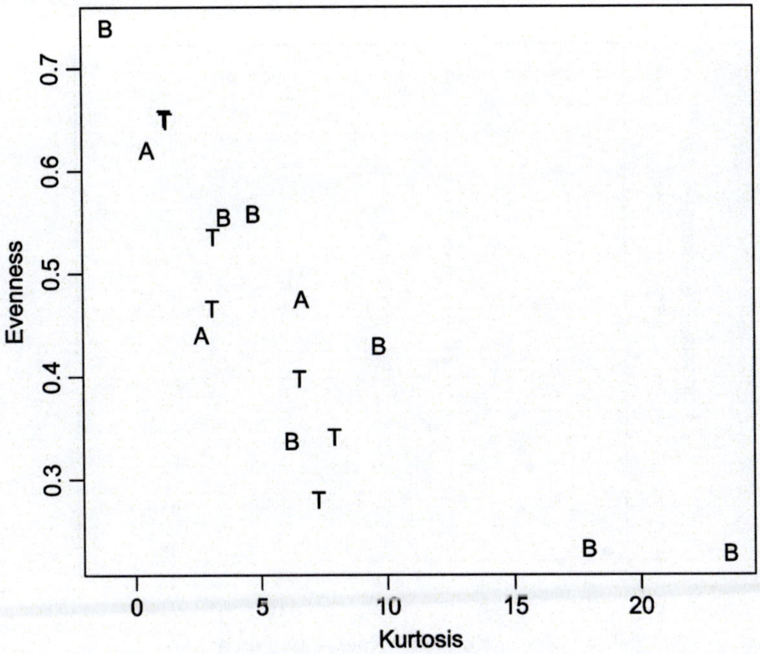

Fig. 8.5 Kurtosis versus evenness for Early Cambrian communities. A comparison of kurtosis values and Buzas and Gibson's evenness. The *letters* refer to the time interval (*T* Tommotian, *A* Atdabanian, *B* Botomian). Note the negative relationship (calculations based on faunal data from Landing (1988))

with the lowest evenness values, kurtosis becomes more sensitive to the community structure by marking communities that have a more peaked RAC (representing higher levels of dominance by one taxon compared to other taxa in the community). The sensitivity of RACs can thereby be used in high stress intervals to determine the nuances of changing community structure that are not revealed by the traditional evenness measures.

When interpreting community data, the abundances may be transformed into percentages to remove perceived biases resulting from variable sample sizes. When the sample size is arbitrary due to a sampling procedure not designed around collecting ecological data, the variations in sample size can introduce bias into an analysis. However, if the data were collected specifically for community analyses (considering the different biases addressed earlier in this chapter), then abundance reflects an important biological or environmental signal. In the case of kurtosis, the use of percentage data makes little difference (Fig. 8.6). Kurtosis is a scale-independent measure, making it ideal for application to ecological data. Even when sample sizes differ by an order of magnitude, the kurtosis of the RAC does not significantly differ, as was tested by bootstrapping a microfossil dataset with an average sample size of 300 subsampled to 50 (Webb et al. 2009). The combination of sensitivity and stability for the kurtosis of RACs makes this relatively new metric a useful tool in the analysis of ecological dynamics.

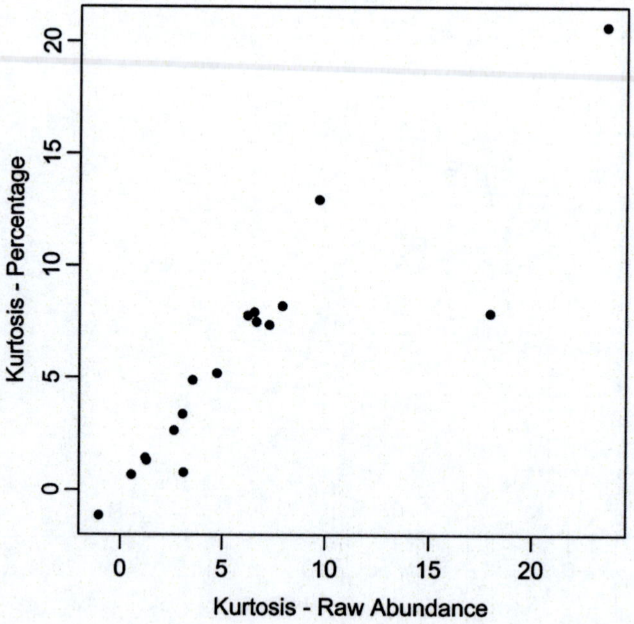

Fig. 8.6 Kurtosis – Raw Abundance versus Percentage Data. A comparison of kurtosis values calculated from raw abundance data and data transformed into percentages (calculations based on faunal data from Landing (1988))

8.5 Implications for Future Work

The results presented here represent the first attempt to quantify the ecological impact of the Botomian extinction. Given the importance of this interval in terms of studying the evolution of metazoan life and the uniquely high rates of both origination and extinction, it is particularly important to investigate the ecological dynamics of Pre-cambrian and Cambrian communities. The elevated levels of stress observed in Botomian communities suggest that there was an extended interval of stress that impacted Early Cambrian communities. The results from the Landing dataset are particularly interesting in that no archaeocyathans were included in the dataset. The Zhuravlev and Wood dataset clearly indicated that although all taxonomic groups were affected by the Botomian extinction, archaeocyathans were by far the most impacted. Therefore, the indication from the Landing dataset that Botomian communities were stressed is actually a conservative indicator of the stress levels during the Botomian. The findings of this study suggest that the evolution of metazoan communities was shaped not only by rapid taxonomic turnover, but also by significant intervals of community stress and extinction events. This preliminary study shows the promise of further work on the ecological dynamics of early metazoan extinctions. Future studies have the opportunity to examine the response of different guilds, the signal from different environments and different regions, all during one of the most critical intervals of evolution and extinction in the history of metazoan life.

The methods presented in this chapter are applicable to any guild, environment, or interval in the fossil record, although different challenges will be faced for each new investigation. In some ways, workers focusing on Pre-Cambrian and Cambrian ecological dynamics face extra challenges. Due to the age of the sediments, exposure and preservation are some of the largest potential stumbling blocks. Erosion and metamorphosis of the relevant strata also make it difficult to accurately reconstruct the paleoenvironment. Beyond challenges with the actual rocks, the taxonomic affiliations and interpretation of life history for different taxa presents further potential problems when investigating ecological patterns. Although these issues may complicate future studies, the preliminary study presented in this chapter proves that it is possible to apply the suggested methods. Often the most challenging projects yield the most rewarding results, and the results of this study suggest that future work on the ecological impact of early extinction events will prove very rewarding.

8.6 Summary

Extinction events provide the ideal opportunity to study ecological dynamics within past systems. The extinction acts as stimuli to the system, resulting in distinct responses that can be detected in the fossil record. This chapter has outlined the unique opportunities available through examining the ecological impact of extinction events as well as introducing the techniques that can be used to pinpoint ecological responses. Using the guidelines presented above, researchers can design a sampling procedure specifically for collecting useful community data. Once a suitable dataset is available, the different methods can be used to detect both taxonomic and ecological signals. Although there are many challenges that must be faced when interpreting ecological data, the rewards of revealing the patterns and processes of extinction events will serve to improve our understanding of evolution, extinction, paleoecology, and macroecology. These insights will be useful not only to paleontologists, but also to modern biologists and conservation workers facing the modern biodiversity crisis which rivals the Permo-Triassic extinction in terms of rate and scale (Wilson 2003).

Acknowledgments We thank William I. Ausich, Matthew Clapham, Simon Darroch, and an anonymous reviewer for insightful comments on this manuscript.

Appendixes

Appendix 8.1 Diversity and Evolutionary Rates for Early Cambrian Taxa

	Nemakit-Daldynian		Tommotian				Atdabanian			
	ND1	ND2	T1	T2	T3	T4	A1	A2	A3	A4
Total	0	0	5	14	19	28	49	81	112	197
T Ext	0	0	0	0	1	2	4	5	4	35
T Orig	0	0	5	9	5	10	23	36	36	89
T Ext Rate	0.0	0.0	0.0	0.0	0.1	0.1	0.1	0.1	0.0	0.2
T Orig Rate	0.0	0.0	1.0	0.6	0.3	0.4	0.5	0.4	0.3	0.5
Archaeocyathids	0	0	4	10	15	20	36	58	88	156
A Ext	0	0	0	0	0	1	2	0	1	32
A Orig	0	0	4	6	5	5	17	24	30	69
A Ext Rate	0.0	0.0	0.0	0.0	0.0	0.1	0.1	0.0	0.0	0.2
A Orig Rate	0.0	0.0	1.0	0.6	0.3	0.3	0.5	0.4	0.3	0.4
Total - Arch.	0	0	1	4	4	8	13	23	24	41
T-A Extinction	0	0	0	0	1	1	2	5	3	3
T-A Originations	0	0	1	3	0	5	6	12	6	20
T-A Ext Rate	0.0	0.0	0.0	0.0	0.3	0.1	0.2	0.2	0.1	0.1
T-A Orig Rate	0.0	0.0	1.0	0.8	0.0	0.6	0.5	0.5	0.3	0.5
Calcitic brachiopods	0	0	0	1	1	3	2	6	6	7
B Exts	0	0	0	0	0	1	0	1	0	0
B Origs	0	0	0	1	0	2	0	4	1	1
B Ext Rate	0.0	0.0	0.0	0.0	0.0	0.3	0.0	0.2	0.0	0.0
B Orig Rate	0.0	0.0	0.0	1.0	0.0	0.7	0.0	0.7	0.2	0.1

(continued)

Appendix 8.1 (continued)

	Nemakit-Daldynian		Tommotian				Atdabanian			
	ND1	ND2	T1	T2	T3	T4	A1	A2	A3	A4
Lingulate brachiopods	0	0	1	2	2	4	6	6	7	17
L Ext	0	0	0	0	0	0	1	2	1	3
L Orig	0	0	1	1	0	2	2	1	3	11
L Ext Rate	0.0	0.0	0.0	0.0	0.0	0.0	0.2	0.3	0.1	0.2
L Orig Rate	0.0	0.0	0.0	0.5	0.0	0.5	0.3	0.2	0.4	0.6
Echinoderms	0	0	0	0	0	0	0	0	0	3
E Ext	0	0	0	0	0	0	0	0	0	0
E Orig	0	0	0	0	0	0	0	0	0	3
E Ext Rate	0.0	0.0	0.0	0.0	0.0	0.0	0.0	0.0	0.0	0.0
E Orig Rate	0.0	0.0	0.0	0.0	0.0	0.0	0.0	0.0	0.0	1.0
Corallimorphs	0	0	0	1	1	1	2	2	2	7
C Ext	0	0	0	0	1	0	0	0	0	0
C Orig	0	0	0	1	0	1	1	0	0	5
C Ext Rate	0.0	0.0	0.0	0.0	1.0	0.0	0.0	0.0	0.0	0.0
C Orig Rate	0.0	0.0	0.0	1.0	0.0	1.0	0.5	0.0	0.0	0.7
Stenothecoids	0	0	0	0	0	0	3	9	9	7
S Ext	0	0	0	0	0	0	1	2	2	0
S Orig	0	0	0	0	0	0	3	7	2	0
S Ext Rate	0.0	0.0	0.0	0.0	0.0	0.0	0.3	0.2	0.2	0.0
S Orig Rate	0.0	0.0	0.0	0.0	0.0	0.0	1.0	0.8	0.2	0.0

	Botomian			Toyonian			Amgan
	B1	B2	B3	Tn1	Tn2	Tn3	Am1
Total	247	188	211	85	61	60	59
T Ext	99	34	133	32	7	12	2
T Orig	85	40	57	7	8	6	11
T Ext Rate	0.4	0.2	0.6	0.4	0.1	0.2	0.0
T Orig Rate	0.3	0.2	0.3	0.1	0.1	0.1	0.2
Archaeo	185	137	161	40	12	6	0
A Ext	79	28	123	28	6	6	0
A Orig	61	31	52	2	0	0	0
A Ext Rate	0.4	0.2	0.8	0.7	0.5	1.0	0.0
A Orig Rate	0.3	0.2	0.3	0.1	0.0	0.0	0.0
Tot - Arch	62	51	50	45	49	54	59
T-A Ext	20	6	10	4	1	6	2
T-A Orig	24	9	5	5	8	6	11
T-A Ext Rate	0.3	0.1	0.2	0.1	0.0	0.1	0.0
T-A Orig Rate	0.4	0.2	0.1	0.1	0.2	0.1	0.2
Calcitic brach	10	8	10	12	13	15	14
B Exts	3	0	1	2	1	2	0

(continued)

Appendix 8.1 (continued)

	Botomian			Toyonian			Amgan
	B1	B2	B3	Tn1	Tn2	Tn3	Am1
B Origs	3	1	2	3	3	3	1
B Ext Rate	0.3	0.0	0.1	0.2	0.1	0.1	0.0
B Orig Rate	0.3	0.1	0.2	0.3	0.2	0.2	0.1
Lingulate brach	24	24	25	21	24	25	32
L Ext	1	0	5	0	0	1	0
L Orig	10	1	1	1	3	1	8
L Ext Rate	0.0	0.0	0.2	0.0	0.0	0.0	0.0
L Orig Rate	0.4	0.0	0.0	0.0	0.1	0.0	0.3
Echinod	9	11	7	7	7	9	11
E Ext	2	4	0	0	0	0	0
E Orig	6	4	0	0	0	2	2
E Ext Rate	0.2	0.4	0.0	0.0	0.0	0.0	0.0
E Orig Rate	0.7	0.4	0.0	0.0	0.0	0.2	0.2
Coralom	10	4	3	1	2	2	0
C Ext	9	2	3	1	0	2	0
C Orig	3	3	1	1	2	0	0
C Ext Rate	0.9	0.5	1.0	1.0	0.0	1.0	0.0
C Orig Rate	0.3	0.8	0.3	1.0	1.0	0.0	0.0
Stenoth	9	4	5	4	3	3	2
S Ext	5	0	1	1	0	1	2
S Orig	2	0	1	0	0	0	0
S Ext Rate	0.6	0.0	0.2	0.3	0.0	0.3	0.3
S Orig Rate	0.2	0.0	0.2	0.0	0.0	0.0	1.0

The diversity and extinction counts presented in this table are taken from Zhuravlev and Wood (1996). Origination counts are calculated from the given diversity and extinction counts. Rates for all non-archaeocyathan taxa (Total – Arch.) are also presented as archaeocyaths comprised the most abundant and diverse group in the analysis. Counts of calcitic and phosphatic small shelly fauna and hyolithomorph hyoliths were excluded as there appeared to be missing data. Note that the term "small shelly fauna" is rather ambiguous, as it has been used to include different taxa by different workers.

Appendix 8.2 Community Data for Early Cambrian Assemblages

Genus	Species	T.a.1	T.a.3	T.a.4	T.b.1	T.b.2	T.b.3	T.b.4	T.b.5
Aldanella	attleborensis	3			1			3	
Allatheca	degeeri					12	5	10	6
Anabarites	korobovi					2	1	1	
Chancelloria	sp.			3					3
Conotheca	mammilata					12			12
Eccentrotheca	kanesia		1	69	33	26		9	180
Fomitchella	infundibuliformis						2	32	
Gracilitheca	sp.					1			
Halkieria	sp.		3			1			

(continued)

Appendix 8.2 (continued)

Genus	Species	T.a.1	T.a.3	T.a.4	T.b.1	T.b.2	T.b.3	T.b.4	T.b.5
Ladathech	cylindrica					2	12	3	5
Lapworthella	laudvigseni					1			
Lapworthella	ludvigseni						56	181	
Plinthokonion	arethion	1	15	10	2				18
Securiconus	minutus	1			1				
Sokolovitheca	sokolovi								1
Sunnaginia	imbricata	1		10		41			6
Torellella	laevigata	7	25	46		28	5		1
Watsonella	crosbyi				3	2			

Genus	Species	A.a.1	A.a.2	A.b.1	B.a.1
Aldanella	attleborensis	29		2	
Allatheca	degeeri	4	25		
Camenella	baltica			2	
Chancelloria	sp.			16	1
Coleoloides	typicalis				1
Conotheca	mammilata			9	
Eccentrotheca	kanesia	1		5	30
Gracilitheca	bayonet		15		
Halkieria	sp.				3
Hyolithellus	micans			2	
Hyolithes	tenuistriatus		83		
Ladatheca	cylindrica	37	12	4	
Linguella	viridis				16
Mellopegma	sp.	1		1	
Pelagiella	sp.	1			1
Plinthokonion	arethion			4	
Rhombocorniculum	cancellatum				278
Stenotheca	acutacosta		2		
Sunnaginia	imbricata			1	8
Torellella	laevigata			54	2

Genus	Species	B.a.1	B.a.3	B.b.1	B.c.1	B.c.2	B.c.3	B.c.4
Chancelloria	sp.	1						
Coleoloides	typicalis	1		1				
Eccentrotheca	kanesia	30	16	8	19	20	41	
Halkieria	sp.	3		1				
Linguella	viridis	16	2	4		6		
Microdictyon	sp. 1		4					
Microdictyon	sp. 2		2					
Pelagiella	sp.	1						1
Rhombocorniculum	cancellatum	278	102	167	18	176	3	19
Rhombocorniculum	sp.		5					
Strenuella	strenua			1				
Sunnaginia	imbricata	8	1		1	4		5
Torellella	laevigata	2	6	3			4	1

(continued)

Appendix 8.2 (continued)

Sample information and lithology		
T.a.1	Sh. 0.7 – Sheldonville	algal wacke- to packstone
T.a.2	Sh. 3.0 – Sheldonville	
T.a.3	Sh. 4.2 – Sheldonville	
T.a.4	Sh. 4.8 – Sheldonville	
T.b.1	HNR-50.7 – Hoppin Resevoir North, Attleboro	wacke- to packstone
T.b.2	HNR-51.2 – Hoppin Resevoir North, Attleboro	
T.b.3	HNR-51.7 – Hoppin Resevoir North, Attleboro	
T.b.4	HNR-52.2 – Hoppin Resevoir North, Attleboro	
T.b.5	HNR-52.5 – Hoppin Resevoir North, Attleboro	
A.a.1	NH-24.5 – Nahant	mud- to wackestone
A.a.2	NH-Sears Collection – Nahant	
A.b.1	PS 0.2 – Pearl Street, Weymouth	wackestone
A.b.2	PS 1.4 – Pearl Street, Weymouth	
B.a.1	HSR-57.5 – Hoppin Resevoir South Attleboro	trilobite wacke- to packstone
B.a.2	HSR-62.0 – Hoppin Resevoir South Attleboro	
B.a.3	HSR-63.5 – Hoppin Resevoir South Attleboro	
B.b.1	HNR-68.9 – Hoppin Resevoir North, Attleboro	wacke- to packstone
B.c.1	ER-3.8 – Ellis Road, Attleboro	trilobite wacke- to packstone
B.c.2	ER-5.1 – Ellis Road, Attleboro	
B.c.3	ER-6.5 – Ellis Road, Attleboro	
B.c.4	ER-9.6 – Ellis Road, Attleboro	

References

Alroy J (2000) New methods for quantifying macroevolutionary patterns and processes. Paleobiology 26:707–733

Alroy J, Aberhan M, Bottjer DJ, Foote M, Fursich FT, Harries PJ, Hendy AJW, Holland SM, Ivany LC, Kiessling W, Kosnik MA, Marshall CR, McGowan AJ, Miller AI, Olszewski TD, Patzkowsky ME, Peters SE, Villier L, Wagner PJ, Bonuso N, Borkow PS, Brenneis B, Clapham ME, Fall LM, Ferguson CA, Hanson VL, Krug AZ, Layou KM, Leckey EH, Numberg S, Powers CM, Sessa JA, Simpson C, Tomasovych A, Visaggi CC (2008) Phanerozoic trends in the global diversity of marine invertebrates. Science 321:97–100

Alroy J, Marshall CR, Bambach RK, Bezusko K, Foote M, Fuersich FT, Hansen TA, Holland SM, Ivany LC, Jablonski D, Jacobs DK, Jones DC, Kosnik MA, Lidgard S, Low S, Miller AI, Novack-Gottshall PM, Olszewski TD, Patzkowsky ME, Raup DM, Roy K, Sepkoski JJJ, Sommers MG, Wagner PJ, Webber A (2001) Effects of sampling standardization on estimates of Phanerozoic marine diversification. Proc Natl Acad Sci USA 98:6261–6266

Ausich WI (1981) Biovolume revisited: a relative density index for paleoecological analyses. Ohio J Sci 81:268–274

Babcock LE, Zhang W, Leslie SA (2001) The Chengjiang biota: record of the Early Cambrian diversification of life and clues to exceptional preservation of fossils. Geological Society of America Today, vol 4–9

Balanda KP, MacGillivray HL (1988) Kurtosis: a critical review. Am Stat 42:111–119

Bambach RK (2006) Phanerozoic biodiversity mass extinctions. Annu Rev Earth Planet Sci 34:127–155

Bambach RK, Knoll AH, Wang SC (2004) Origination, extinction, and mass depletions of marine diversity. Paleobiology 2004:522–542

Bazzaz FA (1975) Plant species diversity in old-field successional ecosystems in Southern Illinios. Ecology 56:485–488

Belaoussoff S, Kevan PG (1998) Toward an ecological approach for the assessment of ecosystem health. Ecosyst Health 4:4–8

Bennington JB, Rutherford SD (1999) Precision and reliability in paleocommunity comparisons based on cluster-confidence intervals: how to get more statistical bang for your sampling buck. Palaios 14:506–515

Bennington JB (2003) Transcending patchiness in the comparative analysis of paleocommunities: a test case from the Upper Cretaceous of New Jersey. Palaios 18:22–33

Brasier MD, Corfield RM, Derry LA, Rozanov AY, Zhuravlev AY (1994) Multiple ∂13C excursions spanning the Cambrian explosion to the Botomian crisis in Siberia. Geology 22:455–458

Briggs DEG, Fortey RA, Wills MA (1992) Morphological disparity in the Cambrian. Science 256:1670–1673

Brown JH (1995) Macroecology. University of Chicago Press, Chicago, p 269

Bulinski KV (2007) Analysis of sample-level properties along a paleoenvironmental gradient: the behavior of evenness as a function of sample size. Palaeogeogr Palaeoclimatol Palaeoecol 253:490–508

Buzas MA, Hayek L-AC (2005) On richness and evenness within and between communities. Paleobiology 31:199–220

Chang Y-M (1967) Accuracy of fossil percentage estimation. J Paleontol 41:500–502

Connell JH (1961) The influence of interspecific competition and other factors on the distribution of the barnacle *Chthalamus stellatus*. Ecology 42:710–723

Darwin C (1904) On the origin of species by means of natural selection, or the preservation of favoured races in the struggle for life (1859). J. Murray, London

Droser ML, Bottjer DJ (1986) A semiquantitative field classification of ichnofabric. J Sed Petrol 56:558–559

Droser ML, Bottjer DJ, Sheehan PM, McGhee GR (2000) Decoupling of taxonomic and ecologic severity of Phanerozoic marine mass extinctions. Geology 28:675–678

Erwin DH (2001) Lessons from the past: biotic recoveries from mass extinctions. Proc Natl Acad Sci USA 98:5399–5403

Erwin DH (1998) The end and the beginning: recoveries from mass extinctions. Trends Ecol Evol 13:344–349

Fatela F, Taborda R (2002) Confidence limits of species proportions in microfossil assemblages. Mar Micropaleontol 45:169–174

Fitzgerald PC, Carlson SJ (2006) Examining the latitudinal diversity gradient in Paleozoic terebratulide brachiopods: should singleton data be removed? Paleobiology 32:367–386

Foote M (2001) Inferring temporal patterns of preservation, origination, and extinction from taxonomic survivorship analysis. Paleobiology 27:602–630

Foote M (2000) Origination and extinction components of taxonomic diversity: Paleozoic and post-Paleozoic dynamics. Paleobiology 26:578–605

Forcino FL, Leighton LR (2010) Determining the smallest sample size required for paleocommunity research: a case-study from the Finish Shale of Texas. In: North Central Geological Society of America meeting abstract, vol 42, p 72

Forcino FL, Stafford ES, Warner JJ, Webb AE, Leighton LR, Schneider CL, Michlin TS, Palazzolo LM, Morrow JR, Schellenberg SA (2010) Effects of data categorization on paleocommunity analysis: a case study from the Pennsylvanian Finis Shale of Texas. Palaios 25:144–157

Gilinksy NL, Bennington JB (1994) Estimating numbers of whole individuals from collections of body parts: a taphonomic limitation of the paleontological record. Paleobiology 20:245–258

Gould SJ, Eldredge N (1993) Punctuated equilibrium comes of age. Nature 366:223–227

Gradstein FM, Ogg JG, van Kranendonk M (2008) On the geologic time scale 2008. Newsl Stratigr 43:5–13

Gray JS (1981) Detecting pollution induced changes in communities using the log-normal distribution of individuals among species. Mar Pollut Bull 12:173–176

Halloy SRP, Barratt BIP (2007) Patterns of abundance and morphology as indicators of ecosystem status: a meta-analysis. Ecol Complex 4:128–147

Harries PJ, Kauffman EG, Hansen TA (1996) Models for biotic survival following mass extinction. Geol Soc Spec Pub 102:41–60

Hayek LC, Buzas MA (1997) Surveying natural populations. Columbia University Press, New York, p 563

Holland SM (1995) The stratigraphic distribution of fossils. Paleobiology 21:92–109

Hough ML, Shields GA, Evins LZ, Strauss H, Henderson RA, Mackenzie S (2006) A major sulphur isotope event at c. 510 Ma: a possible anoxia–extinction–volcanism connection during the Early–Middle Cambrian transition? Terra Nova 18:257–263

Hubbell SP (2001) The unified neutral theory of biodiversity and biogeography. Princeton University Press, Princeton

Hubbell SP (2005) Neutral theory in community ecology and the hypothesis of functional equivalence. Funct Ecol 19:166–172

Joanes DN, Gill CA (1998) Comparing measures of sample skewness and kurtosis. J Roy Statist Soc: Ser D (The Statistician) 47:183–189

Kidwell SM (2007) Discordance between living and death assemblages as evidence for anthropogenic ecological change. Proc Natl Acad Sci USA 104:17701–17706

Kidwell SM (2008) Ecological fidelity of open marine molluscan death assemblages: effects of post-mortem transportation, shelf health, and taphonomic inertia. Lethaia 41:199–217

Kidwell SM, Bosence DWJ (1991) Taphonomy and time-averaging of marine shelly faunas. In: Allison PA, Briggs DEG (eds) Taphonomy: releasing the data locked in the fossil record. Plenum Press, New York, pp 115–166

Kidwell SM, Flessa KW (1995) The quality of the fossil record: populations, species, and communities. Annu Rev Ecol Syst 26:269–299

Kozek AS (1995) A rule of thumb (not only) for gamblers. Stochastic Process Appl 55:169–181

Landing E (1988) Lower Cambrian of Eastern Massachusetts - Stratigraphy and small shelly fossils. J Paleontol 62:661–695

Layou KM (2009) Ecological restructuring after extinction: the Late Ordovician (Mohawkian) of the Eastern United States. Palaios 24:118–130

Leighton LR, Schneider CL (2008) Taxon characteristics that promote survivorship through the Permo-Triassic interval. Paleobiology 34:65–79

Li G, Steiner M, Zhu X, Yang A, Wang H, Erdtmann BD (2007) Early Cambrian metazoan fossil record of South China: generic diversity and radiation patterns. Palaeogeogr Palaeoclimatol Palaeoecol 254:229–249

Lomolino MV, Channell R (1995) Splendid isolation: patterns of geographic range collapse in endangered mammals. J Mammal 76:335–347

Magurran AE, Henderson PA (2003) Explaining the excess of rare species in natural species abundance distributions. Nature 422:714–716

Maheshwari A, Sial AN, Mathur SC (2007) δ13C stratigraphy of the Birmania Basin, Rajasthan, India: implications for the Vendian-Cambrian transition. In: Vickers-Rich P, Komarower P (eds) The rise and fall of the Ediacaran biota, vol 286. Geological Society Special Publication, London, pp 439–441

McCune B, Grace JB (2002) Analysis of ecological communities. MJM Software Design, Gleneden Beach, p 304

McGhee GR, Sheehan PM, Bottjer DJ, Droser ML (2004) Ecological ranking of Phanerozoic biodiversity crises: ecological and taxonomic severities are decoupled. Palaeogeogr Palaeoclimatol Palaeoecol 211:289–297

McGill BJ, Etienne RS, Gray JS, Alonso D, Anderson MJ, Benecha HK, Dornelas M, Enquist BJ, Green JL, He F, Hurlbert AH, Magurran AE, Marquet PA, Maurer BA, Ostling A, Soykan CU, Ugland KI, White EP (2007) Species abundance distributions: moving beyond single prediction theories to integration within an ecological framework. Ecol Lett 10:995–1015

Miller AI (2000) Conversations about Phanerozoic global diversity. Paleobiology 26:53–73

Morris SC (1986) The community structure of the middle Cambrian phyllopod bed (Burgess Shale). Palaeontology 29:423–467

Olszewski T (1999) Taking advantage of time-averaging. Paleobiology 25:226–238

Paine RT (1966) Food web complexity and species diversity. Am Nat 100:65–76

Palmer AR (1984) The biomere problem: evolution of an idea. J Paleontol 58:599–611

Parrish JAD, Bazzaz FA (1982) Competitive interactions in plant communities of differential successional ages. Ecology 63:314–320

Payne JL, Lehrmann DJ, Wei J, Knoll AH (2006) The pattern and timing of biotic recovery from the end-Permian extinction on the Great Bank of Guizhou, Guizhou Province, China. Palaios 21:63–85

Peters SE (2008) Environmental determinants of extinction selectivity in the fossil record. Nature 454:626–629

Peters SE, Bork KB (1999) Species-abundance models: an ecological approach to inferring paleoenvironment and resolving paleoecological change in the Waldron Shale (Silurian). Palaios 14:234–245

Peters SE, Foote M (2001) Biodiversity in the Phanerozoic: a reinterpretation. Paleobiology 27:583–601

Pigliucci M (2008) Sewall Wright's adaptive landscapes: 1932 vs 1988. Biol Philos 23:591–603

R Development Core Team (2009) R: A language and environment for statistical computing. R Foundation for Statistical Computing, Vienna

Raup DM (1976) Species diversity in the Phanerozoic: an interpretation. Paleobiology 2:289–297

Raup DM (1978) Cohort analysis of generic survivorship. Paleobiology 4:1–15

Rowland SM, Hicks M (2004) The early Cambrian experiment in reef-building by metazoans. Paleontol Soc Paper 10:107–130

Saltzman MR, Davidson JP, Holden P, Runnegar B, Lohmann KC (1995) Sea-level-driven changes in ocean chemistry at an Upper Cambrian extinction horizon. Geology 23: 893–896

Schluter D (2000) The ecology of adaptive radiation. Oxford University Press, NY, p 292

Sepkoski JJJ, Bambach RK, Raup DM, Valentine JW (1981) Phanerozoic marine diversity and the fossil record. Nature 293:435–437
Stitt JH (1975) Adaptive radiation, trilobite paleoecology, and extinction, Ptychaspidid biomere, Late Cambrian of Oklahoma. Fossils Strata 4:381–390
Thomas CD, Cameron A, Green RE, Bakkenes M, Beaumont LJ, Collingham YC, Erasmus BFN, de Siqueira MF, Grainer A, Hannah L, Hughes L, Huntley B, van Jaarsveld AS, Midgley GF, Miles L, Ortega-Huerta MA, Peterson AT, Phillips OL, Williams SE (2004) Extinction risk from climate change. Nature 427:145–148
Thomas RC (1995) Cambrian mass extinction ("biomere") boundaries: a summary of thirty years of research. Northwestern Geol 24:67–75
Tucker ME (1992) The Precambrian-Cambrian boundary: seawater chemistry, ocean circulation and nutrient supply in metazoan evolution, extinction and biomineralization. J Geol Soc Lond 149:655–668
Twitchett RJ, Barras CG (2004) Trace fossils in the aftermath of mass extinction events. In: McIlroy D (ed) The application of ichnology to palaeoenvironmental and stratigraphic analysis. Geological Society Special Publications, London, pp 397–418
Twitchett RJ (2006) The palaeoclimatology, palaeoecology and palaeoenvironmental analysis of mass extinction events. Palaeogeogr Palaeoclimatol Palaeoecol 232:190–213
Wagner PJ, Kosnik MA, Lidgard S (2006) Abundance distributions imply elevated complexity of post-Paleozoic marine ecosystems. Science 314:1289–1292
Walker KR, Bambach RK (1971) The significance of fossil assemblages from fine-grained sediments: Time-averaged communities. Geol Soc Am Abstracts with Programs 3:783–784
Watkins R (1996) Skeletal composition of silurian benthic marine faunas. Palaios 11:550–558
Webb AE (2008) Quantifying the ecological response of brachiopods during the Ordovician extinction [M.S. thesis]: San Diego, CA, San Diego State University, 49 p
Webb AE, Leighton LR (2008) Canary in an ordovician coalmine: using rank abundance curves to quantify the ecological impact of the end-ordovician mass extinction on brachiopod communities, Anticosti Island, Quebec. Abstracts with Programs- Geological Society of America
Webb AE, Leighton LR, Schellenberg SA, Landau EA, Thomas E (2009) Impact of the Paleocene-Eocene thermal maximum on deep-ocean microbenthic community structure: using rank–abundance curves to quantify paleoecological response. Geology 37:783–786
Westrop SR, Cuggy MB (1999) Comparative paleoecology of Cambrian trilobite extinctions. J Paleontol 73:337–354
Whittaker RH (1965) Dominance and diversity in land plant communities: numerical relations of species express the importance of competition in community function and evolution. Science 147:250
Wilson EO (2003) On global biodiversity estimates. Paleobiology 29:14
Zhuravlev AY, Wood RA (1996) Anoxia as the cause of the mid-early Cambrian (Botomian) extinction event. Geology 24:311–314
Zhuravlev YA (1996) Reef ecosystem recovery from the Early Cambrian extinction. Geol Soc Spec Pub 102:79–96
Zuschin M, Stachowitsch M, Pervesler P, Kollmann H (1999) Structural features and taphonomic pathways of a high-biomass epifauna in the northern Gulf of Trieste, Adriatic Sea. Lethaia 32:299–317
Zuschin M, Hohenegger J, Steininger FF (2000) A comparison of living and dead molluscs on coral reef associated hard substrata in the northern Red Sea - implications for the fossil record. Palaeogeogr Palaeoclimatol Palaeoecol 159:167–190

Part II
Technological Approaches

Chapter 9
Fossils with Little Relief: Using Lasers to Conserve, Image, and Analyze the Ediacara Biota

Jonathan B. Antcliffe and Martin D. Brasier

Contents

Abstract Fifty years have now passed since the discovery of *Charnia masoni* and *Charniodiscus concentricus* in Charnwood Forest, UK. But what is *Charnia*? And how was it related to the great explosion of animal fossils at the base of the Cambrian that it immediately predates? Recent studies focussing on the growth dynamics and morphology of the group have been greatly aided by the use of innovative photographic and laser scanning techniques that allow the fossils to be analyzed to new levels of detail. Laser methods allow us to build a virtual map of each major fossil

J.B. Antcliffe (✉)
Department of Earth Sciences, University of Bristol, Wills Memorial Building, Queen's Road, Bristol, BS81RJ, UK
e-mail: Jon.antcliffe@bris.ac.uk

M.D. Brasier
Department of Earth Sciences, University of Oxford, South Parks Road, Oxford, OX13AN, UK
e-mail: Martin.Brasier@earth.ox.ac.uk

M. Laflamme et al. (eds.), *Quantifying the Evolution of Early Life*, Topics in Geobiology 36, DOI 10.1007/978-94-007-0680-4_9, © Springer Science+Business Media B.V. 2011

type that can then be viewed and rotated in three dimensions. Laser scanning can be undertaken up to a maximum precision of 1/20 mm. Lasers can be applied in the laboratory to collected specimens, or casts, or alternatively directly in the field to *in situ* specimens when working within fossil conservation areas. Taking laser scans of fossils is completely non-invasive, even unlike casting, the chemicals involved in which can damage delicate fossil surfaces. It is now being seen as a major new tool to aid in the preservation of critical fossil data. Where specimens remain *in situ* in the field, laser scans can be used to monitor the degradation of the fossils. Laser-based conservation of key fossil sites and specimens in England and Newfoundland is now being planned. We have applied this technique to critical fossil cites in Charnwood Forest and to key Ediacaran fossils from the UK, Newfoundland, Australia, and Russia, including *Ivesheadia*, *Fractofusus*, *Charnia*, *Charniodiscus*, *Bradgatia*, and *Dickinsonia*. Further, we shall review the ever expanding application of laser scanning techniques to a variety of paleontological problems exemplified by several diverse case studies, from dinosaur biomechanics to hexapod taxonomy. Laser scanning is increasingly been seen as a game changing technique in paleontology, that can not only conserve fossils and help with their analysis but also to widen access to highly restricted and critically important fossils.

Keywords Ediacara Biota • Lasers • Macrofossils • High resolution scanning • Precambrian

9.1 Introduction

> I think …to… leave out technology misrepresents life as badly as Victorians misrepresented life by leaving out sex.
>
> Kurt Vonnegut (A Man Without a Country, 2005).

The worldwide soft bodied Ediacara biota, c.575–543 Ma ago, is preserved, primarily, as low relief impressions within a variety of fine and coarse sediments (e.g. Ford 1958; Glaessner and Daily 1959; Glaessner and Wade 1966; Wade 1968, 1972; Pflug 1970, 1972; Gehling 1991; Grazhdankin 2003, 2004; Narbonne 2005). Unusually for soft bodied impression fossils, they are highly detailed, reflecting the morphological complexity of the organisms in life, and the high dexterity of Ediacaran sediments for taking fossil imprints (Hagadorn and Bottjer 1997; Gehling 1999; Callow and Brasier 2009; Brasier et al. 2010). In addition, either due to weathering or peculiarities of the preservation process, often counterparts are unavailable (compare typical Phanerozoic soft bodied preservation in nodules and ultra-fine sediments) making the one shallow impression the only source of information. This peculiar set of characteristics makes the imaging of Ediacaran fossils quite unlike dealing with other soft bodied lagerstätte and as such no traditional techniques existed for their detailed analysis. High resolution image analysis of the Ediacara biota is essential for several reasons, where it might potentially be less so for those concerned with other soft-bodied lagerstätten.

Firstly, the specimens are rare and available only in centralized collections close to the particular sections. Secondly, many specimens are in protected sites and remain *in situ*, due to a moratorium in many countries on specimen collection, and so cannot be analyzed easily with modern laboratory techniques (see for example Fedonkin et al. 2009 for a discussion of conservation at critical fossil sites in Russia). Casting can be quite successful for some forms, this allows specimens to be analyzed in more controlled conditions and this has greatly aided many classic studies on the biota (Glaessner 1958, 1966, 1984; Jenkins 1985, 1992; Gehling 1991, 1999; Seilacher 1984, 1985, 1989, 1992; Narbonne 2004). Casting is not permitted, or is being sought to be limited at several sites, however, as the chemicals involved may damage the fossil surface. Thirdly, specimens are highly detailed with complex and enigmatic morphologies that are difficult to record by usual drawing techniques and/or by traditional photographic techniques, such as Camera Lucida and simple sketching by eye. The application of non-invasive high resolution laser scanning techniques uniquely satisfies this complex set of conditions and provides an exciting new means to investigate this important fossil biota as it will others.

Despite the opening quote from Kurt Vonnegut there are indeed laser technologies that shall be left out of this chapter. Particularly, we shall leave con-focal techniques, such as Laser RAMAN Spectroscopy, alone, as the remit of this is somewhat different, being primarily concerned with composition, not morphology, and typically is applied to microfossils. It is also covered in detail in a separate chapter elsewhere in this volume. Those of you interested only in the use of lasers and not their particular application to the Ediacara biota would rightly move directly to Sect. 9.3 of this chapter as the next section details how the Ediacara biota has been imaged by various workers before the advent of laser techniques.

9.2 History of Imaging the Ediacara Biota

The first reconstructions of Ediacaran fossils took place in the 1950s (Ford 1958). These varied greatly, ranging from illustrations of the life habit, to earnest sketches of the fossil material. Ford (1958) provides the first description of material from England and provided simple outline drawings of the holotypes of *Charnia masoni* and *Charniodiscus concentricus* to accompany his taxonomic definitions. These drawings are simple by-eye sketches made whilst working directly with the fossil material. The work of Glaessner, Wade, and Jenkins followed a similar vein through the coming decades (see Glaessner 1958, 1966, 1984; Glaessner and Daily 1959; Glaessner and Wade 1966; Wade 1968, 1972; Jenkins 1985, 1992), though working from casts of the fossils was becoming increasingly common. This may seem to be historical bookkeeping but the history of imaging is to a great extent linked to the history of thought in this field. Jenkins (1985) provides some of the first images of *Rangea schneiderhoehni* (Jenkins 1985; Figures 2–5) which serve to illustrate the structure of his argument. Jenkins' (1985) Figure 3 is an artistic reconstruction of the species *Rangea schneiderhoehni* with interpretative anatomical labels. Jenkins' (1985) Figure 4 is an artistic impression of some parts of this material showing sediment relationships. Jenkins' (1985) Figure 5 shows an artistic reconstruction of *Rangea* as he envisaged it in life habit on the sea floor. Thus interpretation of anatomy and biological affinity is made early in the paper, it is not a conclusion but an assumption, and the following conclusions relate to life habit and ecology. The detailed description of anatomy was a means to an end of illuminating the life modes and habits of these "early animals."

During this time, the use of SLR Photography that allowed careful consideration of the focal plane to be captured was used extensively. Due to the erosion of some *in situ* specimens, some of these early photographic efforts cannot be surpassed today as the degradation of the fossil material provides poorer study. Casting was also widely employed at this time, particularly of the UK material during the 1970s and again because of specimen erosion it has been better to work from casts made in the 1970s than it would be to return to the original *in situ* specimens, or indeed to take new casts. During the 1980s Seilacher produced a series of controversial papers that, none the less, have in many respects endured the test of time (Seilacher 1984, 1985, 1989, 1992). The reconstructions presented by Seilacher during this time have become iconic for several genera. Seilacher's reconstructions are, however, not direct fossil sketches for the most part but are reconstructions of the organism during life and are artistic in nature but here illustrations were used to highlight interpreted organism constructions. It was not until the 1990s that concerted effort was made to document the morphology of the fossils as they actually are, rather than as interpretive reconstructions of the organism in life (cf. Jenkins 1985). Some of the first such work was carried out by Narbonne et al. (1997), who reported new material of Ediacaran age from the sections in Namibia. This new genus was named *Swartpuntia*, after the Swartpunt region in Namibia, and several new specimens were figured and described. The holotype for *Swartpuntia germsi* (Specimen F238-H) was shown photographically (Narbonne et al. 1997; Figure 6) and a "line drawing" was provided (Narbonne et al. 1997; Figure 7). It is unclear how this line-drawing was produced and there is little evidence of the use of multiple lighting or the composition of multiple sketches, though it is clearly a hugely positive step in the right direction regarding actually recording anatomy. Later, Grazhdankin and Seilacher (2002, 2005) re-describe and amend classifications for various species of *Rangea* and *Pteridinium*. In one instance, Grazhdankin and Seilacher (2005; Figure 1c of *Rangea* specimen NESMF541), expressly stated that the drawing of the specimen was carried out using Camera Lucida. There is a clear move in this work to catalogue and document the morphology of the specimens as it actually is and without preconceived notions of what features should be present in primitive animal fossils. This has naturally led to a desire for more sophisticated imaging techniques. Grazhdankin and Seilacher (2005) comment upon the difficulties of trying to actually image these fossils and their attempts to overcome these problems "Upon changing the light direction, however, the furrow pattern in some specimens switch to an opposite inclination."

9.3 The Application of Lasers to the Ediacara Biota

Over the past 5 years several papers have been published that have focussed on the production of high quality fossil illustration aided by high resolution laser analysis (Antcliffe and Brasier 2008; Brasier and Antcliffe 2008, 2009; Liu et al. 2010). The development of these techniques has been necessary as there are unique problems when attempting to image this fossil material for analysis and for publication quality. Firstly, due to the scientific and cultural importance of the fossils the methods must be completely non-invasive. Secondly, the low relief of the impressions means that in direct light the fossil is largely not visible (see Fig. 9.1a). Thirdly, when lighting is not direct (i.e. perpendicular to the surface of the fossil) the fine detail that is preserved in low relief means that shadows often obliterate much of the image, obscuring fossil details. A way to either reduce or utilize shadows is required (see Fig. 9.1b). Fourthly, the large size range (due to high ontogenic variation) of many of the taxa means that methods developed

Fig. 9.1 Illustrating the difficulties of imaging the Ediacara biota. (**a**) *Left*; the low relief of the impressions means that direct light makes the specimens very difficult to see. (**b**) *Right*; this is overcome by using low angle light to cast shadows. However while shadow creates contrast on features perpendicular to light direction it obliterates those features parallel to light direction. Images are laser generated (see below) and both of the *Charnia masoni* holotype from the same view. Specimen is 208 mm long from apex to base

need to be able to image detail in specimens from less than 1 cm in length, in the case of small *Charnia* specimens for instance, to the fine detail in specimens many tens of centimeters in length, in the case of large specimens of *Charnia masoni* syn *grandis* (Antcliffe and Brasier 2008). This macroscopic size invalidates many normal techniques that would usually be applied to analyze such fine morphological detail. Finally, many specimens are still *in situ*, so the methods should be transportable to potentially remote field locations and could be applied to the specimen within a reasonable timeframe for fieldwork.

As would be expected, no one method satisfies all the criteria above (though all must satisfy criteria 1). Two main imaging techniques have been advanced in recent years particularly in their application to macrofossils, that fit the remit discussed above. These two techniques are: Serial Digital Photography and High Resolution Three-Dimensional Laser Scanning. These methods are discussed below, followed by a discussion of producing composite master drawings of fossil morphology given these analyses.

9.3.1 Serially Lit Photograph Sets

With serial photography, successive images are captured using a tripod-mounted digital camera placed above each specimen, with the lens of the camera parallel to the surface of the specimen at a known distance. The direction of lighting can be varied through as many positions around the specimen as required, maintaining a fixed focal length with the camera (see Fig. 9.2; this is similar to the technique used by Hammer et al. (2002)). The choice of lighting position is determined on a

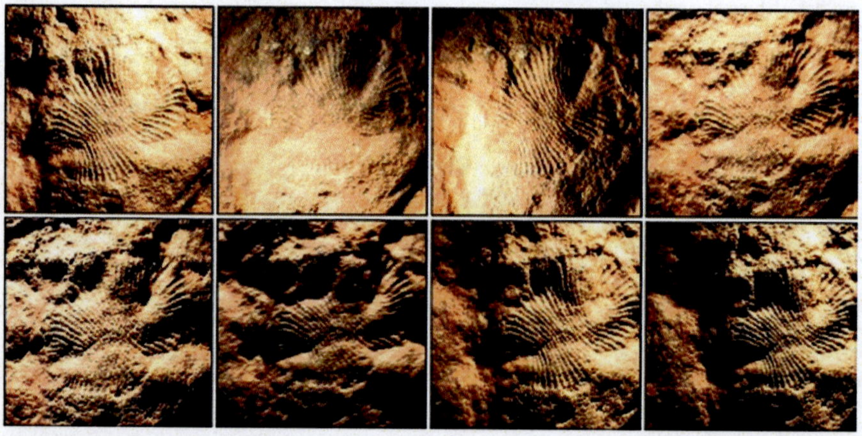

Fig. 9.2 Eight of the source images used in the construction of the drawing of *Dickinsonia costata* specimen OUMNH-AW94. The images are serially lit with the light source moving clockwise round the specimen. Specimen is 15 mm wide by 24 mm long

specimen by specimen basis as other features on the bedding plane and around the fossil will affect the shadows that are cast. If for instance there is a large positive feature elsewhere on the rock surface where the impression fossil is found it may not be possible to light the specimen from several orientations, as the large positive feature casts the specimen into shadow. In these scenarios it is typical to light the specimen from the 180° region that is most accessible. Opposing lighting directions generally show similar features, which appear as a negative to the oppositely lit image. The purpose of this technique when dealing with impression fossils with low relief is to utilize shadows in such a way that they delineate features as distinctly as possible, but whilst trying to prevent the shadow from obscuring other parts of the morphology. Positive ridges in the fossil will cast shadows that are sharpest when the direction of lighting is perpendicular to the orientation of incident light. Thus each lighting direction will display the features in the fossil in a completely unique way that will add to the overall knowledge of the morphology. Given the bespoke way each specimen needs to be lit because of the uniqueness of each fossil surface and because of the variety of sizes of fossils to be imaged it is not possible to have a standardized and automated technique. The images shown (Fig. 9.2) have been taken with a basic seven megapixel digital camera that provides excellent image resolution without producing file sizes that are unmanageable when dealing with many thousands of images. Optical zoom on such images is in focus to approximately 250% of the real fossil size allowing details to be mapped and observed that could not easily be done by the naked eye. A composite Camera Lucida drawing can then be prepared from this series of digital images by plotting all distinct topographic and textural features as outlined below.

9.3.2 *High Resolution Laser Scanning and Composite Master Drawings*

Laser scanning of Ediacaran fossils is an innovative, non-invasive technique that provides a digital topographic data set amenable to image analysis (see Antcliffe and Brasier 2008; Brasier and Antcliffe 2009). The method is portable and can be performed in the field on specimens still *in situ*, in addition to application in more controlled laboratory environments. This method allows quantitative mapping of the surface topography, allowing not only details but cross sectional profiles to be resolved at scales from tens of meters (the whole bedding plane) down to about 100 μm. Each of these three dimensional data arrays can then be freely rotated and/or illuminated from any desired angle, while remaining fully in focus, overcoming many of the difficulties of earlier work.

Laser scanning has been undertaken on the holotypes of both *Charnia masoni* Ford and *Charniodiscus concentricus* Ford, both collected from Charnwood Golf Course in 1957 (Ford 1958) and now catalogued in the Leicester City Museum, plus the holotype of *Charnia grandis* syn *masoni* (Antcliffe and Brasier 2008) and the holotype and three paratypes of *Bradgatia linfordensis* Boynton and Ford, later discovered at a nearby field locality (Boynton and Ford 1995). Each fossil shows sharply focussed profiles, preserved in negative epirelief, on the top surface of a lightly cleaved, gray-green volcanic tuff. The surface containing each fossil was traversed by a laser beam to 1/16th mm resolution to produce a digital data set that was then converted into the master image using Archaeoptics (Copyright) software (Demon 3D; see Figs. 9.3 and 9.4). Rotation of the virtual lighting source across the image of each fossil allowed topographic highpoints to be lit from various directions and plotted graphically using a series of variable viewpoints and allowing composite master drawings to be produced. Multidirectional colored lighting can also be used to show details (Fig. 9.5) and show which lighting direction is illuminating what feature. In addition to laser scanning, the resulting data sets were cross checked with the archive of photos from the field dating back to 1975 and from digital images lit from multiple directions that were produced in the laboratory, using the technique described above. Laser scanning has also been performed on *Dickinsonia costata* specimens (specimen number

Fig. 9.3 High resolution 3-D laser scan of *Charnia masoni* holotype from the Leicester City Museum, UK. The image is resolved with a laser beam accurate to 1/16 mm. This image has been lit with one artificial light source within the laser environment directed from the top left of the specimen. Specimen is 208 mm long

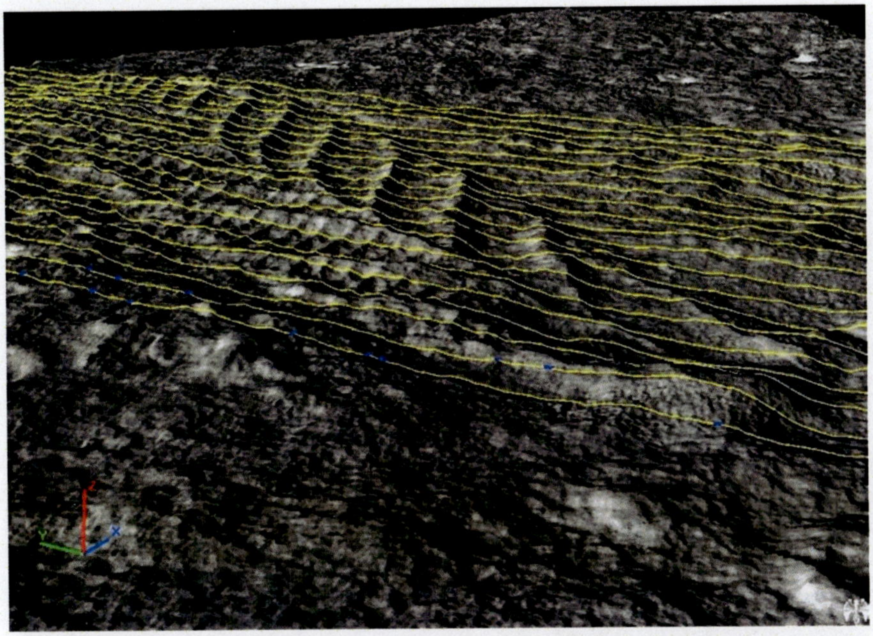

Fig. 9.4 High resolution 3-D laser scan of *Charnia masoni* holotype from Leicester City Museum, UK. This image has been lit with one artificial light source directed from the top left of the specimen and the specimen is then rotated within the virtual environment and viewed from the desired angle. Here also cross-sectional profiles have been added to show the surface topography of the specimen. Specimen is 208 mm long

OUMNH-AW89; Fig. 9.6) from the Roland Goldring collection held in the Oxford University Museum of Natural History.

Figure 9.7 illustrates the basic method of construction of a composite master drawing of *Kimberella quadrata* from the White Sea region of Russia. A composite drawing can be prepared from any combination of serially lit digital images, or higher resolution laser data set. All distinct topographic and textural features are plotted, by light stippling at each step. Shading is later added according to the convention of lighting the specimen from the top left of the image. The drawings can then be checked against original images, montages, laser data sets and, where available, the specimen itself for consistency and accuracy. This provides an objective method for delineating actual fossil features, eliminating ghost features cast by shadows, and checking the veracity of features through a range of independent images. In this sense it is a predictive tool, if a feature is orientated N-S then it would be in sharp focus when lit W-E but would be near invisible when lit from N-S. Thus it is possible to cross check images to distinguish between true fossil impression and features created only by the play of light on the surface of the specimen.

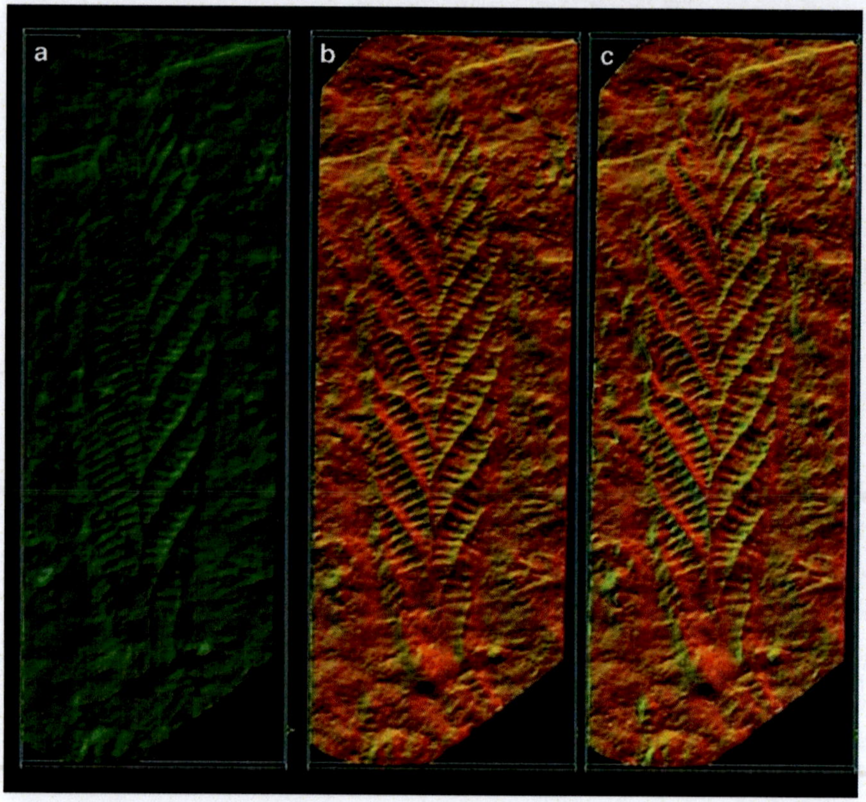

Fig. 9.5 Showing the construction of multiply lit laser scans to reveal maximum morphological detail. High resolution 3-D laser scans of *Charnia masoni* holotype from Leicester City Museum, UK. (**a**) Lit with one artificial light source, *colored green*, within the laser environment directed from the *top left* of the specimen. (**b**) An additional light source *colored red* added in the laser environment and directed from the *top right* of the specimen. (**c**) A further light source, *colored pale blue* added in the laser environment and directed from the *mid left* of the specimen. Specimen is 208 mm long

9.4 Other Recent Applications of Lasers to Paleontology

There is an ever increasing trend for the application of Laser based techniques to paleontological problems. As the biological sciences are continually applying more and more tools from the physicals sciences this is not an unexpected development. There are several fields of study within paleontology where the use of lasers techniques is already sufficiently advanced that they warrant a mention in any review of the methodologies. These can be grouped into four broad classes which are outlined below with a few examples of recent work in each case.

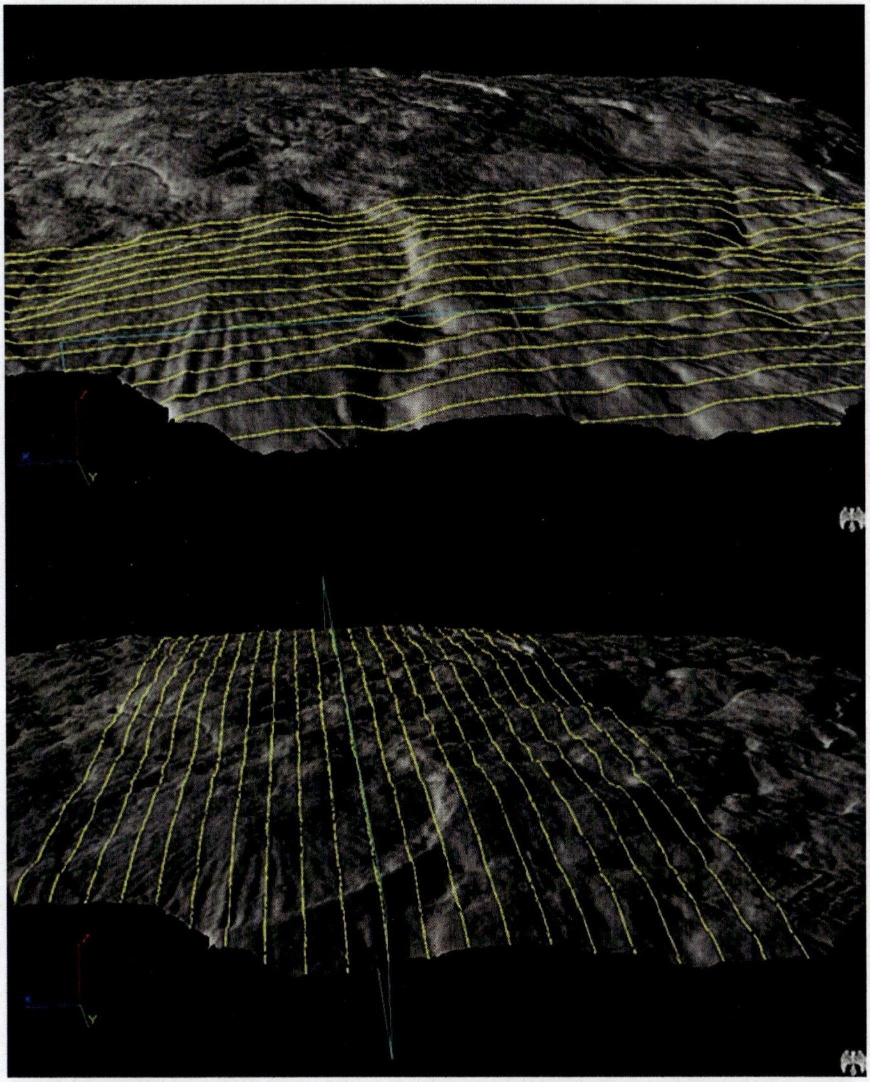

Fig. 9.6 High resolution 3-D laser scan of *Dickinsonia costata* specimen AW89 from the Oxford University Museum of Natural History, UK. The image is resolved with a laser beam accurate to 1/16 mm. This image has been lit with one artificial light source within the laser environment from the left. The image illustrates how the direction of cross-sectional profiles can be controlled within the laser environment to analyze different aspects of the topography of the organism. Specimen is 35 mm wide and 49 mm long

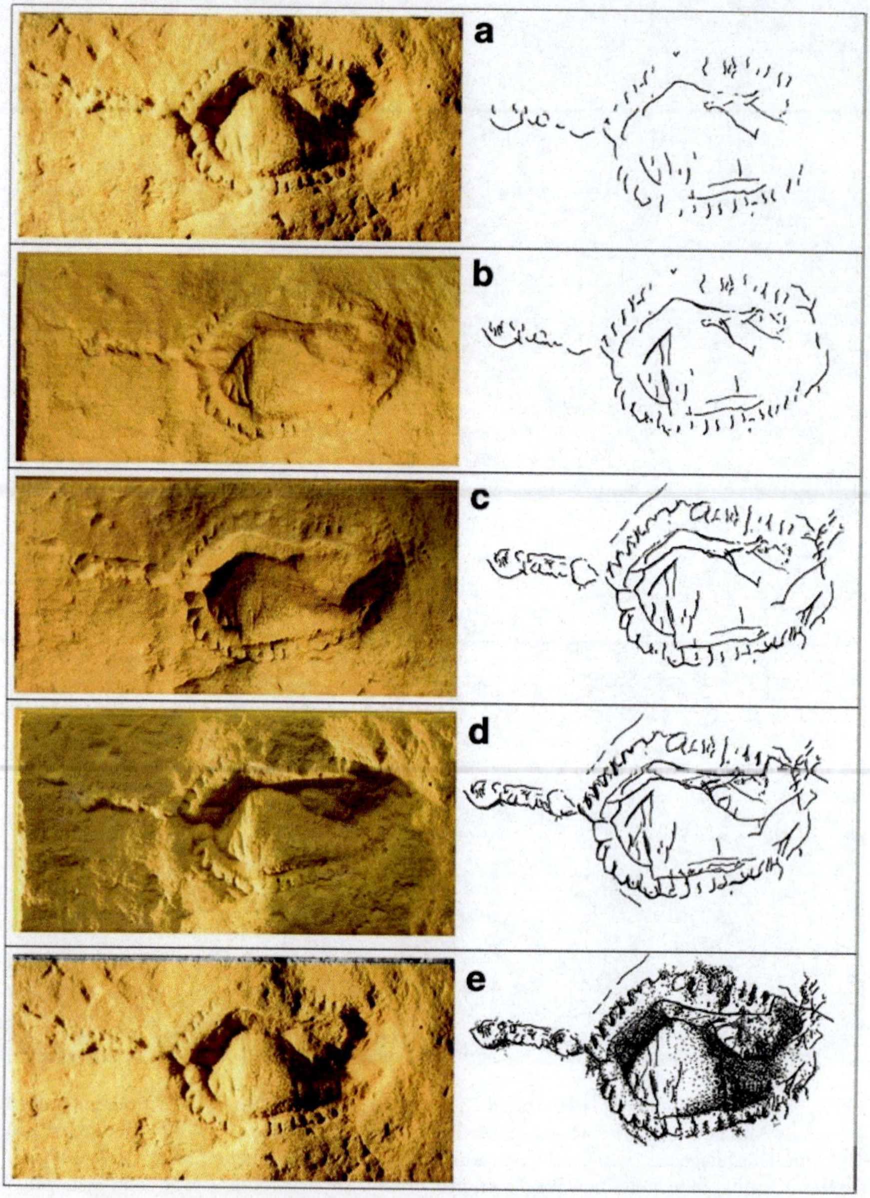

Fig. 9.7 The construction of composite master drawings using a simple sample of four images with lighting from different directions. **(a)** Photograph and overlay of a small high quality cast of *Kimberella quadrata* OUDES-PC701 Erga Formation, Valdai Group. (approximately 3 cm in length) lit from the top left. **(b)** The same but with overlay updated from photograph lit from the right. **(c)** The same but with overlay updated from photograph lit from the bottom. **(d)** The same but with overlay updated from photograph lit from the left. **(e)** The composite master is then produced with the convention of shading the overlay as though it is lit from the top left, thus an image is chosen to match this shading

9.4.1 *Scanning of Whole Large Skeletons for Biomechanics Studies*

Classically three dimensional data arrays captured for biomechanical analysis have used **C**omputer **A**ided **T**omography (CAT; Rayfield et al. 2000) and this has many advantages, specifically high resolution data that Rayfield et al. (2000, 2002) used as a basis for finite element analysis on skulls of theropods. However in recent years there has been an increasing trend to capture such data using **Li**ght **D**etection **A**nd **R**ange (LiDAR; e.g. Bates et al. 2009). This has the advantage of being both quicker and easier than CAT scanning and can be performed on much larger specimens than can be squeezed into even the largest CAT scanner. It also means that scanning can take place on constructed museum vertebrate display specimens which is ideal for biomechanical analysis as the organism is already reconstructed. This particularly allows for much easier biomechanical analysis of large vertebrate organisms when the desired focus is the overall movement of the organism. Typically LiDAR scanners for this application use "a near-infrared laser that is eye safe and requires no additional safety precautions, making it ideal for scanning in museum or public galleries ... able to rapidly acquire dense 3D point data with high accuracy (maximum error of 5 mm) … a range of 800 m, 80° vertical and 360° horizontal fields of view and can be powered by a 24 V or 12 V car battery" (Bates et al. 2009). This makes this a powerful and highly portable tool for the acquisition of high resolution (on the scale of a large vertebrate organism) three dimensional data.

9.4.2 *Scanning Large Bedding Planes, e.g. Vertebrate Track-Ways*

The same specification LiDAR scanner used to capture the three dimensional data array for the dinosaur biomechanics study discussed above (Bates et al. 2009) was also used by the same group to scan a large bedding plane featuring a series of approximately 3,500 dinosaur footprints from the Late Cretaceous Fumanya trackway site in Catalonia (Bates et al. 2008). This approach has the advantage of capturing large data arrays that are very amenable to quantitative analysis of the features on the bedding planes (Breithaupt et al. 2004). For instance the accurate distance between tracks is easily found, as are the depths of the impressions. From such data it is, of course, possible to make assessments of the weight, speed, and biomechanics of the creature that was responsible for the track way. Such an approach also has real and significant potential in the study of paleoecology and community analysis particularly when dealing with life assemblages of invertebrate faunas. Further this could be highly important in analysis of the ecology of the Ediacara biota. The ability to synthesize a large variety of detailed and complex spatial data into one computer file is potentially of staggering importance when it comes to community analysis of any population and no other technique can provide the full 3D scanning over large areas at high resolution like laser scanning. This really has the potential to be a complete game changer in this regard. Again it is important to note that over the coming years computational power will continue to increase making this more widely available without special processing power and technology.

9.4.3 Scanning Cliffs to Map Erosion at Important Fossil Sites

Conservation work at Charmouth, Jurassic Coast, Dorset, UK, has been underway to document the erosion of this part of the UNESCO World Heritage Site that provides an almost continuous section of Mesozoic strata and similar efforts have been undertaken elsewhere. This work has much in common with the traditional mapping applications of LiDAR across the geological sciences and is becoming widespread as a conservation practice at critical fossil and geological sites. This work is often funded by government bodies and/or local councils and focussed primarily on conservation rather than scientific rationale it seems to largely be going unreported in the scientific literature being made the content of internal government reports. This is clearly an expanding and important tool that will find much utility over the coming decades.

9.4.4 High Resolution Scanning of Taxonomically Relevant Characters to Aid Classification and Analysis of Detailed Morphology

The work outlined above (Antcliffe and Brasier 2008; Brasier and Antcliffe 2008, 2009) falls into this category where high resolution scanning is used as a tool to analyze fossil morphology rather than to generate three dimensional data arrays for the purpose of other analyses where the primary focus is not the actual morphology of the organisms but some application of it. Typically such applications use much higher resolution scanners than those concerned with biomechanics or conservation, and this is necessarily so. Béthoux et al. (2004) was the first to apply high resolution laser techniques in a paleontological context, as they analyzed the venation of fossil insect wings which they argue is critical in taxonomy, though others have different opinions about the utility of wing venation in hexapods. Like the Ediacaran fossils, insect wings are very nearly flat and are highly detailed. Consequently they are very difficult to image and analyze and Laser scanning has provided a tool to unlock the secrets of these fossils. It would seem likely that there are many further potential application of this technique in paleontology, where highly detailed and nearly flat surfaces would yield a large amount of taxonomic data but are over all too large, too rare, or too precious to be subject to scanning electron microscopes or other invasive techniques. For instance, fossil specimens preserved in lithographic limestones such as the Solnhofen, Jurassic, Germany (e.g. Kuepp 1977; Barthel 1978; Viohl 1990) where there are mild impressions rather than almost completely flat organic films as in other soft bodied biotas such as the Maotianshan Shales at Chengjiang, Lower Cambrian, China (Hou et al. 2004) would be exciting avenues for the further application of these techniques.

9.5 Concluding Remarks

Laser scanning has the potential to advance a great number of areas of study within paleontology and the history of life. As has been shown above it has already been applied with interesting and innovative results in fields as diverse as the origin of animals, arthropod taxonomy, dinosaur biomechanics, field mapping of vertebrate trackways, and the conservation of critical fossil material. It is the ability of laser scanning to provide microscopic detail to specimens that are (in some cases very) macroscopic that is the real strength of this technique and the reason why it is likely to expand in the field of paleontology over the next decade, applications to the Ediacara biota and arthropod taxonomies clearly illustrate this. There is also a very important future role for laser scanning techniques in the conservation of critical fossils and fossil localities where lasers provide a solution that satisfies so many conservation concerns. Lasers require particularly stable conditions to operate at the highest possible resolution which is not always realistic in the field as such it is usually prudent to scan critical areas of bedding plans at higher resolutions and then stitching these together into their context on a lower resolution scan of the whole bedding plane. Scanning entire bedding planes at the highest possible resolutions (1/20 mm) would be extremely time consuming, and as a result expensive, and produces data sets that are so large so as to be currently out of the operational range of all but the most super of supercomputers. Clearly however the computation limitations will diminish with time making this technique more widely available with the need for special and expensive computation. Scanning for context in the field is very easily achieved however by scanning at lower resolutions (accuracy of c.5 mm or less) is a highly useful and illuminating procedure, that paleontologists are absorbing from geomorphologists, geophysicists, and sequence stratigraphers. The recent work of Bates et al. (2008) clearly shows how laser techniques find utility in providing data and its context when analyzing vertebrate trackways. It is increasingly becoming clear how the geological context of paleontological data is critical in the rational interpretation of fossil material and again laser scanning can provide unparalleled data in this regard. Finally, the fact that laser files are highly portable computer files also offers the tantalizing prospect for greater international exchange of information, in a way that can never happen when dealing with original fossil material (or even casts of original material) that clearly cannot be shipped as easily or with so little regulation. It is well known within the paleontological community how restrictive access to critical fossil material can be. Such restrictions often exist for very good reasons, for instance, the conservation of fossil sites where specimens remain *in situ*, complex access agreements where land is privately owned, or where the interests of students already working on material must be protected. However the scientific endeavor is ultimately about testable hypotheses and it is very difficult to test paleontological hypotheses when fossil material is difficult to access and re-evaluate. The laser scanning of important holotype material across a range of lagerstätte should now be an imminent imperative allowing these three dimensional data to distributed freely and widely in the scientific community even when the actual fossil material could not be so promiscuous in its choice of researcher. Thus published hypotheses could be more widely and rapidly analyzed and the field could progress at a much greater rate. We could be at the beginning of a new age.

References

Antcliffe JB, Brasier MD (2008) *Charnia* at 50: developmental models for Ediacaran fronds. Palaeontology 51:11–26

Barthel KW (1978) Solnhofen. Ein Blick in die Erdgeschichte. Ott, Thun

Bates KT, Rarity F, Manning PL, Hodgetts D, Vila B, Oms O, Galobart A, Gawthorpe RL (2008) High-resolution LiDAR and photogrammetric survey of the Fumanya dinosaur tracksites (Catalonia): implications for the conservation and interpretation of geological heritage sites. J Geol Soc Lond 165:115–127

Bates KT, Manning PL, Hodgetts D, Sellers WI (2009) Estimating mass properties of dinosaurs using laser imaging and 3D computer modelling. PLoSONE 4:e4532. doi:10.1371/journal.pone.0004532

Béthoux O, Mcbride J, Maul C (2004) Surface laser scanning on fossil insects. Palaeontology 47:13–19

Boynton HE, Ford TD (1995) Ediacaran fossils from the Precambrian (Charnian Supergroup) of Charnwood Forest, Leicestershire, England. Mercian Geol 13:165–182

Brasier MD, Antcliffe JB (2008) *Dickinsonia* from Ediacara: a new look at morphology and body construction. Palaeogeogr Palaeoclimatol Palaeoecol 270:311–323

Brasier MD, Antcliffe JB (2009) Evolutionary relationships within the Avalonian Ediacara biota: new insights from laser analysis. J Geol Soc Lond 166:363–384

Brasier MD, Antcliffe JB, Callow RHT (2010) Taphonomy in the Ediacaran interval. In: Allison PA, Bottjer D (eds) Taphonomy: bias and process through time. Springer, Dordrecht

Breithaupt BH, Matthews NA, Noble TA (2004) An integrated approach to three-dimensional data collection of dinosaur tracksites in the Rocky Mountain West. Ichnos 11:11–26

Callow RHT, Brasier MD (2009) A solution to Darwin's Dilemma of 1859: exceptional preservation in Salter's material from the Ediacaran Longmyndian Supergroup, England. J Geol Soc 166:1–4

Fedonkin MA, Ivantsov AYu, Lenov MV, Lipps JH, Serezhnikova EA, Malyutin EI, Khan YV (2009) Paleo-piracy endangers Vendian (Ediacaran) fossils in the White Sea – Arkhangelsk region of Russia. In: Lipps JH, Granier BRC (eds) PaleoParks – The protection and conservation of fossil sites worldwide. Carnets de Géologie/Notebooks on Geology, Brest, Book 2009/03, pp 103–111

Ford TE (1958) Precambrian fossils from Charnwood Forest. Proc Yorks Geol Soc 31:211–217

Gehling JG (1991) The case for the Ediacaran fossil roots to the metazoan tree. Geol Soc Ind Mem 20:181–224

Gehling JG (1999) Microbial mats in terminal Proterozoic siliciclastics: Ediacaran death masks. Palaios 14:40–57

Glaessner MF (1958) New fossils from the base of the Cambrian in South Australia. Trans R Soc S Aust 81:185–188

Glaessner MF (1966) Precambrian palaeontology. Earth Sci Rev 1:29–50

Glaessner MF (1984) The dawn of animal life: a biohistorical study. Cambridge University Press, Cambridge, 256 pp

Glaessner MF, Daily B (1959) The geology and late Precambrian fauna of the Ediacara fossil reserve. Rec S Aust Mus 13:369–410, Plates XLII – XLVII

Glaessner MF, Wade M (1966) The late Precambrian fossils from Ediacara, South Australia. Palaeontology 9:599–628

Grazhdankin D (2003) Structure and depositional environment of the Vendian Complex in the Southeastern White Sea area. Stratigr Geol Correl 11:313–331

Grazhdankin D (2004) Patterns of distribution in the Ediacaran biotas: facies versus biogeography and evolution. Paleobiology 30:203–221

Grazhdankin D, Seilacher A (2002) Underground Vendobionta from Namibia. Palaeontology 45:57–78

Grazhdankin D, Seilacher A (2005) A re-examination of the Nama-type Vendian organism *Rangea schneiderhoehni*. Geol Mag 142:571–582

Hagadorn J, Bottjer DJ (1997) Wrinkle structures: microbially-mediated sedimentary structures common in subtidal silicilastic settings at the Proterozoic-Phanerozoic transition. Geology 25:1047–1050

Hammer Ø, Bengtson S, Malzbender T, Gelb D (2002) Imaging fossils using reflectance transformation and interactive manipulation of virtual light sources. Palaeontol Electron 5. http://palaeo-electronica.org/paleo/2002_1/fossil/issue1_02.htm

Hou XG, Aldrdige RJ, Bergstrom J, Siveter DJ, Siveter DJ, Feng XH (2004) The Cambrian fossils of Chengjiang, China, The flowering of early animal life. Blackwell Science, Oxford, 256 p

Jenkins RJF (1985) The enigmatic Ediacaran (late Precambrian) genus *Rangea* and related forms. Palaeobiology 11:336–355

Jenkins RJF (1992) Functional and ecological aspects of Ediacaran assemblages. In: Lipps JH, Signor PW (eds) Origin and early evolution of the Metazoa. Plenum Press, New York, N.Y. pp 131–176

Kuepp H (1977) Ultrafazies und Genese der Solnhofener Plattenkalke (Oberer Malm, Südliche Frankenalb). Abh Naturhist Ges Nürnberg 37:128

Liu AG, Mcilroy D, Antcliffe JB, Brasier MD (2010) Effaced preservation in the Ediacara biota and its implications for the early macrofossil record. *Palaeontology*. Article published online: 22 DEC 2010. DOI: 10.1111/j.1475-4983.2010.01024.x

Narbonne GM (2004) Modular construction in the Ediacara biota. Science 305:1141–1148

Narbonne GM (2005) The Ediacara biota: Neoproterozoic origin of animals and their ecosystems. Annu Rev Earth Planet Sci 33:421–442

Narbonne GM, Saylor BZ, Grotzinger JP (1997) The youngest Ediacaran fossils from Southern Africa. J Paleontol 71:953–967

Pflug HD (1970) Zur Fauna der Nama-Schichten in Sudwest-Afrika I, II. Palaeontogr. Abt. A, 134 (4–6), 226–262; 135 (3–6), 198–230. [in German]

Pflug HD (1972) The Phanerozoic-cryptozoic boundary and the origin of Metazoa. In: 24th International Geological Congress, Montreal. Section 1: Precambrian geology, pp 58–67

Rayfield EJ, Norman DB, Horner CC, Horner JR, Smith PM, Thomason JJ, Upchurch P (2000) Cranial design and function in a large theropod dinosaur. Nature 409:1033–1037

Rayfield EJ, Norman DB, Upchurch P (2002) Prey attack by a large theropod dinosaur. Nature 416:388–388

Seilacher A (1984) Late Precambrian and Early Cambrian Metazoa: preservational or real extinctions? In: Holland HD, Trendal AF (eds) Patterns of change in earth evolution. Springer, Berlin/Heidelberg, pp 159–168

Seilacher A (1985) Discussion of Precambrian Metazoa. Philos Trans R Soc Lond B 311:47–48

Seilacher A (1989) Vendozoa: organismic construction in the Proterozoic biosphere. Lethaia 22:229–239

Seilacher A (1992) Vendobionta and Psammocorallia: lost constructions of Precambrian evolution. J Geol Soc Lond 149:607–613

Viohl G (1990) Taphonomy of Fossil-Lagerstätten. Solnhofen Lithographic Limestones. In: Briggs DEG, Crowther PR (eds) Palaeobiology: a synthesis. Blackwell Science, Oxford, pp 285–289

Vonnegut K (2005) A man without a country. Seven Stories, New York, 192pp

Wade M (1968) Preservation of soft bodied animals in Precambrian sandstones at Ediacara, South Australia. Lethaia 1:238–267

Wade M (1972) Hydrozoa and Scyphozoa and other medusoids from the Precambrian Ediacara fauna, South Australia. Palaeontology 15:197–225

Chapter 10
Confocal Laser Scanning Microscopy and Raman (and Fluorescence) Spectroscopic Imagery of Permineralized Cambrian and Neoproterozoic Fossils

J. William Schopf and Anatoliy B. Kudryavtsev

Contents

J.W. Schopf (✉) and A.B. Kudryavtsev
Department of Earth and Space Sciences, Institute of Geophysics and Planetary Physics (Center for the Study of Evolution and the Origin of Life), and Molecular Biology Institute, University of California, Los Angeles, CA 90095, USA
and
PennState Astrobiology Research Center, 435 Deike Building, University Park, PA 16802, USA
e-mails: schopf@ess.ucla.edu;Kudryavtsev@ess.ucla.edu

M. Laflamme et al. (eds.), *Quantifying the Evolution of Early Life*, Topics in Geobiology 36,
DOI 10.1007/978-94-007-0680-4_10,

Abstract Among all problems confronting the study of permineralized (petrified) fossils—the most life-like remnants preserved in the fossil record—two stand out, the need for (1) accurate documentation of their three-dimensional morphology, and (2) direct analysis of their chemical composition and that of their embedding mineral matrix. These problems can be addressed effectively by the use of two techniques recently introduced to paleontology: confocal laser scanning microscopy (CLSM), and Raman (and fluorescence) spectroscopic imagery. Both of these techniques, which are non-intrusive and non-destructive, can provide data by which to characterize, *in situ* and at submicron resolution, the cellular and organismal morphology of thin section-embedded organic-walled fossils. The techniques are complementary: CLSM detects laser-induced fluorescence emitted from the carbonaceous kerogen of which such fossils are composed, whereas Raman and fluorescence spectroscopic imagery provide direct analyses of the molecular-structural composition of such kerogen and its embedding mineral matrix as well as providing a means to assess quantitatively the geochemical maturity of the preserved organics. The paleontological usefulness of these techniques is illustrated here by studies of fossils from four Early Cambrian and late Neoproterozoic units: (1) a lowermost Cambrian ctenophore ("comb jelly") embryo from the Kuanchuanpu Formation of Shaanxi Province, China; (2) scale fossils (*Chilodictyon*) from the ~750-Ma-old Lower Tindir Group of Yukon Territory, Canada; (3) acritarchs and cyanobacteria from the ~775-Ma-old Chichkan Formation of southern Kazakhstan; and (4) cyanobacteria from the ~800-Ma-old Bitter Springs Formation of central Australia. Use of these techniques for studies of permineralized fossils can provide information in three dimensions at high spatial resolution about their morphology and cellular anatomy, taphonomy and fidelity of preservation, composition and mode of preservation, and show that they are both indigenous to and syngenetic with the formation of the rocks in which they occur.

Keywords Acritarchs • Ctenophores • Cyanobacteria • Microfossils • Scale fossils

10.1 Introduction

Among all problems confronting paleontologists concerned with the nature of Precambrian life and the rise of multicellular organisms near the Cambrian-Precambrian boundary, two stand out: the need for accurate documentation of the three-dimensional morphology of the microscopic organisms that dominate the Neoproterozoic biota and direct evidence of their chemical composition and that of their embedding mineral matrices. These needs have now been met by the use of two techniques recently introduced to Precambrian paleobiology (Schopf and Kudryavtsev 2005; Schopf et al. 2005, 2006)—confocal laser scanning microscopy (CLSM) and Raman (and fluorescence) spectroscopic imagery—that together can provide data unavailable by any other means about the organismal morphology, cellular anatomy, molecular-structural composition, and taphonomic and preservational history of permineralized organic-walled fossils.

10.1.1 Statement of the Problem

Dating from the 1960s (Barghoorn and Schopf 1965; Barghoorn and Tyler 1965; Cloud 1965; Schopf 1968), studies of permineralized microscopic fossils have played a central role in understanding the composition and evolutionary history of Precambrian life. Of the diverse microorganisms that comprise such assemblages, composed of carbonaceous (kerogenous) cell walls that optically define their three-dimensional form, many are morphologically closely comparable to particular taxa of modern microbes (e.g., Schopf 1968; Schopf and Blacic 1971; Knoll et al. 1975; Knoll and Golubić 1979; Schopf 1999). Although the cellular preservation *in situ* of such microbiotas permits detailed comparison of their components with extant analogues, the life-like preservation afforded by such permineralization also presents problems. Studied in petrographic thin sections, in which the mineralogy of the fossil-embedding matrix and the spatial structure of the preserved community can be deciphered, the coccoidal fossils of chert-permineralized biotas typically occur singly or in three-dimensional colonies of balloon-like quartz-filled bodies, and the filamentous forms are commonly entangled in mat-forming masses that contain few if any specimens aligned parallel to the thin section surface. Because of their minute size, the study of such fossils routinely requires the use of high magnification (e.g., 100×) optical objectives that have a particularly shallow in-focus depth of field (e.g., 0.3 μm). Thus, the study and illustration of the microscopic coccoids of such communities has required the use of multiple photomicrographs acquired at varying focal depths throughout a given cell or three-dimensional colony, and study and illustration of the sinuous filamentous members of such assemblages has typically involved the preparation of montages of photomicrographs acquired at differing optical depths throughout a specimen. Such standard presentations, however, are deficient, both for coccoids and for filaments, since they can depict only a part of the fossil analyzed (in most published illustrations, the thin medial optical plane). In short, standard optical microscopy is incapable of documenting in a single image either the three-dimensional morphology of such fossils or that of their individual cells and overall 3-D form.

Similarly, except in thin sections ≤30 μm thick and by use of a petrographic microscope, optical microscopy is incapable of identifying the various mineral phases of the fossil-associated rock matrix. However, because of the three-dimensional organization of chert-permineralized microbiotas—and, especially, because of the 3-D sinuosity of their commonly predominant filamentous components—most workers prefer to study such communities in 150–250-μm-thick petrographic thin sections (sections prepared at the maximum thickness that permits the transmission of light) in order to maximize the fossil- and community-related information obtained. Since such preparations are not useful for petrologic microscopy, few studies of microscopic fossils have addressed the sequence of mineralogical events that resulted in fossil preservation. Moreover, optical microscopy, the technique preferred by workers in the field worldwide (Schopf and Bottjer 2009), is incapable of definitively identifying the carbonaceous kerogen of which the cell walls of such fossils are composed. Indeed, although such a composition has been assumed since the beginnings of such studies a half-century ago, it has only been in the present decade that the carbonaceous composition of fossil microbes has been firmly established, first for chert-permineralized microorganisms exposed at the surface of thin sections analyzed by secondary ion mass spectrometry (House et al. 2000) and, soon thereafter, of thin section-embedded specimens by use of Raman spectroscopy (Kudryavtsev et al. 2001; Schopf et al. 2002, 2005).

Over the years, the inability to analyze directly the chemical composition of Precambrian microscopic fossils and that of their embedding mineral matrices has led to numerous errors as "biologic-like" mineral grains and other inorganic objects have repeatedly been misinterpreted to be ancient microfossils (Hofmann and Schopf 1983; Schopf and Walter 1983; Mendelson and Schopf 1992). Such errors could have been avoided had there been an established technique by which to document a one-to-one correlation, *in situ* and at micron-scale resolution, of cellular morphology and kerogenous composition, a strong indicator of biogenicity (Schopf et al. 2002, 2005, 2007, 2008; Schopf and Kudryavtsev 2009). Moreover, such analyses in three dimensions of putative microfossils and their enclosing minerals could have provided evidence showing that the objects studied were both indigenous to and syngenetic with the formation of the rock in which they occur (Schopf et al. 2008; Schopf and Kudryavtsev 2009), thereby excluding modern contaminants and endolithic microbes (e.g., Nagy 1974, 1978) such as those mistakenly interpreted as Precambrian microfossils (Hofmann and Schopf 1983; Schopf and Walter 1983; Mendelson and Schopf 1992), while at the same time elucidating their taphonomy and the sequence of mineral-emplacement events that led to their preservation (e.g., Chen et al. 2007).

10.1.2 Focus of This Study

As outlined above, two principal problems have beset studies of Precambrian microorganisms since their inception decades ago: (1) an inability to document accurately the three-dimensional morphology and cellular anatomy of such fossils, and (2) a lack of means and to analyze directly their chemical composition and that of their surrounding mineral matrices. To demonstrate the effectiveness of CLSM and Raman and fluorescence spectroscopic imagery to address these problems, we focus here on fossils permineralized in four deposits ranging in age from lowermost Cambrian (the ~540-Ma-old Kuanchuanpu Formation of Shaanxi, China) to the late Neoproterozoic (the ~750-Ma-old Lower Tindir Group of Yukon Territory, Canada;

~775-Ma-old Chichkan Formation of Kazakhstan; and the ~800-Ma-old Bitter Springs Formation of Australia). Though yet only sparingly applied in paleontology, CLSM and Raman (and fluorescence) spectroscopic imagery are powerful techniques that are broadly useful, not only for the study of microfossils but of megascopic fossils as well (e.g., Schopf et al. 2010).

10.2 CLSM and Raman Imagery

10.2.1 Overview

Confocal laser scanning microscopy (CLSM) affords a means to document and decipher, *in situ* and at submicron-scale resolution, the three-dimensional cellular and organismal morphology of carbonaceous fossils. In contrast, Raman point spectra can be used to analyze the molecular-structural chemistry of such fossils and their associated minerals and, by use of the more advanced techniques of two- and three-dimensional Raman and fluorescence spectroscopic imagery, their composition, morphology and spatial relations to, and the composition of, their embedding matrices. Used in combination, CLSM and Raman and fluorescence spectroscopy can provide telling data about the three-dimensionality and the chemistry of organic-walled fossils, whether they are of micro-, meso- or megascopic size; whether they are permineralized or preserved by compression; and whether the specimens are analyzed in thin sections, palynological macerations, or in cellulose acetate peels (Schopf et al. 2010). Because both techniques are non-intrusive and non-destructive, both can be used to analyze especially precious specimens, such as those archived in museum collections (Schopf et al. 2002, 2007).

10.2.2 Confocal Laser Scanning Microscopy

An instrument available in many university biology departments, the confocal laser scanning microscope was developed in the late-1980s for three-dimensional imaging of the intracellular membranes and organelles of extant organisms (Amos and White 2003). Such confocal microscopes, in which the optical system suppresses the image-blurring effects of out-of-focus planes above and below the thin focal plane studied, can provide images that are appreciably crisper than those attainable by standard optical microscopy. The laser beam of such microscopes excites fluorescence in the material analyzed—fluorescence that in studies of modern cells emanates from introduced fluorescent dyes that bind to cellular components, but that for organic-walled fossils is derived from the interlinked polycyclic aromatic hydrocarbons that comprise their carbonaceous (kerogenous) cell walls (Schopf et al. 2005)—a backscattered laser-induced signal that is then collected by the detector of the instrument to provide a sharply defined image of a single, thin, in-focus plane. By rapidly rastering the laser beam of the system across a defined area of a kerogenous fossil at precisely controlled increasing depths, and then by computer-processing the digitized through-focus series of images acquired, CLSM can produce a three-dimensional image of the specimen studied at high spatial resolution (<0.2 μm).

Although CLSM has provided paleontologically useful information about the three-dimensional structure of Phanerozoic ostracods (Birkmann and Lundin 1996) and radiolarians (O'Conner 1996), most such studies have centered on acid-macerated palynomorphs—spores (Scott and Hemsley 1990), pollen (Foster et al. 1990; Nix and Feist-Burkhardt 2003; Hochuli and Feist-Burkhardt 2004), dinoflagellate cysts (Feist-Burkhardt and Pröss 1999; Feist-Burkhardt and Monteil 2001), micro-algae (Nix and Feist-Burkhardt 2003), and acritarchs (Talyzina 1997). Few CLSM studies have focused on Cambrian (Chen et al. 2007) or Precambrian fossils (Mus and Moczydłowska 2000; Schopf et al. 2006, 2008, 2010 (in press); Schopf and Kudryavtsev 2009).

10.2.3 Raman (and Fluorescence) Spectroscopy and Imagery

Raman spectroscopy is an analytical technique used widely in geochemistry for the identification and molecular-structural analysis of minerals (e.g., McMillan and Hofmeister 1988; Williams et al. 1997; Gaft et al. 2005), including graphite and graphite-like carbonaceous mineraloids (Pasteris and Wopenka 1991, 2003; Jehlička and Beny 1992; Wopenka and Pasteris 1993; Yui et al. 1996; Spötl et al. 1998; Kelemen and Fung 2001; Jehlička et al. 2003). A technique recently introduced to paleontology (Arouri et al. 2000; Kudryavtsev et al. 2001), Raman has not only proven useful in studies of the mineralogy of fossils—whether products of biomineralization, such as the calcite of Paleozoic bimineralic bryozoans (Taylor et al. 2008) and the apatite of Neoproterozoic scale fossils (Cohen et al. 2009), or secondarily emplaced, such as the permineralizing quartz of Precambrian chert-embedded microbiotas (Schopf and Kudryavtsev 2005, 2009; Schopf et al. 2002, 2005, 2007, 2008) and the permineralizing apatite and void-filling calcite shown to have preserved the organic tissues of an Early Cambrian ctenophore embryo (Chen et al. 2007)—but it can be used also to document the molecular structure and geochemical maturity of the kerogen that comprises organic-walled specimens (Schopf et al. 2002, 2005, 2007, 2008).

For many geochemical applications, Raman point spectra—targeted replicate spectral measurements of selected ~1-μm-diameter spots on a specimen—are sufficient to provide the analytical data sought. For analyses of permineralized carbonaceous fossils and associated organic detritus and minerals *in situ* (within petrographic thin sections), however, two additional techniques have proven particularly useful: two- and three-dimensional Raman (and fluorescence) spectroscopic imagery. In such 2-D imagery, whether based on the laser-induced Raman or fluorescence signal emitted by the specimen analyzed, a large number of point spectra are acquired over a defined area of the fossil and its surroundings to provide a micron-scale map that shows the areal distribution both of the carbonaceous matter that comprises the fossil and of the embedding minerals. The resulting 2-D images permit direct correlation of the Raman- and/or fluorescence-based chemical data with fine-scale anatomical features discernable by CLSM and/or optical microscopy (Kudryavtsev et al. 2001; Schopf et al. 2002, 2005, 2007, 2008; Schopf and Kudryavtsev 2005, 2009; Chen et al. 2007; Taylor et al. 2008). Three-dimensional images of such fossils and their enclosing minerals can be prepared by the computerized "stacking" and processing of a through-focus series of 2-D images acquired at sequentially increasing depths throughout a specimen (Schopf and Kudryavtsev 2005; Schopf et al. 2007). Like the 3-D images obtained by CLSM (but unlike optical photomicrographs or 2-D Raman or fluorescence images), computer-processed 3-D Raman images can be manipulated in any direction to reveal features of fossil morphology that may not be discernable by standard optical microscopy. Use of such 3-D Raman imagery has substantiated the biological origin of cellular microbial filaments ~3,465 Ma in age, among the oldest fossils known (Schopf 1992, 1993, 1999, 2006; Schopf et al. 2007) and, combined with carbon isotopic measurements of graphite acquired by secondary ion mass spectroscopy of a microscopic multiphase mineral assemblage ~3,830 Ma old, has provided the earliest hints of life yet discovered (McKeegan et al. 2007).

10.3 Materials and Methods

10.3.1 Materials

The fossils illustrated here are from five geologic units: (1) adult ctenophores ("comb jellies") from the Lower Cambrian (~530-Ma-old) Maotianshan Shale at Mt. Maotian, Chenjiang, Yunnan, China (Fig. 10.1a–c; Chen et al. 2007); (2) an embryonic ctenophore from the Lower Cambrian (~540-Ma-old) Kuanchuanpu Formation at Ningqiang, Shaanxi Province, China (Figs. 10.1d–h and 10.2; Chen et al. 2007); (3) the scale fossil *Chilodictyon* from the Neoproterozoic (~750-Ma-old) Lower Tindir Group of Yukon Territory, Canada (Figs. 10.3 and 10.4; cf. Cohen et al. 2009); (4) filamentous and coccoidal cyanobacteria, and sphaeromorphic acritarchs from the Neoproterozoic (~775 Ma-old) Chichkan Formation of southern Kazakhstan (Figs. 10.5i–x and 10.6a–jj; Schopf and Sovietov 1976); and (5) filamentous cyanobacteria from the Neoproterozoic (~800-Ma-old) Bitter Springs Formation at Ellery Creek, Northern Territory, Australia (Fig. 10.5a–f; Schopf 1968; Schopf and Blacic 1971). The adult ctenophores are compression-preserved whereas all other illustrated fossils are permineralized, the ctenophore embryo by apatite and the cyanobacteria and acritarchs by quartz. The fossils from the Kuanchuanpu Formation occur in bedded phosphorite; those from the Lower Tindir Group, in rounded clasts included in silicified bedded cherts; those from the Chichkan Formation, either in silicified specimens of the columnar stromatolite *Conophyton gaubitza* or in microbially flat- to wavy-laminated stromatolitic chert (cf. *Stratifera*); and those of the Bitter Springs Formation are from flat-laminated *Stratifera*-like chert stromatolites. All specimens were analyzed in unpolished petrographic thin sections (finished by use of a slurry of 600 mesh carborundum) and covered by a ~1-μm-thick veneer of Type FF, fluorescence-free, microscopy immersion oil (R.P. Cargille Laboratories, Inc., Cedar Grove, NJ).

10.3.2 *Optical Microscopy*

Transmitted-light optical photomicrographs were acquired by use of UCLA #0026535 Leitz Orthoplan 2 microscope (Leitz, Wetzlar, Germany) equipped with an Olympus DP12 Microscope Digital Camera (Olympus, Melville, NY).

10.3.3 *Confocal Laser Scanning Microscopy*

CLSM images were obtained by use of an Olympus Fluoview 300 confocal laser scanning biological microscope system equipped with two Melles Griot lasers, a 488 nm 20 mW-output argon ion laser and a 633 nm 10 mW-output helium-neon laser (Melles Griot, Carlsbad CA). The images shown were acquired by use of a 60× oil-immersion objective (numerical aperture 1.4), using the fluorescence-free microscopy immersion oil noted above, with the use of filters in the light-path of the system to remove wavelengths <510 nm (for 488 laser excitation) and <660 nm (for 633 nm laser excitation) from the kerogen-derived fluorescence emitted by the specimens. To maximize the amount of useful data obtained throughout an entire fossil specimen—the detectable backscattered fluorescence signal of which decreases with increasing depth in a petrographic thin section due to its absorption by overlying carbonaceous matter and to light-scattering inhomogeneities in the overlying mineral matrix—the images were obtained by use of the Olympus "Protocol Processor," an add-on software system that increases the voltage and gain of the CLSM detector system linearly with depth. To provide the maximum spatial resolution of such images, the CLSM images have been deconvoluted by use of the computer program Huygens Essential v3.2 (Scientific Volume Imaging Inc., The Netherlands), a process that, compared with standard non-deconvoluted images, results in an improvement in lateral resolution of ~33% and of vertical resolution of ~60% (Schopf and Kudryavtsev 2009). Some images were subsequently processed by use of the VolView v2.0 three-dimensional-rendering computer program (Kitware Inc., Clifton Park, NY) that permits their manipulation in three dimensions.

10.3.4 *Raman (and Fluorescence) Spectroscopy*

Raman and fluorescence spectra were obtained by use of a T64000 (JY Horiba, Edison NJ) triple-stage laser-Raman system having both macro- and micro-capabilities. This system permitted acquisition both of individual point spectra and of Raman and fluorescence images that display the two-dimensional spatial distribution of molecular-structural components of the specimens studied and their associated mineral matrices. Due to the confocal capability of the system, use of a 50× objective (having an extended working distance of 10.6 mm and a numerical aperture of 0.5) provided a horizontal resolution of ~1.5 μm and a vertical resolution of 2–3 μm, whereas use of a 100× objective (having an extended working distance of 3.4 mm and a numerical aperture of 0.8) provided a horizontal resolution of ~0.7 μm and a vertical resolution of 1–1.5 μm. Use in this system of a Coherent Innova (Santa Clara, CA) argon ion laser to provide excitation at 457.9 nm permitted Raman data to be obtained over a range from <500 to >3,100 cm^{-1} by use of a spectral window centered at 1,800 cm^{-1} and, thus, simultaneous acquisition of Raman spectra of the major bands (at ~1,365 and ~1,605 cm^{-1}) and the second-order bands (at ~2,800 cm^{-1}) of kerogen as well as of the major bands of apatite (at ~965 cm^{-1}), of the matrix quartz (at ~465 cm^{-1}), and of associated pyrite (at ~380 cm^{-1}), and the acquisition of fluorescence spectra from <500 to >900 nm.

For 2-D fluorescence (Fig. 10.3j, k) and Raman (Fig. 10.2b–k) imaging, the latter a necessary precursor to preparation of the 3-D Raman images shown here (Fig. 10.5f, o–q), the region of a thin section containing a specimen to be analyzed was covered by a thin (~1-μm-thick) veneer of the fluorescence-free microscopy immersion oil noted above (the presence of which has been shown to have no discernable effect on the Raman spectra acquired; Schopf et al. 2005) and the area of the fossil studied was centered in the path of the laser beam projected through the microscope of the system. The laser power used was ~1–8 mW over a ~1 μm spot, an instrumental configuration well below the threshold resulting in radiation damage to specimens such as those studied here (Schopf et al. 2005). The 2-D fluorescence images shown are maps acquired in windows centered on the major spectral bands of the materials analyzed: for kerogen fluorescence, at ~570 nm (Figs. 10.3j and 10.4); and for the fluorescence of samarium^{+3}, substituting for calcium-I ions in apatite, at ~599 nm (Figs. 10.3k and 10.4). The 2-D Raman maps (Fig. 10.2b–k), including those used for preparation of 3-D Raman images (Fig. 10.5f, o–q), were acquired in windows centered on the major Raman bands of such materials: for kerogen, at ~1,605 cm^{-1}; for apatite, at ~965 cm^{-1}; and for calcite, at ~1,087 cm^{-1}. Both the fluorescence and Raman images, in which the varying intensities correspond to the relative concentrations of the molecular structures detected, show the micron-scale areal distribution of the materials analyzed.

To obtain the through focus sequential series of 10–25 2-D molecular-structural maps needed to construct the 3-D Raman images shown in Fig. 10.5f, o–q, a high precision piezo z-stage [P-721, Physik Instrumente (PI) GmbH & Co. KG] was used. Digitized series of such maps were acquired at precisely defined incremental depths, depending on the dimensions of the specimen analyzed and the microscope objective used for imaging. All 3-D Raman images were processed by use of the VolView three-dimensional-rendering computer program noted above (cf. Schopf and Kudryavtsev 2005).

10.4 Paleontological Uses and Limitations of CLSM and Raman Imagery

Schopf and Kudryavtsev (2009) have recently discussed the strengths and weaknesses of CLSM and Raman imagery as applied to studies of ancient microscopic fossils. Nevertheless, both of these techniques have been introduced only recently to paleontology and, thus, are likely to be new to many members of the paleontological community. Therefore, as requested by the editors of this volume—and because not all permineralized fossils, regardless of age, can be expected to yield the high-quality images shown here—the following is a synopsis of the more extensive discussion of the applicability of these techniques, cited above.

10.4.1 Applicability of the Two Techniques

In their use for studies of organic-walled fossils, CLSM and Raman (and fluorescence) spectroscopic imagery are fundamentally similar.

- Both can be used to analyze flattened compression-preserved specimens (such as those in fine-grained clastic sediments, such as siltstones and shales) and three-dimensionally permineralized fossils (whether the embedding matrix is quartz, phosphate, or calcite).
- Both are non-destructive, non-intrusive and, for fossil embedded in translucent matrices, rock-penetrating.
- Both can provide images at submicron-scale resolution of thin section-embedded fossils and associated organic matter, whether such materials are exposed at the surface of, or are embedded within, the section studied, and whether the section is polished or unpolished.
- Both can yield useful images of rock-embedded fossils to depths within translucent thin sections of ~150 μm.
- Both can provide sharp images of permineralized fossils—in thin sections, acid-macerated preparations or cellulose acetate peels—whether the fossils are of micro-, meso- or megascopic size.

10.4.2 Limitations of the Two Techniques

The two techniques also have limitations:

- Although both can provide useful data from surface-exposed specimens, because the immersion oils used for standard optical microscopy fluoresce with laser-excitation—and because such oil can penetrate into the crevices at mineral grain-boundaries to a depth of a few to several microns (Schopf et al. 2007; Schopf and Kudryavtsev 2009)—data acquired from the uppermost surfaces of previously studied thin sections can be obscured by earlier-used immersion oil (an image-blurring effect that can be countered by deletion of the uppermost slices from CLSM and Raman/fluorescence spectroscopic confocal images and the presence of which is easily discernable from Raman and fluorescence spectra).

- Because the backscattered signal detected by each system (for CLSM and fluorescence spectroscopic images, a strong laser-induced fluorescence signal; for Raman, a very much weaker signal from Raman-shifted backscattered light) is typically diminished and distorted at surficial interfaces, specimens analyzed in thin sections having overlying cover slips generally yield images that are less distinct than those in uncovered sections.
- For both CLSM and Raman (and fluorescence) spectroscopic imagery, the presence of carbonaceous detritus or of fluorescing thin section mounting media in the near-vicinity of an analyzed specimen can blur or obscure its detectable backscattered signal.
- Unlike Raman (and fluorescence) spectroscopic imagery, a technique that can provide molecular-structural data of all mineralic/organic phases of paleontological interest, CLSM (even with the use of a "Protocol Processor" noted above) cannot acquire useful images of organic-walled fossils composed of geochemically highly mature (markedly graphitized) kerogens, such as those preserved in upper greenschist or more severely metamorphosed terrains (Schopf et al. 2005), the weak laser-induced fluorescence of which cannot not recorded by the CLSM detector.

10.4.3 Comparative Strengths and Weaknesses of the Two Techniques

CLSM images of organic-walled fossils can be acquired fairly rapidly (within 20–45-min, depending on the specimen studied), but because various minerals (such as apatite augmented by Sm^{+3}; Fig. 10.4) can also exhibit laser-induced fluorescence (e.g., Gaft et al. 2005), CLSM data do not definitively establish a kerogenous composition. In contrast, Raman (and fluorescence) spectroscopic imagery provides direct, *in situ*, molecular-structural identification of the materials analyzed. However, of the two techniques, Raman/fluorescence spectroscopy affords appreciably less spatial resolution. Because of the optics involved, the maximum resolution obtainable by our Raman/fluorescence system is ~0.7 μm whereas our CLSM system can resolve structures smaller than 0.2 μm (these being, for both systems, near the diffraction-constrained theoretical limits of their application). In addition, Raman imagery is decidedly more time-consuming than imagery by CLSM, acquisition of the hundreds of spectra that make up a useful 2-D image typically requiring about 1 h and the preparation of a 3-D image being a much longer process that involves not only image acquisition-time (e.g., ~50 h for a 25-μm-diameter fossil imaged at 500-nm-intervals) but, also, the time required for computer-stacking and -processing of the images obtained.

10.4.4 Use of CLSM and Raman, Combined

Because CLSM and Raman/fluorescence spectroscopic imagery are complementary—CLSM detecting the laser-induced fluorescence of the kerogenous components of organic-walled fossils and associated organics, and Raman and/or fluorescence spectroscopic imagery identifying the molecular chemistry of such materials and that of their surrounding matrix—and because CLSM is the more rapid of the two techniques, we routinely use CLSM as a preliminary "filter" to determine which of various fossil specimens are the most promising for subsequent Raman/fluorescence spectroscopic studies. Use of the two techniques together, a capability that we regard as especially powerful for paleontology, can provide data unavailable by any other means about the organismal morphology, cellular anatomy, chemistry, and taphonomic and preservational history of carbonaceous microscopic and megascopic fossils, regardless of their geologic age or mode of preservation.

10.5 CLSM and Raman (and Fluorescence) Spectroscopic Imagery of Early Cambrian and Neoproterozoic Fossils

To illustrate the applicability of CLSM and Raman/fluorescence spectroscopy to studies of Early Cambrian and late Neoproterozoic fossils, we present here images of thin section-embedded specimens from ~540 Ma Cambrian strata of China (Figs. 10.1 and 10.2) and from Neoproterozoic, ~750-Ma-old, deposits of Canada (Fig. 10.3); ~775-Ma-old strata of Kazakhstan (Figs. 10.5i–x, and 10.6); and an ~800-Ma-old unit of Australia (Fig. 10.5a–f). Specimens shown include an apatite-permineralized ctenophore embryo (Figs. 10.1 and 10.2) and chert-permineralized scale fossils (Figs. 10.3 and 10.4), filamentous cyanobacteria (Fig. 10.5), sphaeromorph acritarchs (Fig. 10.6a–p), and coccoidal cyanobacteria (Fig. 10.6q–jj).

10.5.1 Early Cambrian Ctenophore Embryo

Ctenophorans, informally referred to as "comb jellies" and represented in the modern biota by fewer than 100 species, are gelatinous, planktonic, typically spheroidal to oblate voracious predators (Ax 1996; Finnerty et al. 1996). Relatively primitive metazoans—as evidenced by their lack of a mesoderm that, as in other coelenterates, gives them a two- rather than the three-tissue-layered organization of more advanced bilaterians—the Ctenophora is a an early-evolved phylum. Despite the unmineralized and seemingly fragile nature of such animals, however, fossils of the group have been reported not only from ~530-Ma-old Early Cambrian Maotianshan Shale of Yunnan, China (Fig. 10.1a–c), but from Middle Cambrian and Early Devonian strata (Chen et al. 2007).

The apatite-permineralized ctenophore egg shown in Figs. 10.1d–h and 10.2—from the ~540-Ma-old Kuanchuanpu Formation of Shaanxi Province,

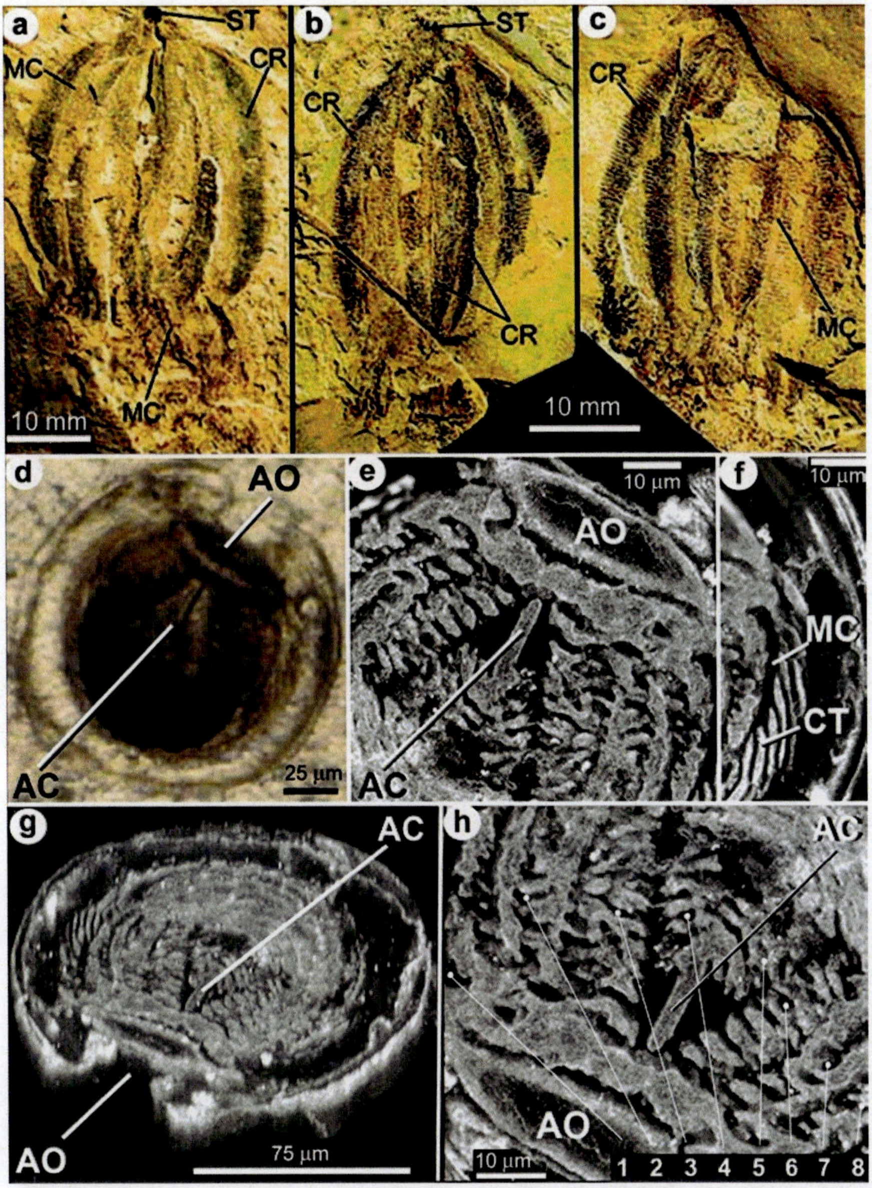

Fig. 10.1 Lower Cambrian adult ctenophores ("comb jellies"), exposed on rock surfaces (**a–c**), and an unnamed ctenophore embryo, in a petrographic thin section (**d–h**). (**a–c**) Two specimens of *Maotianoascus octonarius* from the Lower Cambrian Maotianshan Shale at Mt. Maotian, Chenjiang, Yunnan, China; (**a**) and (**b**), the part and counterpart of the holotype of this species; (**c**), a paratype specimen of this taxon; *CR* comb row; *MC* meridional canal; *ST* statolith. (**d–h**) Optical (**d**) and CLSM images (**e–h**) of a ctenophore embryo from the Lower Cambrian Kuanchuanpu Formation at Ningqiang, Shaanxi Province, China; (**e**) and (**f**), the oral (**e**) and upper-right (**f**) parts of the embryo, showing the fine-scale morphological information provided by CLSM; (**g**), a rotated image of the complete embryo; (**h**), a rotated image, from the same perspective as in (**g**), showing the eight numbered comb rows, characteristic of ctenophores, that overlie meridional canals; *AC* aboral canal, *AO* apical organ, *CT* ctenes, *MC* meridional canal)

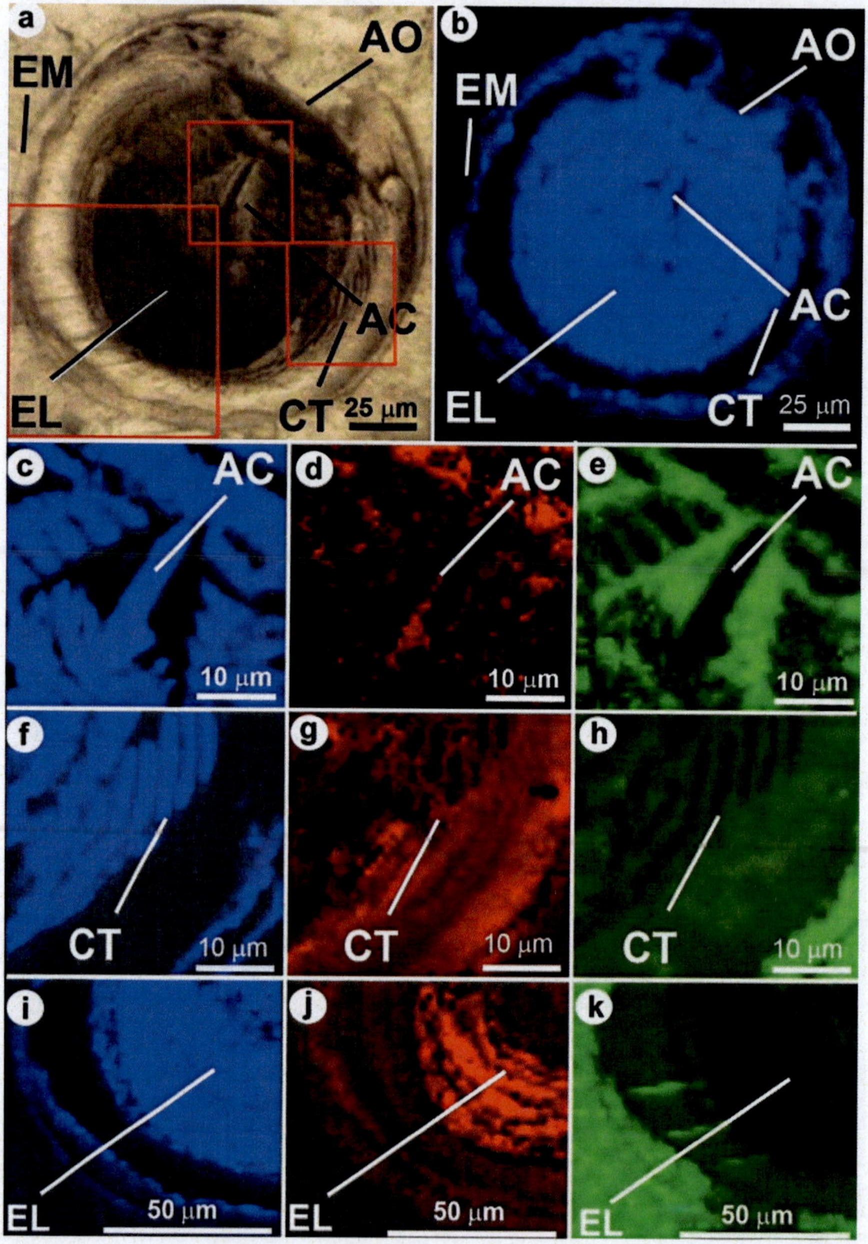

Fig. 10.2 Optical and 2-D Raman images of a thin section-embedded ctenophore embryo from the Lower Cambrian Kuanchuanpu Formation at Ningqiang, Shaanxi Province, China. The color-coded Raman images are maps acquired in spectral windows centered on the major Raman bands of the materials analyzed: blue images (**c**, **f**, and **i**), acquired in a spectral window centered at ~1,605 cm^{-1}, show the spatial distribution of kerogen; red images (**d**, **g**, and **j**), centered at ~965 cm^{-1}, the distribution of apatite; and green images (**e**, **h**, and **k**), centered at ~1,087 cm^{-1}, the distribution of calcite. (**a**) Optical image in which the *central red rectangle* denotes the Raman-imaged area shown in (**c–e**); the *lower-right red rectangle*, that in (**f–h**); and the *lower-left larger red rectangle*, that in (**i–k**). (**b**) Raman image of the complete embryo showing its kerogenous composition. (**c–e**) Raman images

China—enclosing an embryo in late development, before hatching, well illustrates the combined use of optical, CLSM, and Raman imagery. Although the optical images (Figs. 10.1d and 10.2a) of this thin-section embedded specimen show its overall morphology, the CLSM images (Fig. 10.1e–h) establish the presence of the eight comb rows (ctenes) diagnostic of ctenophores (Chen et al. 2007) and provide high-resolution morphological information unavailable by other means. In addition, by documenting the molecular-structural composition of this fossil and the mineral matrix in which it is embedded, 2-D Raman images elucidate the mineralization events by which it was fossilized: Fig. 10.2c, f, i show that the originally soft organic tissues of the specimen are preserved as kerogen; Fig. 10.2d, g, j, that such tissues were permineralized by authigenically emplaced apatite; and Fig. 10.2e, h, k, that voids surrounding the apical canal and the interstices between the ctenes, as well as the region between the embryo and the enclosing egg membrane, were, like the meridional canals and other void spaces of the embryo (Chen et al. 2007), secondarily infilled by calcite. Thus, CLSM documents the fine-structural morphology of this fossil whereas Raman imagery records the sequence of mineralization by which it was preserved.

10.5.2 Late Neoproterozoic Scale Fossils

Biologically controlled mineralization, in both animals and calcareous algae, is typically assumed to have evolved at or near the Cambrian-Precambrian boundary (~542 Ma ago), spurred by the rise of metazoan predation. However, recent discoveries show that biomineralization, at least in protists, evolved appreciably earlier. In particular, as documented by the mutually reinforcing results of optical, CLSM, fluorescence, and Raman analyses, apatite-biomineralized protistan "scale fossils" (*Chilodictyon*)—from the ~750-Ma-old Lower Tinder Group of Yukon Territory, Canada, and first reported by Allison and Hilgert (1986)—pre-date this boundary by some 200 Ma (Cohen et al. 2009).

The morphology of these Tindir scale fossils is distinctive, comprised of an ovate disk-like perforated plate from the center of one surface of which arises a prominent anchor-shaped structure that terminates in four or six flange-like prongs. Figure 10.3a shows an optical image of one such plate of *Chilodictyon*, viewed in face-view, a part of which is obscured by an overlying calcite rhomb both in the

Fig. 10.2 (continued) of the central part of the specimen, showing that its aboral canal (*AC*) is composed of kerogen (**c**) pervaded by fine-grained apatite (**d**), its surrounding interstices having been infilled by calcite (**e**). **(f–h)** Raman images of the lower-right part of the specimen,showing that its embryonic comb plates (ctenes, *CT*) are composed of kerogen (**f**) augmented by apatite (**g**), the adjacent spaces having been infilled by calcite (**h**). **(i–k)** Raman images of the lower-left quadrant of the embryo, showing that its preserved layered tissues are composed of kerogen (**i**) permineralized by apatite (**j**), and that the surrounding matrix is composed of calcite (**k**) (*AC* aboral canal, *AO* apical organ, *CT* ctenes, *EL* embryo layers, *EM* egg membrane)

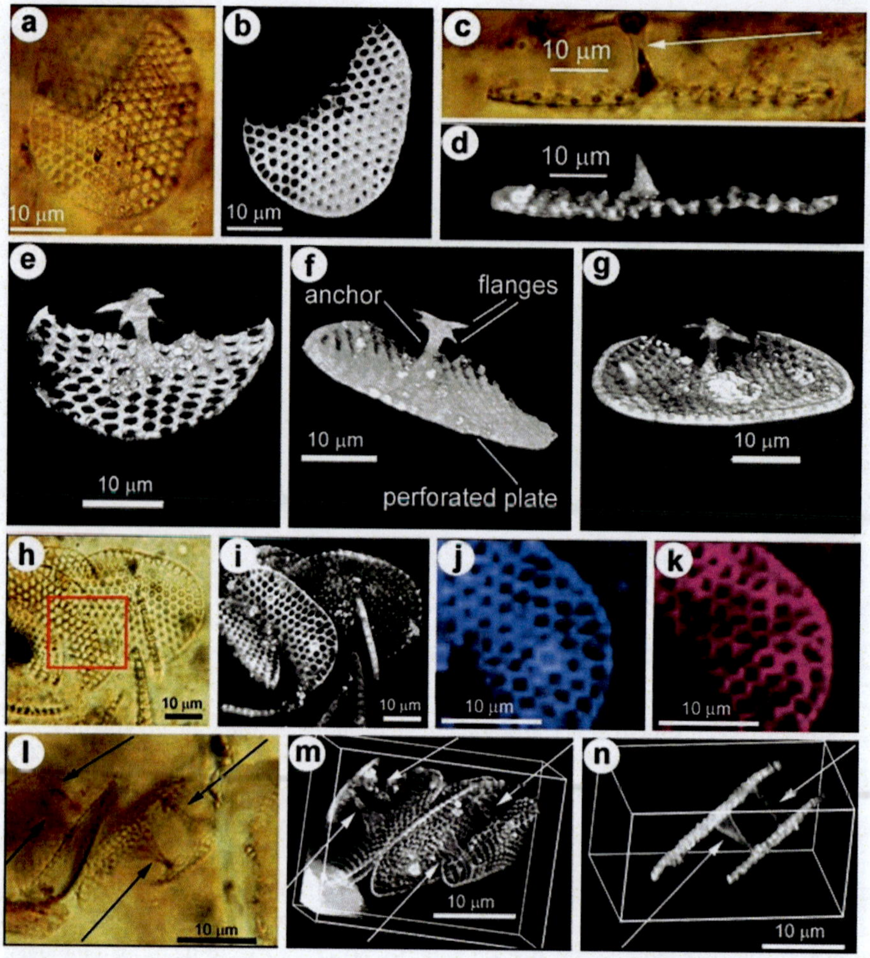

Fig. 10.3 Images of the Neoproterozoic scale fossil *Chilodictyon* in petrographic thin sections of carbonaceous cherts from the ~750-Ma-old Lower Tindir Group of Yukon Territory, Canada. (**a, b**) Optical (**a**) and CLSM (**b**) face-views of the perforated ellipsoidal shield-like plate characteristic of *Chilodictyon* sp; the upper-left quadrant of the specimen is obscured by an overlying calcite rhomb (**a**) that masks the CLSM-detectable fluorescence signal from this part of the scale (**b**). (**c, d**) Optical (**c**) and CLSM (**d**) side-views of *Chilodictyon* showing the distinctive anchor assemblage (*arrow* in **c**) that arises vertically from the center of one surface of the shield-like perforated plate. (**e–g**) CLSM images of three specimens of *Chilodictyon* showing their ellipsoidal shield-like perforated plates and their centrally situated anchor assemblages that terminate in radiating flanges. (**h–k**) Optical (**h**), CLSM (**i**) and fluorescence (**j** and **k**) images of a cluster of specimens, the *red rectangle* in (**h**) denoting the area imaged in (**j**) and (**k**); (**j**), fluorescence image showing the spatial distribution of kerogen (acquired in a spectral window centered on the kerogen fluorescence band at ~570 nm, shown in Fig. 10.4); (**k**) fluorescence image showing the spatial distribution of apatite (acquired in a spectral window centered on the samarium^{+3} fluorescence band at ~599 nm, shown in Fig. 10.4). (**l, m**) Optical (**l**) and CLSM (**m**) images showing two sets of paired, parallel and longitudinally offset specimens that are linked by their anchor assemblages (*arrows*), the flanges of which line up with and are evidently bound to the rib-like struts that define the perforations of the connected plate. (**n**) CLSM side-view of two connected offset *Chilodictyon* plates, linked by their anchor assemblages (*arrows*)

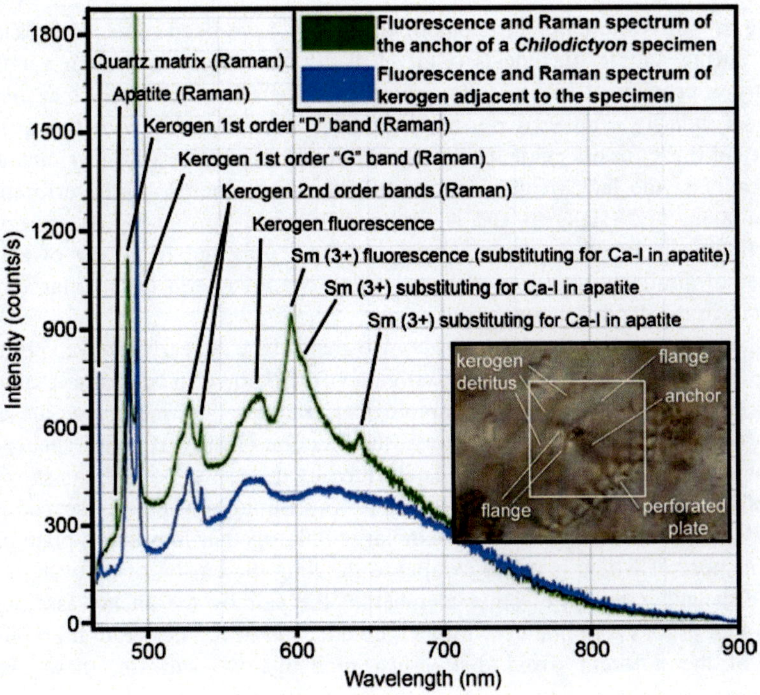

Fig. 10.4 Fluorescence and Raman spectra of a specimen of *Chilodictyon* and of adjacent organic detritus analyzed in a petrographic thin section of carbonaceous chert from the ~750-Ma-old Lower Tindir Group of Yukon Territory, Canada. The photomicrograph (inset) shows a tangential view of the perforated shield-like ellipsoidal scale and its distinctive anchor assemblage; the *white* rectangle encloses fluorescence- and Raman-analyzed areas. Comparison of the fluorescence and Raman spectrum of the anchor with that of adjacent carbonaceous detritus shows that kerogen is present both in the anchor and the detritus (fluorescence and Raman bands); that the anchor also contains apatite (Raman band) in which Sm^{+3} has substituted for Ca-I in the apatite lattice (fluorescence bands); and that apatite is not present in the adjacent kerogen detritus

optical (Fig. 10.3a) and CLSM (Fig. 10.3b) images of the specimen. Side views of another specimen from the assemblage are shown in Fig. 10.3c, d. The overall morphology of such fossils is illustrated especially well by the CLSM images shown in Fig. 10.3e–g.

The composition of these chert-embedded fossils—a mix of apatite and kerogen, permineralized by quartz—is documented by the optical, CLSM, and fluorescence spectroscopic images in Fig. 10.3h–k. Their organic (kerogenous) composition is shown both by their laser-induced CLSM-detectable fluorescence (emitted from the polycyclic aromatic hydrocarbons of which such kerogens are predominantly composed; Schopf et al. 2005) and by their Raman and fluorescence spectra (Fig. 10.4). That they are composed also of apatite, rather than being entirely kerogenous, is similarly documented by such spectra (Fig. 10.4) and by fluorescence spectroscopic images (e.g., Fig. 10.3k), the fluorescence signal emanating not from unaltered apatite (which is non-fluorescent) but from apatite in which the calcium-I ions of the lattice have been replaced by samarium^{+3} during anoxic diagenesis (Gaft et al 2005). Finally, that such apatite is primary, a product of biologically controlled mineralization, rather than having been secondarily emplaced during diagenesis, is consistent with the original rigidity of these fossils evidenced by their fine-structural three-dimensional preservation (i.e., the lack of deformation of plate components, plate perforations, anchor-stems and -flanges) and is further evidenced by the absence of apatite from finely dispersed particulate kerogen (Fig. 10.4) and from any of a large number of similarly analyzed organic-walled cyanobacteria and acritarchs that co-occur in the Tindir cherts.

Although possibly related to prasinophyte green algae (an early-evolved protistan lineage that includes one taxon, *Mesostigma viride*, known to be apatite-biomineralized; Domozych et al. 1991), the biological relations of *Chilodictyon* are enigmatic (Cohen et al. 2009). Nevertheless, the function of the distinctive anchor-like structures of these fossils has been established by the use of CLSM. As shown in the optical (Fig. 10.3l) and CLSM (Fig. 10.3m, n) images of closely spaced pairs of *Chilodictyon* specimens, these assemblages link together imbricated plate-pairs, each member of which is offset parallel to the long dimension of its ovate perforated disk and is linked to other member of the pair by the anchor-assemblage flange-like prongs that line up with and are attached to the perforation-producing struts of the adjacent ovoid disk. Most probably, we surmise, these apatite biomineralized plate-pairs enveloped the external surface of spheroidal protists (cf. the coccoliths of coccolithophorans).

10.5.3 Late Neoproterozoic Filamentous Cyanobacteria

CLSM and Raman imagery are similarly effective for studies of filamentous fossil cyanobacteria such as the helically coiled specimen from the ~800-Ma-old Bitter Springs Formation of central Australia shown in Fig. 10.5b, a CLSM image that displays appreciably more structural detail than the photomontage of the fossil shown in Fig. 10.5a. Oscillatoriacean cyanobacteria, such as the *Spirulina*-like filament shown in Fig. 10.5a, b, are among the most commonly occurring fossils known from the Neoproterozoic. A particularly well preserved example, also from the Bitter Springs Formation, is shown in Fig. 10.5c–f, a tapering cellular oscillatoriacean trichome that plunges sinuously into its embedding mineral matrix from where one of its apices transects the thin section surface (at the upper right in Fig. 10.5c, d) to where the trichome terminates at a depth of ~20 μm. Documentation by standard optical microscopy of inclined microscopic specimens such as this requires the use of photomontages, for this example a composite image comprised of ten high-magnification photomicrographs acquired at progressively deeper focal depths along the length of the trichome (Fig. 10.5c). Because each such photomicrograph records only a thin in-focus plane, and because the acquired images must therefore be especially closely spaced in parts

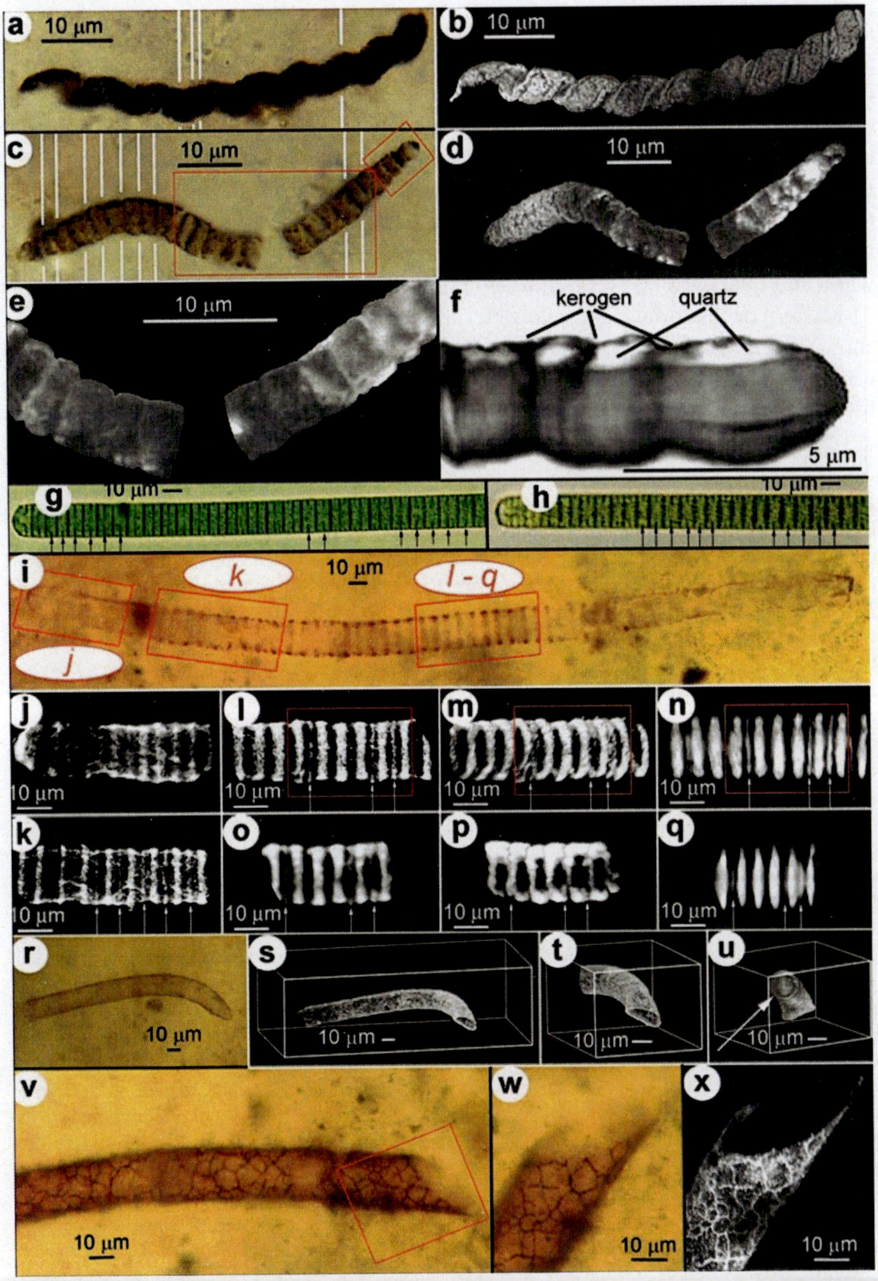

Fig. 10.5 Fossil oscillatoriacean cyanobacteria (**a–f**) in petrographic thin sections of flat-laminated stromatolitic chert (cf. *Stratifera*) from the ~800-Ma-old Bitter Springs Formation at Ellery Creek, west of Alice Springs, Northern Territory, central Australia; modern oscillatoriaceans (**g** and **h**) compared with a morphologically similar fossil trichome (**i–q**) in a thin section of a silicified

(*Caption continued in next page*)

of the trichome that are particularly steeply inclined, the composite optical image depicts only the medial plane of the "photographically flattened" specimen. The distortion produced by such flattening is well illustrated by comparison of the resulting photomontage (Fig. 10.5c) with CLSM images of the specimen (Fig. 10.5d, e) that show its true three-dimensional form, including that of the most steeply inclined part of the trichome (at the lower left in Fig. 10.5d). The three-dimensionality of this fossil is evident also in Fig. 10.5f, a 3-D Raman image of its terminal cells that has been rotated to show their kerogen-defined cell walls and quartz-filled cell lumina where they transect the thin section surface.

Modern oscillatoriacean cyanobacteria, such as the specimens of *Oscillatoria* sp. shown in Fig. 10.5g, h, typically exhibit three diagnostic morphological characters: rounded terminal cells, uniseriate disk-shaped medial cells, and partial septations (denoted by the arrows in Fig. 10.5g, h). Optical (Fig. 10.5i), CLSM (Fig. 10.5j–n),

Fig. 10.5 (continued) columnar stromatolite (*Conophyton gaubitza*) from the ~775 Ma-old Chichkan Formation of southern Kazakhstan; and fossil oscillatoriacean tubular sheaths, also in thin sections of cherty Chichkan specimens of *C. gaubitza* (**r–x**). (**a, b**) Optical montage (**a**), composed of five photomicrographs (denoted by the white lines), and a CLSM image (**b**) of *Heliconema funiculum*, a spirally coiled oscillatoriacean similar to modern *Spirulina* sp. (**c–f**) Optical montage (**c**), composed of ten photomicrographs (denoted by the white lines), and CLSM (**d** and **e**) and a 3-D Raman image (**f**) of a cellular trichome (*Cephalophytarion laticellulosum*, similar to modern *Oscillatoria amoena*) that descends from where it transects the upper surface of the thin section (at the *far right*) to a depth of 20 mm (at the *far left*); the *larger red rectangle* in (**c**) denotes the area shown in (**e**), whereas the *smaller red rectangle* denotes the area shown in (**f**); the 3-D Raman image of its end cells (**f**), rotated to show the flat uppermost surface of the trichome where it transects the thin section surface, demonstrates that the kerogenous cell walls (*gray*) enclose quartz-filled cell lumina (*white*). (**g, h**) Optical images of two specimens of modern *Oscillatoria* sp. showing the rounded terminal cells (*left*), disk-shaped medial cells, and partial septations (*arrows*) characteristic of oscillatoriacean cyanobacteria. (**i**) Optical image of a morphologically similar fossil oscillatoriacean, *Oscillatoriopsis media*, descending into a thin section at a low angle from left to right, shown in a photomontage in which the *red rectangles* denote the areas of the trichome shown in CLSM (**j–n**) and 3-D Raman images (**o–q**). (**j**) The trichome terminus, showing its rounded end-cell and subtending disk-shaped medial cells. (**k**) A part of the trichome situated ~14 mm deeper in the section than the trichome terminus (and ~28 mm below the upper surface of the section) that exhibits partial septations (*arrows*) like those shown in (**g**) and (**h**). (**l–n**) A deeper part of the trichome (~39 mm below the upper surface of the section) that similarly exhibits partial septations (*arrows*), in (**l**) and (**m**) showing the specimen as viewed from above its upper surface [the same perspective as shown in (**i**), but in (**m**) with the trichome tilted slightly to the right to show its interior], and in (**n**) showing the trichome as viewed from its side. (**o–q**) 3-D Raman images (acquired in a spectral window centered in the kerogen "G" band at ~1,605 cm^{-1}) documenting the kerogenous composition of the trichome and its partial septations: (**o**), the part of specimen denoted by the *red rectangle* in (**l**), as viewed from above the trichome; (**p**), the part denoted in (**m**), titled slightly to the left; (**q**), the part denoted in (**n**), showing the specimen from its side. (**r–x**) Optical (**r, v** and **w**) and CLSM (**s–u** and **x**) images of two specimens of *Siphonophycus solidum*, the extracellular tubular sheaths of oscillatoriacean cyanobacteria; (**r**), the shorter of the specimens, shown in a photomontage necessitated by its three-dimensional sinuosity; (**s**), CLSM image of this recurved specimen; (**t**), higher magnification image of a part of the specimen showing its partially flattened end; (**u**), the image shown in (**t**) rotated to show the circular cross-section of the nonflattened sheath (*arrow*); (**v**), a part of a second, broader specimen of *S. solidum* shown in a photomontage; (**w, x**), the part of the specimen enclosed by the *red rectangle* in (**v**) showing its diagenetically produced network-like reticulate surface texture and 3-D Raman images (Fig. 10.5o–q) show that all three of these characters are present in *Oscillatoriopsis media*, a Neoproterozoic analogue of the modern taxon from the ~775 Ma-old Chichkan Formation of southern Kazakhstan. Because of the kerogen-derived fluorescence of the cell walls of this fossil, its rounded terminal cell and uniseriate disk-shaped medial cells are notably better defined in the CLSM images (Fig. 10.5j–n) than in the comparable optical photomicrograph of the specimen (Fig. 10.5i). Similarly, the CLSM images record the presence of partial septations (denoted by the arrows in Fig. 10.5k–n)—submicron-scale structures not evident in the optical photomicrograph (Fig. 10.5i)—thin incipient wall segments arising from the periphery of the fossil trichome that are regularly interspersed between its robust cell-defining walls and that with continued inward growth would have thickened and matured to become the walls of newly formed cells (Pankratz and Bowen 1963; Drews 1973). The 3-D Raman images of this specimen (Fig. 10.5o–q) document the kerogenous composition both of the cell walls and of the interspersed partial septations. The presence of such exceedingly thin partial septations, detectable by CLSM and Raman imagery but not by standard optical microscopy, shows that genetically based cell division in this fossil oscillatoriacean occurred by the same processes as those of modern members of the family.

As exemplified by the specimens shown in Fig. 10.5r–x (from the ~775 Ma-old Chichkan Formation of southern Kazakhstan), the tubular, originally cylindrical extracellular sheaths of oscillatoriaceans are also common components of Neoproterozoic assemblages. As is shown above for the cellular trichomes of such filamentous microorganisms, the use of CLSM for their imagery provides appreciably more information than that available from standard optical microscopy. For example, the optical image in Fig. 10.5r—a photomontage of the type typically presented in paleontological publications—gives the impression that all parts of the specimen are situated within a single focal plane whereas the comparable CLSM image (Fig. 10.5s) shows this fossil to be recurved and, at one extremity, to be partially flattened (Fig. 10.5s, t). Because images acquired by CLSM can be manipulated in three dimensions (by use of processing procedures described above), CLSM permits documentation of the undistorted circularity of the non-flattened end of this sheath (Fig. 10.5u). Moreover, CLSM is especially useful for defining the detailed surface structure of such organic-walled fossils as is shown, for example, by comparison of optical photomicrographs of another Chichkan cyanobacterial sheath (Fig. 10.5v, w) with a corresponding CLSM image (Fig. 10.5x) in which both its three-dimensional morphology and its (diagenetically produced) surface structure are significantly better depicted.

10.5.4 Late Neoproterozoic Spheroidal Acritarchs and Cyanobacteria

Given the foregoing, it is not surprising that CLSM and Raman imagery are useful also for paleontological analyses of Neoproterozoic acritarchs and coccoidal cyanobacteria. Figure 10.6 illustrates their use in studies of such fossils (from the ~775 Ma-old Chichkan Formation of southern Kazakhstan). As shown in

Fig 10.6a–f, the morphology of permineralized acritarchs is decidedly better documented by CLSM than by standard photomicrography. Compare, for example, the optical images of the acritarchs in Fig. 10.6a, b, d, e with their corresponding CLSM images (Fig. 10.6c, f) that contain decidedly more information about their taxonomically important surface texture. As is shown in Fig. 10.6l–p, 3-D Raman imagery of such acritarchs (Fig. 10.6o, p) can identify cell-included non-fluorescing minerals, such as pyrite, that are undetectable by CLSM (Fig. 10.6l).

CLSM is especially useful for studies of three-dimensionally preserved cyanobacterial colonies. Because virtually all coccoidal cyanobacteria are minute, having cells only a few to several microns in diameter, their study by standard optical microscopy requires the use of high magnification objectives that can image only a thin in-focus plane. Thus, documentation of the morphology of such specimens typically requires

Fig. 10.6 Sphaeromorph acritarchs (**a–p**), and few-celled (**q–y**) and many-celled (**z–jj**) cyanobacterial (chroococcalean) colonies in petrographic thin sections of silicified columnar stromatolites (*Conophyton gaubitza*; **a–dd**) and flat-laminated stromatolitic chert (cf. *Stratifera*; **ee–jj**) from the ~775-Ma-old Chichkan Formation of southern Kazakhstan. (**a–f**) Optical (**a, b, d,** and **e**) and CLSM (**c** and **f**) images of two specimens of *Leiosphaeridia crassa*; (**a**) and (**d**), medial focal planes; (**b**) and (**e**), uppermost surfaces; (**c**) and (**f**), the texture of the uppermost surface of each of the two acritarchs. (**g–k**) Optical (**g–i**) and CLSM (**j** and **k**) images of a third specimen of *L. crassa*; (**g**), the uppermost part of the specimen where it transects the upper surface of the section; (**h**), medial focal plane; (**i**), lowermost surface; (**j**), CLSM image from the same perspective as in (**g**) showing the interior surface of the lower part of the specimen; (**k**), the image shown in (**j**) rotated to show the cup-like shape of the transected cell. (**l–p**) CLSM (**l**), optical (**m, n**), and 3-D Raman (**o, p**) images of a specimen of *Leiosphaeridia* sp. that contains a cluster of pyrite crystals; (**l**) CLSM image showing the diagenetically produced surface texture of the kerogenous vesicle wall, but no evidence of the (non-fluorescent, and therefore CLSM-not detectable) included pyrite crystals; (**m**), upper, and (**n**), medial focal planes, showing the included pyrite cubes (*arrows*); (**o**) and (**p**), 3-D Raman images showing the spatial distribution of the kerogen that comprises the fossil-defining vesicle wall [shown in blue in (**p**) and acquired in a spectral window centered on the ~1,605 cm^{-1} kerogen band] and the included pyrite crystals [shown in red in (**p**), acquired in a spectral window centered on a major band of pyrite at ~380 cm^{-1}, and denoted by arrows in (**o**) and (**p**)]. (**q–t**) Optical (**q** and **r**) and CLSM (**s** and **t**) images of a four-celled colony of decussate cells (*Myxococcoides dilutus*); (**q**), upper cell pair; (**r**), lower cell pair; (**s**), CLSM image from the perspective shown in (**q**) and (**r**); (**t**), the image shown in (**s**) rotated to show the underside of the decussate tetrad. (**u–y**) Optical (**u** through **x**) and CLSM (**y**) images of a four-celled colony (*Myxococcoides* sp.), shown in (**u–x**) in a through-focus sequence from the uppermost (**u**) to the lowermost in-focus plane (**x**); (**y**), CLSM image of the colony. (**z–dd**) Optical (**z–cc**) and CLSM (**dd**) images of a thickly ensheathed colony (*Scissilisphaera bistratosa*), shown in (**z–cc**) in a through-focus sequence from its uppermost (**z**) to its lowermost in-focus plane (**cc**); (**dd**), a CLSM image rotated to show the specimen from its underside, demonstrating that the ellipsoidal cell-like structures evident in the optical images (**z–cc**) are actually remnants of originally mucilaginous extracellular envelopes that enclose packets of small spheroidal cells (*arrow*). (**ee, ff**) Optical (**ee**) and CLSM (**ff**) images of an ensheathed colony of spheroidal cells (*Myxococcoides inornata*). (**gg, hh**) Optical (**gg**) and CLSM (**hh**) images of a colony of *M. inornata* enveloped by a diaphanous sheath and composed of cells that contain clumped protoplasmic remnants, evident especially in the CLSM image [*arrows* in (**gg**) and (**hh**)]. (**ii, jj**) Optical (**ii**) and CLSM (**jj**) images of a part of a chroococcalean colony (*M. minor*) showing that thin-walled cells barely discernable by optical microscopy (**ii**) can be defined well by CLSM (**jj**)

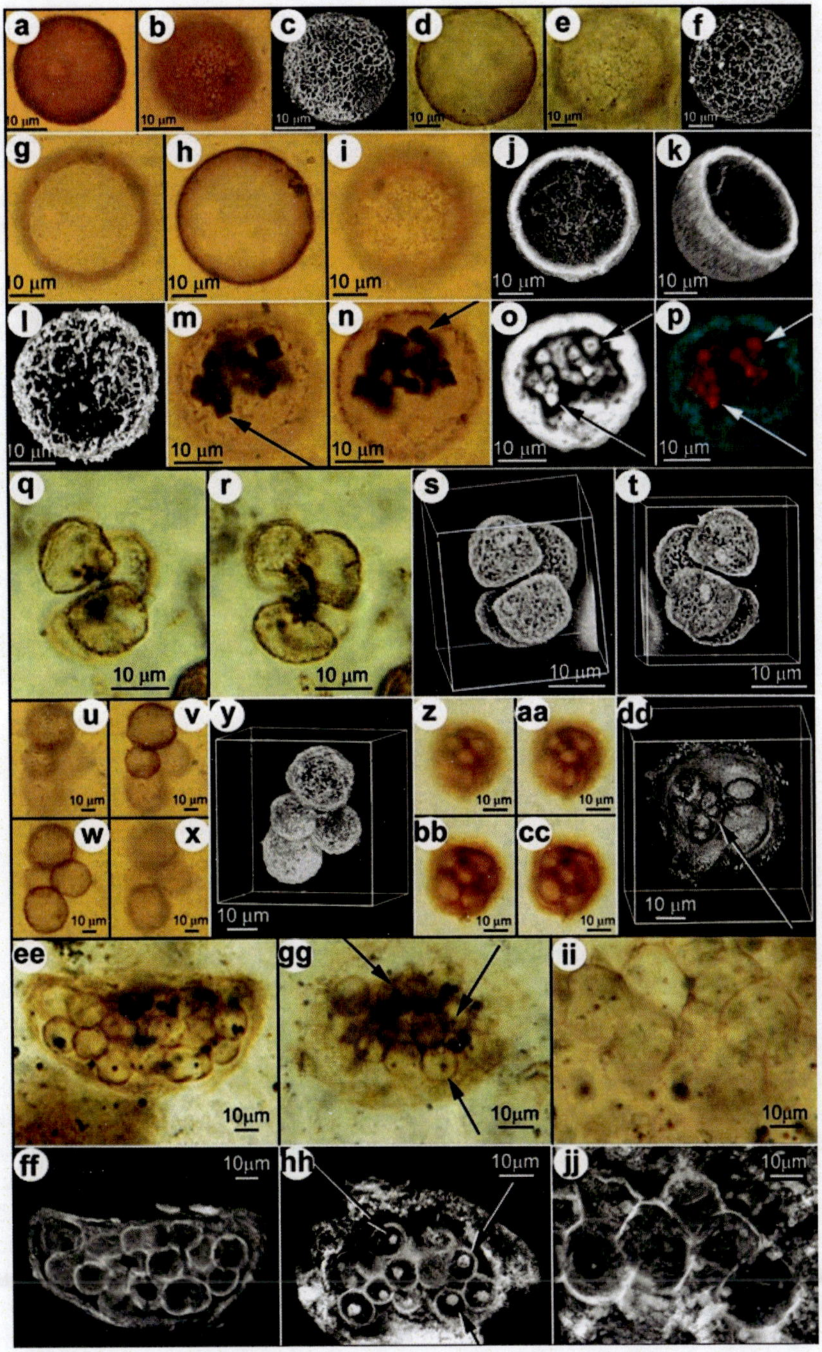
a
b
c
d
e
f
g
h
i
j
k
l
m
n
o
p
q
r
s
t
u
v
w
x
y
z
aa
bb
cc
dd
ee
gg
ii
ff
hh
jj
10 μm

Fig. 10.6

multiple photomicrographs, each showing only the medial optical planes of such specimens (e.g., Fig. 10.6q, r, u–x). In comparison, CLSM provide three-dimensional images of such coccoids at appreciably higher spatial resolution than optical microscopy (compare the CLSM images in Fig. 10.6s, t with the corresponding optical images in Fig. 10.6q, r, and the CLSM image in Fig. 10.6y with the optical images of the same four-celled colony in Fig. 10.6u–x). Moreover, unlike images obtained by standard photomicrography, the CLSM images of such small-celled cyanobacterial colonies can be rotated to show the underlying surfaces of the specimens studied—in the example in Fig. 10.6z–dd showing that the spheroidal cell-like bodies evident in optical photomicrographs (Fig. 10.6z–cc) are actually spheroidal originally mucilaginous envelopes that enclose much smaller cyanobacterial cells (Fig. 10.6dd).

Shown in Fig. 10.6ee–jj are many-celled cyanobacterial colonies from the Chichkan cherts that further illustrate the quality of information provided by the use of CLSM. Comparison of the optical photomicrographs in Fig. 10.6ee, gg with the equivalent CLSM images (Fig. 10.6ff, hh) shows that the use of CLSM well documents the presence of encompassing optically diaphanous organic envelopes; comparison of the optical photomicrograph in Fig. 10.6gg with the corresponding CLSM image in Fig. 10.6hh shows that CLSM can more effectively illustrate the presence of intracellular inclusions of clumped protoplasmic remnants; and comparison of the optical image in Fig. 10.6ii with its CLSM counterpart (Fig. 10.6jj) shows that thin-walled cells, barely discernable by optical microscopy, can be well defined by use of CLSM.

10.6 Geochemical Maturity of the Fossil-Comprising Kerogen

10.6.1 Raman Spectra and the Raman Index of Preservation

The fluorescence and Raman spectra shown in Fig. 10.4 document the identity of the quartz, kerogen, apatite, and Sm^{+3}-substituted apatite that comprise the *Chilodictyon* specimens of the Neoproterozoic Lower Tindir Group discussed above (Fig. 10.3). Comparable data have been acquired from similar analyses of all of the other fossils illustrated here (e.g., Figs. 10.2b–k, 10.5f, o–q and 10.6o, p). Importantly, however, not only can Raman be used to identify the mineralic components of such fossils and to document their morphology and kerogenous composition in three dimensions (Figs. 10.5f, o–q and 10.6o, p), but the Raman spectra of the kerogen on which such images are based can themselves be analyzed to determine quantitatively the geochemical maturity of the preserved organic matter. Shown in Fig. 10.7 are Raman spectra acquired from the kerogenous cell walls of representative fossils from the units studied here together with those of six other Precambrian microfossiliferous cherts. The ten spectra shown—ordered from less (top) to more (bottom) geochemically mature and representative of a much larger suite of kerogen-comprising microfossils for which such data are available (Schopf et al. 2002, 2005, 2007, 2008)—were acquired from microfossils preserved

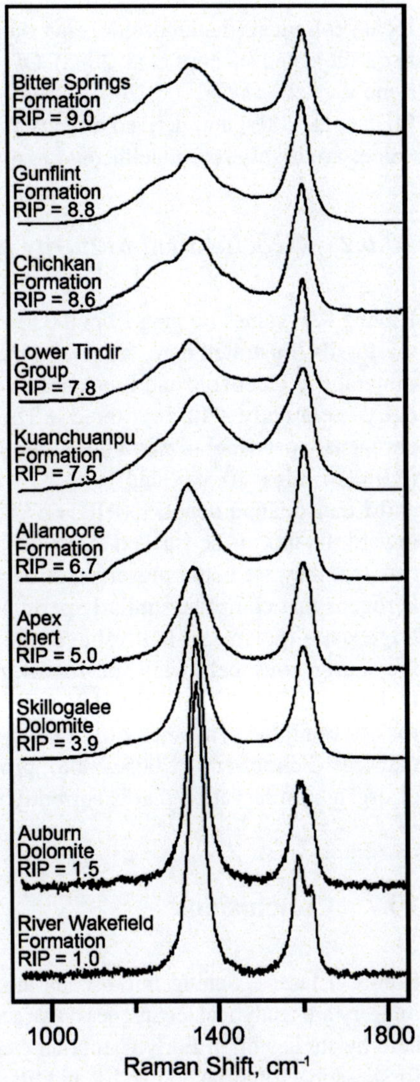

Fig. 10.7 Raman spectra of the kerogen comprising the cell walls of representative microfossils permineralized in cherts of the ~800-Ma-old Bitter Springs, ~1,900-Ma-old Gunflint, and ~775 Ma-old Chichkan Formations, and the ~750-Ma-old Lower Tindir Group; in phosphorite of the ~540-Ma-old Kuanchuanpu Formation; and in cherts of the ~1,050-Ma-old Allamoore Formation, the ~3,465-Ma-old Apex chert, the ~760-Ma-old Skillogalee and ~720-Ma-old Auburn Dolomites, and the ~775-Ma-old River Wakefield Formation (Schopf et al. 2005, 2007), ordered by their RIP values (Schopf et al. 2005) from less (*top*) to more (*bottom*) geochemically mature. The RIP value of the Kuanchuanpu kerogen, determined from a spectrum acquired from the ctenophore embryo shown in Figs. 10.1d and 10.2a, was measured on dehydrated permineralized kerogen which, if rehydrated to more closely approximate its originally preserved condition (Schopf et al. 2005), would yield an RIP value higher than that listed

in rocks ranging from little metamorphosed (top) to middle greenschist facies (bottom). As is evident from these stacked spectra, the two principal Raman bands of kerogen change markedly as its molecular structure, altered primarily by heat, progresses along a geochemical pathway toward graphite: as the carbonaceous matter becomes structurally more ordered, the left-most ("D") band becomes increasingly narrow and more peaked as the right-most ("G") band gradually narrows and, in partially graphitized kerogen, ultimately bifurcates. For each of these ten spectra is listed a Raman Index of Preservation (RIP) value, an objective, quantitative measure of the organic geochemical maturity of the preserved kerogen that reflects the local geological (metamorphic) environment to which the fossil-containing unit has been subjected (Schopf et al. 2005). Of rapidly increasing use in paleontology (e.g., Schopf et al. 2005, 2007, 2008; Chen et al. 2007; Schopf and Kudryavtsev 2009; Igisu et al. 2009) and derived directly from the Raman spectra measured, such RIP values are highly reproducible and easily calculated (Schopf et al. 2005).

10.6.2 Geochemical Maturity of the Analyzed Kerogens

Having RIP values ranging from 9.0 to ≥7.5 (Fig. 10.7), the kerogens that comprise the fossils illustrated here, relative to the kerogenous cell walls of many other permineralized Paleozoic and Precambrian fossils (Schopf et al. 2005), are geochemically relatively little altered. The spectra of these fossil-comprising kerogens—permineralized in the Bitter Springs (RIP = 9.0; Fig. 10.5a–f), Chichkan (RIP=8.6; Figs. 10.5i–x and 10.6), Lower Tindir (RIP = 7.8; Figs. 10.3 and 10.4), and Kuanchuanpu deposits (RIP = ≥7.5; Figs. 10.1d–h and 10.2)—are all situated toward the top of the stacked sequence shown in Fig. 10.7. Their RIP values indicate that they are better preserved or only slightly more geochemically altered than kerogens that comprise similarly permineralized fossils of the ~1,900 Ma Gunflint Lagerstätte (having an RIP value of 8.8) and that they are much less altered than the kerogenous cell walls of fossils permineralized in the ~750 Ma Auburn Dolomite (RIP = 1.5) and River Wakefield Formation (RIP = 1.0). Such data, useful for assessing the syngenicity of organic-walled fossils and their associated mineral matrices (Schopf et al. 2005, 2008), provide a quantitative measure of the quality of organic preservation that is currently unavailable by any other technique.

10.7 Conclusions

Confocal laser scanning microscopy and Raman (and fluorescence) spectroscopic imagery are analytical techniques only recently introduced to paleontology. As shown here by studies of an Early Cambrian ctenophore embryo and Neoproterozoic scale fossils, sphaeromorph acritarchs, and filamentous and coccoidal cyanobacteria, these techniques—whether used individually or in combination—can provide important insight into the morphology, chemistry, and preservational history of permineralized organic-walled fossils. In many respects, the two techniques are similar. Both are rock penetrating, rather than being surface-dependent as are other high-resolution analytical techniques (e.g., scanning electron microscopy, electron microprobe spectroscopy, and secondary ion mass spectrometry), and are non-intrusive and non-destructive (enabling their use for analysis of specimens archived in museum collections). Both are based on confocal optical systems (thereby providing appreciably crisper images than those of standard optical microscopy) and, because the data they provide is based on direct measurement of a specimen analyzed, both avoid the subjectivity necessarily involved in the preparation of optical photomontages. Both are applicable to fossils preserved in the three most commonly occurring permineralizing matrices, apatite (Figs. 10.1 and 10.2), chert (Figs. 10.3–10.6), and calcite (Schopf et al. 2010); both can be used to study standard paleontological preparations (petrographic thin sections, palynological macerations, and cellulose acetate peels), in translucent thin sections to depths of ~150 μm; and both can yield images having submicron spatial resolution—for CLSM <0.2 μm; for Raman (and fluorescence) spectroscopic imagery, ~0.7 μm. Importantly, both can be applied to a single fossil specimen to yield independent but reinforcing lines of evidence about its morphology and molecular-structural composition (e.g., Figs. 10.1d–h, 10.2, 10.3h–k, 10.5i–q and 10.6l–p).

Despite such similarities, however, the techniques are complementary, not redundant. As used here to analyze permineralized organic-walled fossils, CLSM detects the laser-induced fluorescence of the kerogenous components of fossils and their associated organics whereas Raman (and fluorescence) spectroscopy identifies the molecular structure of the preserved organics and that of their embedding minerals. However—and although the high-resolution images of such fossils afforded by CLSM are typically derived from the laser-induced fluorescence emanating from the interlinked polycyclic aromatic hydrocarbons of the kerogen of which they are composed—CLSM-detectable fluorescence can be emitted also from inorganic materials (e.g., Sm^{+3}-containing apatite; Figs. 10.3k and 10.4). Thus, CLSM does not provide unambiguous evidence of chemical composition. In contrast, Raman and fluorescence spectroscopy can provide direct submicron-scale analyses of the molecular-structural composition both of fossil-comprising kerogen and of associated minerals. In addition, the kerogen spectra provided by Raman imagery can themselves be analyzed to determine the RIP value of the preserved organic matter, an objective quantitative measure of its geochemical maturity.

CLSM and Raman/fluorescence spectroscopic imagery address the two outstanding problems that have long-hindered studies of ancient microscopic fossils, namely, the accurate documentation of their three-dimensional morphology and of their chemical composition. Used together, these powerful techniques provide high quality information about the morphology, taphonomy, and fidelity and mode of preservation of permineralized organic-walled fossils and an effective new means to correlate, in three dimensions at high spatial resolution, their micron-scale cellular structure and molecular composition.

Acknowledgments We are grateful to J-Y. Chen (Nanjing Institute of Geology and Palaeontology, China) and N.J. Butterfield (Cambridge University, England) for providing, respectively, the petrographic thin sections containing the fossils illustrated here from the Kuanchuanpu Formation and Lower Tindir Group, and also to V.N. Sergeev (Institute of Geology, RAS, Moscow, Russia) for providing the rock samples of silicified *Conophyton gaubitza* and thin sections of *Stratifera*-like stromatolites in which occur the fossils illustrated from the Chichkan Formation. We are also very appreciative to N.J. Butterfield for first suggesting the possible use of CLSM and Raman to analyze the Tidir Group scale fossils, and to J. Shen-Miller, for a critical review of this manuscript. This work, parts of which are based on material presented at greater length in previous publications (Schopf et al. 2005; Schopf and Kudryavtsev 2005, Schopf and Kudryavtsev 2009; Chen et al. 2007), was supported by UCLA's Center for the Study of Evolution and the Origin of Life and, for a part of the contributions of A.B.K., by the NASA Astrobiology Institute-funded PennState Astrobiology Research Center.

References

Allison CW, Hilgert JW (1986) Scale microfossils from the Early Cambrian of Northwest Canada. J Paleontol 60:973–1015

Amos WB, White JG (2003) How the confocal laser scanning microscope entered biological research. Biol Cell 95:335–342

Arouri KR, Greenwood PF, Walter MR (2000) Biological affinities of Neoproterozoic acritarchs from Australia: microscopic and chemical characterization. Org Geochem 31:75–89

Ax P (1996) Multicellular animals, vol 1. Springer, Berlin/Heidelberg/New York

Barghoorn ES, Schopf JW (1965) Microorganisms from the late Precambrian of central Australia. Science 150:337–339

Barghoorn ES, Tyler SA (1965) Microorganisms from the Gunflint chert. Science 147:563–577

Birkmann H, Lundin RF (1996) Confocal microscopy: potential applications in micropaleontology. J Paleontol 70:1084–1087

Chen J-Y, Schopf JW, Bottjer DJ, Zhang C-Y, Kudryavtsev AB, Tripathi AB, Wang X-Q, Yang Y-H, Gao X, Yang Y (2007) Raman spectra of a ctenophore embryo from southwestern Shaanxi, China. Proc Natl Acad Sci USA 104:6289–6292

Cloud PE (1965) Significance of the Gunflint (Precambrian) microflora. Science 148:27–45

Cohen P, Schopf JW, Butterfield NJ, Kudryavtsev AB, MacDonald FA (2009) Mineralized scales from eukaryotic microorganisms in the ca. 750 Ma Lower Tindir Group, Yukon Territory, Canada. In: Abstracts with program, Geological Society of America Annual Meeting, Portland, 18–21 October 2009, p 31

Domozych D, Wells B, Shaw P (1991) Basket scales of the green-alga, *Mesostigma viride*—chemistry and ultrastructure. J Cell Sci 100:397–407

Drews G (1973) Fine structure and chemical composition of the cell envelopes. In: Carr NG, Whitton BA (eds) The biology of blue-green algae, vol 9, Botanical monographs. University of California Press, Berkeley, pp 99–116

Feist-Burkhardt S, Monteil E (2001) Gonyaulacacean dinoflagellate cysts with multi-plate precingular archaeopyle. Neues Jahrb Geol Paläontol 219:33–81

Feist-Burkhardt S, Pröss J (1999) Morphological analysis and description of middle Jurassic dinoflagellate cyst marker species using confocal laser scanning microscopy, digital optical microscopy and conventional light microscopy. Bull Cent Rech Elf E 22:103–145

Finnerty JR, Master VA, Irvine S, Kourakis MJ, Warriner S, Martindale MQ (1996) Homeobox genes in the Ctenophora: identification of paired-type and Hox homlogues in the atentaculate ctenophore, *Beroe ovata*. Mol Mar Biol Biotechnol 5:249–258

Foster B, Williams VE, Witmer RJ, Piel KM (1990) Confocal microscopy: a new technique for imaging micro-organisms and morphology in three-dimensions. Palynology 14:212

Gaft M, Reisfeld R, Panczerer G (2005) Modern luminescence spectroscopy of minerals and materials. Springer, Berlin/Heidelberg/New York

Hochuli P, Feist-Burkhardt S (2004) An early boreal cradle of angiosperms? Angiosperm-like pollen from the Middle Triassic of the Barents Sea (Norway). J Micropalaeontol 23:97–104

Hofmann HJ, Schopf JW (1983) Early Proterozoic microfossils. In: Schopf JW (ed) Earth's earliest biosphere, its origin and evolution. Princeton University Press, Princeton, pp 321–360

House CH, Schopf JW, McKeegan KD, Coath CD, Harrison TM, Stetter KO (2000) Carbon isotopic composition of individual Precambrian microfossils. Geology 28:707–710

Igisu M, Ueno Y, Shimojima M, Nakashima S, Awramik SM, Ohta H, Maruyama S (2009) Micro-FTIR spectroscopic signatures of bacterial lipids in Proterozoic microfossils. Precambrian Res 173:19–26

Jehlička J, Beny C (1992) Application of Raman microspectrometry in the study of structural changes in Precambrian kerogens during regional metamorphism. Org Geochem 18:211–213

Jehlička J, Urban A, Pokorny J (2003) Raman spectroscopy of carbon and solid bitumens in sedimentary and metamorphic rocks. Spectrochim Acta A59:2341–2352

Kelemen SR, Fung HL (2001) Maturity trends in Raman spectra from kerogen and coal. Energy Fuels 15:653–658

Knoll AH, Golubić S (1979) Anatomy and taphonomy of a Precambrian algal stromatolite. Precambrian Res 10:115–151

Knoll AH, Barghoorn ES, Golubić S (1975) *Paleopleurocapsa wopfneri* gen. et sp. nov.: a late Precambrian alga and its modern counterpart. Proc Natl Acad Sci USA 72:2488–2492

Kudryavtsev AB, Schopf JW, Agresti DG, Wdowiak TJ (2001) *In situ* laser-Raman imagery of Precambrian microscopic fossils. Proc Natl Acad Sci USA 98:823–826

McKeegan KD, Kudryavtsev AB, Schopf JW (2007) Raman and ion microscopic imagery of graphite inclusions in apatite from older than 3830 Ma Akilia supracrustal rocks, West Greenland. Geology 35:591–594

McMillan PF, Hofmeister AM (1988) Infrared and Raman spectroscopy. Rev Mineral 18:99–159

Mendelson CV, Schopf JW (1992) Proterozoic and selected early Cambrian microfossils and microfossil-like objects. In: Schopf JW, Klein C (eds) The Proterozoic biosphere, a multidisciplinary study. Cambridge University Press, New York, pp 865–951

Mus MM, Moczydłowska M (2000) Internal morphology and taphonomic history of the Neoproterozoic vase-shaped microfossils from Visingsö Group, Sweden. Norsk Geol Tidsskr 80:213–228

Nagy LA (1974) Transvaal stromatolite: first evidence for the diversification of cells about 2.2 x 10^9 years ago. Science 183:514–516

Nagy LA (1978) New filamentous cystous microfossils, 2, 300 M.Y. old, from the Transvaal sequence. J Paleontol 52:141–154

Nix T, Feist-Burkhardt S (2003) New methods applied to the microstructure analysis of Messel Oil Shale: confocal laser scanning microscopy (CLSM) and environmental scanning electron microscopy (ESEM). Geol Mag 140:469–478

O'Conner B (1996) Confocal laser scanning microscopy: a new technique for investigating and illustrating fossil Radiolaria. Micropaleontology 42:395–402

Pankratz HS, Bowen CC (1963) Cytology of blue-green algae. I. The cells of Symploca muscorum. Am J Bot 50:387–399

Pasteris JD, Wopenka B (1991) Raman spectra of graphite as indicators of degree of metamorphism. Can Mineral 29:1–9

Pasteris JD, Wopenka B (2003) Necessary, but not sufficient: Raman identification of disordered carbon as a signature of ancient life. Astrobiology 3:727–738

Schopf JW (1968) Microflora of the Bitter Springs Formation, late Precambrian, central Australia. J Paleontol 42:651–688

Schopf JW (1992) Paleobiology of the Archean. In: Schopf JW, Klein C (eds) The Proterozoic biosphere, a multidisciplinary study. Cambridge University Press, New York, pp 25–39

Schopf JW (1993) Microfossils of the early Archean Apex chert: new evidence of the antiquity of life. Science 260:640–646

Schopf JW (1999) Cradle of life, the discovery of Earth's earliest fossils. Princeton University Press, Princeton

Schopf JW (2006) Fossil evidence of Archaean life. Philos T Roy Soc B 361:869–885

Schopf JW, Blacic JM (1971) New microorganisms from the Bitter Springs Formation (late Precambrian) of the north-central Amadeus Basin, central Australia. J Paleontol 45:925–960

Schopf JW, Bottjer DJ (2009) Preface: world summit on ancient microscopic fossils. Precambrian Res 173:1–3

Schopf JW, Kudryavtsev AB (2005) Three-dimensional Raman imagery of Precambrian microscopic organisms. Geobiology 3:1–12

Schopf JW, Kudryavtsev AB (2009) Confocal laser scanning microscopy and Raman imagery of ancient microscopic fossils. Precambrian Res 173:39–49

Schopf JW, Sovietov YK (1976) Microfossils in *Conophyton* from the Soviet Union and their bearing on Precambrian biostraigraphy. Science 193:143–146

Schopf JW, Walter MR (1983) Archean microfossils: new evidence of ancient microbes. In: Schopf JW (ed) Earth's earliest biosphere, its origin and evolution. Princeton University Press, Princeton, pp 214–239

Schopf JW, Kudryavtsev AB, Agresti DG, Wdowiak TJ, Czaja AD (2002) Laser-Raman imagery of Earth's earliest fossils. Nature 416:73–76

Schopf JW, Kudryavtsev AB, Agresti DG, Czaja AD, Wdowiak TJ (2005) Raman imagery: a new approach to assess the geochemical maturity and biogenicity of permineralized Precambrian fossils. Astrobiology 5:333–371

Schopf JW, Tripathi AB, Kudryavtsev AB (2006) Three-dimensional optical confocal imagery of Precambrian microscopic organisms. Astrobiology 1:1–16

Schopf JW, Kudryavtsev AB, Czaja AD, Tripathi AB (2007) Evidence of Archean life: stromatolites and microfossils. Precambrian Res 158:141–155

Schopf JW, Tewari VC, Kudryatsev AB (2008) Discovery of a new chert-permineralized microbiota of the Proterozoic Buxa Formation of the Ranjit Window, Sikkim, N.E. India, and its astrobiological implications. Astrobiology 8:735–746

Schopf JW, Kudryavtsev AB, Tripathi AB, Czaja AD (2010) Three-dimensional morphological (CLSM) and chemical (Raman) imagery of permineralized fossils. In: Allison PA, Bottjer DJ (eds) Taphonomy: process and bias through time. Springer, Berlin/Heidelberg/New York, pp 457–486

Scott AC, Hemsley AR (1990) A comparison of new microscopical techniques for the study of fossil spore wall ultrastructure. Rev Palaeobot Palynol 67:133–139

Spötl C, Houseknecht DW, Jaques RC (1998) Kerogen maturation and incipient graphitization of hydrocarbon source rocks in the Arkoma Basin, Oklahoma and Arkansas: a combined petrographic and Raman study. Org Geochem 28:535–542

Talyzina NM (1997) Fluorescence intensity in early Cambrian acritarchs from Estonia. Rev Palaeobot Palynol 100:99–108

Taylor PD, Schopf JW, Kudryavtsev AB (2008) Calcite and aragonite in the skeletons of bimineralic bryozoans as revealed by Raman spectroscopy. Invertebr Biol 127:87–97

Williams KPJ, Nelson J, Dyer S (1997) The Renishaw Raman database of gemological and mineralogical materials. Renishaw Transducers Systems Division, Gloucestershire, England

Wopenka B, Pasteris JD (1993) Structural characterization of kerogens to granulite-facies graphite: applicability of Raman microprobe spectroscopy. Am Mineral 78:533–557

Yui T-F, Huang E, Xu J (1996) Raman spectrum of carbonaceous material: a possible metamorphic grade indicator for low-grade metamorphic rocks. J Metamorph Geol 14:115–124

Chapter 11
X-Ray Microanalysis of Burgess Shale and Similarly Preserved Fossils

Patrick J. Orr and Stuart L. Kearns

Contents

P.J. Orr (✉)
UCD School of Geological Sciences, University College Dublin, Belfield, Dublin 4, Ireland
e-mail: patrick.orr@ucd.ie

S.L. Kearns
Department of Earth Sciences, University of Bristol, Queen's Road, BS8 1RJ Bristol, UK
e-mail: stuart.kearns@bristol.ac.uk

M. Laflamme et al. (eds.), *Quantifying the Evolution of Early Life*, Topics in Geobiology 36, DOI 10.1007/978-94-007-0680-4_11, © Springer Science+Business Media B.V. 2011

Abstract Non-biomineralized fossils from the Burgess Shale Formation are typically preserved as near two dimensional carbonaceous remains of the original tissues associated with films of aluminosilicate minerals; the host lithologies have experienced a long, complicated, diagenetic and metamorphic history. Such 'layered' substrates exhibit lateral and vertical variation in sample chemistry over micron-scale distances. X-ray microanalysis of these presents a particular challenge to the analyst, as the results are highly sensitive to the choice of operating conditions.

In a scanning electron microscope an electron beam generates an interaction volume at and extending below the sample's surface; various signals, including x-rays, are generated from different parts of this volume. X-ray microanalysis is therefore not an analysis of the surficial chemistry of a sample. The lower the mean atomic number of the area impacted, and the higher the accelerating voltage, the further the electron beam will penetrate, the effects of which on the generated and emitted signals can be readily visualized via computer-based simulations.

The emitted X-rays can be collected via either a wavelength or an energy dispersive spectrometer and provide compositional information. Results are presented as (a) quantitative analysis; (b) an x-y plot of x-ray energy against counts (an X-ray spectrum); (c) a visual image of abundance of an element over a user-defined area (an element map).

Specimen preparation prior to analysis is designed to eliminate sample charging (most geological samples are dielectric) and topography, both of which impact negatively on X-ray microanalysis. When sample integrity must be maintained, unprepared samples may be used. Grounding part of the sample will minimize the likelihood of specimen charging; in 'variable pressure' SEMs the presence of gas in the specimen chamber negates the tendency for any charge to accumulate on the surface, but reduces markedly the resolution of analyses (the 'beam skirt' effect). Topographic artifacts are inevitable. Surface irregularity alters absorption path length; the longer such a path and the lower the energy of the X-ray the less likely it is to be emitted. Compositional analyses of unprepared samples are best considered a qualitative assessment of a sample's chemistry.

Keywords Variable pressure • Low vacuum • Scanning electron microscope (SEM) • Electron beam • Interaction volume • Energy dispersive • X-ray spectrum • Accelerating voltage • Charging • Topography

11.1 Introduction

Scanning electron microscopy (SEM) is now used routinely in paleobiological studies. It offers high resolution imaging of samples that often can, depending on the configuration of the microscope, be integrated with precise analyses of their chemical composition. SEM-based imaging of fossils has a long history. SEM-based analyses of fossil composition is a more recent phenomenon, but has developed rapidly over the past two decades. Pioneering studies used fully or semi-destructive techniques to acquire samples; samples were thin-sectioned, or small pieces ('chips') were picked from their surface, typically using scalpels and needles. The latter were either resin-embedded, sectioned and polished before analysis or simply mounted directly onto SEM stubs. Recent technological advances, in particular the development of variable pressure scanning electron microscopes (VP-SEM), allow routine imaging and analyses of entire specimens with minimal, even no, preparation. The obvious advantage this offers is that the composition of rare or valuable material, such as that from the Burgess Shale Formation (Briggs et al. 1994) and other exceptional biotas, can be determined while maintaining sample integrity.

The general principles of scanning electron microscopy are well served by texts such as Goldstein et al. (2003); others, such as that by Reed (2005), treat the principles and methods involved in analysis of the composition of materials. In summary, in a SEM a primary electron beam (with energy between a few hundred electron volts and 40 keV) striking a surface (the 'spot') will yield various signals from the surface of, and within the sample. This is unlike an optical photograph, where light is reflected from the surface of a sample. SEM signals are therefore derived from within what is termed the interaction volume (Fig. 11.1). The relevant emitted signals are: (a) secondary electrons (SE); (b) backscattered electrons (BSE); and (c) electromagnetic radiation, most importantly X-rays. These are the signals used to generate images of the specimen's appearance (its surface topography) or its composition, either as an element (X-ray) map, or a mean atomic number (z-contrast) image (more commonly referred to as a BSE image). The excitation volume is that part of the interaction volume from which a specific signal is generated and, possibly, emitted. It is important to note that each signal has a different excitation volume (Fig. 11.1). In addition, chemical compositions can be

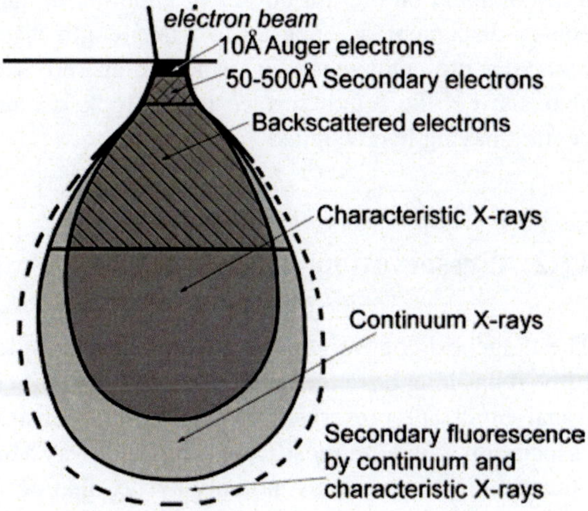

Fig. 11.1 Schematic cross-section of the interaction volume showing different areas from which electrons and X-rays are emitted

derived from X-ray analysis of small (micron-sized) volumes of the sample. More recent developments include the application of charge contrast imaging (CCI), cathodoluminescence imaging (CL-imaging) and Electron Backscatter Diffraction (EBSD) analysis to paleontological materials.

In this chapter we review first recent improvements in our understanding of the diagenesis of Burgess Shale fossils, We do not, however, attempt to resolve outstanding issues among the various taphonomic models that have been developed. Instead, we use these to illustrate the potential challenges fossils preserved in this manner present to the analyst. Of the various SEM techniques available, charge contrast imaging (Kearns and Orr 2009) (Fig. 11.2a) and, to a greater extent, backscattered electron imaging (see, for example, Conway Morris 1990; Caron et al. 2006; and Butterfield et al. 2007) (Fig. 11.2b) have both been used in the study of Burgess Shale fossils. Non-biomineralized Burgess Shale fossils are almost invariably preserved in minimal relief (Briggs and Williiams 1981). While useful for detailed imaging of the textures of different diagenetic mineral phases, secondary electron imaging, used mainly to visualize the topography of fossils, has not been widely employed. To the best of our knowledge, CL-imaging has not been undertaken previously (Fig. 11.2c). This chapter focuses on X-ray microanalysis, the principal SEM investigative technique that has been used to date (Fig. 11.2d). We discuss briefly what constraints the nature of the instrumentation and samples may place on a study, and the use of computer based modeling of electron beam interaction with solids, a methodology we have found particularly useful (see, for example, Orr et al. 2002, 2009; Maletz et al. 2005), and feel is under-utilized. We then discuss the most fundamental concept that underpins any SEM-based study: how the size and shape of the interaction volume will impact on the results. Finally, we discuss the principles of X-ray generation and the nature of the output. Our focus is on the methodologies involved in the production of X-ray, or element, maps that show spatial variation in element abundance as images. We also comment on the difficulties of undertaking more quantitative analyses using energy-dispersive, as opposed to wave-length dispersive, spectrometers, and the particular problems what we term 'unprepared' samples present in X-ray microanalysis; i.e. the rough surface of the rock is analyzed without any conductive coating having been applied.

11.2 Preservation of Burgess Shale Fossils

The diagenesis of Burgess Shale fossils is complex and, as yet, not understood fully. It has long been recognized that non-biomineralized fossils from this exceptional biota are preserved as near-two dimensional carbonaceous residues in association with aluminosilicate films (Conway Morris 1990) (Fig. 11.2). Various other authigenic minerals, notably pyrite, also occur. High fidelity replication of decay-prone tissues in authigenic minerals is, however, limited; the mid-gut glands of various arthropods are replicated in calcium phosphate, often in

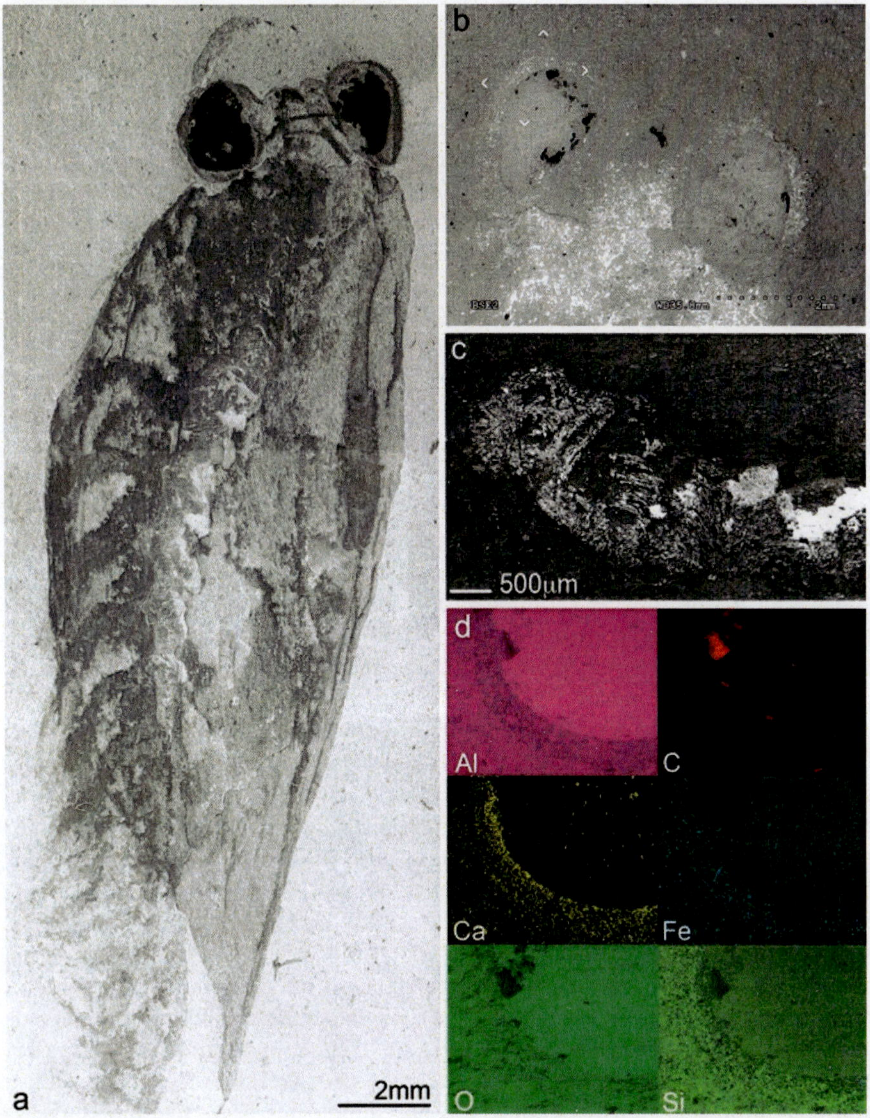

Fig. 11.2 Aspects of the techniques used to elucidate the preservation of Burgess Shale fossils illustrated using a specimen of *Isoxys* sp. (ROM930912a, b). (**a**) Montage of fifteen charge contrast images of the part. (**b**) BSE image of anterior of counterpart, showing variable z-contrast between carbon, authigenic minerals, and sedimentary matrix. (**c**) CL-image revealing detail of the mid-gut glands. (**d**) Element maps of the area indicated in part (**b**)

exquisite detail (Butterfield 2002a) (Fig. 11.2c). There is general consensus that the carbonaceous remains represent original tissues (see, for example, Gaines et al. 2008); whether highly labile (decay-prone) volatile (proteinaceous) cellular tissues are preserved as carbonaceous remains in addition to recalcitrant cuticular tissues (as for example argued by Conway Morris 2008, but see Butterfield 1990) has yet to be agreed upon.

There is some consensus as to the origin of the aluminosilicate films associated with the original tissues. Elemental mapping by Orr et al. (1998) demonstrated that the composition of these aluminosilicate films differed significantly from that of the sedimentary matrix surrounding the fossils, and thus were of diagenetic origin. There now appears little support for the aluminosilicate films being of sedimentary origin (with perhaps rare exceptions; see Butterfield (1990, Fig. 4)) the result of sediment being forcibly injected inside the fossils during transport within depositing event beds (Butterfield 1990) or by sediment leaking into specimens to infill voids created during early decay (Towe 1996). The aluminosilicate films have been considered as being of early diagenetic origin, later metamorphosed, and to have formed predominantly inside the fossils (e.g. Orr et al. 1998; see also Orr et al. 2008). Alternatively, Page et al. (2008) favored their having originated later in the taphonomic history: reduction of the fossils to near-two dimensional carbonaceous substrates during diagenesis, followed by volatilization of these during metamorphism, and growth of minerals in the voids created; differences in the chemistry of the aluminosilicates between and within fossils reflects variation in the timing of volatilization of different carbonaceous tissues. Butterfield et al. (2007) envisaged the aluminosilicates having two origins. Some represent replacement of carbonaceous substrates associated with volatilization of the latter during metamorphism; others represent complete replacement of a precursor, carbonate, phase that precipitated inside the fossils during early diagenesis. It has also been posited that the aluminosilicates are external to the fossils, an early diagenetic sedimentary imprint produced as the buried carcass decayed, and forming an envelope around the fossil (Butterfield 2002b, p. 242). Finally, the possibility that metamorphic minerals, including aluminosilicates, have grown on, not replaced, Burgess Shale fossils, inside pressure shadows should also be borne in mind (see discussion by Orr et al. 2009).

It has proved instructive to consider the diagenesis of other, similarly preserved, fossils. We also include here data from the Solite biota, an exceptionally preserved biota of vertebrates and insects from the Late Triassic of Virginia, USA (Fraser et al. 1996). The insects are preserved as carbonaceous residues locally coated by thin aluminosilicate films ('biotite'); the host lithologies have, like the Burgess Shale, been metamorphosed.

In summary, fossils from the Burgess Shale and other deposits can be modeled as a complex, near two-dimensional, arrangement of carbonaceous and various mineral substrates that are each microns, or less, thick. These we term 'layered substrates' herein. The majority of early to middle Cambrian Burgess Shale-type fossils analyzed to date are mineralogically less complex than those from the Burgess Shale itself. However, as the primary preservational pathway (Burgess Shale-type preservation) is as near two-dimensional carbonaceous remains (Gaines et al. 2008), the various techniques described herein, including their limitations, also apply to X-ray analysis of these.

11.3 Instrumentation

11.3.1 'High Vacuum' and 'Variable Pressure' SEMs

For paleontological materials, it is convenient to define scanning electron microscopes as one of two types: 'high vacuum' and 'variable pressure'. As the term 'high vacuum' implies, such microscopes operate only with the specimen chamber and column evacuated to a pressure of typically 10^{-3} Pa or lower. In contrast, many modern SEM systems include the capability to operate under variable pressure (VP-SEM); typically the pressure inside the specimen chamber would be in the range 1–1,300 Pa. Some authors consider the terms environmental SEM (ESEM) where chamber pressures may rise to 4,000 Pa, and low vacuum SEM (LV-SEM) synonymous with VP-SEM; see Thiel and Toth (2005) for discussion of the subtle differences between each. The principle common to each is that the SEM column, maintained at high vacuum, is decoupled from the chamber by a pressure limiting aperture. Higher pressure (lower vacuum) inside the chamber is maintained by leaking atmospheric air or another gas (for example, water vapor or nitrogen) into the chamber at a controlled pressure (Moncrieff et al. 1979). Which system is available impacts most on how specimens are prepared, in particular the steps required to inhibit specimen charging.

11.3.2 Specimen Charging

The vast majority of geological samples, including fossils, are dieletrics. Rarely, graphitic rocks may exhibit some degree of conductivity but the rule of thumb is to assume that samples are electrically insulating. This is important as the negative charge of the electron beam will accumulate on the surface of the specimen unless a path to ground potential is available. This, sample charging, will distort the image and introduce streaking, highlights and other aberrant effects which render the image largely useless, and compromise any chemical analyses. The likelihood of charging depends on the nature of the sample and the operating conditions of the microscope; the effect of each is difficult to predict in advance. The composition and topography of the sample are important; samples with a high degree of surface roughness are likely to charge more than planar surfaces. The accelerating voltage of the beam, beam current, scan speed and working distance can be adjusted by the operator and will exacerbate or reduce charging. Operating at a higher accelerating voltage, e.g. 20 as opposed to 10 keV, offers improved resolution and better high energy X-ray ionization, but will exacerbate any charging. Experimental adjustment to optimize the analytical conditions is likely to be required. It is often necessary to compromise between the benefits of operating at a high accelerating voltage and the requirement that the specimen does not charge during analysis.

Under VP-SEM conditions the surface charge is dissipated because the positively ionized gas atoms in the specimen chamber counterbalances the negative charge on the sample surface. Increasing the gas pressure enhances charge neutralization, but can induce deleterious effects on results, e.g. loss of signal, and slightly poorer resolution.

Traditionally, under high vacuum conditions, charge neutralization is accomplished by the application of a thin (on the order of 1–20 nm) coating of a conductive material (e.g. gold or carbon). Additionally, the coated sample must be grounded to the sample holder. Specimens that must remain uncoated present a particular challenge under high vacuum conditions. The only practical solution to the issue of sample charging is to find a compromise between operating at as high an accelerating voltage as permitted by the sample, yet maintain acceptable image quality. In practice, this means, for most geological materials, operating at less than 5 keV. The problem may be circumvented, to some extent, by using backscattered electrons, not secondary electrons, for imaging.

11.3.3 Alternative SEM Systems

For the past few decades, most SEM systems have used a tungsten filament as the electron source. Electrons are emitted when the tungsten is heated to 2,700 K and accelerated by an anode to form the primary beam. In recent years field emission gun (FEG) sources have become increasingly popular. While more expensive, the FEG-SEM offers improved beam stability and reliability, and is emerging as the instrument of choice for many researchers. In a FEG, electrons are extracted from the tip by a primary excitation voltage applied to an anode, and then regulated by a second anode to the accelerating voltage required (Goldstein et al. 2003). This yields a higher beam current density than a tungsten filament at the same accelerating voltage, as a result of which images have higher resolution, especially at lower accelerating voltages and higher magnifications

Alternative versions of what is generally considered to be a SEM are available. Tabletop tungsten-source scanning electron microscopes are a relatively new phenomenon that offer limited functionality (e.g. small chamber size and fixed keV) but ease of use, portability, and relatively low acquisition costs. "Electron microprobes" (EPMA) are configured specifically for the analysis of x-rays by both energy dispersive (ED) and multiple wavelength dispersive (WD) spectrometers. Their primary role, high precision, high accuracy, quantitative chemical analyses, precludes the addition of other detectors and necessitates stages with limited motion, in order to optimize acquisition of the x-ray signal. Whilst of some benefit to the specific study of elemental composition of fossil specimens these can be regarded as more specialized EM instruments and not particularly suited to the routine analysis of fossils.

11.4 Simulation of Electron Beam-Sample Interactions

Herein, we simulate the effects of electron beam interaction with samples using the Monte Carlo simulation program Electron Flight Simulator™. Other software packages are available. The results are illustrated by synthetic interaction and excitation volumes and ED spectra. All simulations used common operating conditions (Super Ultra Thin Window (SUTW) on a silicon detector; x-ray modeling using a generic phi-rho-z algorithm). Simulations are run for 32,000 trajectories and assume a zero tilt angle on the sample. The detector has a 30° take off angle and a nominal resolution of 130 eV FWHM (full width at half maximum) at MnKα (5.9 kV). Note that no absolute value illustrated (for example depth of beam penetration) should be extrapolated to other operating conditions. In particular, the presence or absence of a window on the front end of the relevant detector, and the thickness of any window, will significantly alter the signal received. SUTW allow the detection of light elements, but are fragile; if a more robust, beryllium, window is present atomic numbers less than 11 (Na) cannot be detected; of these carbon is the most significant in analysis of fossil material.

11.5 Size and Shape of the Interaction Volume

All signals are emitted from parts of an interaction volume the size and shape of which is controlled by two variables: the density of the material; the accelerating voltage applied to the electron beam. For a specimen held in a constant orientation, the higher the accelerating voltage and the lower the density of the material the further the beam will penetrate. For geological samples, analyzed at typical accelerating voltages (10–20 keV) interaction volumes extend to depths of, at most, a few microns. It may therefore seem pedantic to emphasize the complexity of how the signals are generated. There are, however, contexts in which the size of the interaction volume, and thus the position from which the different signals are generated within the sample, is critical. The most significant are when materials are mineralogically heterogeneous (most natural samples), including where discrete layers of different density and thickness are juxtaposed (typical of highly compressed carbonaceous compression fossils). A second, related, variable is that the different emitted signals are sourced from different parts of the interaction volume.

11.6 Selection and Preparation of Samples for X-Ray Analysis

11.6.1 Destructive Preparation Techniques

Optimal conditions for X-ray analysis are, of course, if the specimen is presented in the form of a flat, highly polished, surface. If suitable fractured surfaces through the specimen are not available, sectioning will be necessary to examine internal detail. Standard petrological techniques can be used to produce either thin sections or polished blocks. It may be useful to embed such materials in resin beforehand, particularly if small and/or fragile. Poorly consolidated or porous samples may require impregnation of resin under vacuum (i.e. resin not only encases the sample but penetrates to its interior) prior to cutting or polishing. Most such resins are organic-based; i.e. strongly carbonaceous.

It can be useful to 'spike' the embedding medium with a compound unlikely, ideally known not, to occur in the sample (e.g. metallic powder). Alternatively, the resin itself may contain an exotic element (e.g. chlorine) that is unlikely to occur in the fossil material. Thus any carbon in the sample itself can be distinguished from that of the embedding medium. The surface of samples is then coated with a thin conductive layer prior to examination; for X-ray analysis carbon is preferable to gold or other metals. The coating reduces the likelihood of charge build up on the specimen surface. Rough surfaces of specimens can be similarly coated; various methods to remove coatings from unpolished samples exist (see Leslie and Mitchell 2007 and references therein). Additionally, a thin line of conductive paint (carbon or silver dag) or tape between the sample and the mount will ensure that the sample is grounded.

The nature of the fossil sample will determine whether such destructive techniques are warranted. The scientific value of a specimen must not be compromised by the analysis or any component of the sample preparation. It is highly likely that future technological advances will push the boundaries of what can be achieved and it is therefore necessary to protect as much of the sample as possible. However, it is equally valid to argue that the scientific value of a sample can be increased as a result of the analysis to be performed. Two possible methodologies that often are a suitable compromise are non-destructive analytical techniques e.g. VP-SEM, and, secondly, micro sampling from a larger host if possible, including the use of Focused Ion Beam Milling of extremely small areas to generate sections through specimens that can be imaged and analyzed (see Chap. 13). This can often be a persuasive argument in securing the loan of samples for analytical study. We focus on the first of these here.

11.6.2 Selection and Preparation of Paleontological Specimens for VP-SEM

Selection of specimens depends, most obviously, on their size relative to that of the SEM specimen chamber. Movement in the x, y and z directions may be limited for larger specimens, thus restricting examination to only the more central areas. This problem can be minimized for samples that have an irregular shape (e.g. elongate specimens) by repeating the examination with the specimen in different orientations. It may be useful to prioritize those samples for which part and counterpart are available; in many cases the plane of splitting passes through rather than around the external surface of fossils preserved as carbonaceous compressions.

Obvious impediments to analysis can be introduced during recovery, preparation and long term curation of material, most obviously, coating of samples' surfaces by lacquers. The surficial layers of dust that can accumulate over time on curated samples can prove difficult to remove, If sample surfaces require cleaning, suitable brushes, careful use of compressed air or a low pressure airball cleaning puffer are alternatives. Ultrasonic cleaning should be used with caution as it may damage materials, especially friable samples.

The obvious advantage of a VP-SEM is that no other sample preparation is required. A sample that is not grounded is susceptible to charge build up. Areas that are not to be analyzed can be wrapped in metal foils to assist the charge neutralization process. As a result of doing so, it may be possible to operate the VP-SEM at lower chamber pressures; this will improve the intensity of the emitting signals. The sample is grounded using techniques as for coated samples. Conductive paint, however, can be difficult to remove from specimens subsequently. To minimize the length of time required to evacuate the specimen chamber, air can be removed from inside porous specimens in advance, using a vacuum desiccator or other vacuum chamber, such as a gold- or carbon coater. Subsequent to this, the specimen should be inserted in the SEM chamber as soon as possible.

11.7 X-Ray Imaging and Microanalysis

The emitted X-ray signal can be used to either image the distribution of elements over a specific user-defined area (termed elemental or X-ray mapping) or to analyze the composition of specific points or areas. In both cases the emitted signal, comprising X-rays that vary in their energy, is collected by either an energy-dispersive (ED), or wavelength dispersive (WD) spectrometer. In WD systems data are collected in turn; more than one spectrometer is therefore typically deployed to expedite analysis. As such, WD systems are typically found mounted on 'electron microprobes', the principal function of which is to provide high precision, high accuracy, chemical analyses of samples presented as highly-polished surfaces and carbon coated; see, for example, Boyce et al. (2001). WD systems can be, and are, coupled with SEMs, but it is more usual that an ED system will be present. We therefore focus here principally on the applications, and limitations, of analysis using ED systems. In analysis using an ED system different x-ray energies are sampled simultaneously, and can be displayed in 'live time'; the analysis is essentially a 'snapshot' of the chemistry of the sample. Output generated by the controlling software is in the form of an X-ray spectrum that plots intensity (number of counts), essentially a measure of abundance, versus X-ray photon energy (Fig. 11.3). The energy resolution on an ED system is considerably less than on a WD system, and results in the phenomenon of 'peak overlap', and potential misidentification of mineral phases. An ED X-ray detector is optimized for a specified working distance; other working distances result in geometrical misalignment between samples and detector that will reduce X-ray intensity and compromise quantitative accuracy.

11.7.1 Principles of X-Ray Generation

The atom comprises a cloud of negatively charged electrons bound to the nucleus by electromagnetic forces (the Rutherford-Bohr model of the atom). The electron cloud can be modeled as a series of shells orbiting around the nucleus (Fig. 11.4). If an incoming electron causes the ejection of a tightly bound inner shell electron (inner shell ionization) then the atom is left in an excited state. The atom returns to normal

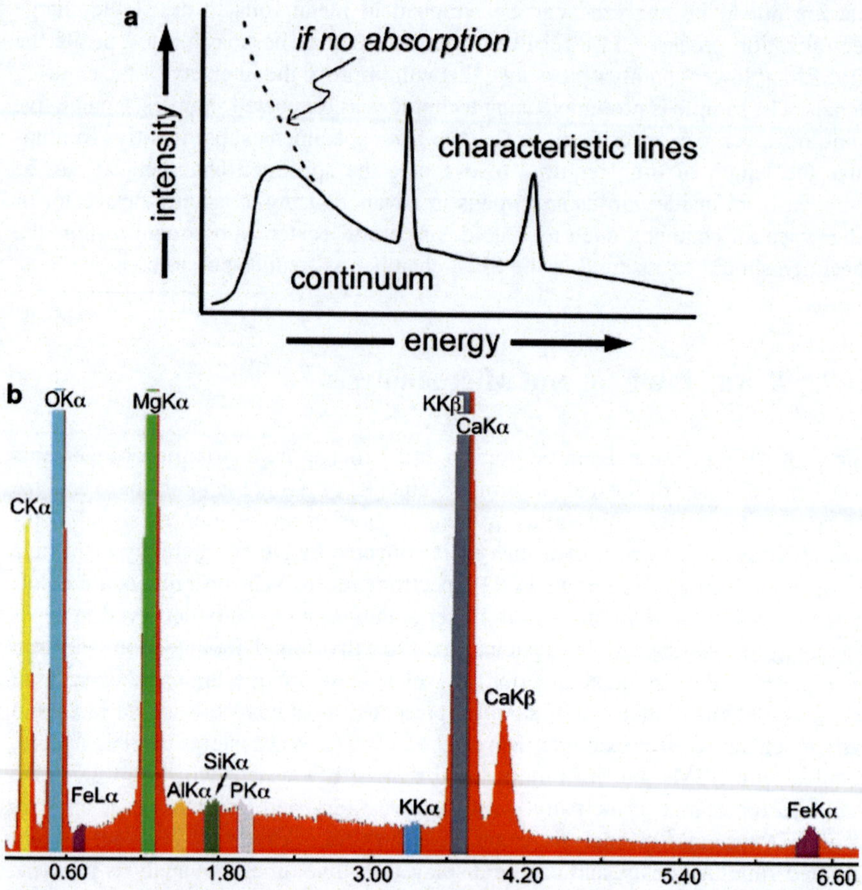

Fig. 11.3 Key components of an ED spectrum. (**a**) Schematic diagram, illustrating distinction between characteristic lines and continuum. (**b**) ED spectrum of dolomite with regions of interest (ROI) marked prior to element mapping. Note the near-identical positions of the CaKα and KKβ characteristic lines (3.69 and 3.59 kV, respectively); the ROI corresponds to the former as there is no peak in the spectrum corresponding to the position of the KKα characteristic line, and one corresponding to the position of CaKβ characteristic line

(ground) either via production of an Auger electron or an electron from an outer shell falling into the inner shell, in the process of which X-rays are emitted. X-rays are mainly designated K, L or M depending on which inner shell is ionized (other, more distal, shells exist; these are the three closest to the nucleus), and α, β or γ (the so-called Siegbahn nomenclature) depending on which outer shell or sub-shell supplies the electron to return to the ground state. The latter designation reflects the intensity of the transition (the difference in energy of the atom between its initial and final state). For the most part, the Siegbahn system denotes whether the replacement electron has been sourced from the adjacent outer shell, or one more distal. Thus, Kα and Kβ indicate that the K shell has been ionized, and usually that the replacement

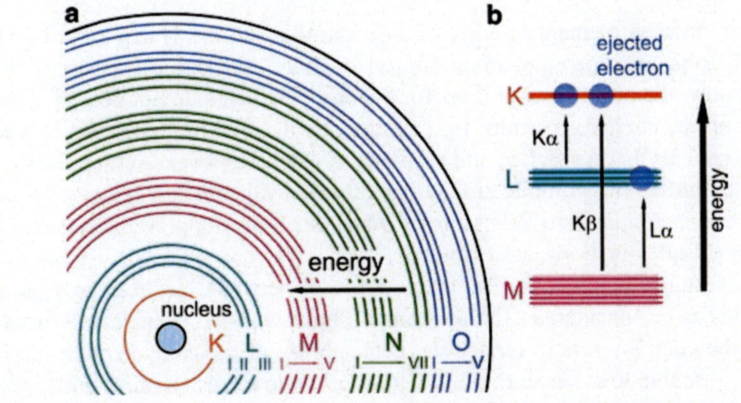

Fig. 11.4 Schematic diagrams illustrating X-ray production. (**a**) Distribution of *K*, *L* and *M* shells around the nucleus. (**b**) Notation used to define characteristic X-ray lines comprises two main parts. The prefix, *K* and *L* in this example, denotes which inner shell is ionized. The suffix, α and β in this example, indicates the source of the electron that returns the atom to the ground state. The characteristic X-ray energy (the x-axis on an ED spectrum) is equal to the difference in energy between the levels involved in the transition. After Reed (2005) and various other sources

electron has been sourced from the L shell, and M shell, respectively. Further designation is possible, as each shell comprises a number of sub-shells, typically labeled with the suffixes 1, 2, 3, etc., but are not discussed further (see Reed 1993, appendix 2 for a full discussion of origin and nomenclature of characteristic x-rays). There are two principal points derived from this summary that are relevant to the SEM user. Incident electrons from the electron beam require energy greater than that binding the electron in the shell to the nucleus to 'knock out' the latter; this is known as the critical excitation energy. Whether or not the critical excitation energy is exceeded depends on the accelerating voltage applied to the incident electron beam. Ideally, to generate enough X-rays ('counts') for analysis the overvoltage ratio should be greater than two. Typically, at accelerating voltages of 15 keV and above, X-rays analyzed will have energies of 10 kV and less. Related to this, the energy of the emitted X-ray, which is what will be analyzed, reflects which shell is ionized and which outer shell supplies the electron; for example, FeKα has an energy of 6.4 kV, and FeLα, 0.7 kV. Thus at an accelerating voltage of 15 keV, FeKα and FeLα x-rays will both be generated from Fe, but at 5 keV, only the latter as the critical excitation energy for FeKα has not been exceeded. Reed (1993) provides further details.

11.7.2 Structure of the ED Spectrum

As X-rays are produced via one of two mechanisms, there are two components to the ED spectrum (Fig. 11.3). The characteristic lines (also referred to as 'peaks') extend above the smooth continuous spectrum (the 'continuum' or 'bremsstrahlung'); The characteristic lines result from electron transitions between energy levels that are specific to each element (Sect. 11.7.1); from this summary, it is clear that the X-ray spectrum for a given element comprises more than one characteristic line. Unfortunately, the characteristic lines for different elements do not occupy unique positions on the energy spectrum; for example, the Kα line for Titanium is essentially the same as that for BaLα, and similarly SKα shares a very similar energy to MoLα and PbMα. This complication is exacerbated by the relatively low resolution of ED systems compared to WD systems where such overlapping interferences in many, but not all, situations, are resolved.

The continuum is generated by the interaction between electrons and atomic nuclei, and can be considered as 'noise' on the signal we wish to analyze. A notable effect of the continuum is to reduce the detectability of the characteristic lines of elements present in low concentrations. The continuum intensity varies with atomic number, thus is different between mineral phases. Finally, the spectrum falls off at low energies owing to absorption, for example that by the sample itself and the detector window (Sect. 11.8.2). Consequently, the height of low energy characteristic X-ray lines is suppressed. For example, Fig. 11.5 shows simulated ED spectra for a material containing equal concentrations of Oxygen, Calcium and Iron at accelerating voltages of 10 keV and 15 keV. The CaKα characteristic line is the most intense peak in each spectrum. There is less ionization of the most energetic X-ray (FeKα) (Sect. 11.7.1) and greater absorption of the lowest energy X-ray (OKα). Between the two spectra, the intensity of a given peak is not constant, a function of changing the accelerating voltage. For example, FeKα is more intense at 15 keV than it is at 10 keV; the reverse is true for FeLα. For this detector configuration, the characteristic line for OKα is significantly more prominent at 10 keV. At 15 keV more X-rays are generated, but as this occurs deeper within the sample many are absorbed hence the decrease in intensity.

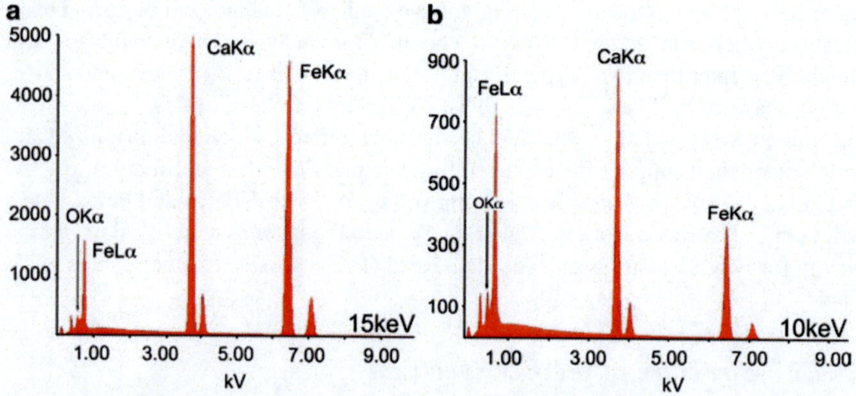

Fig. 11.5 Simulated ED spectrum acquired at 15 keV (**a**) and 10 keV (**b**) for a theoretical material containing *equal* abundances of oxygen, calcium and iron. In each spectrum the Kα characteristic line for each element has a different intensity. The relative intensity of the FeKα and FeLα characteristic lines varies between spectra acquired at the two accelerating voltages. OKα is less intense at the higher of the two accelerating voltages, due to greater absorption

11.8 Element, or X-Ray, Mapping

Following pioneering studies (e.g. Aldridge and Armstrong 1981; Martill et al. 1992) this technique is now widely used on both specimens prepared as polished blocks and thin sections (e.g. Wilby et al. 1996; Orr et al. 2000; Butterfield et al. 2007), as well as unprepared, and often uncoated, samples (e.g. Orr et al. 1998; Loydell et al. 2004; von Bitter et al. 2007; Gaines et al. 2008; Page et al. 2008; Moore and Lieberman 2009). The output is a series of maps showing the spatial distribution of elements selected by the user in advance over a pre-defined area.

Modern systems are capable of storing the entire spectrum for each pixel (termed spectral imaging). This has the benefit that the user is not required to specify in advance which elements to record but can interrogate the system after acquisition. If spectral imaging is not available, 'regions of interest' (ROI) or 'windows' must be defined on the spectrum before analysis (Fig. 11.3b). These ROI correspond to the energy of the characteristic lines for specific elements.

The width of these ROI can be defined automatically by the system, or the user, either by drawing on-screen their starting and end points on the x-axis, or by entering specific values for each. In either case, care should be taken to both position and size the ROI accurately, as the software will acquire the total number of counts within the window. The possibility that the ROI may encompass overlapping lines of more than one element should always be borne in mind (Sect. 11.7.2). Careful definition of the width of the ROI should eliminate the possibility that two separate elements with characteristic lines that have near-equivalent, i.e. closely spaced, energy values are inadvertently recorded as the same element. However, it is more difficult when these characteristic lines occupy essentially the same place on the energy spectrum (e.g. ZnLα and NaKα). It should not be assumed that automatic labeling of characteristic lines by the system software is infallible. Examination of the spectrum as a whole will, in many cases, and depending on the accelerating voltage, reveal the presence of other characteristic lines for one, but not the other, of two alternatives. The counts acquired for a characteristic line will also include a proportion derived from the continuum (Fig. 11.3).

The specimen is then stepped through a pre-defined x–y grid by moving either the specimen stage or beam depending on size of the area being analyzed. Counts acquired within each ROI are recorded at each step ('pixel' in the resultant X-ray map) for a specified dwell time. For each window, the range of pixel counts is then converted such that the tone in the X-ray map is brighter in those areas where the element is more abundant. Typically different color images are produced for each element (Fig. 11.1d).

The content of elemental maps is extremely sensitive to various parameters, including the system configuration (in particular the type of detector window), sample topography, the mineralogical complexity, and what proportion of the counts acquired is derived from the continuum. These are considered in turn, below. Note that although discussed in terms of the production of element maps, these same principles also apply to assessment of the chemistry of specific parts of the sample.

11.8.1 Contribution of the Continuum to X-Ray Maps

The intensity of the continuum is a function of atomic number, and thus not constant between different mineral phases. The impact on the generation of X-ray maps is usually small, and can be observed by placing a ROI over a peak-free part of the spectrum. The resulting map is thus a record of spatial variation in the continuum intensity.

11.8.2 Importance of the Detector Window

Although what are termed 'windowless' detectors exist, it is more usual that the front of the X-ray detector will have an interface that decouples the vacuum conditions inside the detector from those of the chamber. These 'windows' are one of two types: either a 'super ultra-thin window' (SUTW), composed of polymer foils, or a thicker (typically 10–15 μm) window of beryllium. The latter inhibits, to a much greater extent, the detectability of low energy X-rays; they are absorbed by the window itself. It is only possible to analyze for elements with an atomic number less than that of sodium (11) using a SUTW or a windowless detector.

11.8.3 Operator-Controlled Variables

Elemental maps are highly sensitive to several variables relating to the operating conditions, particularly accelerating voltage. The volume from which X-rays are emitted is a function of the accelerating voltage of the electron beam and the characteristic X-ray energy (Fig. 11.6). There are two aspects to consider.

1. For a given composition, more of the emitted X-ray signal is sourced from deeper within the sample at higher accelerating voltages.
2. The generation of an X-ray within a sample does not imply that it will be emitted, and thus potentially detected. The likelihood of an X-ray being emitted is dependent on the amount of absorption it undergoes on the way to the surface and then to the detector. In detail, the relationship between X-ray energy and absorption is more complex and, significantly, non-linear (Reed 2005). In general, a high energy X-ray (e.g., FeKα, 6.4 kV) will be much less readily absorbed than a low energy X-ray (e.g., CKα, 0.3 kV). It follows that, in practice, we can assume that low energy X-rays will generally come from nearer the surface whereas high energy X-rays will come from a greater thickness, potentially from depths up to that to which the beam penetrates.

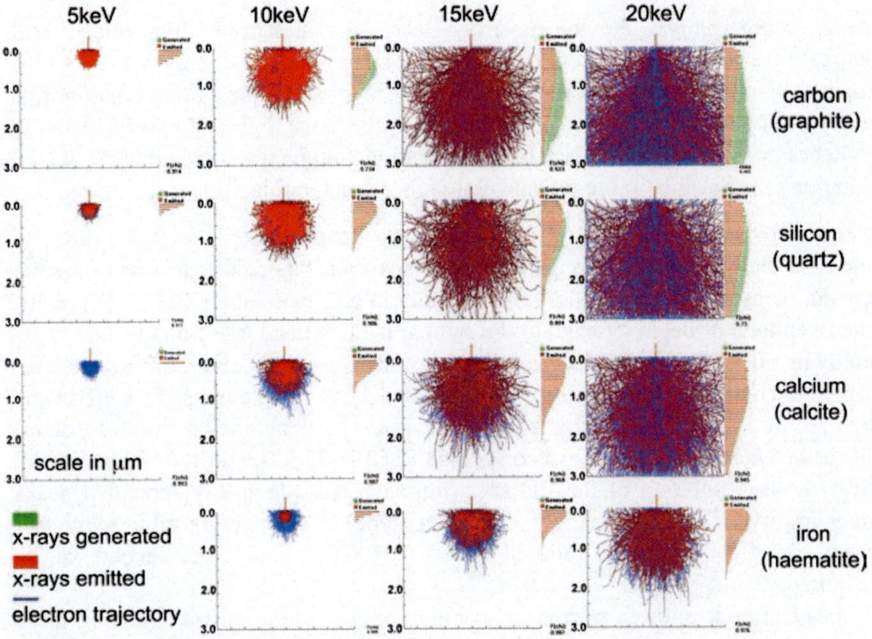

Fig. 11.6 Simulations of the generation and emission of Kα characteristic lines for C, Si, Ca and Fe at the accelerating voltages indicated. Note the variation in two key parameters: depth of electron penetration, and thus excitation volume; changes with depth in the proportion of X-rays that are generated and emitted. After Orr et al. (2009, Fig. 3)

11.8.4 Resolution of X-Ray Maps

Element mapping is inevitably a slower process than electron imaging. Total acquisition time of an X-ray map depends on the number of pixels that make up the image and the dwell time per pixel. For example, a map of 1024 × 1024 pixels with a dwell time of 100 ms will take about 30 h. To reduce the total time to say 3 h would require a reduction in dwell time to 10 ms, but each pixel will contain only one tenth of the number of counts. If the number of pixels is reduced to 512 × 512 the total acquisition time would be circa 7.5 h at a 100 ms dwell time, or 45 min at 10 ms per pixel.

The objective is to acquire as high resolution a map as possible in the least amount of time. The resolution of X-ray maps can be considered both in terms of what the smallest observable feature is (spatial resolution), and the abundance of that feature's component elements. For a specific set of operating conditions, the first depends on the number of steps and the distance between successive steps. The abundance of a given element is derived from the number of counts recorded for each pixel; its statistical precision is improved by increasing the number of counts, by moving the detector closer to the sample and/or adjusting some or all of accelerating voltage, beam current and dwell time, although some caveats apply.

Accelerating voltage. For the most part increasing the accelerating voltage will increase the count rate. However, this may not be true for low energy X-rays, which as most X-ray generation is now occurring deeper within the sample, may not be emitted. The spatial resolution of the map acquired at the higher of two accelerating voltages may be less, as the size of the excitation volume has increased (Fig. 11.6). Charging may also be more prevalent at higher accelerating voltages.

Beam Current. Count rate is proportional to beam current. It is tempting to increase the beam current to a maximum. However, this results in loss of spatial resolution as the spot size is larger, especially in tungsten-source SEMs. Secondly, the likelihood of beam damage to the sample is increased. Additionally, and especially in older ED systems, increasing beam current can lead to dead time problems where the output count rate drops off with increasing input counts. This effect can be to some extent overcome by reducing the pulse-process time (aka time constant) of the detector to a few microseconds (see Reed 2005). This in turn however will degrade the resolution of the ED spectrum such that previously separated peaks may now overlap. Modern silicon drift (SDD-type) EDS detectors suffer much less from this effect and count rates in excess of 100,000 counts per second can be acquired.

In practice it is often necessary to compromise resolution by adjusting these parameters, in order to acquire X-rays maps in an acceptable period of time. The distance between successive steps need not, of course, be smaller than the diameter of the excitation volume; typically, a spatial resolution of about 500 nm is adequate in most circumstances. Maintaining spatial resolution over large areas inevitably generates large image sizes, and extends total acquisition time significantly. It is therefore often prudent to generate lower resolution X-ray maps quickly to test the potential outcome, before committing to lengthy analysis.

11.8.5 X-Ray Analysis of 'Layered Substrates'

Natural samples, including many paleontological specimens, are likely to be mineralogically complex, and involve various phases. One particular configuration merits further consideration, that of a 'layered substrate'. Herein we demonstrate this concept using a thin surficial layer of low atomic number (carbon) in the simulations. This configuration simulates examples where non-biomineralized tissues (for example arthropod cuticles) are preserved as carbonaceous compression fossils. Typically, the plane of splitting passes through such fossils, rather than around their external surface, to leave the cuticle exposed as thin films on the surface of both part and counterpart. This has the effect of reducing further the thickness of the carbonaceous layer; such films can have thicknesses on the scale of microns, often considerably less. Critically, it is possible to render such surficial layers essentially invisible to the X-ray detector (Fig. 11.7). As accelerating voltage is increased, proportionally more of the emitted X-ray signal is sourced from below the surficial layer until a critical threshold is reached where the characteristic line

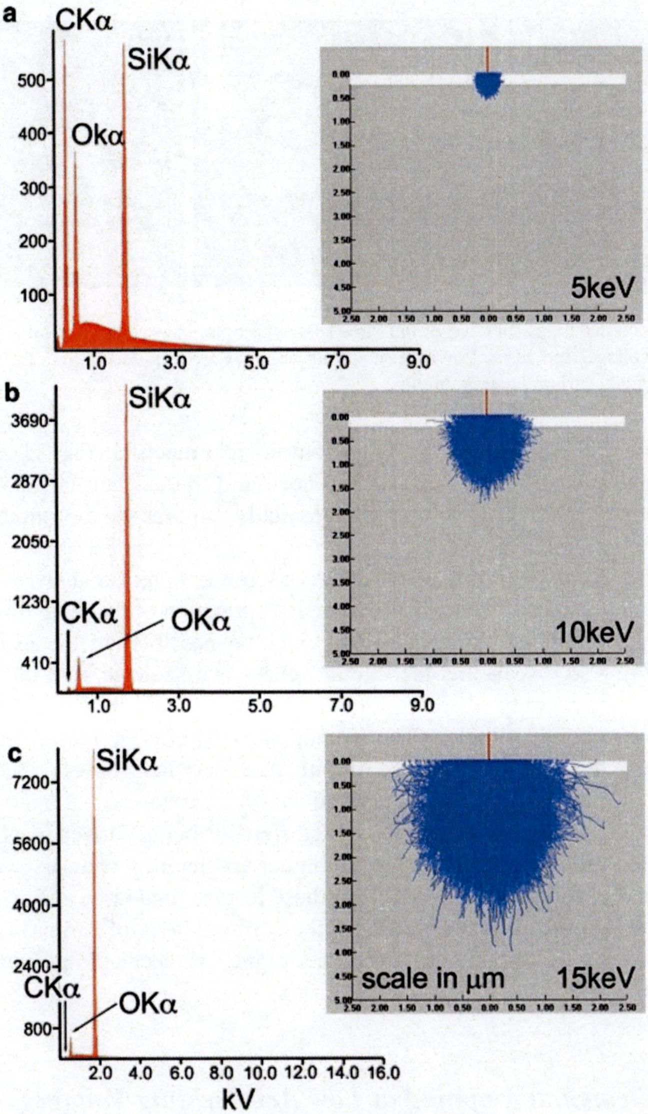

Fig. 11.7 Simulated ED spectra at the accelerating voltages indicated for a thickness of surficial carbon overlying a silica substrate. At accelerating voltages of 15 keV or greater the contribution of the surficial carbon layer to the total emitted X-rays is negligible and the CKα characteristic line is effectively subsumed within the continuum

for the surficial film is suppressed, and effectively subsumed within the continuum. Assuming all other operating parameters are held constant the threshold depends on the thickness of the surficial layer as well as the accelerating voltage. The effect will be more pronounced the lower the mean atomic number of any surficial layer, hence its particular significance for fossils preserved as carbonaceous remains.

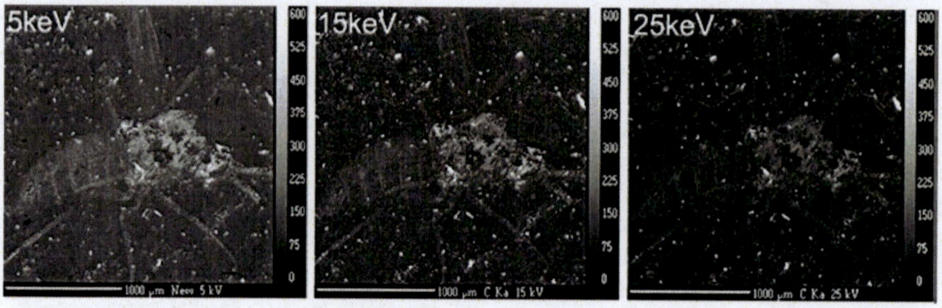

Fig. 11.8 Elemental maps for CKα of the same specimen of an insect from the Solite biota at the accelerating voltages indicated. Lower accelerating voltages are required to map for those areas where the surficial carbon layer is thinner

This concept is also illustrated using paleontological material (Fig. 11.8). In this, the surficial carbon film is thicker in the head and thorax than in the abdomen; increasing the accelerating voltage preferentially suppresses the emitted X-ray signal for carbon in the latter area.

In practice, it can be useful, albeit time-consuming, to repeat analyses at different accelerating voltages. In doing so, some interpretation of the three dimensional vertical structure of the preserved fossil may be possible. In the examples in Fig. 11.8, doing so reveals that the 'biotite' phase is associated with the head and thorax of the insect only; there is none concealed beneath the surficial layer of carbon in the abdominal region. The methodology is, however, not akin to serial slicing, in which the data is produced for only one specific level below the surface. The emitted X-ray signal is always derived over a discrete vertical interval; i.e. at low accelerating voltages most of the emitted signal is being derived at, and a short distance below, the sample surface, but at higher accelerating voltages *some* of the emitted signal is still being sourced from these levels. Note how, in Fig. 11.7, the characteristic line for CKα (derived from the surficial layer of carbon) is present, albeit decreasing in amplitude, over the range of accelerating voltages of 5–15 keV.

11.8.6 Elemental Mapping at Low Accelerating Voltages

At higher accelerating voltages (typically 10 keV or above) it is usual to map using the Kα radiation as these are the most intense characteristic lines for the majority of elements in common rock-forming minerals. It can, however, be desirable to map at lower accelerating voltages, for example if a thin surficial layer is present (Sect. 11.8.5). In such cases it is possible to map using the low energy characteristic lines of such elements; for example, at 5 keV it is necessary to use the Lα characteristic line to map for Fe (0.705 kV) instead of the Kα characteristic line (6.4 kV). In addition, the Kα lines of light elements (e.g., oxygen and carbon) can be mapped at lower accelerating voltages.

The advantage of operating at low accelerating voltages is that the interaction volume is reduced, thus more of the emitted signal from the excitation volume is from the surficial layers of the sample. The negative aspect of doing so is, of course, that count rates will be lower as absorption is greater at the lower energy end of the spectrum (Sect. 11.7.2). Overlapping peaks between elements can also pose a problem, particularly at energies less than 2 kV.

11.8.7 Specimen Topography

Surface roughness has the effect of extending or reducing the absorption path length of the X-ray photon (Fig. 11.9). Increasing the absorption path length reduces the likelihood of the X-ray photon reaching the detector; this is, of course, exacerbated the lower the energy of the X-ray. A low energy X-ray map is therefore more likely to contain a significant topographic component than a high energy X-ray map. This is illustrated in Fig. 11.10 in two ways: for the same characteristic line of two elements (OKα (0.523 kV) and FeKα (6.4 kV)); for different characteristic lines of the same element (FeKα and FeLα (0.705 kV)). The lower energy X-ray map of each pair is more sensitive to topography.

There are two methods by which the topographic component of the X-ray map can be, to some extent, suppressed.

Composite mapping. The effect of absorption is sensitive to the orientation of the specimen with respect to the detector; re-orientating the specimen through 90° and then repeat mapping three times results in four images displaying different sets of artifacts (Fig. 11.11a–d). A cumulative map generated from these suppresses much of the effects of the surface roughness (Fig. 11.11e). The prominent feature remaining represents a deep hole, where all X-rays photons were always absorbed and no

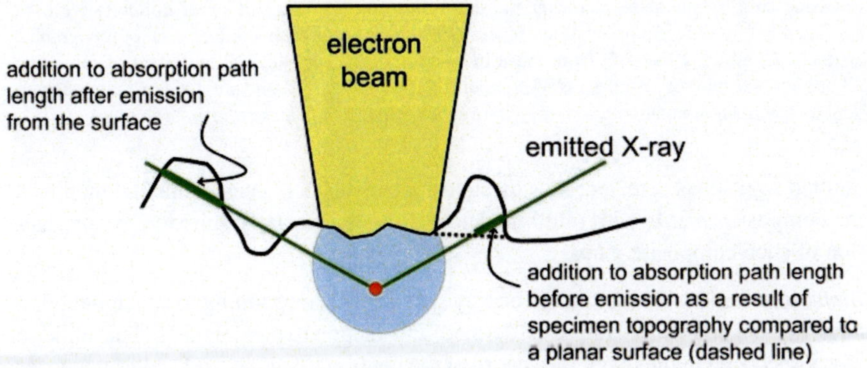

Fig. 11.9 Schematic diagram illustrating the effect of surface roughness on the absorption path length of the X-ray photon. Note that if the absorption path length is lengthened, the likelihood of the X-ray photon reaching the detector is reduced, by having to travel a longer distance within the sample before emission, and/or encountering an impediment after emission

Fig. 11.10 X-ray maps for the characteristic lines indicated illustrating that the effects of specimen topography are more pronounced for lower energy X-rays

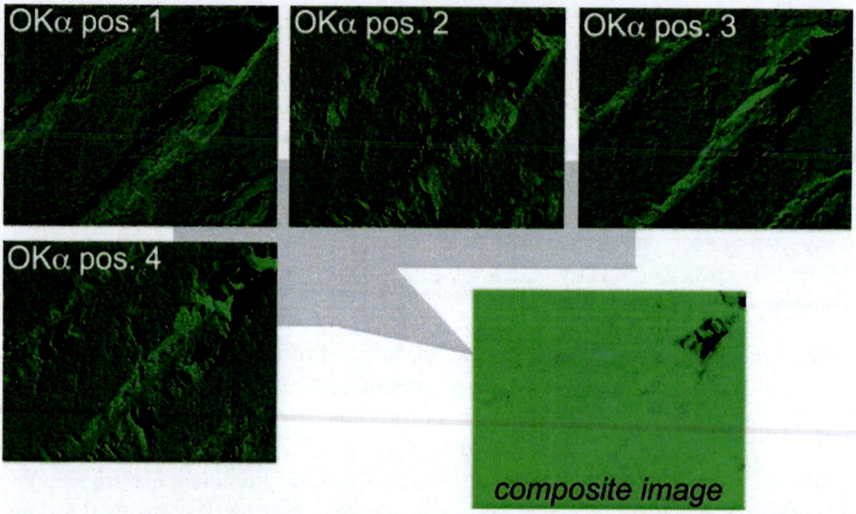

Fig. 11.11 Reducing topographic artifacts in X-ray maps I: composite images. Repeat mapping of the same area for the same characteristic line (OKα) on a rough surface of dolomite with the specimen in four different orientations generates four maps each with a different set of topographic artifacts. Combining the data from these in one composite map (e) has the effect of removing all artifacts except those from areas of pronounced topography (*a deep hole*) from which, independent of specimen orientation, no X-rays were emitted (hence its *black color* in each map)

emitted signal was detected; thus the area appears black in each elemental map, and the composite. SEMs with multiple detectors are particularly suited to the production of such composite maps.

Blank subtraction. A ROI on the energy spectrum corresponding to an element that does not occur in the sample is defined and mapped (the 'blank'). The emitted X-ray spectrum in the blank is a record of the continuum alone, and shows whatever topographic artifacts are present on an unprepared specimen. Visual comparison of the blank to the maps for characteristic lines is often sufficient to identify artifacts in the latter; compare maps for PKα (effectively a blank as it is not present as a characteristic line) and SiKα in Fig. 11.12. A more sophisticated technique involves

Fig. 11.12 (**a**) Reducing topographic artifacts in X-ray maps II: blank subtraction. ROIs for key elements are defined as normal (e.g. a, SiKα). One or more ROIs are defined at energy ranges on the spectrum in which no characteristic lines are present, in this case using the position on the spectrum occupied by PKα, as calcium phosphate minerals are not present. (**b**) These, the 'blanks', record spatial variation in the continuum intensity over the area mapped, the most significant effect on which is derived from topography. The grayscale tones in the blank are inverted and incrementally added to the corresponding pixels in maps of the characteristic lines. The result is a map in which topographic effects can largely, if not fully, be eliminated, leaving only spatial variation in chemistry expressed by variation in the grayscale tone. (**c**) Specimen of *Marrella splendens* from the Burgess Shale (ROM53348)

what we term 'blank subtraction', whereby the pixel count in the blank is inverted, and then added incrementally to the corresponding map for the characteristic line, until the background is effectively normalized and rendered topography free (see, for example, Bengtson 2000, on how this can be achieved using commercially available image processing software). Strictly, as the intensity of the continuum varies across the energy range the energy of the blank ROI should be close to that of the characteristic line. For example, in Fig. 11.12 the PKα blank is centered at an energy of 2.1 kV, slightly higher than that of SiKα.

11.8.8 Interpretation of Element Abundance from X-Ray Maps

In many SEM systems the abundance of an element is displayed in the x-ray map over the entire greyscale range (0–255 for an 8-bit map); this is usually the case even where the data itself is stored as number of counts per pixel. Unless the analytical system is calibrated against a standard and background intensities subtracted (i.e., quantitative mapping; see Reed (2005)) then it cannot be assumed that similar greyscale tones in two or more maps, even those acquired at the same time, imply each element is equally abundant. The same element, with identical abundance, can have a different greyscale tone in two maps, depending on what other mineral phases are present. For example, in a map for SiKα, the *same* biotite crystal would appear brighter than the surroundings if embedded in, for example, calcite, but darker if embedded in quartz. The brightness and contrast of any map can also be

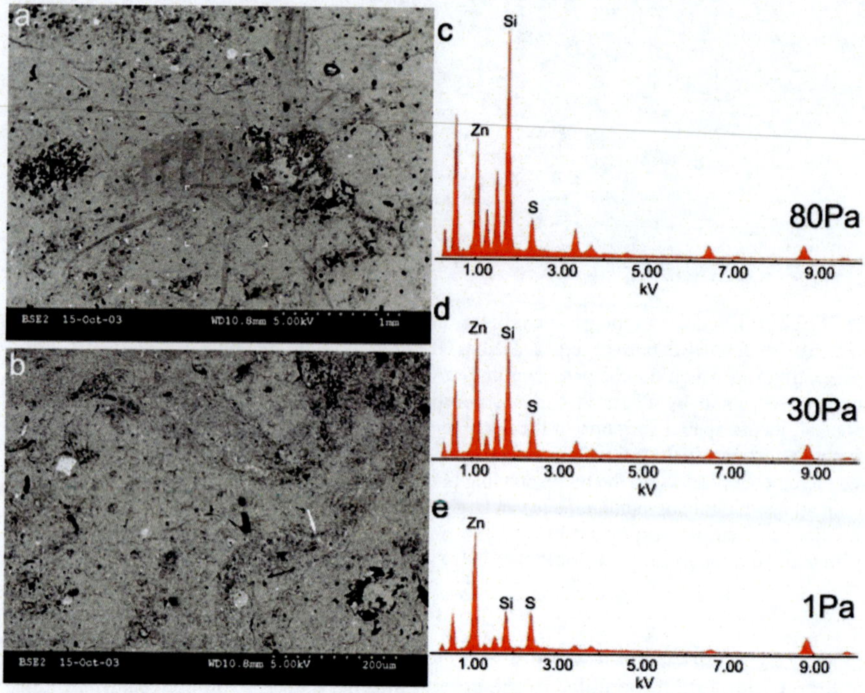

Fig. 11.13 The beam skirt effect in VP-SEM, a consequence of the presence of gas in the specimen chamber, reduces the accuracy of any analysis. **(a, b)** Backscattered electron image of an insect from the Sollte biota (VMNII1372b); area indicated in part a shown enlarged in **(b)**. Position of sphalerite crystals indicated by *arrow*. **(c–e)** ED spectra of the results of spot analysis of the same sphalerite crystal at the different chamber pressures indicated. Note that each spectrum includes an X-ray signal from the surrounding siliceous matrix (most obvious as a peak for SiKα), the scale of which is reduced as chamber pressure is reduced (i.e. when the chamber is at higher vacuum)

altered using image processing software. It should not be automatically assumed that the relative abundance of elements can be inferred by comparing elemental maps visually, even those acquired under identical operating conditions. In those systems that allow the number of counts per pixel to be saved, this data can be normalized between two or more maps in which case the greyscale tone in those maps indicates the relative abundance of each element.

In every map, the counts recorded include a contribution from the continuum below the characteristic line (Fig. 11.3). This is not constant across the energy spectrum and thus varies between elements. More significantly, two or more phases of different mean atomic number that contain the same element in equal abundance will generate different characteristic line intensities for that elements; the contributions from the continuum are different and the absorption characteristics will depend on the remaining compositional components.

11.9 Determining Composition Under VP Conditions: The Beam Skirt Effect

X-ray analyses acquired under VP conditions on uncoated materials are, for the most part, subject to the same parameters as maps of coated materials, acquired under high vacuum, including, ideally a charge-neutralization of the surface (see Carlton et al. 2004). At low vacuum (high pressure), however, there is an additional problem (Bolon 1991), termed the beam skirt effect. The presence of a gas in the specimen chamber results in primary beam scattering. The result is a skirt of low current density with a diameter of up to several hundred microns surrounding the focused central probe spot (Thiel and Toth 2005). This impacts on imaging samples to only a limited, typically negligible, extent, but does have significant implications for the interpretation of any x-ray analysis. For example, point analysis of a sample will yield an x-ray spectrum representative of a much larger bulk than would be expected. It is even possible that this X-ray spectrum could include a contribution from beyond the margins of the sample. The beam skirt diameter varies in response to the gas pressure in the specimen chamber, and the gas path length, the distance which the electron beam travels through the gas before impacting the sample. This can be demonstrated by spot analysis of a sphalerite (ZnS) crystal emebbed within carbonaceous insect cuticle in a siliceous sedimentary matrix (Fig. 11.13); as the gas pressure is reduced the diameter of the beam skirt reduces and the contribution to the ED spectrum from the siliceous matrix diminishes. Caution should be exercised in using any x-ray analysis acquired under VP-SEM conditions (see Newbury 2004).

The beam skirt effect can result in soft-looking images, including X-ray maps, in which the edges of features become blurred. In practice, there is little discernible problem in images acquired at a vacuum of 50 Pa or higher.

11.10 Selection of ED Detectors

For the past few decades the standard ED detector has been a Si(Li) detector. More rarely, high purity germanium has been used. For details on detector types see Goldstein et al. (2003). The detector and its preamplifier are connected to a rod cooled to liquid nitrogen temperatures. To the frustration of all laboratory managers, the nitrogen source – typically a 10 l dewer mounted above the detector requires constant replenishment. With correct care and usage the detector remains stable for many years. A recent alternative, the silicon drift detector (SDD), has the very significant advantage of being able to operate at much higher count rates with only a relatively small loss of resolution. Additionally they are now manufactured to sizes of up to 80 mm^2 compared to the standard 10 mm^2; this allows efficient use at relatively low beam currents (Newbury 2006). The net advantage is that mapping is significantly quicker as higher count rates that are not compromised by extended dead time are possible. Finally, and to the delight of laboratory managers, Peltier (thermo-electric) cooling is sufficient.

11.11 Quantitative Analysis of the Composition of Mineral Phases Using X-Rays

Fully quantitative X-ray microanalysis of major and minor element composition is widely employed in petrological studies of minerals using EPMA on carbon-coated and highly polished samples, including paleontological materials (e.g. Orr et al. 2000). The quantification is determined by comparison of the count rate for a given element in the sample with that from a set of calibration standards. The two cannot be compared directly. The effects of variable atomic number, absorption and fluorescence are combined into a correction algorithm; ZAF or the phi-rho-z function are two common variants thereof. Several texts detail these techniques (Reed 1993; Goldstein et al. 2003). Materials prepared using the same rigorous methods can yield reasonably accurate quantitative results if examined using SEM-based ED systems calibrated against appropriate standards. Low vacuum conditions render quantitative analysis of these more challenging; see Carlton et al. (2006) for further discussion, including a summary of the correction methods developed to counter the effects of beam skirt. The logistical problems are compounded hugely if samples are unprepared. Most ED systems in use today contain an automated quantification routine. These results are not calibrated against known standards (thus termed standardless), in the absence of which their accuracy is hard to evaluate. In our opinion, such ED spectra, especially those from unprepared specimens, are best considered a qualitative analysis of composition and phase identification. Typically two options are available. In spot analysis, a small volume is analyzed, typically of a single mineral phase. In area analysis the ED spectrum is of the bulk composition over a rastered area the size of which is controlled by the user.

11.12 Summary

(a) X-ray microanalysis is a widely employed technique by which Earth Scientists, including paleobiologists, acquire information on the composition of samples. Spot and area analyses are possible; element maps, one variant of the latter, show the distribution and abundance of selected elements over user-defined areas.

(b) Destructive preparation of specimens allows samples to be presented as polished, carbon-coated surfaces and analyzed under high vacuum; X-ray microanalysis of such yields optimal results. Fully quantitative compositional analyses are possible, and typically acquired on an EPMA system using WD spectrometers. Analysis on SEM-based ED systems is more robust when the results are calibrated against appropriate standards.

(c) Unprepared samples are more challenging, but offer one significant advantage: specimen integrity is wholly or largely retained. There is usually little to be gained from analyzing such material using a WD as opposed to an ED detector, although there are exceptions.

(d) Sample charging will have deleterious effects on both image quality and analyses of composition. Charging can be minimized by using various non-permanent methods to ensure the sample is well grounded, reducing accelerating voltage (if appropriate) and/or operating a VP-SEM under low vacuum conditions. The presence of a gas in the specimen chamber induces a 'beam-skirt' effect, that reduces the resolution of any analysis.

(e) Specimen topography inevitably introduces analytical error, most obviously, apparent variation in the spatial distribution of an element in an element map. ED spot and area analyses are also compromised. The effects are more pronounced the lower the energy of the X-ray.

(f) The depth to which the electron beam penetrates the sample can impact significantly on the results that are obtained, as X-rays are emitted from within, not only at the surface of the sample. This is particularly relevant to 'layered substrates'.

(g) It is often instructive to compare visually the results of x-ray microanalysis (in particular element maps) with SE, CCI and, especially, BSE, images of the same area, with the following caveat. The different signals are emitted from different parts of the interaction volume.

(h) Proprietary software that simulates electron beam-sample interactions provides a means of 'ground-truthing' the results of any analysis. In particular, simulation provides insight as to whether, and how, lateral and vertical variation in sample chemistry, and changing various operator-controlled variables, will impact on results.

(i) As a matter of good practice, the user should consider how the following will potentially impact on the results of any analysis: (1) the chemistry of the sample, and especially any heterogeneity, for example, , as exemplified by Burgess Shale fossils, whether the chemistry varies over short vertical distances; (2) the operating conditions selected, i.e. accelerating voltage, beam current, dwell time and set-up of the detector (its position and pulse process time).

The goals of undertaking an analysis may be diverse and disparate. It is rare that a particular set of operating conditions will be ideal for all aspects of the analysis. It may be possible to define a set of compromise operating conditions that yield acceptable results. Analysis is best viewed as an iterative process and there are situations where there is no alternative but to repeat analyses having altered one or more operating conditions (e.g. accelerating voltage) in the expectation of improved results.

Acknowledgments We thank Nick Butterfield, Bob Gaines and James Schiffbauer for their extremely helpful comments, and the Virginia Museum of Natural History (VMNH) and the Royal Ontario Museum (ROM) for the loan of materials. This is a contribution to the Royal Ontario Museum Burgess Shale Research Project number 31.

References

Aldridge RJ, Armstrong HA (1981) Spherical phosphatic microfossils from the Silurian of North Greenland. Nature 292:531–533

Bengtson S (2000) Teasing fossils out of shales with cameras and computers. Palaeontol Electronica, 3: article 4: 14 pp., 7.7 MB. http://palaeo-electronica.org/2000_1/fossils/issue1_00.htm

Bolon RB (1991) X-ray microanalysis in the ESEM. In: Bailey GW (ed) Microbeam analysis-1991. San Francisco Press, San Francisco, pp 199–200

Boyce CK, Hazen RM, Knoll AH (2001) Nondestructive, in situ, cellular-scale mapping of elemental abundances including organic carbon in permineralized fossils. Proc Natl Acad Sci 98:5970–5974

Briggs DEG, Williiams SH (1981) The restoration of flattened fossils. Lethaia 14:157–164

Briggs DEG, Erwin DH, Collier FJ (1994) The fossils of the Burgess Shale. Smithsonian Institution Press, Washington, DC

Butterfield NJ (1990) Organic preservation of non-mineralizing organisms and the taphonomy of the Burgess Shale. Paleobiology 16:272–286

Butterfield NJ (2002a) *Leanchoilia* guts and the interpretation of three-dimensional structures in Burgess Shale type fossils. Paleobiology 28:155–171

Butterfield NJ (2002b) Perimineralization vs. compression: disparate modes of exceptional preservation in the Burgess Shale and their palaeobiological significance. In: DeRenzi M, Pardo Alonso MV, Belinchón M, Peñalver E, Montoya P, Márquez-Aliaga A (eds) Current topics on taphonomy and fossilization. Ajuntament de Valencia, Valencia, pp 241–246

Butterfield NJ, Balthasar U, Wilson LA (2007) Fossil diagenesis in the Burgess Shale. Palaeontology 50:537–543

Carlton RA, Lyman CE, Roberts JE (2004) Charge neutralization in the ESEM for quantitative X-ray microanalysis. Microsc Microanal 10:753–763

Carlton RA, Lyman CE, Roberts JE (2006) Accuracy and precision of quantitative energy-dispersive x-ray spectrometry in the environmental scanning electron microscope. Scanning 26:167–174

Caron JB, Scheltema A, Schander C, Rudkin D (2006) A soft-bodied mollusc with radula from the Middle Cambrian Burgess Shale. Nature 442:159–163

Conway Morris S (1990) Burgess Shale. In: Briggs DEG, Crowther PR (eds) Palaeobiology, a synthesis. Blackwell, Oxford, pp 270–274

Conway Morris S (2008) A redescription of a rare chordate, *Metaspriggina walcotti* Simonetta and Insom, from the Burgess Shale (Middle Cambrian) British Columbia, Canada. J Paleontol 82:434–430

Fraser NA, Grimaldi DA, Olsen PE, Axsmith B (1996) A Triassic Lagerstätte from eastern North America. Nature 38:615–619

Gaines RR, Briggs DEG, Yuanlong Z (2008) Cambrian Burgess Shale–type deposits share a common mode of fossilization. Geology 36:755–758

Goldstein JI, Newbury DE, Echlin P, Joy DC, Lyman CE, Lifshin E, Sawyer L, Michael JR (2003) Scanning electron microscopy and X-ray microanalysis, 3rd edn. Kluwer Academic/Plenum Publishers, New York

Kearns SL, Orr PJ (2009) Charge contrast imaging of exceptionally-preserved fossils. Palaeontology 52:673–680

Leslie SA, Mitchell JC (2007) Removing gold coating from SEM samples. Palaeontology 50:1459–1461

Loydell DK, Orr PJ, Kearns SL (2004) Preservation of soft tissues in Silurian graptolites from Latvia. Palaeontology 47:503–513

Maletz J, Steiner M, Fatka O (2005) Middle Cambrian pterobranchs and the question: what is a graptolite? Lethaia 38:73–85

Martill DM, Wilby PR, Williams N (1992) Elemental mapping: a technique for investigating delicate phosphatized fossil soft tissues. Palaeontology 35:869–874

Moncrieff DA, Barker PR, Robinson VNE (1979) Electron scattering by gas in the scanning electron microscope. J Phys D Appl Phys 12:481–488

Moore RA, Lieberman BS (2009) Preservation of early and middle Cambrian soft-bodied arthropods from the Pioche Shale, Nevada, USA. Palaeogeog Palaeoclimatol Palaeoecol 277:57–62

Newbury DE (2004) Assessing charging effects on spectral quality for X-ray microanalysis in low voltage and variable pressure/environmental scanning electron microscopy. Microsc Microanal 10:739–744

Newbury DE (2006) The new X-ray mapping: X-ray spectrum imaging above 100 kHz output count rate with the silicon drift detector. Microsc Microanal 12:26–35

Orr PJ, Briggs DEG, Kearns SL (1998) Cambrian Burgess Shale animals replicated in clay minerals. Science 281:1173–1175

Orr PJ, Briggs DEG, Siveter DJ, Siveter DJ (2000) Three-dimensional preservation of a non-biomineralized arthropod in concretions in Silurian volcaniclastic rocks from Herefordshire. Engl J Geol Soc Lond 157:173–186

Orr PJ, Kearns SL, Briggs DEG (2002) Backscattered electron imaging of fossils exceptionally preserved as organic compressions. Palaios 17:110–117

Orr PJ, Briggs DEG, Kearns SL (2008) Taphonomy of exceptionally preserved crustaceans from the Upper Carboniferous of southeastern Ireland. Palaios 23:298–312

Orr PJ, Kearns SL, Briggs DEG (2009) Elemental mapping of exceptionally preserved 'carbonaceous compression' fossils. Palaeogeogr Palaeoclimatol Palaeoecol 277:1–8

Page A, Gabbott SE, Wilby PR, Zalasiewicz JA (2008) Ubiquitous Burgess Shale–style "clay templates" in low-grade metamorphic mudrocks. Geology 36:855–858

Reed SJB (1993) Electron microprobe analysis, 2nd edn. Cambridge University Press, Cambridge

Reed SJB (2005) Electron microprobe analysis and scanning electron microscopy in geology, 2nd edn. Cambridge University Press, Cambridge

Thiel BL, Toth M (2005) Secondary electron contrast in low-vacuum / environmental scanning electron microscopy of dieletrics. J Appl Phys 97:1–18

Towe KM (1996) Fossil preservation in the Burgess Shale. Lethaia 29:107–108

von Bitter PH, Purnell MA, Tetreault DK, Stott CA (2007) Eramosa Lagerstätte – exceptionally preserved soft-bodied biotas with shallow-marine shelly and bioturbating organisms (Silurian, Ontario, Canada). Geology 35:879–882

Wilby PR, Briggs DEG, Riou B (1996) Mineralization of soft-bodied invertebrates in a Jurassic metalliferous deposit. Geology 24:847–850

Chapter 12
Ultrastructural Approaches to the Microfossil Record: Assessing Biological Affinities by Use of Transmission Electron Microscopy

Sebastian Willman and Phoebe A. Cohen

Contents

S. Willman (✉)
Department of Earth Sciences, Palaeobiology, Uppsala University, Villavägen 16, Uppsala SE-752 36, Sweden
e-mail: Sebastian.Willman@geo.uu.se

P.A. Cohen
Department of Earth and Planetary Sciences, Harvard University, 26 Oxford Street, Cambridge, MA 02138, USA
and
Massachusetts Institute of Technology, NASA Astrobiology Institute, Cambridge, MA, USA
e-mail: pacohen@fas.harvard.edu

M. Laflamme et al. (eds.), *Quantifying the Evolution of Early Life*, Topics in Geobiology 36,
DOI 10.1007/978-94-007-0680-4_12,

Abstract One of the major technological advances in biological research was the invention and development of the transmission electron microscope, which enables high resolution and high magnification studies of cross-sections of specimens. As such, it has proved to be a useful tool to describe ultrastructural features of taxonomic and phylogenetic importance in modern organisms. Here we discuss how to extend the use of transmission electron microscopy (TEM) to the fossil record, with emphasis on acritarchs (organic-walled microfossils of unknown affinity). Microfossils are traditionally studied by use of transmitted light microscopy, a method that reveals details of external morphology only. TEM however, gives an additional level of detail and reveals structures that can greatly aid in interpretation of taxonomic affinity, and thus can reveal further detail on the origination and diversification of myriad eukaryotic groups in the fossil record. In this chapter we describe the preparation procedure, show advantages and shortcomings of the technique, and discuss how to interpret the results from a geobiological perspective.

Keywords TEM • Ultrastructure • Preparation • Acritarch • Biological affinities

12.1 Introduction

Microscopy has fundamentally changed our understanding of the natural world by enabling us to observe objects at a scale that is impossible to see with our eyes alone. One such microscopic technique is called transmission electron microscopy (TEM). In a transmission electron microscope, an electron beam is accelerated by high voltage and passed through a sample that has been previously cut into ultra-thin slices, forming an image of the sample much like a slide projector displays figures on a screen. The major advantage of TEM over other techniques is the high level of resolution achievable, which in theory reaches down to the molecular level.

TEM has long been an important analytical tool in a range of fields including the physical and biological sciences, and its use in the field of geobiology is growing. The method is used to study many different aspects of a sample, including morphology, crystallography, and chemical composition. In this contribution we discuss how to successfully prepare and interpret biological structures with the use of TEM, focusing on organic-walled microfossils, with special emphasis on acritarchs – a group of microfossils with unknown and likely polyphyletic affinities (Evitt 1963).

12.2 The Transmission Electron Microscope

The first transmission electron microscope was built in 1931 and is credited to two German scientists; Max Knoll and his doctoral student, and later Nobel laureate, Ernst Ruska. Their invention was based on the simple understanding that the order of magnification obtainable by a microscope is limited by wavelengths, and microscopes using electrons rather than visible light could provide a more detailed picture of an object, since beams of electrons have wavelengths 1,000 times shorter than visible light. Some years after the first microscope was built, Knoll (1935) published the first images obtained by scanning and transmission electron beam microscopy. Since Knoll and Ruska's first attempt, these microscopes have been greatly refined, with the development of the scanning electron microscope (SEM) by Manfred von Ardenne in the late 1930s as another major breakthrough in high resolution microscopy. Despite major technical advancements, the fundamentals of TEM remain the same (see McMullan (1995) for a detailed historical review).

In theory, a TEM works in the same way as an optical light microscope (LM). However, instead of visible light as the illuminating system, a TEM is equipped with a gun that emits a beam of electrons (Fig. 12.1a–d). This beam is accelerated towards a sample using an electrical charge and passes through a set of magnetic lenses that focus the beam. The electron beam passes through the sample, which has been previously cut into ultra-thin (~50–90 nm) sections. Given that the lenses are magnetic, it is possible to vary the electrical current that passes through the coil, which in turn allows for variations in intensity and focal length to be made. As electrons pass

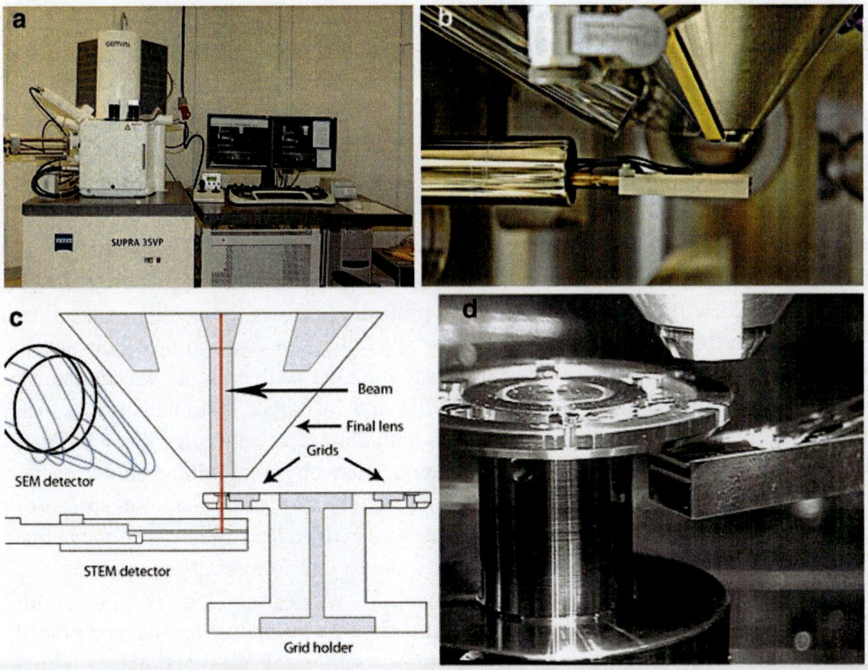

Fig. 12.1 (**a**) Equipment setup, scanning-transmission electron microscope (STEM) to the *left* and control panel to the *right*. (**b**) Showing setup of the beam accelerator and the STEM and SEM detectors inside the microscope. (**c**) Schematic view of the technical setup, showing grid holder, beam accelerator and detectors. (**d**) Actual setup inside the microscope with the grid holder with the sample in place

through the sample, they interact with the specimen and scatter. Scattered electrons are focused again using a magnetic lens that allows the scatter to appear as a high resolution image of the sample (Reimer and Kohl (2008) and Williams and Carter (2009) both give authoritative and informative aspects of TEM including the technical features, but see also Egerton (2005)).

There are many factors that affect the resulting TEM image, of which preparation is most important and will be dealt with extensively in this contribution. Preparation success aside, the final results also depend on the technical settings such as acceleration voltage (high voltage gives better resolution but less contrast, whereas low voltage gives better contrast but lower resolution), objective aperture (a large aperture gives better resolution but less contrast, whereas a small aperture gives better contrast but lower resolution) and how well one is able to correct for beam astigmatism. Consequently, in order to achieve good results TEM requires some basic training by an experienced technician. In this chapter, we aim instead to provide the reader with a thorough background in sample preparation, the appropriate lens through which to interpret results, and keys to identifying probable artifacts.

12.3 Preparation for TEM

TEM is a technique suitable for the study of structures in a range of disparate samples; however, the scope of this chapter is to describe its potential in analyzing and interpreting biological structures of fossil organisms, with a focus on organic-walled microfossils. There are several TEM preparation protocols available for biological material, each customized to best suit a specific organism (e.g., Kennaway et al. (2008) gives a detailed protocol for dinoflagellate preparation). In preparing specimens for TEM, virtually every step of the procedure affects the quality of the final product. Therefore, it is important to follow prescribed preparation protocols and to understand how each process affects the sample. We will present two alternative methods for the preparation of specimens, the first for larger specimens (more than 300 μm in diameter), and one for specimens with diameters between 50 and 300 μm. For even smaller samples, preparation techniques used by microbiologists for bacterial cells can also be used (see Bozzola and Russel 1999 for examples). Regardless of preparation method, the first step in the procedure is to select suitable specimens that can easily be picked from suspension under a stereo microscope using a pipette.

The major difference between the two methods we describe here is how to contain the samples during preparation. In the first method, the fossils are placed individually in a microcentrifuge tube (a small, cylindrical plastic container with a conical bottom and an integral snap cap, often called Eppendorf tube) where the specimen settles through gravity in the pointed tip of the tube. It is important at this point to identify the specimen selected and remove unwanted organic debris. All the subsequent steps in the preparation procedure take place with the specimen in the tube. The advantage of a conical tube is that the fossil is in a known location even if it is not visible to the naked eye. Fluids can be removed with a pipette from the top more easily and the location of the specimen is known when the time comes to make the final microtome cuts. The second, more elaborate method is described in detail in Sect. 12.3.5. There are many ways to prepare samples for TEM but the methods described here have been tried successfully for Precambrian and Cambrian acritarchs, Cenozoic dinoflagellates, modern prasinophyte algae phycoma, and modern metazoan resting egg stages. The method below follows a standard protocol from fixation of the sample to the sectioning. Allow a few days for the whole procedure.

12.3.1 Primary Fixation

Fixatives are by definition hazardous to live tissue and should therefore be used with great caution under a fume hood and using protective gloves, eyewear, and other necessary protection. The purpose of the fixation is to stop ongoing biological processes and to preserve structural and chemical relations in the specimens. Although fossils are not live tissue, fixation may increase the mechanical strength of treated specimens, which in turn helps to preserve morphology. There are many types of fixatives on the market today, each directed towards a specific set of analyses. One standard method is to fix the specimen with 2.5% glutaraldehyde in buffer solution, for at least 1 h (to several days) at +4°C. The buffer solution used here and in all other steps is PBS (phosphate buffered solution) at pH 7.4. After the first fixation the specimen is washed with buffer solution three times for at least 10 min.

12.3.2 Secondary Fixation

After the first washing we have used 1–2% osmium tetroxide (OsO_4) as the fixative (for 2 h at +4°C). In general, specimens should be as small as possible to allow rapid and complete penetration (particularly true when using OsO_4), however, the samples described here are microscopic, and so size is not an issue. Osmium tetroxide reacts primarily with double bonds and sulfhydryl groups of proteins, causing major conformational changes in the protein structure. It stabilizes and stains lipids – preferentially unsaturated fatty acids – which is necessary for membrane fixation and improved contrast in TEM. Note that OsO_4 is toxic, an irritant, and very volatile, so precautions need to be followed.

After the second fixation the specimen is again washed with buffer solution three times for at least 10 min.

12.3.3 Dehydration

Water should normally be avoided in high vacuum electron microscopy since it will prevent stabilization of the specimen (i.e., make the specimen less tolerant to further treatments and conditions in the microscope), disturb the vacuum, or even destroy the specimen during evaporation. Consequently it is important to dehydrate the specimens. This is often done by washing the specimen in buffer solution followed by increasing concentrations of ethanol – allowing 10 min at each concentration (e.g., 20%, 50%, 70%, 80%, 90%, 95%, and two times 100% for 20 min each) and finish with washing in acetone twice. As alternatives, acetone or DMP (2,2-dimethoxypropane) can be used in all or some of the steps.

12.3.4 *Infiltration and Embedding*

After dehydration, the acetone is removed from the tube using a pipette and microfossils are infiltrated with a mixture of TAAB 812 epoxy resin (or similar epoxy, e.g., Agar 100) and acetone in equal proportions and left overnight. In the next step the excess fluids are removed using a pipette and replaced with 100% epoxy and left overnight. Finally, the microfossils are transferred to, and embedded in silicon plates (or gelatine capsules) using a pipette. The plates are subsequently left to polymerize at 50°C for 48 h. The hardness of the embedding media is important and it is best if the epoxy is of a similar hardness as the sample; this greatly reduces artifacts that occur during the microtoming processes if the knife blade encounters hardness contrasts during cutting. However, it is often impossible to know beforehand how recalcitrant a certain sample will be, and similar fossils can differ in hardness from locality to locality. Thus, some degree of experimentation is required. If the amount of material, or the fossil itself, is small it is possible to centrifuge the samples between the steps above. Speed and time depends on the fragility of the material but 2,000 rpm for 10 min can be used as guideline.

12.3.5 *A Modified Preparation Technique for Small Specimens*

The above technique works well with larger specimens (more than 300 μm in diameter), however, there are additional methods that, while more time intensive, can make the final sectioning and analysis much easier and more fruitful.

As opposed to dehydrating and infiltrating samples in microcentrifuge tubes as in the first method, this method calls for making small baskets out of Beem capsules and a mesh with a pore size smaller than the sample size. The top of the Beem capsule is removed and a small hole punched in the cap. The bottom of the capsule is then cut off with a razor blade and the cap, with a circle of mesh laid over the top, is fitted over the bottom, leaving a layer of mesh slightly elevated above the base of the basket (Fig. 12.2). The sample is then pipetted directly into the basket so that it sits on the mesh layer. This basket can then be immersed in the progressive stages of dehydration and infiltration as described above. This is easily done by using a 12-well cell culture plate.

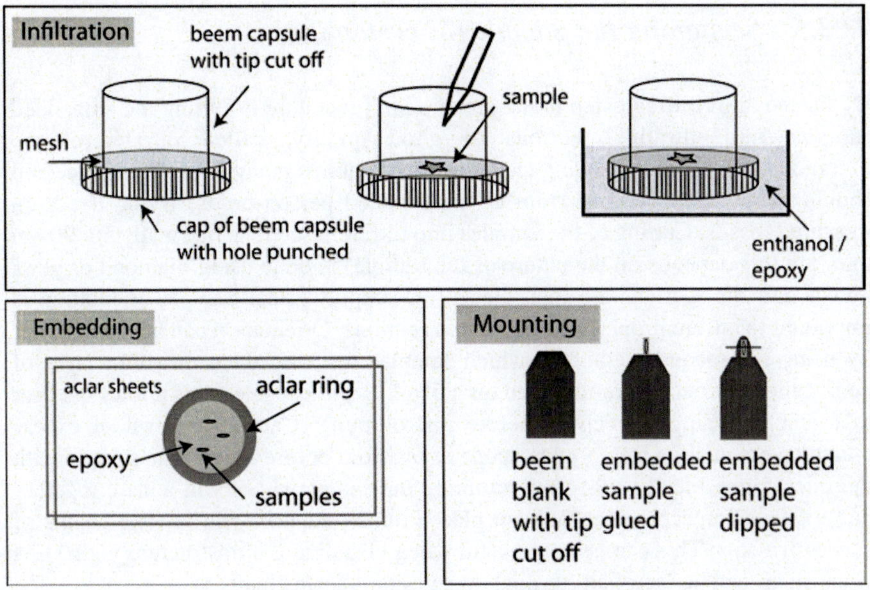

Fig. 12.2 Schematic drawing of the modified preparation procedure discussed in Sect. 12.3.5

Once the sample is in the final stage of 100% epoxy, it is embedded in a thin disc. This is best done using sheets of aclar film, which easily peel away from set epoxy. Cut two rectangles of aclar as well as a ring about 1 cm in diameter with a 3–5 mm edge thickness. Adhere the ring to one of the rectangles of aclar using a tiny dab of epoxy. Then deposit the samples inside the ring, and add a few additional dabs of epoxy to fill the ring. Then place the second rectangle of aclar over the ring assemblage and set according to the epoxy instructions.

Once the epoxy is set, peel away the top layer of aclar. The samples should be oriented parallel to the aclar sheets. Under a stereo microscope, cut out small rectangles of epoxy containing your sample and remove them with tweezers. To keep track of them it is useful to have a piece of double-sided tape right next to the epoxy ring on which the samples can be placed. At this point, the rectangles are mounted on blank Beem capsules, i.e., Beem capsules that have been filled with epoxy, hardened, and then removed from the plastic housing. File down the tip of a blank so that there is a small flat surface. Then, apply a small dab of a strong two-part glue to the flat surface. Orient the epoxy rectangle containing the sample so that subsequent microtome cuts will be appropriate for your analysis and place the sample on the tip of the blank. Let the glue harden according to its instructions. At this point, it is useful to dip the sample in additional epoxy in order to get a coating around the sample and blank for ease when microtoming. To do this, dip the tip of the sample and the blank in a dab of epoxy and let dry upside down in an oven. Do this a few times until an adequate coating is formed around the sample. At this point, the sample should be oriented so that when microtomed (see Sect. 12.3.6) the knife edge will be perpendicular to the sample, allowing for the best thin sections possible.

12.3.6 Sectioning the Sample (Microtoming)

By far the most difficult step in the preparation procedure is cutting the embedded samples into ultrathin sections of good quality, called ultramicrotomy. Ultramicrotomy is a demanding technique that requires many hours of practice and should therefore be learned from an experienced person before trying it out on treasured fossils. Cutting of the samples into ultrathin sections (normally 50–90 nm thin, but this depends on the nature of the sample) is done using diamond or glass knives and the sections are then placed on support grids. Sample orientation is important to ensure properly arranged thin sections. Orientation can be made easier by using the second method in which samples are embedded in a thin layer of epoxy, then cut out and re-mounted on a blank Beem capsule. This allows the user to orient the sample precisely before microtoming. Careful removal of excess epoxy layers under a stereo microscope is essential before cutting the sample with the microtome. Ideally, when microtoming the sample slices will adhere together due to static forces, so that they are placed on the support grid as long chains of successive cuts. This can be very useful when visualizing ultrastructure with TEM since slices can be 'scrolled' through to see small-scale changes.

12.3.7 Support Grid

The support grid is the electron microscope equivalent of a glass slide used in light microscopy, i.e., it is where your sample is placed for study. The support grid can be made up of Cu, Au or nylon grids with a thin (10 nm) plastic film. See Bozzola and Russel (1999) for a detailed overview.

12.3.8 Staining

In order to obtain higher contrast of a sample it is often useful to stain the specimen using uranyl acetate and lead citrate. Staining can be done pre- or post-embedding but it is often easiest to stain samples once they are already on the support grids. The first step is to saturate the sample with uranyl acetate. One effective way of doing this is to place drops of the liquid on a small square of parafilm. Then, each grid can be placed sample-side down on the drops, which effectively coats the sample. First, place each grid on a drop of uranyl acetate, and leave for 10 min. Pick up the grid and wash it gently with deionized water. Then place the grid sample-side down on a drop of lead citrate for another 10 min, followed by a final wash. Touch the edge of the support grid gently to a paper towel or lab tissue to draw away any excess water before storing the grid in a specimen box. Make sure to dispose of both the uranyl acetate and lead citrate properly.

Note: If you wish to halt the preparation processes you can do that after washing and during the dehydration step at 70% and 95% ethanol (Sect. 12.3.3).

Equipment: The studies were made using an environmental STEM Zeiss Supra 35VP and JEOL 2100 TEM microscopes and using an LKB Ultramicrotome for sectioning. A STEM (scanning transmission electron microscope) is a modified microscope which produces a transmitted image as a TEM does but uses a scanning electron beam as opposed to a transmitted beam.

12.4 How to Interpret TEM Structures

One of the major issues in TEM microscopy of fossils is distinguishing between original features and artifacts. One must be constantly aware of the fact that diagenetic factors such as mineral overgrowths and precipitation, compression, and bacterial degradation can all affect the final ultrastructure of a fossil (Grey and Willman 2009). Additionally, knife marks, embedding, staining, and folding of the microtomed thin section can also create artifacts that can be misinterpreted as original structures and textures. Artifacts can also be introduced during the analysis of a sample due to interactions between the electron beam and the sample material. Most of the radiation energy produced by the TEM electron beam is converted to heat that may alter the appearance of a sample. The rise in temperature caused by the beam is limited to the illuminated area and is proportional to the charge. The electron excitation of organic molecules causes bond rupture and loss of mass and crystallinity (Reimer and Kohl 2008) and consequently it is best to keep the radiation of samples to a minimum to prevent melting and distortion of the sample material For example, it is usually best to focus and align the electron beam on a feature away from the area of your sample you wish to examine. Figure 12.3 shows folds, chatter marks and other common examples of artifacts. Most are easily distinguished from original textures when examined in the appropriate context.

12.4.1 Strengths of the Technique

The advantage of TEM is that it allows one to see details and textures that are impossible to observe with other techniques. Such features as the cross-sections of cell walls can, for example, give unique insight into taxonomic distinctions. TEM is especially useful when attempting to determine the taxonomic affinity of fossils that may have morphologically simply exteriors, such as smooth-walled acritarchs. In such instances, TEM can provide a valuable additional set of data with which to compare fossils to possible modern analogs, since convergence of external structures can make identification based solely on external morphology difficult or even impossible, while ultrastructural characters may indicate specific taxonomic affinities. If such taxonomic characters are absent or convergent, studies of wall ultrastructure can at least

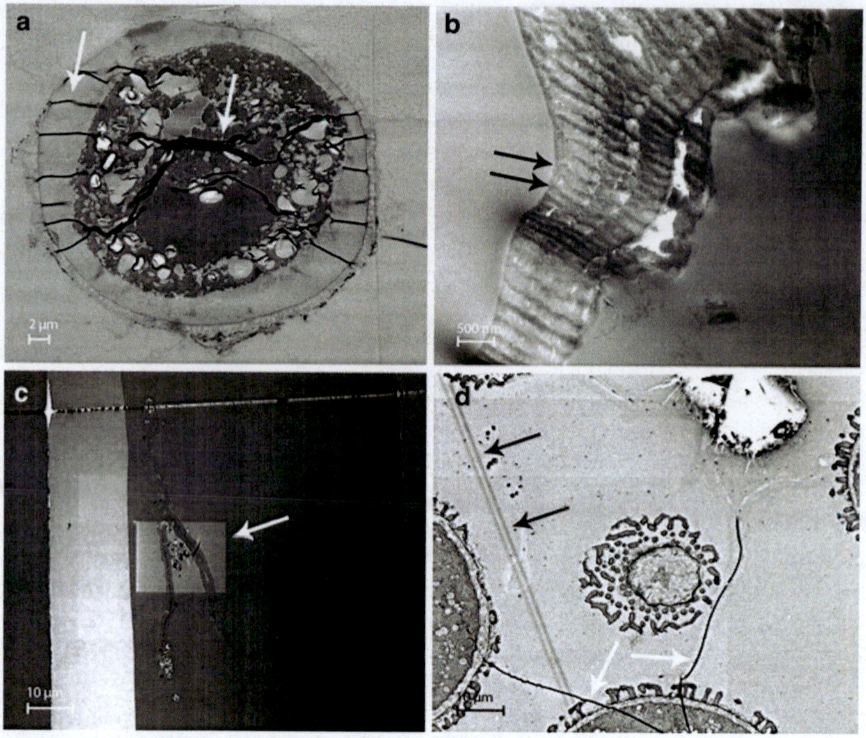

Fig. 12.3 Artifacts produced during preparation. (**a**) Folds in the thin section (*arrows*). Note how the folds, due to varying mechanical strength, appear different in different parts of the fossil; *Spiniferites* sp., a dinoflagellate cyst from the Late Eocene, Florida. (**b**) Folds (*arrows*) in the cell wall of an acritarch introduced during cutting. (**c**) *White arrow* shows the area of destruction that is the result of heat produced by the radiation energy. If the beam is focused in one spot too long, the area will become bleached (the contours of a faint *larger rectangle* is the result of another area of focus). (**d**) So-called chatter marks (*black arrows*) can occur during microtomy when the diamond knife "jumps" and are easily distinguished from folds (*white arrows*); Recent semitectate pollen grain

separate morphologically similar acritarchs (*sensu* Javaux et al. 2004) and potentially identify acritarchs in different stages of their life cycle (Moczydłowska et al. 2010). Below we outline research using TEM which has helped to elucidate taxonomic affinity in some organic-walled microfossils. This work has great importance for our understanding of the diversity and affinities within this major group of fossils, which are the most abundant and diverse record of eukaryotic life during the Precambrian.

12.5 TEM in Geobiology

TEM has proven to be an important analytical tool, especially in many areas of biology where it is now considered indispensable. However, the technique is still in its infancy in the field of geobiology, despite being used for more than 50 years.

The pioneering work in studying microfossils with the use of TEM started in the late the 1960s and continued sporadically into the 1980s (e.g., Wall 1962; Jux 1968, 1969a, b, 1971, 1977; Kjellström 1968; Martin and Kjellström 1973; Peat 1981). These early studies demonstrated the advantages of TEM over LM. Early TEM work showed that it is possible to recognize details of fossilized cell walls that are poorly visible under LM, such as radial channels in the walls of the acritarch genera *Leiosphaeridia* and *Tasmanites*, which indicate that these taxa may be related to prasinophycean green algae such as *Pachysphaera pelagica* (Jux 1968). Talyzina and Moczydłowska (2000) revitalized the use of TEM in their study of early Cambrian acritarchs from Estonia. They revealed four different types of cell wall ultrastructure within six acritarch genera, of which the most interesting ultrastructure was found in *Leiosphaeridia* spp. This morphologically simple, unornamented acritarch displayed a multilayered structure including an outer trilaminar sheath structure (TLS) that is indicative of certain types of green algae (e.g., Allard and Templier 2000; Hagen et al. 2002; Damiani et al. 2006). The acanthomorphic (process-bearing) acritarchs revealed single-layered structures; electron-tenuous and fibrous in *Archaeodiscina umbonulata*; electron-dense and homogeneous in *Globosphaeridium cerinum, Comasphaeridium brachyspinosum* and *Skiagia compressa*, and also single-layered, electron-dense, homogeneous and perforated by pore canals in the leiosphaeric *Tasmanites tenellus* (see Moczydłowska and Willman 2009 for a discussion of these and other recent results).

In a series of more recent papers Javaux and co-authors discussed the difficulties in recognizing eukaryotes in the early fossil record and showed the potential of TEM to address such questions (Javaux et al. 2001, 2003, 2004). As an example, Javaux et al. (2004) established a eukaryotic affinity of some morphologically simple leiospheres (e.g., *Shuiyousphaeridium macroreticulatum*) from the Mesoproterozoic (ca. 1,500–1,400 Ma) Roper Group in Australia and the coeval Ruyang Group in China. This interpretation was based solely on TEM studies that revealed a complex ultrastructure in the specimens, a complexity unknown in prokaryotes. Consequently they interpreted *S. macroreticulatum* and other acritarchs as eukaryotes, a conclusion that would have been impossible to make based on external morphology alone. In addition, *Leiosphaeridia crassa* was described as having partially preserved trilaminar sheath structure (TLS) indicating a putative green algal affinity (possible TLS is also reported in leiosphaeric acritarchs from the Paleoproterozoic Chuanlinggou Formation; Peng et al. 2009). Accordingly, if their interpretation of complex ultrastructure is correct, TEM has provided us with some of the best examples of early eukaryotic evolution (Knoll et al. 2006).

12.5.1 Current Research in Geobiology Using TEM

In recent years the use of TEM in the study of microfossils has been revived. Investigations of the ultrastructure of acritarchs, spores, and pollen have shown its growing importance (see review by Wellman et al. (2009) and references therein).

Work on acritarchs has been especially insightful. Willman and Moczydłowska (2007) studied the wall ultrastructure of the Ediacaran acritarch taxon *Gyalosphaeridium pulchrum* using scanning- and transmission electron microscopy. Their study revealed a four-layered, complex wall ultrastructure, consistent with a eukaryotic organization (Fig. 12.4a, c). Although not conclusive with regard to the exact systematic position of *G. pulchrum*, affinities to green algae and dinoflagellates were discussed. More recently a paper by Willman (2009) aimed to discuss and compare the wall ultrastructure of Ediacaran leiosphaeric (or sphaeromorphic) and acanthomorphic (again, *G. pulchrum*) acritarchs. Both *G. pulchrum* and *Leiosphaeridia* sp. have wall ultrastructures that vary from single- to multilayered, which may indicate either that the specimens belong to different clades but are morphologically convergent; or that the specimens represent different stages in the life cycles of a single taxon (Fig. 12.4b, d).

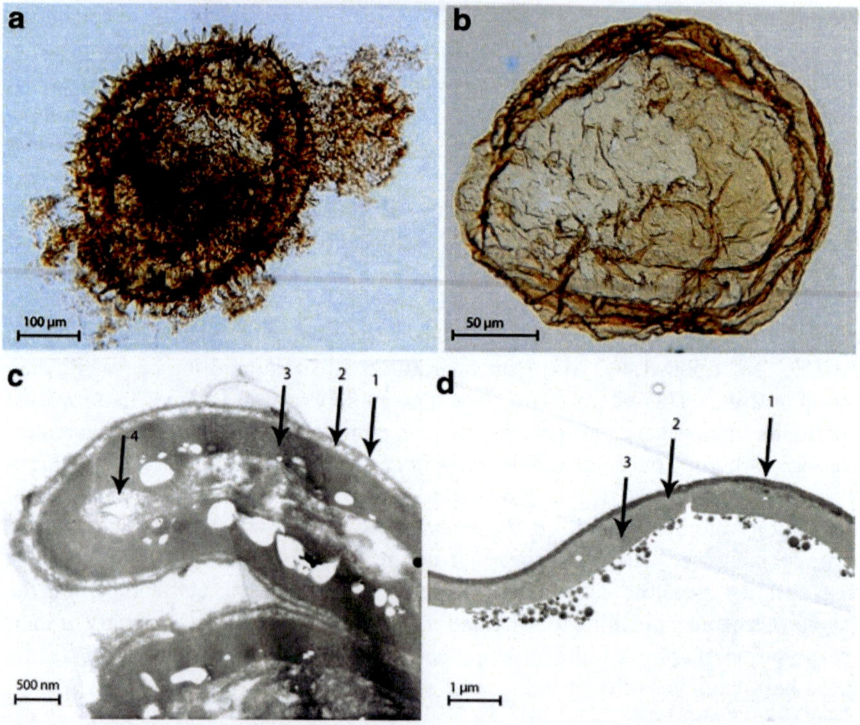

Fig. 12.4 Ultrastructural comparison between two morphological end-members from the Ediacaran of Australia. (**a**) Acanthomorphic acritarch *Gyalosphaeridium pulchrum*; Murnaroo 1 borehole, Dey Dey Mudstone, 230.4 m +45 3, England finder coordinates B13. (**b**) Leiosphaeric acritarch *Leiosphaeridia* sp.; Giles 1 borehole, Tanana Formation, 445.0 m .2 +45, England finder coordinates D47. (**c**) Cross-section of the vesicle wall of a *G. pulchrum* showing four different and clearly distinguished layers (*arrows*). (**d**) Cross-section of the vesicle wall of a *Leiosphaeridia* sp. showing a three-layered ultrastructure (*arrows*)

The possibility that some acritarchs are metazoans was discussed (Willman 2009) and further elaborated in Cohen et al. (2009). A possible metazoan affinity for some organic-walled microfossils was for example suggested by van Waveren and Marcus (1993), who emphasized the morphological similarities between some of these fossils and diapause egg cysts produced by copepods and other animals. Work on potential metazoan affinities for some organic walled microfossils is motivated by the fact that acritarchs are by definition polyphyletic, but are most often interpreted as belonging to an algal group. However, many non-algal organisms form resting stages with preservation potential, including in most major clades of animals. Such recalcitrant structures include sponge gemmules, cnidarian podocysts, and the egg and diapause cysts of diverse bilaterian metazoans, especially marine crustaceans (Cáceres 1997; Hill and Shepard 1997; Buckland-Nicks and Hodgson 2000; Onoue et al. 2004). Convergence on similar external morphologies among resting and reproductive stages of green algae, dinoflagellates and animals make it challenging to assign a definitive taxonomy based solely on external morphology. Thus, ultrastructure as viewed through TEM is essential to further testing taxonomic hypotheses in organic walled microfossils.

For example, the walls of some prasinophyte phycomata, including the genera *Pachysphaera*, *Halosphaera*, and *Cymbomonas*, have a distinctive ultrastructure (Fig. 12.5a, c) marked by radially oriented punctae (Wall 1962; Jux 1969a; Inouye et al. 2003) that have also been recorded in large sphaeromorphic microfossils from Ediacaran rocks (Arouri et al. 2000). However, other Ediacaran sphaeromorphs preserve ultrastructural features that do not clearly conform to a prasinophyte affinity (Fig. 12.5b, d).

The wall ultrastructure of the taxon *Gyalsosphaeridium* sp. from the Ediacaran Officer Basin of Australia has been studied and described in several publications (Willman and Moczydłowska 2007; Cohen et al. 2009; Willman 2009) and provides an excellent example of the level of detail obtainable by TEM and the conclusions that can be drawn based on TEM data. The study by Cohen and co-authors revealed a wall ultrastructure consisting of an electron dense outer layer that extends outward to form processes (Fig. 12.6a–c), a fibrous and electron tenuous middle layer, and a thin, electron dense inner layer. Such ultrastructure is different from that of punctate phycomata, and does not closely resemble the trilaminar ultrastructure of green algal walls. Therefore, they compared the ultrastructure with that of the diapause cysts of the modern brine shrimp *Branchinella longirostris* which is morphologically similar to many large ornamented Ediacaran microfossil (LOEM) taxa (Fig. 12.6d). Under TEM, the cyst wall exhibits a thin, electron dense outer layer from which processes arise (Fig. 12.6e, inset), a thicker, more electron tenuous and fibrous middle layer (Fig. 12.6f), and an inner layer, similar in electron density to the outer wall but thinner. Comparable ultrastructure has been documented for other arthropods (Couch et al. 2001). Like its external morphology, the ultrastructure of *B. longirostris* compares closely with those of some observed LOEM taxa. Thus, a metazoan affinity for some Ediacaran organic-walled microfossils is supported by the additional information obtained through TEM.

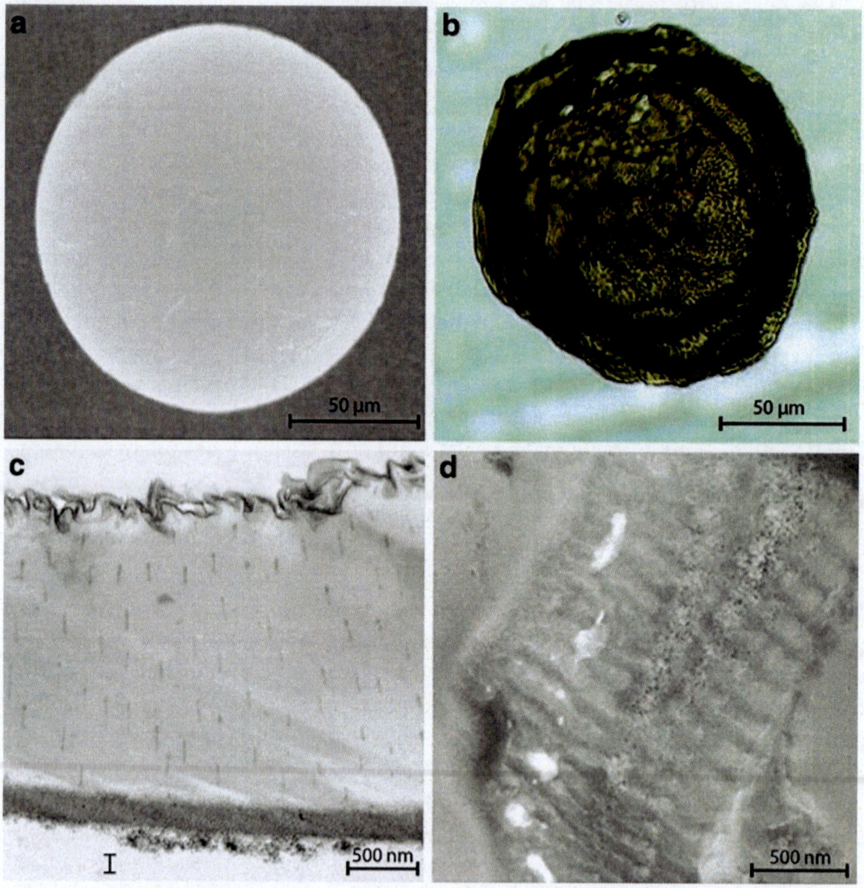

Fig. 12.5 Comparison of a modern algal analog and Ediacaran smooth-walled microfossil. (**a**) Modern *Halosphaera* sp. phycoma from Puget Sound, north-western USA. (**b**) Ediacaran *Leiosphaera* sp. from Australia showing similar simple external morphology. (**c**) TEM section of *Halosphaera* sp. phycoma (I indicates inside of phycoma). (**d**) TEM section of *Leiosphaera* sp. Diagonal lineations running from *top left* to *bottom right* in (**d**) are the result of the differing density of the embedding material and the fossil, and are not primary. The distinctive pores seen in (**c**) are not found in (**d**), but have been documented by Arouri et al. (1999)

As the TEM dataset of both fossil and modern forms grows, old results can be interpreted in the light of new knowledge. Modern green microalgae (such as e.g., *Haematococcus pluvialis*, Volvocales, Chlorophyta) go through a series of developmental stages in aplanospore (resting stage) formation during, for example, adverse environmental conditions. This morphogenesis from flagellate to aplanospore involves, among others things, discarding of flagellae, formation of a multilayered cell wall including a TLS (e.g., Hagen et al. 2002). Unrecognized in the initial ultrastructural study by Kempe et al. (2005) this type of structure was recently interpreted in 650 Ma leiosphaeric microfossils from the Chichkan Formation by

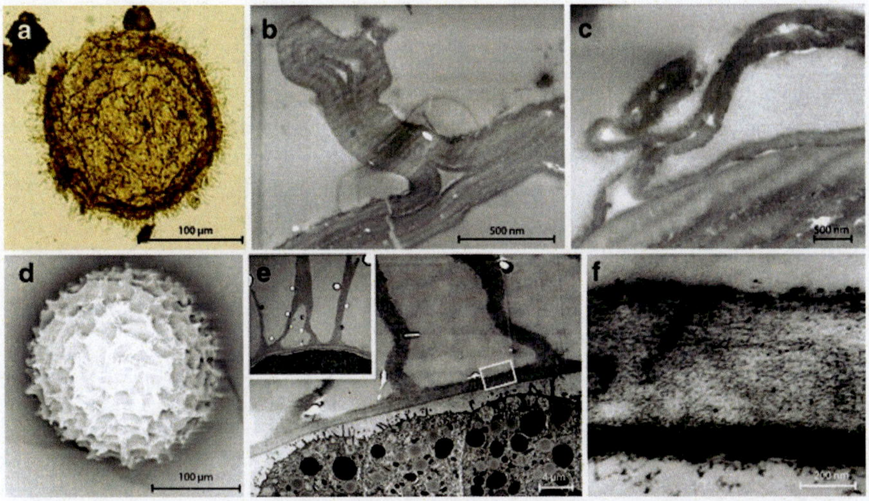

Fig. 12.6 Comparison of a LOEM fossil and a modern crustacean analog. (**a–c**) Light and transmission electron micrographs of *Gyalosphaeridium* sp. (**a**) Light micrograph showing abundant processes. (**b**) TEM showing an electron dense outer layer, a fibrous and electron tenuous middle layer, and a thin, electron dense inner layer. Diagonal lineations running from *top left* to *bottom right* are the result of the differing density of the embedding material and the fossil, and are not primary. (**c**) TEM showing detail of a processes extending out from the outermost electron dense layer and a closer view of the fibrous inner layer. (**d–f**) Scanning- and transmission electron micrographs of the modern crustacean *Branchinella longirostris*, (**d**) SEM of the external morphology of a *B. longirostris* cyst. (**e**) TEM showing the outer wall of the cyst and some of the internal contents, including black-stained lipid deposits. Inset showing hollow process. (**f**) TEM detail of outer wall showing an electron dense outer layer, a fibrous and electron tenuous middle layer, and a thin, electron dense inner layer (*rectangle* in (e))

Moczydłowska et al. (2010) showing that the Chlorophyceae may have deep phylogenetic roots in the Proterozoic.

Apart from identifying true morphological features, the major challenge in interpreting TEM images of microfossils is that there is a paucity of modern analogs with which to compare them. TEM studies on modern organisms tend to focus on the cell contents (such as the nucleus or the mitochondria) and thus only occasionally document the ultrastructure of walls that might conceivably be preserved in the fossil record. Additionally, taphonomic and diagenetic processes must be taken into consideration when comparing modern and fossil TEM images, making direct comparisons challenging, though not fruitless. There is a need for increased TEM studies on both fossil and modern cell walls but references in Sect. 12.5 gives an overview of what has been published thus far in the literature.

While most geobiological research using TEM has, to this point, focused on organic-walled microfossils, the technique can provide additional insight into other earth history questions as well. Among the earliest structures interpreted as biological fossils are a group of carbonaceous structures preserved in a local chert unit of the ~3.5 Ga old Apex Basalt in Western Australia (e.g., Schopf 1993), however, their biogenicity have been seriously questioned (e.g., Brasier et al. 2002). A similar set of structures that are more or less accepted as true biological microfossils come from the ~1.9 Ga old Gunflint Formation in Canada (e.g., Barghoorn and Tyler 1965). Similar Raman spectra of the carbon in the Apex chert have been interpreted both as indicating biological affinity as well as indicating an abiotic origin (Schopf et al. 2002; Brasier et al. 2002, respectively). Moreau and Sharp (2004) used TEM to study the relationship and distribution of silica and carbonaceous material in the microfossils from the Gunflint chert and concluded a biological origin to the structures. Most importantly, the kerogen in the Gunflint chert does not display crystallinity, which would be the case if the structures were abiotically produced. With regard to the biogenicity of the Earth's oldest putative microfossils De Gregorio et al. (2009) used similar TEM techniques to show that Apex carbonaceous matters appeared structurally amorphous and similar to the structure of the Gunflint fossils. Along with other lines of evidence such as the macromolecular hydrocarbon structure, carbon bonding, and functional group chemistry, they supported a biogenic origin for the Apex microfossils.

A recent discovery of organic-walled microfossils collected from siliciclastic deposits in the 3.2 Ga Moodies Group, South Africa (Javaux et al. 2010) shows another use of TEM. It is known that abiotic processes can form morphologies that resemble microfossils and that geochemical alteration can produce organic compounds with depleted carbon isotope values. Consequently several lines of evidence are needed to prove a biological origin of structures, especially those that push back the age of origin of major groups. Using TEM, Javaux and colleagues showed that the organic structures collected from the Moodies Group are hollow organic-walled vesicles, with cell lumen visible between the compressed cell walls, and not large abiotically produced kerogen particles. Together with other data including geochemical analyses and general morphology, TEM helped to show that the structures are in fact bona fide fossils.

12.6 Discussion

TEM has been utilized extensively in the study of fossil spores and pollen, and the additional information provided by the technique has helped to clarify phylogenetic relationships in plants (see for example chapters in Jones and Rowe (1999) or the special issue of *Review of Palaeobotany and Palynology* edited by Grauvogel-Stamm et al. (2009)). Similarly, the continued use of TEM in combination with microchemical analyses will improve our understanding of microfossil taxonomic affinity. This endeavor has important implications for our understanding of the early evolution of the biosphere, as organic-walled microfossils represent the most abundant and morphologically diverse group of Precambrian fossils. The Proterozoic fossil record consists largely of hundreds of species of unornamented, simple leiospheres that are described using parameters such as size and wall thickness that may have little taxonomic significance. Consequently, distinguishing among taxa is difficult at best and it is likely that these divisions have little phylogenetic meaning. TEM studies of the ultrastructure of some leiospheres show that they differ in terms of number of layers and types of cell walls and this *is* indicative of different phylogenetic relationships. Thus, TEM can help us clarify our view of the Proterozoic fossil record of simple organic-walled microfossils. Additionally, TEM can help us resolve the taxonomic affinities of more complex organic-walled microfossils that appear in the Neoproterozoic. Determining which forms are algal versus which forms are potentially metazoan has important implications for how we view the timing and patters of diversification of eukaryotes. Further ultrastructural work on both modern analogs and the Precambrian and Phanerozoic fossil record will reveal further information on the taxonomic affinity and function of the diverse organic-walled microfossil record.

Although TEM provides us with a unique kind of information, there are other methods that obtain similar results. One such microscopic technique similar to TEM is focused ion beam electron microscopy (FIB-EM; Schiffbauer and Xiao 2009). FIB-EM has the advantage that it requires much less sample preparation than TEM does, and is more site specific. However, despite the somewhat lengthy preparation process, TEM allows the user to detect variability over the whole vesicle at once with relative efficiency both of time and cost, relative to FIB-SEM. Ideally, FIB-EM and TEM can be used as complementary techniques, as each provides a unique view of the microstructure of complex samples.

12.7 Conclusions

The true potential of TEM imaging lies in the added dimensions of information it provides over traditional light microscopy and scanning electron microscopy, which are both surface techniques. TEM is most valuable when combined with other lines of evidence such as external morphology, microchemistry, and stratigraphic relationships (Arouri et al. 1999, 2000; Marshall et al. 2005, 2006; Javaux and Marshal 2006; Cohen et al. 2009). It is also essential to remember that the images produced by TEM are rarely a perfect representation of the sample. There is always a transfer function between the object and the image that leads to a degradation of detail. In addition, and especially with biological and fossil material, the preparation procedure can alter the final results. With these caveats in mind, TEM can reveal detail of samples indistinguishable with other techniques, and, when combined with a comprehensive understanding of modern analogs, can help to shed light on the taxonomic affinities of organic-walled microfossils and other enigmatic fossilized forms. Future applications to the fossil record will help to broaden our understanding of the biology, physiology, and phylogenetic relationships of myriad fossil taxa.

Acknowledgments Gary Wife, Anette Axén and Stefan Gunnarsson at the Microscopy and Imaging unit at EBC, Uppsala University, are thanked for their expertise regarding preparing and sectioning the samples and microscope work. Margaret Coughling at the Harvard Medical School provided help and inspiration to PAC in developing a new preparation method and provided help and advice with microtoming.

References

Allard B, Templier J (2000) Comparison of neutral lipid profile of various trilaminar outer cell wall (TLS)-containing microalgae with emphasis on algaenan occurrence. Phytochemistry 54:369–380

Arouri K, Greenwood PF, Walter MR (1999) A possible chlorophycean affinity of some Neoproterozoic acritarchs. Org Geochem 30:1323–1337

Arouri K, Greenwood PF, Walter MR (2000) Biological affinities of Neoproterozoic acritarchs from Australia: microscopic and chemical characterisation. Org Geochem 31:75–89

Barghoorn ES, Tyler SA (1965) Microorganisms from the Gunflint Chert. Science 147:563–577

Bozzola JJ, Russel LD (1999) Electron microscopy: principles and techniques for biologists, 2nd edn. Jones & Bartlett, Sudbury, 670 pp

Brasier MD, Green OR, Jephcoat AP, Kleppe AT, Van Kranendonk MJ, Lindsay JF, Steele A, Grassineau NV (2002) Questioning the evidence for Earth's oldest fossils. Nature 416:76–81

Buckland-Nicks J, Hodgson A (2000) Fertilization in *Callochiton castaneus* (Mollusca). Biol Bull 199:59–67

Cáceres CE (1997) Dormancy in invertebrates. Invert Biol 116:371–383

Cohen PA, Knoll AH, Kodner RB (2009) Large spinose microfossils in Ediacaran rocks as resting stages of early animals. PNAS 106:6519–6524

Couch K, Downes M, Burns C (2001) Morphological differences between subitaneous and diapause eggs of *Boeckella triarticulata* (Copepoda: Calanoida). Freshw Biol 46:925–933

Damiani MC, Leonardi PI, Pieroni OI, Caceres EJ (2006) Ultrastructure of the cyst wall of *Haematococcus pluvialis* (Chlorophyceae): wall development and behaviour during cyst germination. Phycologia 45:616–623

De Gregorio BT, Sharp TG, Flynn GJ, Wirick S, Hervig RL (2009) Biogenic origin for Earth's oldest putative microfossils. Geology 37:631–634

Egerton RF (2005) Physical principles of electron microscopy. An introduction to TEM, SEM, and AEM. Springer, New York, 202 pp

Evitt WR (1963) A discussion and proposals concerning fossil dinoflagellates, hystrichospheres, and acritarchs. Proc Natl Acad Sci USA 49(158–164):298–302

Grauvogel-Stamm L, Guignard G, Wellman CH (eds) (2009) Spore/pollen fine structure in living and fossil plants. Rev Palaeobot Palynol 156:1–262

Grey K, Willman S (2009) Taphonomy of Ediacaran (late Neoproterozoic) acritarchs: significance for taxonomy and biostratigraphy. Palaios 24:239–256

Hagen C, Siegmund S, Braune W (2002) Ultrastructural and chemical changes in the cell wall of *Haematococcus pluvialis* (Volvocales, Chlorophyta) during aplanospore formation. Eur J Phycol 37:217–226

Hill R, Shepard W (1997) Observations on the identification of California anostracan cysts. Hydrobiologia 359:113–123

Inouye I, Hori T, Moestrup Ø (2003) Ultrastructural studies on *Cymbomonas tetramitiformis* (Prasinophyceae). I. General structure, scale microstructure, and ontogeny. Can J Bot 81:657–671

Javaux EJ, Marshal CP (2006) A new approach in deciphering early protist palaeobiology and evolution: combined microscopy and microchemistry of single Proterozoic acritarchs. Rev Palaeobot Palynol 139:1–15

Javaux EJ, Knoll AH, Walter MR (2001) Morphological and ecological complexity in early eukaryotic ecosystems. Nature 412:66–69

Javaux EJ, Knoll AH, Walter MR (2003) Recognizing and interpreting the fossils of early eukaryotes. Orig Life Evol Biosph 33:75–94

Javaux EJ, Knoll AH, Walter MR (2004) TEM evidence for eukaryotic diversity in mid-Proterozoic oceans. Geobiology 2:121–132

Javaux EJ, Marshall CP, Bekker A (2010) Organic-walled microfossils in 3.2-billion-year-old shallow-marine siliciclastic deposits. Nature 463:934–938

Jones TP, Rowe NP (eds) (1999) Fossil plants and spores – modern techniques. Geological Society Publishing House, Bath

Jux U (1968) Über den Feinbau der Wandung bei *Tasmanites* Newton. Palaeontogr Abt B 124:112–124

Jux U (1969a) Über den Feinbau der Zystenwandung von *Pachysphaera marshalliae* Parke, 1966. Palaeontogr Abt B 125:104–111

Jux U (1969b) Über den Feinbau der Zystenwandung von *Halosphaera* Schmitz, 1878. Palaeontogr Abt B 128:48–55

Jux U (1971) Über den Feinbau der Wandungen einiger paläozischer Baltisphaeidiacean. Palaeontogr Abt B 136:115–128

Jux U (1977) Über die wandstrukturen sphaeromorpher acritarchen: *Tasmanites* Newton, *Tapajonites* Sommer & Van Boekel, Chuaria Walcott. Palaeontogr Abt B 160:1–16

Kempe A, Wirth R, Altermann W, Stark RW, Schopf JW, Heckl WM (2005) Focussed ion beam preparation and in situ nanoscopic study of Precambrian acritarchs. Precambrian Res 140:35–54

Kennaway GE, Eaton GL, Feist-Burkhardt S (2008) A detailed protocol for the preparation and orientation of single fossil dinoflagellate cysts for transmission electron microscopy. Palynology 32:1–15

Kjellström G (1968) Remarks on the chemistry and ultrastructure of the cell wall of some Palaeozoic leiospheres. Geol Fören Stockh Förh 90:118–221

Knoll M (1935) Aufladepotentiel und Sekundäremission elektronenbestrahlter Körper. Z techische Phys 16:467–475

Knoll AH, Javaux EJ, Hewitt D, Cohen P (2006) Eukaryotic organisms in Proterozoic oceans. Philos Trans R Soc B 631:1023–1038

Marshall CP, Javaux EJ, Knoll AH, Walter MR (2005) Combined micro-Fourier transform infrared (FTIR) spectroscopy and micro-Raman spectroscopy of Proterozoic acritarchs: a new approach to Palaeobiology. Precambrian Res 138:208–224

Marshall CP, Carter EA, Leuko S, Javaux EJ (2006) Vibrational spectroscopy of extant and fossil microbes: relevance for the astrobiological exploration of Mars. Vib Spectrosc 41:182–189

Martin F, Kjellström G (1973) Ultrastructural study of some Ordovician acritarchs from Gotland, Sweden. Neues Jahrbuch für Geologie und Paläontologie Monatshefte 1:44–54

McMullan D (1995) Scanning electron microscopy 1928–1965. Scanning 17:175–185

Moczydłowska M, Willman S (2009) Ultrastructure of cell walls in ancient microfossils as a proxy to their biological affinities. Precambrian Res 173:27–38

Moczydłowska M, Schopf JW, Willman S (2010) Micro- and nano-scale ultrastructure of cell walls in Cryogenian microfossils: revealing their biological affinity. Lethaia 43:130–136

Moreau JW, Sharp TG (2004) A transmission electron microscopy study of silica and kerogen biosignatures in 1.9 Ga Gunflint microfossils. Astrobiology 4:196–210

Onoue Y, Toda T, Ban S (2004) Morphological features and hatching patterns of eggs in *Acartia steueri* (Crustacea, Copepoda) from Sagami Bay, Japan. Hydrobiologia 511:17–24

Peat CJ (1981) Comparative light microscopy, scanning electron microscopy and transmission electron microscopy of selected organic walled microfossils. J Microsc 122:287–294

Peng Y, Bao H, Yuan X (2009) New morphological observations for Paleoproterozoic acritarchs from the Chuanlinggou Formation, North China. Precambrian Res 168:223–232

Reimer L, Kohl H (2008) Transmission electron microscopy. Physics of image formation, 5th edn. Springer, New York, 590 pp

Schiffbauer JD, Xiao S (2009) Novel application of focused ion beam electron microscopy (FIB-EM) in preparation and analysis of microfossil ultrastructures: a new view of complexity in early eukaryotic organisms. Palaios 24:616–626

Schopf JW (1993) Microfossils of the early Archean Apex chert: new evidence of the antiquity of life. Science 260:640–646

Schopf JW, Kudryavtsev AB, Agresti DG, Wdowiak TJ, Czaja AD (2002) Laser-Raman imagery of Earth's earliest fossils. Nature 416:73–76

Talyzina NM, Moczydłowska M (2000) Morphological and ultrastructural studies of some acritarchs from the Lower Cambrian Lükati Formation, Estonia. Rev Palaeobot Palynol 112:1–21 van Waveren IM, Marcus NH (1993) Morphology of recent copepod egg envelopes from Turkey Point, Gulf of Mexico, and their implications for acritarch affinity. Spec Pap Palaeontol 48:111–124

Wall D (1962) Evidence from recent plankton regarding the biological affinities of *Tasmanites* Newton 1875 and *Leiosphaeridia* Eisenack 1958. Geol Mag 99:353–362

Wellman CH, Grauvogel Stamm L, Guignard G (2009) Studies of spore/pollen wall ultrastructure in fossil and living plants: a review of the subject are and the contribution of Bernard Lugardon. Rev Palaeobot Palynol 156:2–6

Williams DB, Carter CB (2009) Transmission electron microscopy: a textbook for materials science, 2nd edn. Springer, New York, 760 pp

Willman S (2009) Morphology and wall ultrastructure of leiosphaeric and acanthomorphic acritarchs from the Ediacaran of Australia. Geobiology 7:8–20

Willman S, Moczydłowska M (2007) Wall ultrastructure of an Ediacaran acritarch from the Officer Basin, Australia. Lethaia 40:111–123

Chapter 13
Paleobiological Applications of Focused Ion Beam Electron Microscopy (FIB-EM): An Ultrastructural Approach to the (Micro)Fossil Record

James D. Schiffbauer and Shuhai Xiao

Contents

Abstract Coupled dual-beam focused ion beam electron microscopy (FIB-EM) has gained popularity across multiple disciplines over the past decade. Widely utilized as a stand-alone instrument for micromachining and metal- or insulator-deposition in numerous industries, the sub-μm-scale ion milling and integrated electron imaging capabilities of such FIB-based systems are well documented in the materials science literature. These capacities make FIB-EM a powerful tool for *in-situ* site-specific preparation of ultrathin foils for transmission electron microscopy. Recent advancements in the field-emission guns of FIB-EM systems have

J.D. Schiffbauer (✉) and S. Xiao
Department of Geosciences, Virginia Polytechnic Institute and State University, 4044 Derring Hall, Blacksburg, VA 24061, USA
e-mail: jdschiff@vt.edu; xiao@vt.edu

M. Laflamme et al. (eds.), *Quantifying the Evolution of Early Life*, Topics in Geobiology 36, DOI 10.1007/978-94-007-0680-4_13, © Springer Science+Business Media B.V. 2011

provided spatial resolution comparable to that of many high-grade scanning electron microscopes, providing enhanced imaging capacities with material-deposition and material-removal capabilities. Recently, FIB-EM preparation techniques have been applied to geological samples to characterize mineral inclusions, grain boundaries, and microfossils. We here provide a summary of recent paleobiological studies that use FIB-EM methodology for the examination of fossils. Additionally, we demonstrate a novel method for analyzing the three-dimensional ultrastructure of microfossils (reported previously by Schiffbauer and Xiao [Palaios 24: 616–626, 2009]). This method, FIB-EM nanotomography, consists of sequential ion milling, or cross-sectioning, and concurrent SEM imaging, a technique that provides three-dimensional data of precise sites at high spatial resolution, yielding new insight into fossil ultrastructure. We here illustrate the use of FIB-EM nanotomography by studies of herkomorphic and acanthomorphic acritarchs (organic-walled microfossils) extracted from the ≥999 Ma Mesoproterozoic Ruyang Group of North China. The three-dimensional characteristics of important but controversial acritarch features such as extravesicular processes and vesicularly enclosed central bodies are described. Taken together, these case studies demonstrate that FIB-EM instruments are powerful and useful tools for investigating the three-dimensionality of microfossil ultra- and nanostructures.

Keywords Acritarch • Focused ion beam electron microscopy • Microfossil ultrastructure • Nanotomography

13.1 A Brief History of the Focused Ion Beam Electron Microscope

As alluded to in Chaps. 11 and 12, advanced microbeam instrumentation has truly revolutionized our view of the natural world by allowing living things, present or past, to be observed at the micro- (or nano-) scale. Most microbeam instruments, including transmission electron microscopy (TEM) and scanning electron microscopy (SEM), were designed for the physical, chemical, and materials sciences, but their applicability to biological (and paleobiological) sciences cannot be ignored. Indeed, shortly after the SEM became commercially available in 1965, researchers were applying the instrument and methods to biological and paleobiological samples (e.g., Karnovsky 1965; Seligman et al. 1966; Sandberg and Hay 1967). However, as electron microbeam tools became more readily available, methods of preparing novel samples for investigation became a focal point of development of instrument accessories and/or method development for existing instruments. For example, TEM requires samples to be sliced into very thin sections (also known as ultrathin foils), typically less than 200 nm in thickness, to allow electrons to pass through the sample (also known as electron transparency; see Chap. 12 for details on TEM sample preparation).

This type of sample preparation requires specialized equipment; the most common of which include: (1) ultramicrotomy, developed in the in the late eighteenth century as a means of preparing reproducible sections for light microscopy, is a process for TEM sample preparation in which epoxy-embedded specimens are physically sliced with a precisely controlled diamond or glass knife (common in the preparation of biological samples; Quekett 1848); (2) electropolishing, borrowed from the metal-finishing industry, is an electrochemical process where samples, hosted in an electrolyte bath, are thinned by electrical current-induced oxidation (common in the preparation of metals and alloys; Vander Voort 2004); (3) physical lapidary polishing methods such as wedge (or tripod) polishing, which are highly common in materials sciences; and (4) ion milling, a process that uses an ionized inert gas (commonly argon) that sputters material from the sample surface. Ion milling techniques became popular in the 1950s as a means of thinning samples to electron transparency for TEM analysis. Because many geological materials are inherently brittle or friable, other standard TEM preparation techniques proved inadequate or produced poor ultrathin foils. Thus, argon ion milling grew to be the principal methodology for preparing TEM samples for Earth science research and revolutionized the microstructural study of rocks and minerals (Castaing and Labourie 1953; Barber 1970, 1999; Heaney et al. 2001 and references therein).

Since the 1990s, however, there has been an increase in the use of a different type of tool—focused ion beam (FIB) systems. While the terminology "focused ion beam" technically incorporates several instruments that employ micrometer-scale (μm) spatial resolution (or better) ion beams—such as secondary ion mass spectrometry (SIMS) and scanning helium ion microscopy (HeIM or SHIM)—the FIB acronym is used in the literature to describe instruments that use high current density ion beams from a field emission liquid metal ion source (LMIS), most frequently gallium (Ga^+), to ablate and deposit materials in controlled patterns. Interestingly, the development of liquid metal ion sources came from the field of aerospace engineering, where charged metal droplets were investigated as potential rocket thruster sources (Krohn 1961; Phaneuf 1999)—so the innovation of FIB instruments did indeed take rocket science.

The most common FIB systems are either single beam systems, that utilize the ion beam for both material addition/excavation and secondary ion imaging, or dual-beam systems, that combine a LMIS into a field emission scanning electron microscope (FE-SEM; Fig. 13.1a, b). These dual-beam FIB-EM (also abbreviated as FIB-SEM and FIB/SEM) workstations couple high-resolution, high-magnification secondary electron imaging capabilities with ion-beam-induced material modification, and are the focus of this chapter. Recent advancements in imaging technologies have resulted in FEGs capable of subnanometer-scale resolution (very high screen magnifications ≥1,000,000×), technologies now commonly incorporated into FIB-EM workstations. The integrated SEM in dual-beam systems allows visual control on where and how the sample is modified by the ion beam, whereas single-beam FIB systems rely on the ion beam for both sample modification and imaging, which is intrinsically detrimental to the sample. The principal advantages of dual-beam systems, therefore, are that they not only allow for simultaneous electron imaging and ion milling but also

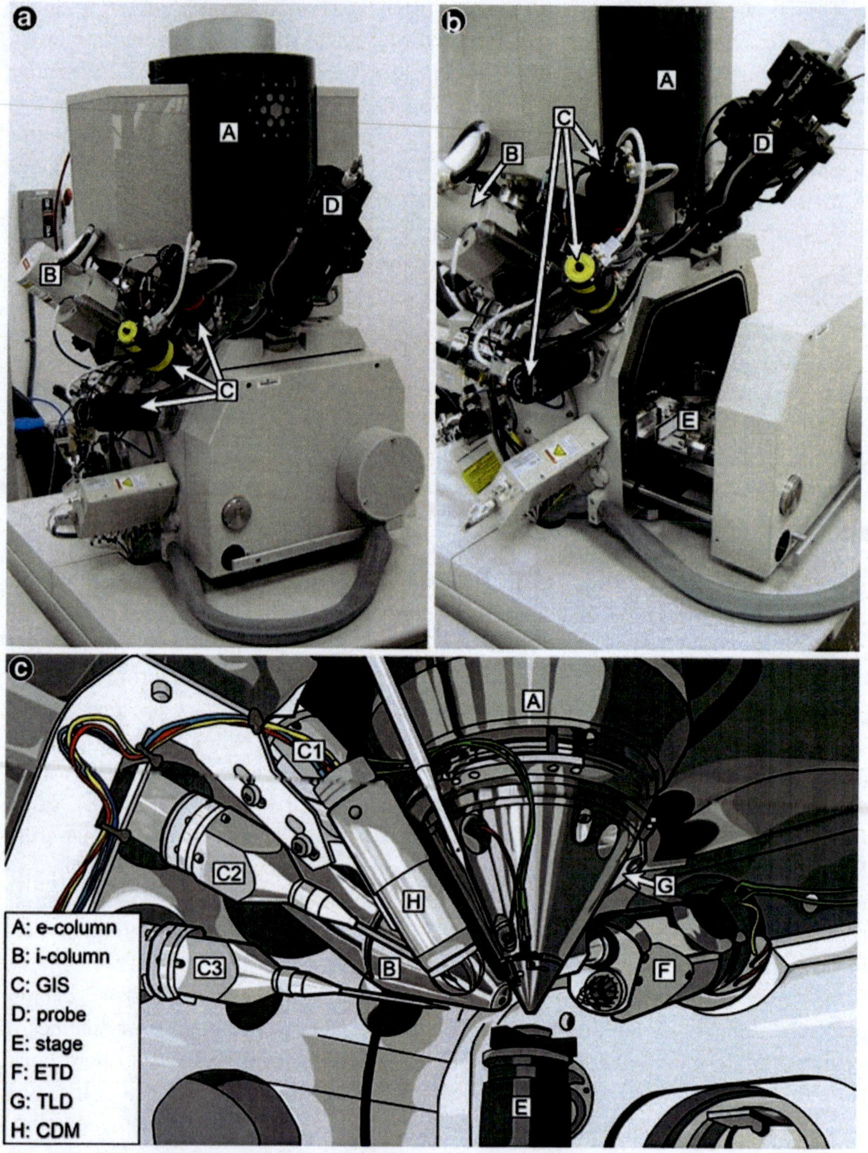

Fig. 13.1 FIB-EM instrument photographs and sample chamber schematic with labeled components. (**a**) Photograph of FEI Company's DualBeam™ Helios 600 NanoLab™ with e-column (*A*, field emission source), i-column (*B*, liquid metal ion source), three-nozzle gas injection system (*C*), and micromanipulator probe controller (*D*) labeled; (**b**) Photograph of Helios 600 NanoLab with chamber door ajar to illustrate sample stage (*E*) and size of sample chamber (*A–D* additionally labeled); (**c**) Graphic representation of sample chamber interior with e-column (*A*, field emission gun), i-column (*B*, liquid metal ion gun), gas injection system nozzles (*C*), micromanipulator probe needle (*D*), sample stage (*E*), and image detectors (*F*, ETD = Everhart-Thornley detector; *G*, TLD = in-lens detector; *H*, CDM = continuous dynode electron multiplier detector) labeled

circumvent surficial damage to the sample caused by extended ion imaging in single-beam systems. Although not commonly accessible in research institutions, triple-beam FIB systems are available, merging the FE-SEM and Ga^+ LMIS capabilities of dual-beam FIB systems with an argon ion column (which is less damaging than gallium due to its comparatively low atomic mass) for high precision sample finishing. Additionally, dual-beam environmental FIB-EM systems (typically abbreviated as ESEM/FIB) are available as well, incorporating the capabilities of dual-beam FIB-EM systems into a variable pressure system. Sample analysis can be performed in low and extended vacuum modes that circumvent the need for a conductive charge-control coating (such as carbon or gold-palladium), although the ion milling capabilities function only in high vacuum mode.

While the semiconductor industry has dominated the FIB-EM-use market for the construction and modification of prototype microdevices, FIB-EM instruments have expanded into industrial and research laboratories and have vastly diversified the types of materials examined. The primary functions of FIB-EM instruments are the dissection and deposition of material at the micro- to nanometer scale using precisely controlled ion beam rastering. Because the ion beam can excavate sample material *in-situ* while under secondary electron observation, TEM ultrathin foil preparation became one of the primary research uses for FIB-EM instruments. These systems have been successfully used to prepare TEM ultrathin foils of a large variety of specimens, ranging from metals, alloys, and ceramics, to biological samples, earth and planetary materials, and fossils (e.g., Giannuzzi et al. 1998, 1999; Heaney et al. 2001; Weiss et al. 2002; Lee et al. 2003; Floss et al. 2004; Wirth 2004, 2009; Benzerara et al. 2005; Kempe et al. 2005; Westall et al. 2006; Bernard et al. 2007, 2009; Cavalazzi 2007; Marko et al. 2007).

Ultrathin foil preparation with FIB-EM instruments has drastically changed the data that can be acquired from TEM analyses as compared to samples prepared by traditional ultrathin foil preparation approaches. And FIB-EM systems also offer numerous advantages over conventional ion milling, ultramicrotomy, or lapidary ultrathin foil preparation techniques: (1) live-time secondary electron imaging with spatial resolutions at ≤5 nm during sample preparation (this, in and of itself, is the greatest advantage of FIB-EM foil preparation as all other methods rely on optical microscopy for visualization of sample preparation); (2) *in-situ* sample excavation from polished or unpolished material; (3) site specific material extraction at the nm-scale; (4) introduction of few sample deformation artifacts as compared to mechanical preparation methods; (5) ease of sample preparation and rapid processing as compared to methods that demand sample embedding; and (6) useable on a broad range of sample types and compositions, some of which may not be possible with conventional methods (Phaneuf 1999; Heaney et al. 2001). Nevertheless, these instruments should not only be considered as TEM ultrathin foil preparation tools: FIB-EM workstations are comprehensive micromachining platforms capable of housing numerous SEM application packages (as described in Sect. 13.2.1: Technical description and operational capabilities) for carrying out integrated, three-dimensional sample analyses. FIB-EM systems are now being use as stand-alone analytical instruments and are becoming increasingly accessible.

13.2 FIB-EM Systems

13.2.1 Technical Description and Operational Capabilities

The FEI Company's DualBeam™ Helios 600 NanoLab™ (Fig. 13.1a, b), housed at the Virginia Tech Institute of Critical Technology and Applied Science Nanoscale Characterization and Fabrication Laboratory (VT-ICTAS-NCFL), is built on a high-resolution field emission scanning electron microscope platform. The large vacuum chamber of this instrument can accommodate samples up to 15 cm in diameter and 5 cm high. The primary components of this workstation are the following: a high-resolution, high-magnification Elstar™ Schottky FEG for scanning electron microscopy (SEM); multiple electron detectors for image acquisition, including a through-the-lens detector (TLD, or in-lens) for ultra-high resolution secondary electron detection (sample surface topography), and an Everhart-Thornley detector (ETD) for conventional secondary electron detection (which can also be used for backscattered electron detection [Z-contrast compositional information] by reversing the charge applied to the detector's Faraday cage); and a high-resolution Sidewinder™ focused Ga^+ ion beam column for precisely controlled nanoscale-material addition and excavation using a Continuous Dynode Electron Multiplier detector (CDM detector) for ion imaging (see Fig. 13.1c for a generalized sample chamber diagram).

The electron column and ion column are oriented in the workstation so that the electron and ion beams (*e*-beam and *i*-beam, respectively) intersect at a 52° angle at their coincident point (also known as the eucentric point or eucentric height). Where the sample also intersects the coincident beam point, it can be tilted perpendicular to either beam without changing focal position (see Fig. 13.2a, b for corresponding electron images perpendicular to the *e*-beam and *i*-beam). The *e*-beam is normally used for secondary electron imaging, whereas the *i*-beam is used to volatize material in an operator-defined pattern. Both the electron column and ion column can be operated at a broad range of beam accelerating voltages (that determines the energy and wavelength of the beam) and currents (the quantity of electrons or ions interacting with the sample surface), allowing for optimal imaging and milling capacities. The Elstar™ electron column is capable of 1,000,000× screen magnification (256 nm horizontal field width), providing subnanometer-resolution image detail (0.9 nm resolution at 15 keV). The Sidewinder™ ion column can also provide high-resolution images with 5.0 nm resolution (at 30 kV). Both beams can operate at lower beam energies, allowing for sample charge control. Operating the *i*-beam at lower energy is useful in reducing surface amorphization during prolonged periods of ion imaging. The piezoelectrically controlled sample stage permits precise positioning in the x- and y-directions. Stage tilting (Fig. 13.2a, b), as well as rotational and z-axis movement, are motorized so that at eucentric height of the sample a precise point of interest on the sample surface can be positioned coincident to the e- and *i*-beams for simultaneous and synchronized electron imaging and ion milling.

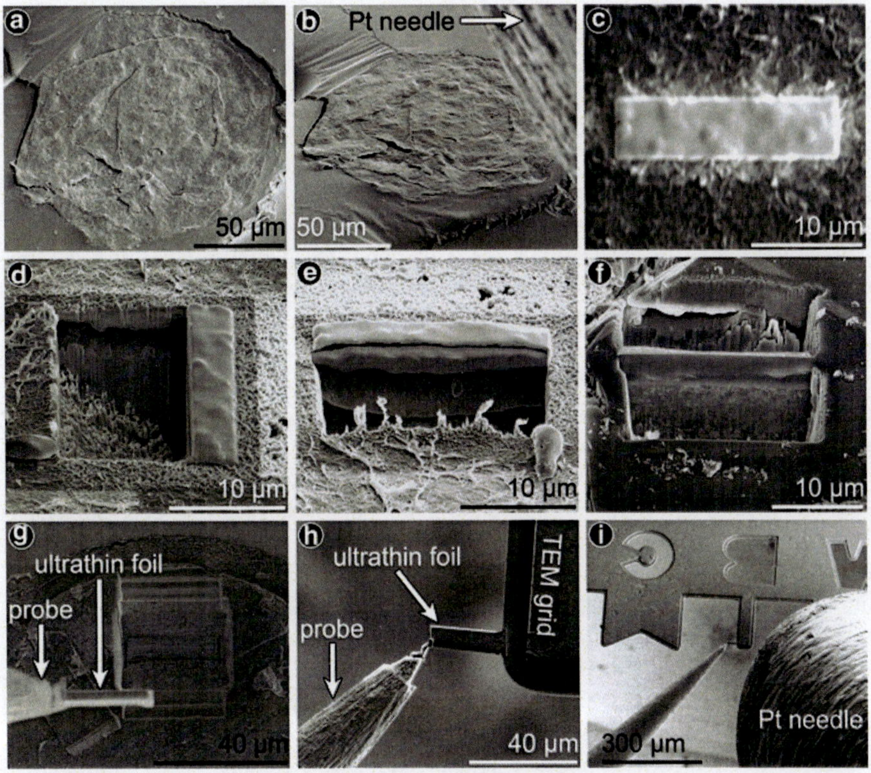

Fig. 13.2 Electron and ion micrographs of FIB-EM sample processing. (**a–e**) *Dictyosphaera delicata* (Ruyang Group) acritarchs; (**f–i**) Jingangku graphitized acritarchs. (**a**) Overview of uncut *D. delicata* specimen perpendicular to *e*-beam; (**b**) Overview of same *D. delicata* specimen as shown in (**a**) oriented perpendicular to *i*-beam and imaged with *e*-beam, platinum GIS needle visible in upper right corner of frame of view (*arrow*); (**c**) Deposited platinum strip (~20 µm width) on surface of *D. delicata* surface; (**d**) Side view of step-cut into *D. delicata* specimen with Pt strip shown in (**c**) on right; (**e**) Step-cut shown in (**d**) rotated 90° for illustration of standard cross-section viewing via *e*-beam imaging—this view also demonstrates sample surface orientation at 142° (52° below horizontal) to the *e*-beam (see Fig. 13.3 for graphic representation of beam and image orientation); (**f**) TEM lift-out preparation of Jingangku specimen, consisting of two opposing step-cuts juxtaposed to central Pt protective strip; (**g**) Micromanipulator probe (*left arrow*) attached to TEM ultrathin foil (*right arrow*), which was cut free at its base and is now being "lifted-out"; (**h**) Micromanipulator probe with attached TEM foil lift-out being attached via Pt deposition to a standard copper TEM grid; (**i**) Lower magnification view of process shown in (**h**) to show Pt GIS needle and more complete view of TEM sample grid

This FIB-EM instrument is capable of nanoscale lithography, material deposition (Fig. 13.2c) and removal (Fig. 13.2d, e), and application of tomographic techniques. It houses an integrated Omniprobe™ tungsten micromanipulator probe needle (Fig. 13.2g–i) that can be used for precision probing, straining, moving and placing of nanometer-sized objects cut from or deposited on larger samples. This system also houses an integrated three-nozzle gas injection system (GIS) that can support organometallic gases for material/insulator deposition and various other material-specific assist gases to enhance specialized ion-milling processes. These GIS nozzles are in essence 100 μm-diameter hypodermic needles that can be inserted to within 100 μm of a sample surface for delivery of accessory gases in direct proximity to a region of interest. For example, the organometallic platinum gas is injected into the sample chamber via a GIS nozzle (Fig. 13.2b) and platinum is then deposited by controlled *i*-beam volatization of the organic component of the precursor gas (Fig. 13.2c). This deposited platinum layer protects the original sample surface from superfluous *i*-beam sputtering during milling procedures. Additionally, platinum can also be used as a binding material; for instance, the micromanipulator probe needle can be attached via platinum deposition to the sample for material biopsies or TEM foil lift-outs (Fig. 13.2g–i). In addition to the deposition of metals, decorative gases can be used. For example, such accessory gases as xenon difluoride (XeF_2) and magnesium sulfate heptahydrate ($MgSO_4 \cdot 7H_2O$) can be injected in proximity to the sample surface to aid, respectively, in selectively etching silicates and oxides or carbon/organic materials. The XeF_2 accessory gas, which is decomposed by interaction with the *i*-beam, etches the sample surface much like hydrofluoric acid (HF), but with microscopic visual control under live-time secondary electron imaging. Similarly, the $MgSO_4 \cdot 7H_2O$ accessory gas volatizes to water vapor with *i*-beam interaction for selective carbon milling.

To aid its use as a stand-alone analytical instrument, the Helios 600 system is equipped with a Pegasus XM 4 integrated energy dispersive x-ray spectrometer (EDS or EDX) with a Genesis EDS x-ray microanalysis system, an Electron Backscatter Diffraction (EBSD) package, and a Hikari EBSD camera. When coordinated with material-removal via the *i*-beam, these analytical accessories allow for 3D reconstructions of the internal structure, chemistry, and orientation of objects a few nanometers in size.

13.2.2 Principal Uses

The principal tasks conducted by FIB-EM workstations are *in-situ* material deposition (Fig. 13.2c), nanometer-scale ion milling, and site-specific sample cross-sectioning (Fig. 13.2d–f)—all of which can be monitored in real-time via concurrent electron imaging (see Fig. 13.3 for graphic representation of beam orientation, sample milling, and imaging). The FIB-EM's most common use is TEM ultrathin foil preparation, which consists of cutting an electron-transparent foil through ion material excavation and subsequently moving and attaching the sample to a TEM grid via an integrated micromanipulator probe needle (Fig. 13.2g–i). Such sub-μm capabilities well illustrate the advantages of FIB-EM workstations in TEM ultrathin foil preparation. Additional advantages over standard ultramicrotomy include the preservation of sample structural integrity, that would be otherwise damaged by ultramicrotomy, and avoidance of physical artifacts such as fissures (microcracks), epoxy resin shatter, and knife chatter. However, it must be

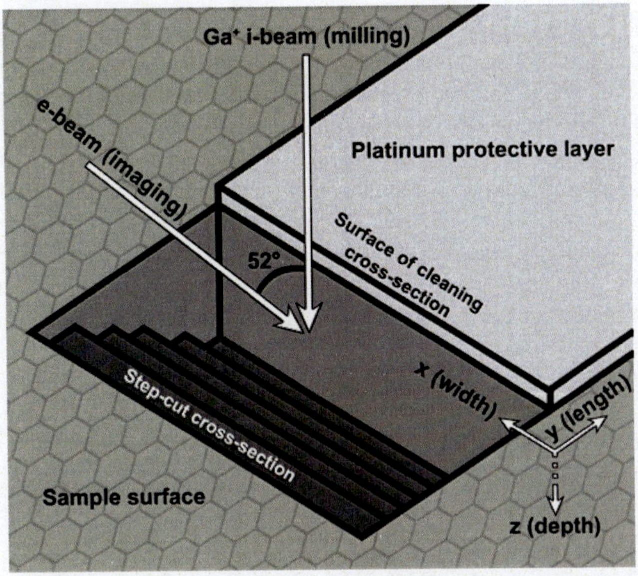

Fig. 13.3 Graphic representation of reticulated acritarch sample processing for FIB nanotomography, illustrating beam orientation during sample milling (sample surface perpendicular to *i*-beam, *e*-beam intersecting *i*-beam and milled cross-section at 52° and sample surface at 142°). X, Y, and Z sample axes are labeled at *lower right*

noted that FIB-EM material excavation may itself introduce artifacts during ion imaging and cross-sectioning, such as sample surface amorphization and consequent loss of microstructural detail, cross-section curtaining (i.e., the appearance of gathered curtains on the cross-section surface), and Ga^+ impregnation (Prenitzer et al. 2003; Rubanov and Munroe 2004). Many of these problems can be easily resolved or circumvented. For instance, broad surface amorphization can be avoided with careful selection of beam accelerating voltage and current, minimal usage of *i*-beam imaging, and deposition of a protective platinum layer.

A major advantage over other systems of the FIB-EM technique is that it can cut nearly any material with minimal damage. Because material removed by the *i*-beam is volatized rather than sheared by physical means, the FIB-EM can be used for preparation of challenging samples that are not easily prepared by microtome techniques. Moreover, this instrument is highly applicable to *in-situ*, site-specific, nanometer-scale tomographic analyses. FIB-EM nanotomography (also known as FIB nanotomography or FIB tomography, e.g. Inkson et al. 2001; Holzer et al. 2004, 2006, 2007; Kubis et al. 2004; Groeber et al. 2006) is a three-dimensional electron microscopy technique that is accomplished by slice and view—or more specifically, constructing (via *i*-beam) and imaging (via *e*-beam) sequential cross-sections of a given region of interest.

For the following sections of this chapter, some caveats should be noted. FIB-EM techniques are applicable for the examination of structures across a broad range of material or sample types, e.g., polycrystalline materials, photonic crystal fibers, ceramics, silicon wafers, alloys, metals and metal powders, biological materials, and so on, and preparation techniques and protocols may vary dependent on the material type to be analyzed. Therefore, because various samples differ in composition, no specific beam currents or accelerating voltages are specified here. Efficient and effective material removal will inevitably be sample-dependent, and further will be contingent on the specific combination of beam current, accelerating voltage, size of milling patterns, use of material-specific assist gases, etc. Because the purpose of this chapter is to describe methodologies and instrument potential for the examination of Earth science materials, focusing specifically on paleobiological specimens, consultation of the relevant materials sciences literature has focused on samples having similar compositions to the fossils of interest. However, because the procedures illustrated here can be used to analyze minute volumes of sample, numerous combinations of instrument conditions can be applied to a single specimen to find the optimal settings for milling.

13.2.3 Overview of the FIB-EM "Lift-Out" Method for Preparation of Electron-Transparent Ultrathin Foils

One of the major benefits of using FIB-EM systems for ultrastructural analysis is its relative ease of sample preparation. While use of FIB-EM to prepare TEM ultrathin foils can be time-consuming, there are a few points to consider: (1) preparing a sample for FIB-EM analysis or for FIB-EM TEM foil preparation is the same as preparing an SEM stub; (2) during the construction a TEM foil lift-out, data can be collected on the sample via secondary electron imaging, EDS, EBSD, etc.; and (3) because the sample must be embedded in epoxy for standard ultramicrotomy of a TEM foil, analytical approaches that are applicable for the remainder of the sample (known as the potted butt) are limited. If samples are relatively limited in number and conservation of material is a priority, the FIB-EM ultrathin foil preparation technique damages only a small portion of the sample, and additionally provides unembedded material for further analyses—which can be especially important in reference to paleobiological samples.

For initial sample preparation, specimens should be placed on an adhesive mount, such as carbon- or copper-tape, and affixed to an aluminum SEM stub. Prior to mounting, no other preparation is required, as polished or unpolished host-rock-embedded fossils can be mounted directly to adhesive stubs (although extracted microfossils, or fossils hosted within petrographic/paleontologic thin sections are just as easily used). It is important to note, however, that care must be taken when using samples with large topographic variation as movement of the sample may be difficult or even damaging to the instrument with GIS nozzles or micromanipulator apparatuses in place. The shape of the specimen is also not a concern as even spherical fossils can be prepared or analyzed via FIB-EM. After such samples are mounted, they can be gold-palladium or carbon sputter-coated if charge control (e.g., if the sample is not conductive) is a concern during electron imaging.

Alternatively, lower beam accelerating voltages/currents may be used in an attempt to avoid sample charging on uncoated specimens.

Following sample-mounting and sputter-coating, such stubs are individually secured to the sample stage and inserted into the FIB-EM vacuum chamber. Once under vacuum, the *i*-beam and GIS warm-up sequences are initiated and the sample surface is viewed using the *e*-beam. After a region of interest is determined and the LMIS source is warmed, the sample surface is adjusted to its eucentric height where the *e*-beam and *i*-beam are in coincident focus (the region of interest on the sample being centered in the sample chamber universe). Ion beam imaging is damaging to a sample, but is necessary to ensure that both beams are focused on the same point of the specimen. Accordingly, several additional procedures during ion imaging are recommended to minimize marring of the sample surface. First, lower *i*-beam currents may be used for imaging purposes, which allow for longer live-viewing times with less surficial damage than do higher beam currents. This is still destructive, albeit minimally. In addition, changing *i*-beam currents may require imaging adjustments (such as focus, brightness, contrast, beam shift, etc.). Second, *i*-beam snapshots may be used at the currents selected for milling purposes to avoid the potential need to re-tune the beam; but even with very short viewing times, this technique may result in significant surficial damage. Third, the organometallic platinum gas may be injected into the system without a deposition pattern in place; thus, when scanning the sample in *i*-beam mode (even at higher currents), a very thin layer of platinum is deposited and the sample surface remains undamaged. This gas-injection technique is probably the most efficient if surficial damage needs to be prevented across the entire sample surface. Finally, if the sample surface is relatively flat, or there is a reasonably large surface area of the sample that is of less significance, the *i*-beam and *e*-beam can be brought to coincident focus outside of the region of interest. The stage can then be positioned back to the specific region of interest with the beams in coincident focus and surficial damage to this area can be avoided.

Once a region of interest is identified, the eucentric height is reached, and both beams are focused, preparation of a TEM foil lift-out begins with the deposition of a thin platinum strip (Fig. 13.2c). Other metals can be used, but platinum is most commonly used for sectioning purposes. The *i*-beam has a Gaussian (parabolic) shape and therefore would unevenly erode the sample surface during milling and cleaning/ion-thinning procedures if not for this platinum strip, which functions to define and protect the surface of the sample. The platinum GIS needle is inserted to within 100 μm of the sample surface and the organometallic gas is injected into the sample chamber atmosphere in direct proximity to the region of interest. Once the platinum deposition pattern is defined (typically with a length ranging between 10 and 20 μm and a width of <10 μm) the *i*-beam is rastered within the boundaries of this pattern to decompose the organic component of the organometallic platinum precursor gas. The platinum is deposited specifically where it interacts with the *i*-beam, for TEM foil preparation usually to a vertical thickness of approximately 1–2 μm. The platinum, protecting the sample surface, also provides rigidity to the ultrathin foil during lift-out procedures as well as during the final thinning of the foil to electron transparency. If using automated milling functions (available on some FIB-EM software interface packages), reference markers, usually in the form of X patterns, may be milled on both sides of the platinum strip to serve as guides to compensate for beam shift or sample drift.

Following the deposition of platinum, two step-cut milling patterns are positioned on either side of the protective strip, with the deepest portion of the step-cuts (usually designated between 5 and 15 μm) placed nearest to the strip (Fig. 13.2f). The step-cut patterns are typically milled with higher *i*-beam currents for optimization of cutting time. These cuts allow direct secondary or backscattered electron-viewing of the sample cross-section as well as access to make the vertical side and horizontal bottom cuts to free the lift-out from the remainder of the specimen.

Prior to the final cutting lines, the sample stage must be angled away from the *i*-beam (perpendicular to the *e*-beam) for maximum exposure of the sample cross-section, and the micromanipulator probe needle must be attached to the foil. Once the probe needle has been brought to within 1 μm of the platinum strip on the foil surface, the organometallic platinum GIS is again used, this time as a binding agent. A small *i*-beam scanning pattern, of only a few micrometers in both width and length, is defined directly on the junction between the micromanipulator probe needle and platinum protective strip. The platinum deposited in this pattern joins the tip of the probe needle to the TEM foil platinum strip. Following probe needle mounting, a horizontal cutting line and two vertical cutting lines are positioned at the base and each side of the foil, cutting lines that serve to disconnect the foil from the bulk of the specimen. The TEM foil can then be lifted out directly by use of probe needle controls (Fig. 13.2g). In the finishing stages of the lift-out procedure, the foil needs to be thinned to electron transparency with the *i*-beam at a lower current in an effort to prevent unnecessary sample damage. TEM-ready foils can then be attached to 3-mm-sized copper TEM sample grids in the FIB-EM sample chamber using the platinum gluing procedure previously used to attach the foil to the probe needle (Fig. 13.2h–i). The last step prior to TEM analysis requires a line cutting pattern to separate the foil from the probe needle in order to leave it attached to the TEM sample grid—although the electron transparency thinning process or thinning touch-ups can also be conducted once the foil has been attached to the TEM grid .

While typical length and depth measurements have been described here for TEM foil preparation, it is important to note that these measurements are highly dependent on the structure of interest or the sample itself. The prime measurement to be considered is the thickness of the foil, as it must be electron transparent for TEM analysis. The goal, therefore, is to produce a TEM-ready foil ≤200 nm thick for optimum electron transparency.

13.2.4 Overview of the FIB-EM Nanotomography Method: A Three-Dimensional View of Microstructure

FIB-EM nanotomography is accomplished by sequential ion sectioning and acquisition of scanning electron micrographs of each serial section of the structure of interest. This technique has been widely used in materials sciences to study materials interfaces in, for instance, ceramics, alloys, composites, cements, and particulates (Inkson et al. 2001; Holzer et al. 2004, 2006, 2007; Kubis et al. 2004; Groeber et al. 2006). The interslice spacing of sequential sections in the studies of these materials typically ranges from 10 to 300 nm, the selection of which is dependent on the size of the features of interest. Most commonly, increments of 100–250 nm have been used, although FIB-EM nanotomography of 100 nm-sized particles require sectioning as fine as 10 nm for preparation of three-dimensional (3D) reconstructions (Holzer et al. 2004). The application of FIB-EM nanotomography allows for real-time, three-dimensional ultrastructural analysis and compositional mapping with precise site selectivity.

Following the same introductory sample preparation procedures outlined above for the TEM lift-out section, and after a region of interest is selected and coincident beam focus is attained, a platinum deposition pattern is outlined. For FIB-EM nanotomography, however, the entire region of interest is covered with a thin ≤ 1 μm layer of platinum rather than the narrow platinum protective strip used in TEM foil preparation. A single step-cut pattern is used at the leading edge of the region of interest or platinum cover to permit easy viewing of the cross-sections. A second step-cut pattern is unnecessary; all subsequent ion-cutting patterns are simply very narrow strips, or cleaning cross-sections. After each cut, the newly exposed section is imaged and the cutting pattern is then advanced sequentially to obtain successive cross sections of the region of interest (for more examples of FIB-EM nanotomographic preparation and data, see Figs. 13.5–13.7 and the corresponding case study presented in Sect. 13.3). The computer interface and concurrent *e*-beam imaging provide nanometer-scale positional accuracy for sequential cuts. As noted above, the spacing of the cleaning cross-section should be chosen to be appropriate for the features of interest and for time-efficiency. Perhaps most importantly for paleontological samples, determination of the specific protocol to be used for FIB-nanotomography benefits from preliminary investigation of the samples to be examined, whether from published ultrastructural studies based on the use of more traditional methods or from exploratory FIB cuts outside the specific region of interest, in order to provide a baseline for the size range of features of interest. If nanometer-scale features are to be investigated, the advancing cutting distance may be as small as a single nanometer, but if larger regions are to be imaged, the cutting pattern can be increased to μm-scale. To ensure that all of the leading edge material is cleanly and completely removed with each cleaning cross-section, the patterns generated should be greater than the length (y-axis) of the advancement distance. For instance, a 150–200 nm y-axis length pattern can be used for 100-nm-length sections; this pattern would be advanced only 100 nm for each sequential section, which provides additional *i*-beam rastering/material removal overlying the previously cut region. Although sectioning patterns may be advanced in 1 nm increments, sequential sections of <100 nm can produce surface amorphization complications. Because of the Gaussian shape of the *i*-beam, the wider upper portion of the beam can unevenly erode the platinum protective layer above exterior of the uncut sample. Therefore, increasing the number of tightly spaced sequential cross-sections results in a progressively thinner protective platinum layer and eventually causes sample surface amorphization.

If three-dimensional modeling or rendering is a goal of the use of this technique, the characteristics of the electron imaging data to be provided should be taken into account. Because the sample is oriented at an angle of 52° below horizontal (to achieve perpendicularity to the *i*-beam), the electron micrographs of sequential ion sections are oriented 142° (rather than perpendicular) to the surface of the sample in order to obtain the best available view of the cross-section. Therefore, the resultant image data inherently contain a y-axis length component and an image thresholding-procedure is required to reduce the three-dimensionality of the sample image to two dimensions. Thresholding procedures are included in numerous image-manipulation software packages, such as Adobe Photoshop®, Scion Image, and ImageJ. Such procedures reduce the image to (commonly black and white) binary representations that include only the pixels corresponding to the flat-sectioned face of the sample studied. This alteration removes unwanted image data, such as sections of the mounting tape and platinum protective layer. The resultant 2D data can be organized into image stacks and then be imported to three-dimensional rendering software packages such as ParaView or Amira. These software packages have multiple algorithms available for 3D renderings, one of the more common being the "marching cubes" algorithm (Lorensen and Cline 1987) which extracts a polygonal mesh of an isosurface (a surface that represents points of a constant value) from the 2D image stacks. The individual polygons are then incorporated into 3D surface renderings of the structures studied. Furthermore, by reducing the opacity of the surface contours, which helps elucidate internally contained structures, these 3D surface models can be converted to 3D volume renderings.

13.3 FIB-EM Methods in the Paleobiological Literature

For the past half century, electron microscopy has played an important role in the examination of microfossils (see, for example: Sandberg and Hay 1967; Oehler 1976, 1977; Schopf and Oehler 1976; Xiao et al. 1998, 2007; Arouri et al. 1999, 2000; Talyzina and Moczydłowska 2000; Javaux et al. 2001, 2003, 2004; Moreau and Sharp 2004; Javaux and Marshal 2006; Schiffbauer et al. 2007; Willman and Moczydłowska 2007; Xiao and Schiffbauer 2008; Cohen et al. 2009; Moczydłowska and Willman 2009). One might even say that electron microscopy has played an integral role in understanding and deciphering the biogenicity, structure, and taxonomic affinity of microfossils throughout the record of life on Earth. Because of their capabilities, FIB-EM systems are ideal for analyzing intricate details of microfossils—whether the fossils are embedded in their host rock or extracted. As discussed below, within the last 5 years, several papers have reported the use of FIB-EM instruments and corresponding ultrastructural methods for studies of microfossils (and macrofossils). The following discussion section provides a brief survey of FIB-EM methods used to date in paleobiological research.

The earliest uses of FIB-EM for fossil analyses are those of Wirth (2004) and Kempe et al. (2005), both of which used FIB-EM to produce TEM-ready ultrathin foils of fossil specimens—in the Wirth (2004) study, the anchoring spicule of a deep-sea hexactinellid sponge, *Monorhaphis chuni*, and in the Kempe et al. (2005) report, permineralized leiospheric acritarchs in carbonaceous-cherty *Conophyton gaubiza* stromatolites.

Anchoring spicules of *M. chuni* are the largest biosilica structures known, up to 3 m in length and 8.5 mm in diameter (Müller et al. 2007). Starting with a polished cross-section of a *M. chuni* spicule, Wirth (2004) used the GIS with XeF_2 accessory gas in the FIB-EM sample environment to aid in etching the polished spicule surface, revealing a previously-described radial arrangement of amorphous silica and finely-layered organic-rich material. A TEM lift-out was then produced from the etched spicule, which was subsequently examined under TEM bright-field imaging and further elementally characterized by electron energy-loss spectroscopy (EELS). From the FIB-EM and TEM analyses, Wirth (2004) was able to characterize the silica-organic material composite nanostructure of the *M. chuni* anchoring spicule and describe the potential structural integrity that this configuration provides.

The leiospheric acritarch samples, from the ~650 Ma Chichkan Formation of southern Kazakhstan and analyzed by Kempe et al. (2005), were hosted in thin sections approximately 300 μm thick. The thin sections were hand-ground with aluminum oxide powder to expose the carbonaceous material of the acritarch cell walls, and were then polished with decreasing grain-size diamond powders (to 0.5 μm) to reduce surface variation of the thin section. The samples were then etched in a 5% HF bath to dissolve the embedding rock matrix and thus expose, in bas relief, the organic acritarch walls. Raman spectroscopy and atomic force microscopy (AFM) were used, respectively, to determine the geochemical thermal maturity of the carbonaceous acritarch walls and to produce surface topography maps. The thin sections were then sputter-coated with a thin gold layer, followed by a thin carbon layer, for electron charge-dispersal under secondary electron imaging in the FIB-EM. Because of the site-specific milling capabilities of the FIB-EM instrument, there was no need to macerate the rock sample to remove these potentially fragile microfossil specimens. Kempe et al. (2005) prepared TEM thin foils of the acritarch walls directly from the thin sections using FIB-EM lift-out procedures (producing foils 15–20 μm in width, 10 μm in height, and 150 nm in thickness). The foils, once attached to TEM grids, were examined via analytical electron microscopy (AEM) as well as TEM bright-field imaging and EELS. In addition to other techniques used in this study, Kempe et al. (2005) noted the site-specific capabilities of the FIB-EM instrument and suggested that this instrument should provide new information on fossilization processes and nanoscopic structures of microfossils. Moreover, after their work had shown that FIB-EM-conducted TEM lift-outs provided complete and solid foils across a range of cell preservations, Kempe et al. (2005) further stated that the FIB-EM ultrathin foil preparation method is applicable even to the study of poorly preserved fossils, such as those common in the Precambrian, where maceration techniques may produce only bits and pieces of fossils, rather than intact specimens, and where taphonomic details must be thoroughly considered to assess the putative biogenicity of specimens.

Bernard and colleagues, in 2007 and 2009, used FIB-EM to analyze the ultrastructure of Triassic plant spore walls: lycophyte megaspores preserved in blueschist-grade metasediments of the Roc de la Pêche-Chanrossa in the Vanoise massif, Western Alps, France, and well-preserved lycopsid spores from the Lettenkohle Formation of the Wasselonne Quarry, France (Bernard et al. 2007, 2009). Using multidisciplinary/multiple analytical approaches (including Raman spectroscopy, energy resolution near-edge X-ray absorption fine structure spectroscopy [NEXAFS], and scanning transmission X-ray microscopy [STXM] for high spatial spectroscopic analyses of subsurface structures), the FIB-EM instrument was used as a preparative tool for *in-situ* construction of ultrathin foils from fossils hosted in petrographic thin sections (Fig. 13.4a–c). In addition to such studies of fossil specimens, Benzerara et al. (2005) have shown that FIB-EM foils prepared for TEM study are applicable to other geobiological problems, such as the investigation of mineral-microbe interactions—which undoubtedly have implications in fossil preservation.

Although FIB-EM was used essentially in the above described studies as a sample preparation tool for specimens to be analyzed by other techniques, Westall et al. (2006) and Cavalazzi (2007) have shown that the capabilities of FIB-EM make it a useful stand-alone workstation. In 2006, Westall and colleagues investigated ~3,400 Ma microbial mats from the Barberton Greenstone Belt, South Africa, in an effort to decipher the effects of ultraviolet radiation on the early Earth. Instead of using FIB-EM as only a TEM foil-preparation tool, they used the FIB-EM to excavate fossil mat material for direct observation of the subsurface structures (Fig. 13.4d–f). The direct cuts were investigated by integrated secondary and backscatter electron imaging as well as EDS elemental analysis within the FIB-EM system, studies showing that the mat surfaces and associated evaporitic halides were coated by a layer of silica. Further, the direct FIB-EM cuts illustrated that the interior organic components of the fossil mats were not preserved as individual filaments but, rather, degraded into alveolar textures, which, in conjunction with the observation of well-preserved silica-coated exteriors of the mat structures, suggest that silicification was rapid (Westall et al. 2006). In 2007, Cavalazzi (2007) expanded on the paleobiological applicability of directly observed FIB-EM-produced cross-sections, shown by Westall et al. (2006). By directly observing *in-situ* cross-sections of putative filamentous microfossils (and their host rock) from the Devonian Hollard Mound, Morocco, Cavalazzi (2007) demonstrated the use of FIB-EM in assessing the biogenicity of objects having microbial-like morphologies.

While multidisciplinary/multi-instrument approaches are rapidly becoming the *modus operandi* in paleobiology (especially in the Precambrian; e.g., see brief review by Cady and Noffke 2009), the studies noted above document the capabilities of FIB-EM for performing *in-situ* analyses of the surficial structures and subsurface structures of fossils and their taphonomic relation to their host

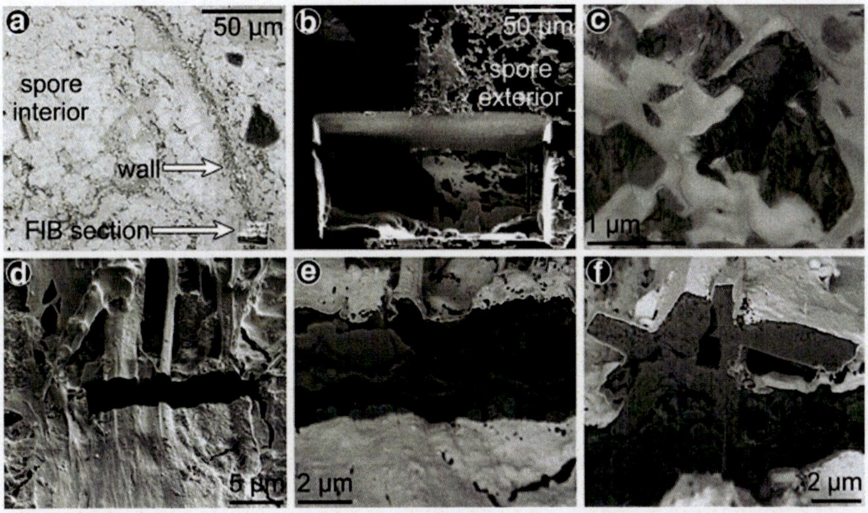

Fig. 13.4 Representative FIB-EM sample preparation images from published paleontological literature. These representatives were chosen for their differing FIB-EM techniques and analyses, and more importantly to illustrate the applicability of FIB-EM techniques across a range of paleontological samples. (**a–c**) FIB-EM preparation and TEM analysis of fossil lycophyte megaspore (Bernard et al. 2007; reproduced with permission): small (~250–300 μm diameter) *in-situ* samples in polished petrographic thin section; (**d–f**) FIB-EM preparation and direct SEM cross-sectional imaging of fossil microbial mat (Westall et al. 2006; reproduced with permission): broad unpolished surfaces with numerous fossil microbial structures (variable sizes, typically structures of interest <5 μm). (**a**) FIB-EM preparation of lycophyte spore wall TEM ultrathin foil (Modified from Fig. 3A), step-cuts visible in lower right corner (*arrow and label*); (**b**) direct cross-sectional photomicrograph of ultrathin foil shown in (**a**) prior to micromanipulator-assisted extraction (Modified from Fig. 3B); (**c**) TEM photomicrograph of spore wall organic material and graphitic carbon (Modified from Fig. 6A); (**d**) Secondary electron photomicrograph of FIB-EM cut through unpolished microbial mat structure (imaged perpendicular to cut; Modified from Fig. 11A); (**e**) Backscattered electron photomicrograph of FIB-EM cut shown in (**d**), darker regions in cross-section correspond to amorphous kerogenous material with lighter colored fine intervening authigenic minerals (imaged with a stage tilt of 52°; Modified from Fig. 11B); (**f**) Backscattered electron photomicrograph of FIB-EM cut through evaporate-coated microbial mat, darker regions correspond to kerogeneous masses with lighter colored evaporitic minerals dispersed throughout the structure (imaged with a stage tilt of 52°; Modified from Fig. 11D)

rock—and, perhaps more importantly, their morphological and chemical (elemental) three-dimensionality. The following sections of this contribution describe a study that we conducted that uses a novel FIB-EM tomographic method to reveal the three-dimensionality of acritarch ultrastructures, such as the external processes (or spines) of acanthomorphic acritarchs.

13.4 FIB-EM Nanotomography: An Acritarch-Based Case Study

The case study presented in the following sections of this chapter using FIB-EM nanotomography was originally published in Palaios (Schiffbauer and Xiao 2009); the discussion below is an abbreviated version of the original publication that focuses specifically on the method used and the results obtained.

13.4.1 Examining Acritarch Ultrastructure: A Brief Introduction

In this study, we used an FEI Company Helios 600 Nanolab FIB-EM system to examine the ultrastructures of Mesoproterozoic acritarchs from the Beidajian Formation (likely 1,300–1,400 Ma, but poorly constrained to <1,625 Ma and ≥999 Ma) of the Ruyang Group, southern Shanxi, North China (Xiao et al. 1997). Widely known from Proterozoic deposits, acritarchs are as old as 1,600–1,800 Ma (Yan 1982, 1991, 1995; Luo et al. 1985; Zhang 1986, 1997; Sun and Zhu 2000). Similar fossils are known from older sediments as well: recent findings of graphitized vesicles in high-grade metamorphic rocks of Australia and China (<2,500 Ma) and less thermally altered organic-walled vesicles from the Moodies Group (~3,200 Ma), South Africa, suggest that acritarchs may have been well established by the Mesoarchean to Neoarchean (Schiffbauer et al. 2007; Zang 2007; Javaux et al. 2010). The term "acritarch" is used as a trash-can grouping for organic-walled vesicular microfossils that cannot be placed with confidence into any existing biologic group (Evitt 1963). Acritarchs undoubtedly have diverse biological affinities and, therefore, should be considered a polyphyletic grouping with no true taxonomic status or rank (Servais 1996). The identification of taxonomically and phylogenetically useful features in acritarchs can be a challenging task because of their potential diversity. Indeed, individual acritarch taxa have been proposed to represent diapause egg cases of early metazoans (Yin et al. 2007), resting cysts of prasinophyte chlorophytes (Arouri et al. 1999), dinoflagellates (Moldowan and Talyzina 1998; Arouri et al. 2000; Meng et al. 2005; Willman and Moczydłowska 2007), and various prokaryotes (Javaux et al., 2003). Traditionally, some acritarchs have been interpreted as eukaryotic fossils based on a large vesicle size (Schopf 1992). However, size may not be an adequate indicator of taxonomic standing for single-celled organisms, as some such prokaryotes may reach large sizes, such as the 750 μm-diameter *Thiomargarita namibiensis* (Schulz et al. 1999). In light of this, microstructures and ultrastructures are often sought to elucidate taxonomic interpretations of acritarchs (Arouri et al. 1999, 2000; Talyzina and Moczydłowska 2000; Javaux et al. 2004). For example, micro- and ultrastructural characterization of ~1,500 Ma acanthomorphic (i.e., process-bearing) acritarchs, using both SEM and TEM, has provided the earliest unambiguous evidence for the presence of a cytoskeleton, unquestionable evidence for Mesoproterozoic (1,600–1,000 Ma) eukaryotes (Javaux et al. 2001, 2003, 2004).

13.4.2 Sample Description and Preparation

The siliciclastic host rocks containing the acritarchs used for this study were collected at the Shuiyougou Section (SYG 6) of the Beidajian Formation (Xiao et al. 1997). Shales and siltstones of the Beidajian Formation contain highly abundant and well-preserved acritarchs, including *Shuiyousphaeridium macroreticulatum*, *Dictyosphaera delicata*, *Valeria lophostriata*, and *Tappania plana*. This study focused on *D. delicata* and *S. macroreticulatum*, the two most abundant species from the Shuiyougou acritarch assemblage.

Although the analytical method detailed here is applicable to (perhaps better-preserved) microfossils of younger geologic ages as well as modern microorganisms, acritarchs of this antiquity (~1,400–1,300 Ma) and from this locality were chosen because detailed descriptions and relevant geochemical analyses are available (Xiao et al. 1997; Marshall et al. 2005). Importantly, the ultrastructures of conspecific and morphologically similar acritarchs have been documented by TEM (Javaux et al., 2001, 2003, 2004), data available from this well-established technique that provide a basis for comparison with ultrastructural findings using FIB-EM.

The vesicles of *D. delicata* and *S. macroreticulatum* range in diameter from 50–300 μm. Their vesicle walls consist of interlocking polygonal (mostly hexagonal) plates. Although *D. delicata* is herkomorphic and lacks processes, *S. macroreticulatum* is acanthomorphic with unevenly distributed extravesicular processes. Excystment structures or medial splits may occur in these taxa, but need further confirmation. The hexagonal plates are 1.5–3 μm in maximum width, and the plate boundaries consist of 100–300 nm thick, 100–200 nm high raised ridges. The processes on *S. macroreticulatum* are flared or conspicuously branched, and on average are 10–15 μm long with diameters of 2–3 μm. The processes have been described as hollow, appear to be open at the distal end, and flare outward at both the vesicular attachment and outer tip (Xiao et al. 1997; Javaux et al. 2004).

To extract specimens of such acritarchs from their encompassing matrix, ~2 g of Shuiyougou carbonaceous shale was dissolved using standard hydrofluoric acid-digestion techniques and the resulting carbonaceous macerates were rinsed with distilled deionized water and then vacuum-sieved using a 10 μm nylon screen (Vidal 1988). During inspection of the sieved macerates under stereomicroscopy, numerous individual *S. macroreticulatum* and *D. delicata* vesicles were isolated, lightly rinsed to remove any attached debris, placed on a glass slide, and photomicrographed. The specimens chosen for analysis ranged in color from orange-brown to brown, and have a corresponding thermal alteration index (TAI) of 3–4, using the TAI scale published by Batten (1996) and consistent with the results reported by Marshall et al. (2005). *S. macroreticulatum* specimens that contained vesicularly enclosed central bodies, interpreted as possible nuclear or other organelle remnants as well as vestiges of degraded cytoplasm (Knoll and Barghoorn 1975; Oehler 1976, 1978; Golubic and Barghoorn 1977; Hagadorn et al. 2006), were specifically selected for analysis in an effort to elucidate their ultrastructure. In addition, initial optical microscopy showed that some of the *S. macroreticulatum* specimens exhibited a discontinuous outer membrane supported by extravesicular processes, and that some such processes appeared to have rounded or bulbous terminations, interesting targets for FIB-EM nanotomography.

After they were imaged by optical microscopy, the Shuiyougou shale acritarchs were mounted to adhesive tape affixed to a standard SEM stub. Because removal of a standard depth of material with each *i*-beam cross-section could excavate into the adhesive tape due to the variable vesicle thickness of the acritarch specimens, copper tape mounts (rather than carbon tape) were used in order to maximize the differences in composition between the carbonaceous acritarchs and the mounting medium. After mounting the acritarchs, the aluminum stubs were sputter-coated with gold-palladium to a thickness of 20 nm with a Cressington 208 HR (equipped with a MTM-20 thickness-controller) high-resolution sputter coater. Thinner Au-Pd coatings could have been applied; the 20 nm thickness was chosen as a preventative measure to avoid surface amorphization during *i*-beam tuning.

13.4.3 FIB-EM Nanotomography of Shuiyougou Acritarchs

Using a trial and error approach, we achieved successful results using variable width, 3–5 μm depth (z-axis) cleaning cross-section patterns advanced from 100 to 250 nm at a time, the most common thicknesses reported in the materials literature.

The relatively large depth selection (z-axis) of our cross-sectioning patterns ensured that the entire thickness of the fossil, in addition to the 20 nm gold-palladium coating and the ~1 μm thick platinum protective surface, was completely sectioned. Typically, deeper cutting patterns of up to ~10 μm were used for the construction of TEM lift-outs on rock hosted specimens, but the cutting pattern-depth for extracted specimens was determined by the fossil thickness, expected depth of structures of interest, and/or the time efficiency of the ion milling process. Because the volume of material excavated via the *i*-beam is directly proportional to the time of *i*-beam rastering, the selection of 100 or 250 nm cleaning-section advancements depended on the total width (x-axis) of the sectioning pattern. For instance, in cases where the cutting pattern was narrow (≤10 μm), 100-nm-length sections were used for optimal spatial resolution of potential 3D ultrastructures, whereas 250-nm-length sections were used for wider cross-sections (>10 μm) to maximize both spatial resolution and beam-time efficiency.

13.4.4 Case Study Results

13.4.4.1 FIB-EM Nanotomography of *Dictyosphaera delicata*

In order to refine the FIB-EM nanotomography technique for potentially more ultrastructurally detailed acanthomorphic *S. macroreticulatum* specimens, a total of ten *D. delicata* specimens were used as initial test subjects. From these specimens, numerous FIB ultra-thin sections were constructed to examine the capabilities and establish proper settings for *i*-beam material excavation, to observe potential *i*-beam deformation artifacts, and to establish a 2D baseline for ultrastructures observed using the FIB-EM sectioning technique. Once these conditions had been evaluated, 90 sequential cross-sections were constructed at a spacing of 100 nm and width of 8.2 μm, covering a total surficial area of 73.8 μm^2, on one specimen. Significant aspects of the ultrastructure of *D. delicata* vesicles are shown by FIB-EM. Vesicle walls vary in thickness (up to 2.5 μm in some cases), but are typically constrained between approximately 200–750 nm (Fig. 13.5a, b). The vesicle walls are typically multilamellar (alternating light and dark bands, or layers, which likely correspond to higher and lower electron density [z contrast], respectively; (Fig. 13.5a–c), although these laminations are less distinct in some cross-sections and may appear to be composed of homogenous unilayered walls (e.g., Fig. 13.5f). The vesicle architecture consists of reticulate polygonal (mostly hexagonal) plates, which are raised on the exterior of vesicle and either beveled or raised in the vesicle interior (Fig. 13.5d–j). This vesicle wall morphology is very similar to the multi-layered reticulate vesicle walls of *S. macroreticulatum* reported in previous TEM ultrastructural analyses (compare Fig. 13.5g–j with Fig. 13.5k, l [modified from Javaux et al. 2004]). The innermost layer, where visible, which shows up in a lighter grayscale with both the ETD and TLD, is approximately 150 nm thick. With sequential *i*-beam sectioning, it was observed that the two vesicle walls contain numerous smaller flattened (Fig. 13.5m) to rounded chambers (Fig. 13.5n; the maximum diameter observed in rounded chambers ranging from 460 to 548 nm) distinct from the large vesicle cavity. Additionally, the vesicle walls merge and separate in multiple places (Fig. 13.5o).

13.4.4.2 FIB-EM Nanotomography of *Shuiyousphaeridium macroreticulatum*

In the more complex vesicles of *S. macroreticulatum*, two features were of primary interest for use of FIB-EM nanotomography: the 3D ultrastructures of extravesicular processes and vesicularly-enclosed central bodies. After study by standard optical microscopy, six specimens having identifiable central bodies were prepared for and analyzed by FIB-EM nanotomography. Under ion imaging, these specimens showed topographically raised regions on the vesicle wall that corresponded to the location of the central bodies (Fig. 13.6a–d). Particularly useful for this study, one specimen contained what we interpreted to be an extravesicular process situated on the outer vesicle wall at the edge of the central body (Fig. 13.6d inset, extravesicular process ultrastructures shown in Fig. 13.7). Seventy cross-sections, having 250 nm spacing and 15.4 μm widths, were prepared of this specimen (occupying a surficial area of 265.65 μm^2). As described below, the nanotomography of this specimen serves as the focal-point case study for both the central body and extravesicular process ultrastructures.

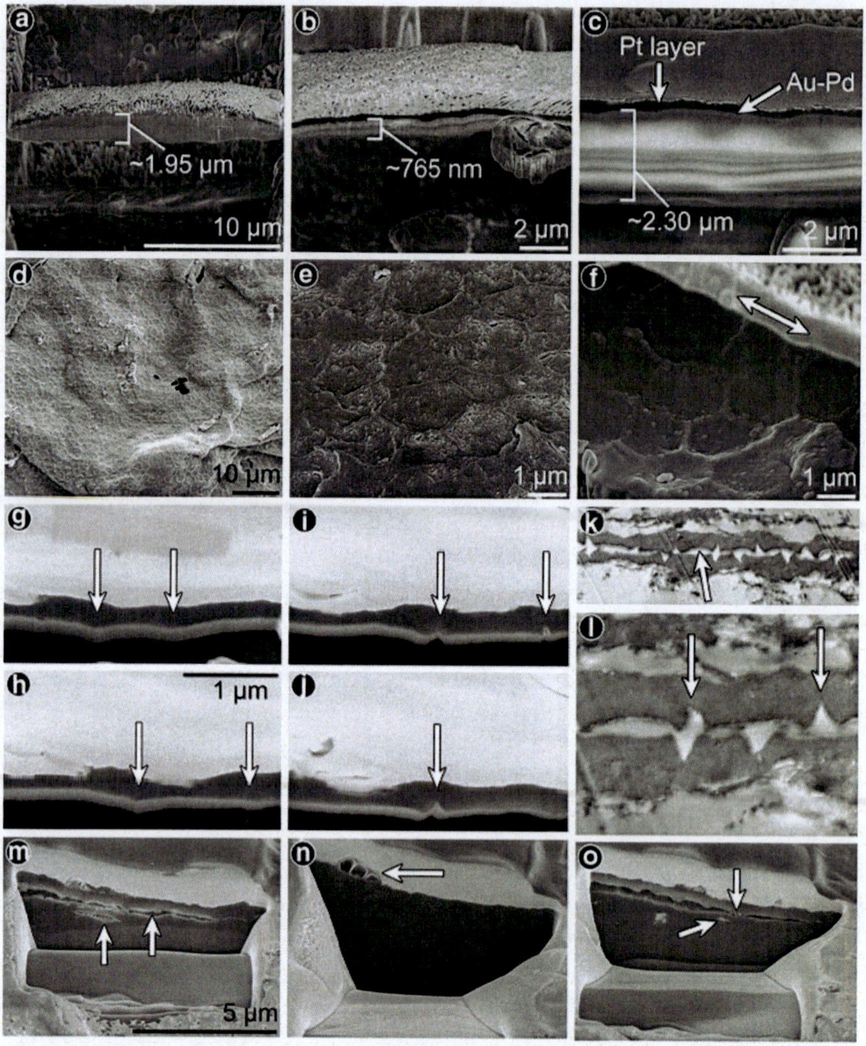

Fig. 13.5 Secondary electron photomicrographs of FIB-EM prepared *Dictyosphaera delicata* specimens and standard ultramicrotomy-preparation/TEM analysis of *Shuiyousphaeridium macroreticulatum* for comparison of techniques. (**a–c**) Standard FIB-EM ultrathin foil lift-out preparation of *D. delicata*; (**d–f**) FIB-EM surface and subsurface analyses of *D. delicata* vesicle walls via secondary electron imaging; (**g–j**) FIB-EM nanotomographic technique for three-dimensional analysis of *D. delicata* vesicle wall ultrastructures, secondary electron imaging of sequential FIB-EM sections with an interslice spacing of 100 nm, scale bar in (**h**) applicable to (**g–j**); (**k–l**) TEM analysis of vesicle wall ultrastructure of *S. macroreticulatum* (Javaux et al. 2004, reproduced with permission); (**m–o**) Vesicle wall chambers and merge points observed from FIB-EM nanotomography of *D. delicata*, scale bar in (**m**) applicable to (**m–o**). (**a**) Lift-out preparation of *D. delicata* illustrating ~1.95 μm-thick multilamellar vesicle wall (darker outer layer, lighter inner layer); (**b**) Lift-out preparation of *D. delicata* illustrating ~765 nm-thick multilamellar (darker central wall band between two lighter layers) vesicle wall, polygonal attachment at right edge of FIB-EM cut is a single organic plate (multilamellar structure continues in plate); (**c**) Higher magnification view of *D. delicata* vesicle

As was expected for this specimen from its topographic observation using ion imaging, sequential-ion sectioning that moved into the vesicularly enclosed central body was accompanied by an overall thickening of the fossil, from a relatively constant total vesicle thickness of 300–500 nm—individual vesicle wall thickness ranging from ~150 to ~250 nm (Fig. 13.6e)—to a total thickness of approximately 1,500 nm (Fig. 13.6f, g). Throughout the analyzed region of the central body, there was no distinction between the vesicle walls and the central body (i.e., no gaps exist between the central body and the vesicle walls). Given that the vesicle walls of such compressed acritarchs were shown to merge in the study of *D. delicata*, summarized above, it was not unexpected in this specimen to observe a merging of the compressed vesicle wall and central body. No direct measurement could be made on the thickness of the central body, though it is estimated to be ~1,000 nm by assuming vesicle wall thicknesses of ~250 nm—the maximum individual wall thickness outside of the central body region. In comparison with the acritarch cell wall, this central body is distinguished by its high incidence of nanometer-scale pores (Fig. 13.6f–i). These pores (some of which are outlined by a lighter color in cross-section than the surrounding darker vesicle material) are predominantly localized to the central-body region. To evaluate these findings, this assessment was tested by adding a hypothetical vesicle wall having an estimated wall thickness of 250 nm to the image slices obtained (Fig. 13.6h, i). At any given image slice of the

Fig. 13.5 (continued) cross-section (~2.30 μm-thick) illustrating numerous layers—alternating light and dark layers—within wall structure (also note *fine light gray layer* at junction of *darker platinum protective layer* cross-section, which corresponds to the 20 nm-thick Au-Pd sputter-coating for charge control); (**d**) Overview of *D. delicata* vesicle surface reticulation; (**e**) Higher magnification view of surface reticulation with raised organic plate junctions; (**f**) Higher magnification view of internal, raised organic plate boundaries able to be observed through FIB-EM material excavation, note *dual-sided arrow* at *upper right* illustrating outer, unilayered vesicle wall; (**g**) Initial cross-section in FIB-EM nanotomography of *D. delicata*, *arrows* indicate raised organic plate boundaries on interior side of multilayered vesicle wall (*darker outer layer*, *lighter inner layer*; this pattern persists though frames (**h–j**); (**h**) 200 nm from cut shown in (**g**) with *arrows* illustrating lateral migration of raised interior plate boundaries; (**i**) 100 nm from cut shown in (**h**) illustrating beveled plate boundary junctions at same location of raised boundary junctions in previous cut; (**j**) 100 nm from cut shown in (**i**) with slight lateral movement of mid-image plate boundary incision and termination of right edge plate boundary incision; (**k**) TEM image of ultra-microtomy-prepared *S. macroreticulatum* specimen with highly comparable vesicle wall structures, including inner plate boundary incisions and multilayered wall structure (lighter electron-tenuous inner plate lining with homogenous, darker, electron-dense plate layer) with central vesicle cavity indicated by *arrow* (Modified from Fig. 5J); (**l**) Higher magnification view of *S. macroreticulatum* wall ultrastructure from TEM analyses, *arrows* indicate inner vesicle plate boundary incisions (Modified from Fig. 5H); (**m**) FIB-EM cross-section of *D. delicata* with broad central vesicular cavity (*right arrow*), multilayering of vesicle walls, and smaller, compressed intravesicular wall chamber (*left arrow*); (**n**) FIB-EM cross-section of same *D. delicata* specimen shown in (**m**) with merged vesicle walls (no central cavity), disappearance of multilayering (appearing unilayered), and three chambers at outer edge of vesicle wall (*arrow*); (**o**) FIB-EM cross-section of same *D. delicata* specimen shown in (**m** and **n**) with central vesicular cavity, point of vesicle wall merging (*upper arrow*), and small, compressed intravesicular wall chamber (*lower arrow*)

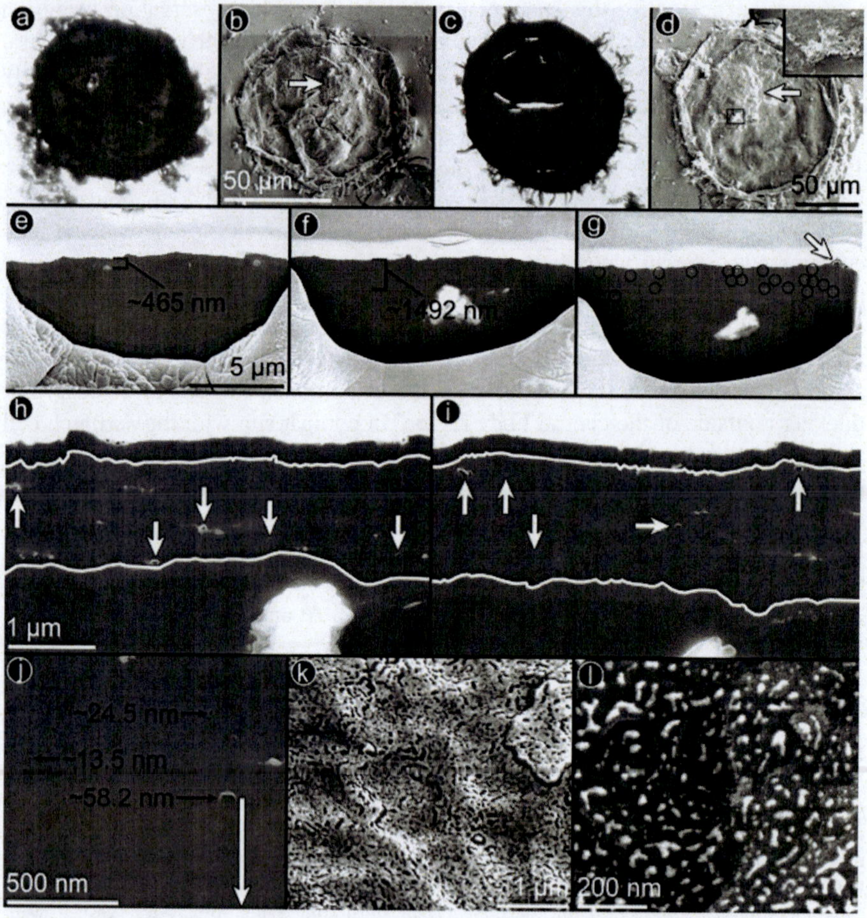

Fig. 13.6 Representative light, electron, and ion photomicrographs of FIB-EM nanotomography-prepared *Shuiyousphaeridium macroreticulatum* specimens highlighting central body ultrastructures. (**a–d**) Light, electron, and ion photomicrographs of *S. macroreticulatum* central bodies; (**e–g**) Cross-sections showing overall vesicle thickening corresponding to central body region; (**h–j**) High magnification views of central body ultrastructures with nanoporous structures (*arrows* in **h** and **i**), hypothetical outer vesicle wall boundaries (*white lines* in **h** and **i**), and nanopore (or potential tube) dimensions (*labels* in **j**); (**k–l**) FE-SEM photomicrographs of interpreted nanoporous structures on *Dictyosphaera delicata* vesicle surfaces (Kaufman and Xiao 2003, reproduced with permission). (**a**) Light micrograph of *S. macroreticulatum* with enclosed central body; (**b**) Corresponding ion micrograph of specimen in (**a**) with topographically distinct central body region (*arrow*); (**c**) Light micrograph of *S. macroreticulatum* with enclosed central body; (**d**) Corresponding ion micrograph of specimen in (**c**) with topographically distinct central body region (*arrow*) and conspicuous extravesicular process (*box*; inset in *upper right corner* shows secondary electron photomicrograph of extravesicular process, ultrastructures shown in Fig. 13.7); (**e**) Unilayered merged vesicle walls of specimen shown in (**c–d**), cross-section outside of central body region; (**f**) Unilayered merged vesicle walls and central body of specimen shown in (**c–d**), note overall thickening of slightly more than 1 μm; (**g**) Unilayered merged vesicle walls and central body of specimen shown in (**c–d**) with labeled nanoporous structures (*black circles*) and chamber at outer edge of vesicle wall (*arrow*); (**h**) Higher magnification view of central body region illustrating numerous nanoporous structures (*arrows*)

central body, up to 25 nanopores can be identified within an area of ~5 μm^2 (5,120 nm, the horizontal field width at 25k × system magnification × 1,000 nm, the total estimated central body thickness). These nanopores vary over an order of magnitude in size, ranging from 5.8 to 58.2 nm (although what appear to be larger amalgamations of nanopores can be hundreds of nanometers in dimension; Fig. 13.6j), and are notably similar to previously illustrated *D. delicata* nanopores observed on both the exterior and interior vesicle surfaces (Fig. 13.6k, l, modified from Kaufman and Xiao 2003). In one example, a relatively large nanopore exhibits what appears to be a tangential extension downward. This extension extends for 575.7 nm, nearly ten times the diameter of the nanopore, possibly indicating that this nanopore may be a cylindrical nanotube, or alternatively a single curtaining artifact (Fig. 13.6j). A large pore or enclosed grain of different density than the surrounding acritarch wall could create enough scatter of the *i*-beam to affect the local milling rate, thus resulting in the curtaining effect. However, while curtaining artifacts may be superficially similar to the proposed nanotubular structure, other large pores of similar or even greater diameters did not show the same effect.

The structure interpreted in *S. macroreticulatum* as an extravesicular process examined with FIB-EM nanotomography was located on the vesicle surface, near the outer boundary of the raised central body region. Given the site-specificity capability of this system, we sectioned through the length of this structure—a length of ~4.5 μm acquired in 19 cross-sections. FIB-EM nanotomography shows that this extravesicular process is composed of a complex three-dimensional sub-structure having multiple chambers, rather than being a hollow or cylindrical projection (see Fig. 13.7 for representative sections through the extravesicular process). Indeed, two distinct but conjoined structures were observed during analysis of this extravesicular process: a larger bulbous structure (Fig. 13.7a–l) 3.23 μm broad and ~3.0 μm long, present in 13 sections, and a smaller stalk-like columnar structure (Fig. 13.7d–p) 1.51 μm wide and ~3.5 μm long, present in 15 sections. The larger bulbous structure has a rounded central vacuole or cavity, up to 1,494 nm broad, and as many as eight smaller, radially adjacent, elliptical to reniform chambers (maximum dimension = 1,255 nm). These reniform chambers extend through the cross-sectioning images, continuing proximally to distally and perpendicular to the plane of the cross-sections. The smaller columnar structure is extended perpendicular to the plane of sectioning. It flanks and adjoins the larger bulbous structure, but it shows a different chamber arrangement—as many as four axially arranged, predominantly elliptical,

Fig. 13.6 (continued) predominantly confined to the central body, *white lines* show hypothetical individual vesicle wall thicknesses of ~250 nm; **(i)** 750 nm from cross-section shown in **(h)** showing persistence of nanopores (*arrows*) in central body region, *white lines* show hypothetical individual vesicle wall thicknesses of ~250 nm; **(j)** High magnification view of nanopores with maximum dimension labeled, nanopore with tangential extension (*downward vertical arrow* runs parallel to extension here, *horizontal arrow* in **(i)** indicates nanopore with tube-like extension); **(k)** Nanopores (*darker color*, lower electron density) on outer surface *D. delicata* specimen (Modified from Fig. 1D); **(l)** Nanopores (*lighter color*, higher electron density) on interior surface of *D. delicata* as viewed through pustular opening (Modified from Fig. 1F)

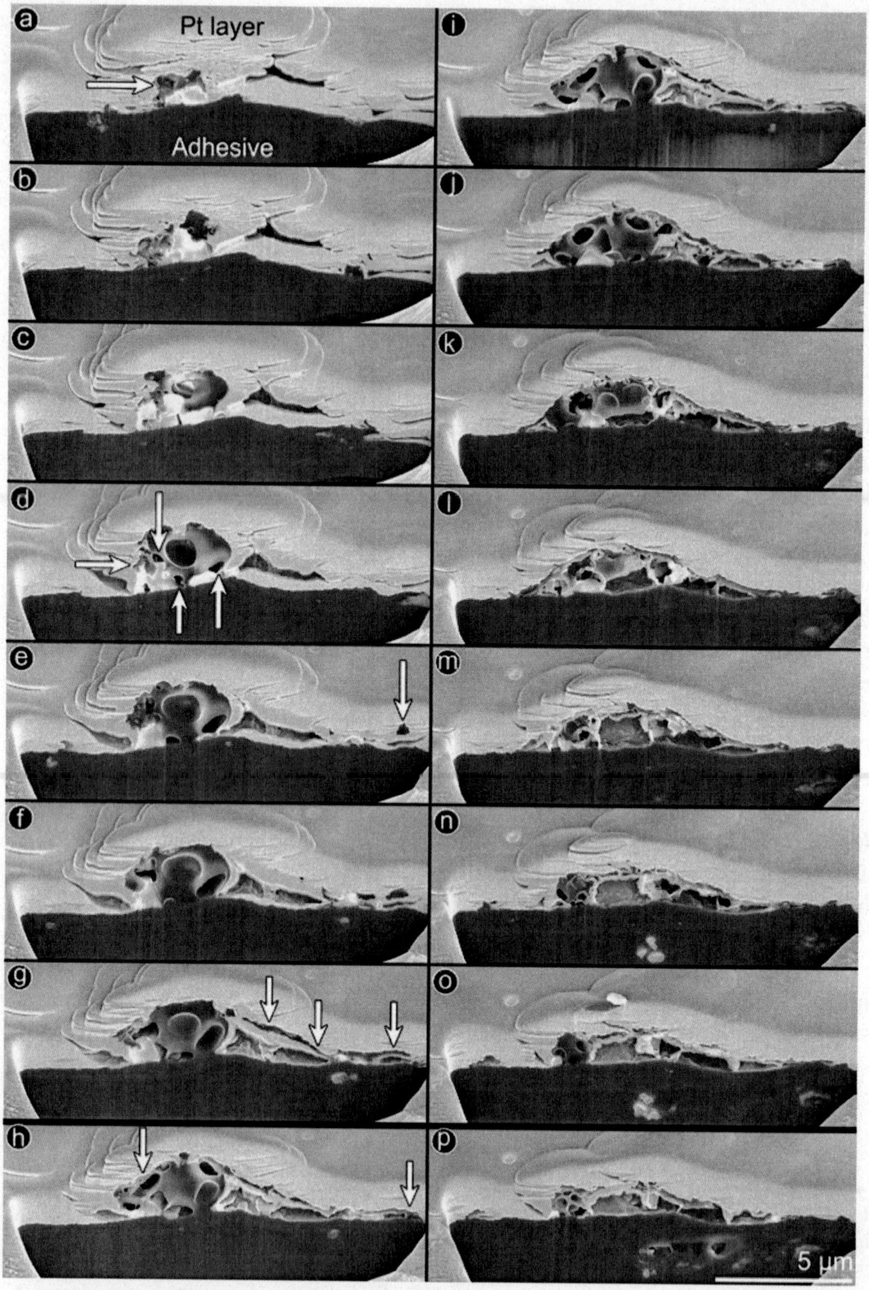

Fig. 13.7 FIB-EM nanotomographic sequential cross-section (250 nm cross-section increments) electron photomicrographs of *Shuiyousphaeridium macroreticulatum* extravesicular process and outer membrane shroud. **(a–l)** Sequential photomicrographs of bulbous extravesicular process termination; **(d–p)** Sequential photomicrographs of stalk-like portion of extravesicular process; **(e–p)** Sequential photomicrographs of outer membrane shroud. Scale bar in **(p)** applicable to all frames.

intermittent chambers (maximum dimension = 539 nm) that disappear and reappear suddenly in the cross-sections without tapering through the sequential sectioning. The bulbous and columnar structures are attached: the columnar structure joins with the bulbous structure above its midpoint, at only ~1.0 μm into the structure, and continues for an additional ~1.5 μm past the end of the bulbous structure. Both of these structures maintain attachment to the outer vesicle surface for nearly their entire lengths but show no evidence of being connected to vesicle cavity. Furthermore, a shroud of organic material covers this complex set of structures through nearly the entire length of the extravesicular process (Fig. 13.7e–p). This shroud begins to appear in the advancing sequential cross-sections and seems to be attached to the bulbous structure at nearly its midpoint, some 250 nm below the top of the columnar structure, as well as being loosely attached to the vesicle wall at about 6.0–6.5 μm laterally away from the edge of the extravesicular process (see arrow in Fig. 13.7e and right-most arrow in Fig. 13.7h). Sequential cross-sections of the shroud occur over a distance of ~4.5 μm, extending past the termination of the stalk-like portion of the extravesicular process by ~1.25 μm. At the termination of the stalk-like portion of the extravesicular process (only one section below, or 250 nm), this shroud becomes contiguous to, although not merging with, the vesicle surface. Conceptual diagrams of the full living view of a *S. macroreticulatum* vesicle, as well as tentative reconstructions of the process-structure both in-life and as sectioned from the fossil representative, are shown in Fig. 13.8. While these diagrams are speculative, they incorporate key characteristics observed in previously examined specimens and information elucidated by FIB-EM nanotomography and, therefore, provide a comprehensive interpretation of available surficial and ultrastructural data.

13.4.5 Case Study Discussion

FIB-EM nanotomography of the Ruyang Group specimens shown here and in Schiffbauer and Xiao (2009) illustrate numerous similarities to previously published ultrastructures detailed from TEM examination of conspecific and morphologically similar Ruyang (China) and Roper Group (Australia) acritarchs, such as multilayered vesicle walls comprised of reticulate organic plates (Javaux et al., 2001, 2003, 2004).

Fig. 13.7 (continued) **(a)** Beginning of bulbous structure (*arrow*) with bright Pt protective layer and dark adhesive labeled, vesicle wall corresponds to *lighter gray* between dark adhesive and bright Pt; **(b–d)** Sequential cross-sections of the bulbous structure showing central cavity (**c** and **d**), beginning of radially adjacent reniform chambers (3 *vertical arrows* in **d**), and stalk-like portion of process (*horizontal arrow* in **d**); **(e)** Beginning of outer membrane shroud (*arrow*) and increase in number of radially adjacent reniform chambers; **(f–g)** Growing continuity of outer membrane shroud (*arrows* in **g**); **(h–i)** Increase in number of reniform chambers, termination of central cavity, and more definition of stalk-like structure as well as connection to bulbous structure (*left arrow* in **h**), loose attachment of membrane shroud to vesicle surface (*right arrow* in **h**); **(j–k)** Reduction in size of bulbous structure; **(l)** Termination of bulbous structure; **(m–p)** Continuation of outer membrane shroud and stalk-like structure with axially-arranged intermittent chambers

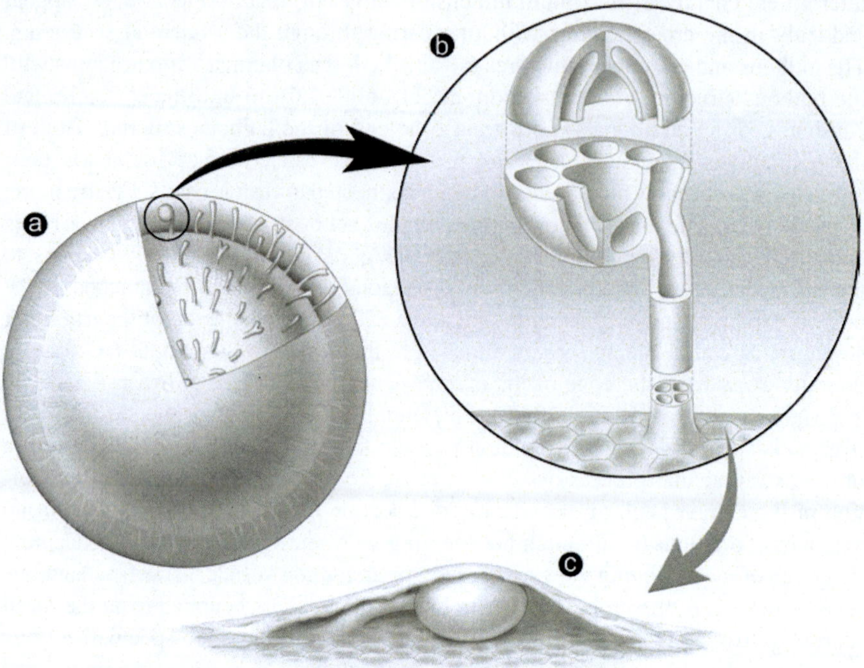

Fig. 13.8 Conceptualization of process ultrastructure and fossil observations. (**a**) Living full vesicle depiction with outer membrane, upper slice removed to show process distribution, morphologies, and vesicle surface reticulation; (**b**) Close-up view and hypothetical structural/ultrastructural representation of bulbous-tipped process as it may have appeared during the life of the acritarch; (**c**) Conceptual reconstruction of fossil bulbous-tipped process and outer membrane shroud as sectioned during FIB-EM analyses

Such similarities demonstrate that FIB-EM nanotomography is a useful method for examining microfossil ultrastructure.

The biological origin of the chambers within the vesicle walls of *D. delicata* is debatable, as they may be products of taphonomic or thermal alteration, similar to the amalgamation and separation of compressed vesicle walls in the same specimens. Nonetheless, it is important to note that similar chambers, described as rounded voids—maximum diameter of approximately 3 μm, nearly an order of magnitude larger than those observed here—have been observed from TEM ultrastructural examination of younger (~580 Ma) and less thermally mature (TAI = 2–3, Willman et al. 2006) leiosphere acritarchs from the Officer Basin, South Australia (Willman 2009). FIB-EM nanotomography of *S. macroreticulatum* specimens, on the other hand, bring to light complexities unknown from previous examinations. First, the nanoporous structures present in the central bodies may represent nanotubular structures of biological significance. These structures seem intriguingly similar to nanoporous structures observed on the vesicle wall surfaces (both exterior and interior) of *D. delicata* (Kaufman and Xiao 2003), but those observed here are restricted to the central body region and show no evidence of continuing through the outer vesicle walls. Unfortunately, the removal of 250 nm of material between sequential sections subtracts a fairly large volume of potential data from such specimens because of their small size. It is not possible to determine, therefore, whether these nanopores are a network of interconnected nanotubes. The two interconnected process-like structures, if they are indeed part of what is commonly recognized as an external process, are far more intricate than the hollow cylindrical processes as previously described from these fossils (Xiao et al. 1997). Their relatively consistent fusion with the vesicle wall may draw the process-interpretation into question, but this connection throughout nearly the entire length of the process is likely taphonomic—similar to the amalgamation of compressed vesicle walls in *D. delicata*. As is shown by our conceptualized representation of these structures, we interpret the larger bulbous structure to be a process-termination attached to the vesicle surface by the stalk-like columnar structure. Potential bulbous-tipped extravesicular processes, such as the one detailed here, were observed on a number of the analyzed specimens—although exceedingly rare in comparison to the typical furcated process terminations. We do not feel that the bulbous process termination is a common form, though it may be more frequent than observed if it is a structure easily damaged or disarticulated from the processes during maceration and handling. Another feature of note is the shroud of organic material that covers the majority of the process-like structure of *S. macroreticulatum*. We suggest that this shroud may represent a fragment of an outer membrane, which has previously been described as a thin organic veil supported by the extravesicular processes (Xiao et al. 1997). The surficial features observed by electron imaging, therefore, likely reflects this shroud rather than the extravesicular process observed in FIB-EM nanotomography. If so, the process is obscured underneath and, thus, protected by the outer membrane shroud. Further, one apparent detail that can be resolved from the process ultrastructural reconstruction is the lack of communication between the process—specifically the stalk in our conceptualized view—and the inner vesicle cavity. Finally, the segmented chambers viewed in the smaller columnar structure and the vacuolated larger bulbous structure illustrate a level of ultrastructural complexity that was not previously documented in Ruyang Group acritarchs and, therefore, should encourage further microfossil ultrastructural study via FIB-EM nanotomography.

13.5 Strengths of FIB-EM Instrument and Final Remarks

One of the chief strengths of FIB-EM rests on the instrument's capability to perform *in-situ* analyses of the micro- and ultrastructures as well as microchemistry of rock-hosted microfossils, with no need for fossil-extraction techniques. Further, FIB-EM examination of rock-embedded microfossils can demonstrate spatial and chemical (elemental) relationships of the fossil and the host rock on the μm-to-nm scale, which can provide insight into microfossil preservation and the fossilization processes. Although even with extracted microfossils, as shown in the above case study, the FIB-EM nanotomographic technique provides potential for gaining precise, site-specific information on the nature of particular microfossil structures and presents an important opportunity to examine the three-dimensionality and continuity of microfossil ultrastructures and microchemistries. As pointed out in Chap. 12, it is important to note that FIB-EM and TEM should ideally be used as complementary techniques in order to gain the best ultrastructural and microchemical perspectives of complex organisms. In addition, FIB-EM should be (and has been, as described in Sect. 13.3) used in conjunction with other investigative techniques, such as STXM, EELS, AEM, Raman spectroscopy, and micro-Fourier transform infrared (FTIR) spectroscopy (e.g., Marshall et al. 2005), in order to obtain an integrated ultrastructural and chemical analysis of fossil organisms. Finally, the field of paleobiology has much to gain by popularizing the use of FIB-EM, and the analyses it provides, as it holds promise for solving taxonomic puzzles that surround problematic microfossils, leading to new interpretations of microfossil structures or exceptionally-preserved macrofossil structures, providing taphonomic and preservational information on rock-hosted specimens, and furthering insights into the history and evolution of early life.

Acknowledgments Research was supported by NASA Exobiology and Evolutionary Biology Program, the Virginia Tech Institute for Critical Technology and Applied Science (VT-ICTAS), and the Virginia Space Grant Consortium. We thank J. McIntosh and S.R.F. McCartney (VT-ICTAS Nanoscale Characterization and Fabrication Laboratory) for technical assistance; P. Shinpaugh (VT-CAVE, Visualization and Animation Group of ICTAS) for assistance with the construction of 3D renderings; T.A. Dexter and P.J. Voice (Virginia Tech) for critical comments and suggestions for improvement, and J. Norton for conceptual graphical representations of the *Shuiyousphaeridium macroreticulatum* process. A shortened version of this paper dealing specifically with the case study presented in Sect. 13.4 is available in Schiffbauer (2009), published by SEPM Society for Sedimentary Geology; and thank the editorial staff, including Palaios co-editor S.T. Hasiotis, managing editor J. Hardesty, associate editor B. Granier, and two anonymous reviewers, for greatly improving the quality of the first version of this report. We are additionally grateful to reviewers Bradley De Gregorio, Emmanuelle Javaux, and J. William Schopf for constructive and thoughtful comments and suggestions that enhanced this newest adaptation of our work. FIB-EM analyses were conducted at the VT-ICTAS Nanoscale Characterization and Fabrication Laboratory under the supervision and operation of J.D. Schiffbauer and J. McIntosh.

References

Arouri KR, Greenwood PF, Walter MR (1999) A possible chlorophycean affinity of some Neoproterozoic acritarchs. Org Geochem 30:1323–1337

Arouri KR, Greenwood PF, Walter MR (2000) Biological affinities of Neoproterozoic acritarchs from Australia: microscopic and chemical characterisation. Org Geochem 31:75–89

Barber DJ (1970) Thin foils of non-metals made for electron microscopy by sputter-etching. J Mater Sci 5:1–8

Barber DJ (1999) Development of ion-beam milling as a major tool for electron microscopy. Microsc Anal 36:5–8

Batten DJ (1996) Palynofacies and petroleum potential. In: Jansonius J, McGregor DC (eds) Palynology: Principles and Applications: American Association of Stratigraphic Palynologists Foundation. College Station, Texas

Benzerara K, Menguy N, Guyot F, Vanni C, Gillet P (2005) TEM study of a silicate-carbonate-microbe interface prepared by focused ion beam milling. Geochim Cosmochim Acta 69(6):1413–1422

Bernard S, Benzerara K, Beyssac O, Menguy N, Guyot F, Brown GE Jr, Goffé B (2007) Exceptional preservation of fossil plant spores in high-pressure metamorphic rocks. Earth Planet Sci Lett 262:257–272

Bernard S, Benzerara K, Beyssac O, Brown GE Jr, Grauvogel Stamm L, Duringer P (2009) Ultrastructural and chemical study of modern and fossil sporoderms by scanning transmission X-ray microscopy (STXM). Rev Palaeobot Palynol 156:248–261

Cady SL, Noffke N (2009) Geobiology: evidence for early life on Earth and the search for life on other planets. GSA Today 19:4–10

Castaing R, Labourie P (1953) Examen direct des métaux par transmission au microscopie électronique. Compte Rendus Académie Sci 237:1330–1332

Cavalazzi B (2007) Chemotrophic filamentous microfossils from the Hollard Mound (Devonian, Morocco) as investigated by focused ion beam. Astrobiology 7:402–415

Cohen PA, Knoll AH, Kodner RB (2009) Large spinose microfossils in Ediacaran rocks as resting stages of early animals. Proc Natl Acad Sci USA 106:6519–6524

Evitt WR (1963) A discussion and proposals concerning fossil dinoflagellates, hystrichospheres, and acritarchs. Proc Natl Acad Sci USA 49: 158–164, 298–302

Floss C, Stadermann FJ, Bradley J, Dai ZR, Bajt S, Graham G (2004) Carbon and nitrogen isotopic anomalies in an anhydrous interplanetary dust particle. Science 303(5662):1355–1358

Giannuzzi LA, Prenitzer BI, Drown-MacDonald JL, Brown SR, Irwin RB, Stevie FA, Shofner TL (1998) Advances in the FIB lift-out technique for TEM specimen preparation: HREM lattice imaging. Microstruct Sci 26:249–253

Giannuzzi LA, Prenitzer BI, Drown-MacDonald JL, Shofner TL, Brown SR, Irwin RB, Stevie FA (1999) Electron microscopy sample preparation for the biological and physical sciences using focused ion beams. J Process Anal Chem 4:162–167

Golubic S, Barghoorn ES (1977) Interpretation of microbial fossils with special reference to the Precambrian. In: Flügel E (ed) Fossil algae: recent results and developments. Springer, Berlin

Groeber MA, Haley BK, Uchic MD, Dimiduk DM, Ghosh S (2006) 3D reconstruction and characterization of polycrystalline microstructures using a FIB–SEM system. Mater Charact 57: 259–273

Hagadorn JW, Xiao S, Donoghue PCJ, Bengtson S, Gostling NJ, Pawlowska M, Raff EC, Raff RA, Turner FR, Yin C, Zhou C, Yuan X, McFeely MB, Stampanoni M, Nealson KH (2006) Cellular and subcellular structure of Neoproterozoic embryos. Science 314:291–294

Heaney PJ, Vicenzi EP, Giannuzzi LA, Livi KJT (2001) Focused ion beam milling: a method of site-specific sample extraction for microanalysis of Earth and planetary materials. Am Mineral 86:1094–1099

Holzer L, Indutnyi F, Gasser P, Münch B, Wegmann M (2004) Three-dimensional analysis of porous BaTiO3 ceramics using FIB nanotomography. J Microsc 216:84–95

Holzer L, Münch B, Wegmann M, Gasser P, Flatt RJ (2006) FIB-nanotomography of particulate systems—Part I: particle shape and topology of interfaces. J Am Ceram Soc 89:2577–2585

Holzer L, Gasser P, Kaech A, Wegmann M, Zingg A, Wepf A, Münch B (2007) Cryo-FIB-nanotomography for quantitative analysis of particle structures in cement suspensions. J Microsc 227:216–228

Inkson BJ, Mulvihill M, Möbus G (2001) 3D determination of grain shape in a FeAl-based nanocomposite by 3D FIB tomography. Scr Mater 45:753–758

Javaux EJ, Marshal CP (2006) A new approach in deciphering early protist paleobiology and evolution: combined microscopy and microchemistry of single Proterozoic acritarchs. Rev Palaeobot Palynol 139:1–15

Javaux EJ, Knoll AH, Walter MR (2001) Morphological and ecological complexity in early eukaryotic ecosystems. Nature 412:66–69

Javaux EJ, Knoll AH, Walter MR (2003) Recognizing and interpreting the fossils of early eukaryotes. Orig Life Evol Biosph 33:75–94

Javaux EJ, Knoll AH, Walter MR (2004) TEM evidence for eukaryotic diversity in mid-Proterozoic oceans. Geobiology 2:121–132

Javaux EJ, Marshall CP, Bekker A (2010) Organic-walled microfossils in 3.2-billion-year-old shallow-marine siliciclastic deposits. Nature 463:934–938

Karnovsky MJ (1965) A formaldehyde-glutaraldehyde fixative of high osmolality for use in electron microscopy. J Cell Biol 27:137A

Kaufman A, Xiao S (2003) High CO_2 levels in the Proterozoic atmosphere estimated from analyses of individual microfossils. Nature 425:279–282

Kempe A, Wirth R, Altermann W, Stark RW, Schopf JW, Heckl WM (2005) Focussed ion beam preparation and in situ nanoscopic study of Precambrian acritarchs. Precambrian Res 140(1–2):36–54

Knoll AH, Barghoorn ES (1975) Precambrian eukaryotic organisms: a reassesment of the evidence. Science 190:52–54

Krohn VE (1961) Progress in astronautics and rocketry. Academic, New York

Kubis AJ, Shiflet GJ, Dunn DN, Hull R (2004) Focused ion-beam tomography. Metall Mater Trans A 35A:1935–1943

Lee MR, Bland PA, Graham G (2003) Preparation of TEM samples by focused ion beam (FIB) techniques: applications to the study of clays and phyllosilicates in meteorites. Mineral Mag 67(3):581–592

Lorensen WE, Cline HE (1987) Marching cubes: a high resolution 3D surface construction algorithm. ACM SIGGRAPH Comput Graph 21:163–169

Luo Q, Zhang Y, Sun S (1985) The eukaryotes in the basal Changcheng system of Yanshan ranges. Acta Geol Sin 1985(1):12–16

Marko M, Hsieh C, Schalek R, Frank J, Mannella C (2007) Focused-ion-beam thinning of frozen-hydrated biological specimens for cryoelectron microscopy. Nat Meth 4:215–217

Marshall CP, Javaux EJ, Knoll AH, Walter MR (2005) Combined micro-Fourier transform infrared (FTIR) spectroscopy and micro-Raman spectroscopy of Proterozoic acritarchs: a new approach to palaeobiology. Precambrian Res 138:208–224

Meng F, Zhou C, Yin L, Chen Z, Yuan X (2005) The oldest known dinoflagellates: morphological and molecular evidence from Mesoproterozoic rocks at Yongji, Shanxi Province. Chin Sci Bull 50:1230–1234

Moczydłowska M, Willman S (2009) Ultrastructure of cell walls in ancient microfossils as a proxy to their biological affinities. Precambrian Res 173:27–38

Moldowan JM, Talyzina NM (1998) Biogeochemical evidence for dinoflagellate ancestors in the early Cambrian. Science 281:1168–1170

Moreau JW, Sharp TG (2004) A transmission electron microscope study of silica and kerogen biosignatures in 1.9 Ga Gunflint microfossils. Astrobiology 4:196–209

Müller WEG, Eckert C, Kropf K, Wang X, Schloßmacher U, Seckert C, Wolf SE, Tremel W, Schröder HC (2007) Formation of giant spicules in the deep-sea hexactinellid *Monorhaphis chuni* (Schulze 1904): electron-microscopic and biochemical studies. Cell Tissue Res 329:363–378

Oehler DZ (1976) Transmission electron microscopy of organic microfossils from the late Precambrian Bitter springs formation of Australia: techniques and survey of preserved ultrastructure. J Paleontol 50(1):90–106

Oehler DZ (1977) Pyrenoid-like structures in late Precambrian algae from the Bitter springs formation of Australia. J Paleontol 51(5):885–901

Oehler DZ (1978) Pyrenoid-like structures in late Precambrian algae from the Bitter springs formation of Australia. J Paleontol 51(5):885–901

Phaneuf MW (1999) Applications of focused ion beam microscopy to materials science specimens. Micron 30:277–288

Prenitzer BI, Urbanik-Shannon CA, Giannuzzi LA, Brown SR, Irwin RB (2003) The correlation between ion beam/material interactions and practical FIB specimen preparation. Microsc Microanal 9(3):216–231

Quekett J (1848) Microtomes and microtome knives, A practical treatise on the use of the microscope. Hippolyte Bailliere, London

Rubanov S, Munroe PR (2004) FIB-induced damage in silicon. J Microsc 214(3):213–221

Sandberg PA, Hay WW (1967) Study of microfossils by means of the scanning electron microscope. J Paleontol 41:999–1001

Schiffbauer JD, Xiao S (2009) Novel application of focused ion beam electron microscopy (FIB-EM) in preparation and analysis of microfossils ultrastructures: a new view of complexity in early eukaryotic organisms. Palaios 24:616–626

Schiffbauer JD, Yin L, Bodnar RJ, Kaufman AJ, Meng F, Hu J, Shen B, Yuan X, Bao H, Xiao S (2007) Ultrastructural and geochemical characterization of Archean-Paleoproterozoic graphite particles: implications for recognizing traces of life in highly metamorphosed rocks. Astrobiology 7(4):684–704

Schopf JW (1992) Proterozoic prokaryotes: affinities, geologic distribution, and evolutionary trends. In: Schopf JW, Klein C (eds) The Proterozoic biosphere: a multidisciplinary study. Cambridge University Press, Cambridge

Schopf JW, Oehler DZ (1976) How old are the eukaryotes? Science 193:47–49

Schulz HN, Brinkhoff T, Ferdelman TG, Mariné MH, Teske A, Jørgensen BB (1999) Dense populations of a giant sulfur bacterium in Namibian shelf sediments. Science 284:493–495

Seligman AM, Wasserkrug HL, Hanker JS (1966) A new staining method for enhancing contrast of lipid-containing membranes and droplets in osmium tetroxide-fixed tissue with osmiophilic thiocarbohydrazide (TCH). J Cell Biol 30:424–432

Servais T (1996) Some considerations on acritarch classification. Rev Palaeobot Palynol 93:9–22

Sun S, Zhu S (2000) Paleoproterozoic eukaryotic fossils from northern China. Acta Geol Sin 74(2):116–122

Talyzina NM, Moczydłowska M (2000) Morphological and ultrastructural studies of some acritarchs from the Lower Cambrian Lükati Formation, Estonia. Rev Palaeobot Palynol 112:1–21

Vander Voort GF (ed) (2004) Chemical and electrolytic polishing. In: ASM handbook: metallography and microstructures. ASM International, Materials Park

Vidal G (1988) A palynological preparation method. Palynology 12:215–220

Weiss BP, Vali H, Baudenbacher FJ, Kirschvink JL, Stewart ST, Shuster DL (2002) Records of an ancient Martian magnetic field in ALH84001. Earth Planet Sci Lett 201(3–4):449–463

Westall F, de Ronde CEJ, Southam G, Grassineau N, Colas M, Cockell C, Lammer H (2006) Implications of a 3.472–3.333 Gyr-old subaerial microbial mat from the Barberton greenstone belt, South Africa for the UV environmental conditions on the early Earth. Phil Trans Roy Soc B 361:1857–1875

Willman S, Moczydłowska M, Grey K (2006) Neoproterozoic (Ediacaran) diversification of acritarchs: A new record from the Murnaroo 1 drillcore, eastern Officer Basin, Australia: Review of Palaeobotany and Palynology 139:17–39

Willman S, Moczydłowska M (2007) Wall ultrastructure of an Ediacaran acritarch from the Officer Basin, Australia. Lethaia 40:111–123

Willman S (2009) Morphology and wall ultrastructure of leiosphaeric and acanthomorphic acritarchs from the Ediacaran of Australia. Geobiology 7:8–20

Wirth R (2004) Focused Ion Beam (FIB): A novel technology for advanced application of micro- and nanoanalysis in geosciences and applied mineralogy. Eur J Mineral 16:863–876

Wirth R (2009) Focused Ion Beam (FIB) combined with SEM and TEM: advanced analytical tools for studies of chemical composition, microstructure and crystal structure in geomaterials on a nanometre scale. Chem Geol 261:217–229

Xiao S, Schiffbauer JD (2008) Microfossil phosphatization and its astrobiological implications. In: Seckbach J, Walsh M (eds) From fossils to astrobiology: cellular origin, life in extreme habitats and astrobiology. Springer, New York

Xiao S, Knoll AH, Kaufman AJ, Yin L, Zhang Y (1997) Neoproterozoic fossils in Mesoproterozoic rocks? Chemostratigraphic resolution of a biostratigraphic conundrum from the North China platform. Precambrian Res 84:197–220

Xiao S, Zhang Y, Knoll AH (1998) Three-dimensional preservation of algae and animal embryos in a Neoproterozoic phosphorite. Nature 391:553–558

Xiao S, Hagadorn JW, Zhou C, Yuan X (2007) Rare helical spheroidal fossils from the Doushantuo Lagerstätte: Ediacaran animal embryos come of age? Geology 35:115–118

Yan Y (1982) *Schizofusa* from the Chuanlinggou Formation of Changcheng System in Jixian County. Bull Tianjin Inst Geol Min Res 6:1–7

Yan Y (1991) Shale-facies microflora from the Changzhougou Formation (Changcheng System) in Pangjiapu Region, Hebei, China. Acta Micropalaeontol Sinica 8(2):183–195

Yan Y (1995) Shale facies microfloras from lower Changcheng System in Kuancheng, Hebei, and comparison with those of neighboring areas. Acta Micropalaeontol Sinica 12(4):349–373

Yin L, Zhu M, Knoll AH, Yuan X, Zhang J, Hu J (2007) Doushantuo embryos preserved inside diapause egg cysts. Nature 446:661–663

Zang W-L (2007) Deposition and deformation of late Archean sediments and preservation of microfossils in the Harris Greenstone Domain, Gawler Craton, South Australia. Precambrian Res 156:107–124

Zhang Z (1986) Clastic facies microfossils from the Chuanlinggou Formation (1800 Ma) near Jixian, North China. J Micropalaeontol 5(2):9–16

Zhang Z (1997) A new Palaeoproterozoic clastic-facies microbiota from the Changzhougou Formation, Changcheng Group, Jixian, north China. Geol Mag 134(2):145–150

Chapter 14
Reconstructing Deep-Time Biology with Molecular Fossils

Christian Hallmann, Amy E. Kelly, S. Neal Gupta, and Roger E. Summons

Contents

C. Hallmann (✉) and R.E. Summons
Department of Earth, Atmospheric and Planetary Sciences, Massachusetts Institute of Technology, 77 Massachusetts Ave, Cambridge, MA 02139, USA
e-mail: hallmann@mit.edu; rsummons@mit.edu

A.E. Kelly
School of Earth and Space Exploration, Arizona State University, 85281-1404, Tempe, AZ, USA
e-mail: kellya@alum.mit.edu

S.N. Gupta
Department of Earth, Atmospheric and Planetary Sciences, Massachusetts Institute of Technology, 77 Massachusetts Ave, Cambridge, MA 02139, USA
and
Indian Institute of Science Education and Research, Mohali, India
e-mail: ngupta@ciw.edu

M. Laflamme et al. (eds.), *Quantifying the Evolution of Early Life*, Topics in Geobiology 36,
DOI 10.1007/978-94-007-0680-4_14,

Abstract Molecular fossils can be a useful source of information about past organismic physiology and an effective tool for reconstruction of paleoenvironmental conditions. This is particularly the case with Precambrian sediments, which only rarely contain visible fossils. However, since these most ancient of rocks are frequently thermally altered and contain low contents of organic matter, analytical uncertainties and concerns surround the syngeneity of biomarkers. Analysts must be vigilant in recognizing and avoiding contamination by hydrocarbons derived from more recent organisms. Here we present an overview of the current state of Precambrian molecular paleontology and provide a theoretical and practical entrée into the field of organic geochemistry. We discuss improved strategies and analytical methodologies including a guide to interpreting data as well as recognizing and avoiding contamination.

Keywords Organic geochemistry • Molecular palaeontology • Lipids • Biomarkers • Deep time • Precambrian • Evolution

14.1 Introduction

14.1.1 Deep Time Phylogeny

Body fossils have been studied for many years with the goal of understanding past biota, their evolutionary emergence, disappearance, paleogeographical distributions and the reasons behind these phenomena. However, only a vanishingly small fraction of all organisms is typically preserved in sedimentary rocks. Furthermore, a well-known issue concerns preservation biases, depending on location, sedimentology, local redox conditions and most of all, the composition of organisms (Briggs 2003). In contrast to very resistant biotic remains such as chitin, e.g. from insects, or lignin from vascular plants, the majority of marine and terrestrial primary producers – protists and bacteria – bear a minuscule fossilization potential. Even when preserved, morphology only allows for a tentative phylogenetic classification of microbes.

The redox structure of the surface of the Earth, as we know it today, had been largely established by the beginning of the Phanerozoic eon (~542 Ma BP) (e.g. Des Marais et al. 1992; Canfield and Teske 1996). Before reaching this state however, Earth's ocean and atmosphere system witnessed a succession of biogeochemical transitions. The Hadean, Archean and Proterozoic eons, comprising more than 85% of Earth's history, saw not only the origin of life, but also key physiological innovations such as the advent of anoxygenic photosynthesis, oxygenic photosynthesis and a progressive increase in the complexity of organisms. There is no doubt that Earth's biota played a significant role in shaping our planet through a complex reciprocal maze of physical, chemical and biological processes. It is our aim to improve the understanding of the *when* and the *why* of these events. In the near-absence of macroscopic morphological remains, traditional paleontology is limited to the analysis of acritarchs, resistant organic microfossils and probable remains of microalgae and primitive metazoa that diversify throughout time, but which evade conclusive phylogenetic classification.

Physiology is encoded into an organism's genome. The sequence of nucleotides in deoxyribonucleic acid (DNA) determines not only morphology, which forms the basis of traditional paleontology, but also biochemical pathways, the proteome (sum of all proteins), and metabolome (sum of all intermediary metabolites, including lipids). Highly conserved stretches of DNA coding for ribosomal ribonucleic acid (rRNA) have been shown to be a reliable tool by which the phylogenetic affinities of all extant organisms can be deduced. Further, the properties of DNA, RNA and proteins of known organisms suggest that all life on Earth has descended from a last universal common ancestor (LUCA). Hence, a 'Tree of Life' depicting these relationships was first constructed by Woese and colleagues (Woese et al. 1990). In this respect, DNA is almost *the* perfect fossil. Unfortunately however, DNA is unstable in the environment. Even though it can be preserved in sediments for several thousands of years under exceptional conditions (Coolen and Overmann 2007), DNA typically decays very rapidly after cell death and bears no long-term fossil preservation potential. The individual DNA fragments progressively lose taxonomic value and this is, in some ways, analogous to the breakdown processes that act on entire organisms. While partially or wholly complete they can carry morphological information, but after bacterial decomposition, only the most recalcitrant molecular fragments remain. In 1936, Alfred Treibs reported the presence of tetrapyrroles in a Triassic shale (Treibs 1936), thereby suggesting that certain molecules necessary for specific physiological processes, and thus carrying information (in Treibs' case, photosynthesis), can be preserved for millions of years. This finding was the foundation of modern biomarker geochemistry and molecular paleobiology (Peterson et al. 2007).

14.1.2 Sedimentary Organic Matter

An organic compound is classified as a biomarker when its structure can be confidently linked to a molecular precursor that is present in a single organism, or in a limited number of extant organisms (Peters et al. 2005a, b). Assuming the conservation of biochemical pathways over time, biomarkers in ancient sediments can be used as molecular fossils to reconstruct the composition of past microbial communities and, by inference, the paleoenvironmental conditions favored by them (e.g. Gaines et al. 2009).

14.1.2.1 Organic Matter Preservation

While biomarkers can, in theory, be preserved for billions of years, this requires a set of specific conditions to be met during deposition of their precursor organisms within sediments. Deceased organisms contain a vast array of different compounds, ranging from small functional molecules such as fatty acids, amino acids, alcohols, and sterols, to larger macromolecules that include triglycerides, phospho- and glycolipids, proteins and nucleic acids (Killops and Killops 2005). The initial disassembling of biomacromolecules into smaller fragments via enzymatic, photolytic and other chemical processes takes place in the water column or upper sediment (Fig. 14.1). The decomposition pathways and the fate of the products depend on the environmental conditions. For example, under oxic conditions, the majority of compounds will be completely respired to CO_2 and thereby lost forever (e.g. Canfield 1994). A small fraction of the inventory of compounds may be bound to the surfaces of clay particles and inaccessible to microbial attack (Hedges and Keil 1995). Another small fraction escapes after major modification.

Under specific depositional conditions, the preservation of organic matter (OM) may be enhanced. The most important factor herein is the concentration of molecular oxygen. Anoxic (<0.1‰ dissolved O_2) and suboxic (0.1–0.5‰ O_2) conditions are hostile to most larger eukaryotic detritivores and decomposers, which account for a significant proportion of OM remineralization and also for the aeration of anoxic sediments by their burrowing (bioturbation) activities (Revsbech et al. 1980). Furthermore, such conditions preclude the oxidation of organic matter by aerobic microbes. Anaerobic microbial respiration is energetically less efficient (Konhauser 2007) and thus, generally speaking, consumes organic matter at a slower pace. However, it should be noted that anaerobic metabolism continues as long as there is a supply of electron acceptors and anaerobic microbes can thrive in deep subsurface sediments (Gold 1992; Schippers et al. 2005; Biddle et al. 2006), where they only seem to be limited by the upper temperature limit for life of ~110°C (Wilhelms et al. 2001; Head et al. 2003; Hallmann et al. 2008). While sediments readily become anoxic with depth, an anoxic water column can also be created when the rate of supply of organic matter exceeds the re-supply of molecular oxygen. This can occur during and after seasonal algal blooms or when other environmental parameters, such as nutrient availability, promote the expansion of primary producers.

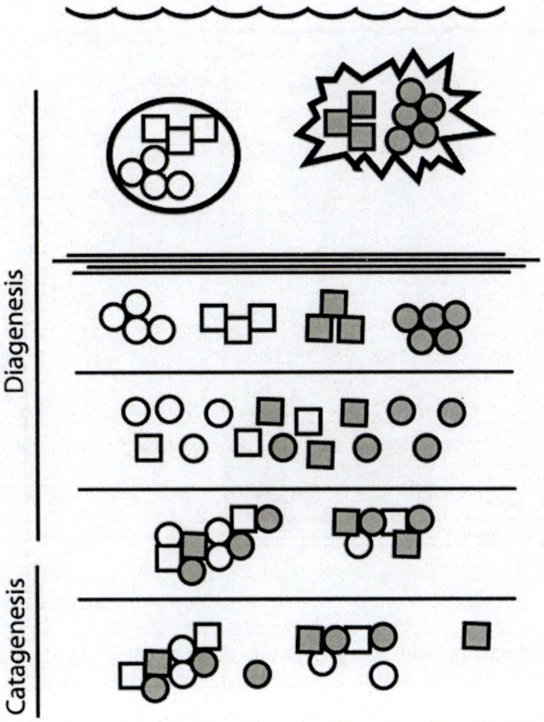

Fig. 14.1 Origin and fate of sedimentary organic matter. (*I*) Two hypothetical organisms (*circle* and *spike*) containing biomacromolecules (1x poly-circle, 1x poly-block, each) expire in the water column and settle with other sinking particles. Cell lysis in the water column (*I*) and upper sediments (*II*) releases the still-intact large biomacromolecules. (*III*) Bacterial enzymatic attack will break these polymers into their constituent monomers. Most of this sedimentary biological material is recycled by heterotrophic bacteria, respired as CO_2 and lost to the record. (*IV*) Some of the surviving compounds recombine and, together with undegraded biopolymers, form a modified macromolecular structure, kerogen. Cross-linking by incorporation of reduced sulfur compounds can be an important component of the process but many details remain to be elucidated (de Leeuw et al. 2006; Vandenbroucke and Largeau 2007). Note, however, that some of the individual building blocks that were present in the living organisms are incorporated intact. The macromolecular character of kerogen causes it to be immobile and insoluble in organic solvents. (*V*) Constant burial brings the sedimentary kerogen into a realm of elevated pressures and temperatures and small molecules are cleaved from the main network. These molecules, which are mobile and soluble in organic solvents are termed bitumen or extractable organic matter (EOM). While many of the bitumen components are so altered as to bear no unequivocal similarity their precursors, a select few can be directly linked to specific biological precursors. These are termed 'biomarkers' or 'molecular fossils'

Depending on the intensity and persistence of reducing conditions and, in the case of marine environments, the presence of sulfide, organic compounds will experience characteristic modes of alteration (Fig. 14.2) that can often be used to deduce the depositional conditions. A second important factor in the preservation of OM is the presence of clay minerals. Labile OM is typically enriched in shales and

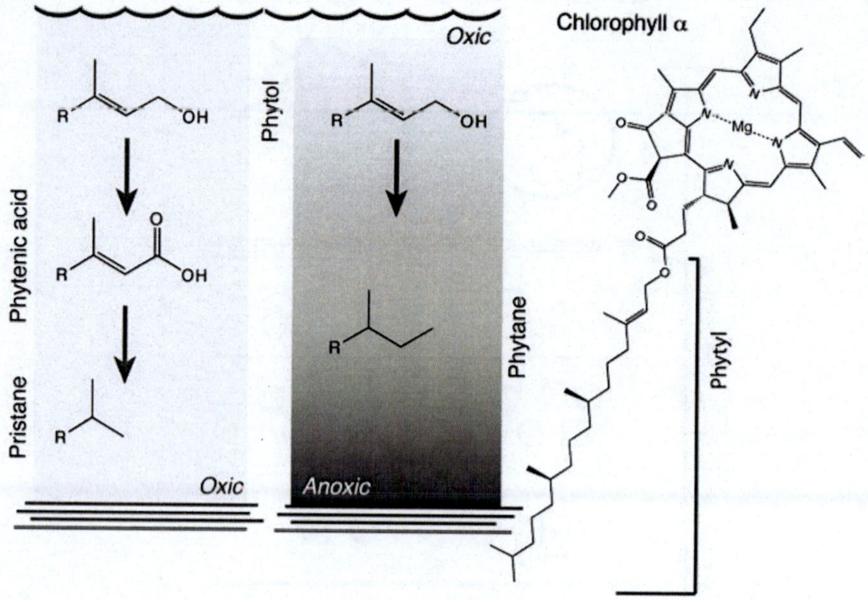

Fig. 14.2 Depositional environment. The phytyl side chain of chlorophyll molecules is readily cleaved to become the C_{20} alcohol phytol. Its subsequent fate depends on redox conditions of the depositional environment. Under aerobic conditions (high Eh) it undergoes oxidation to form phytenic acid, which can decarboxylate to form the C_{19} hydrocarbon pristene and its saturated counterpart pristane. Under anoxic and low Eh conditions, however, phytol is preferentially reduced to the C_{20} hydrocarbon phytane. Pristane (Pr) and phytane (Ph) both have a geological preservation potential. The Pr/Ph ratio of bitumens recovered from ancient sediments can thus be employed as an indicator of depositional redox conditions (Didyk et al. 1978)

carbonates, but rarely in sand- or siltstones. It was shown that sorption of organic matter to mineral surfaces in marine sediments stabilizes the molecules, thereby significantly slowing remineralization rates (Keil et al. 1994; Hedges and Keil 1995), and that the majority of labile organic matter in marine environments is protected from degradation by the inorganic matrix of sinking particles (Hedges et al. 2001). The proportion of sedimentary clay minerals thus plays a significant role in the survival and preservation of organic matter.

The sedimentary preservation of individual labile organic compounds is significantly enhanced by polymerization reactions that fuse small individual molecules together to form large molecular networks. If these networks are extensive enough, the product is insoluble and is termed kerogen. Although the molecular reactions involved in kerogen formation are manifold and their individual relevance is still debated, it has been found that the presence of reduced sulfur species often plays a significant role (e.g. Kok et al. 2000; Schouten et al. 1993; Hebting et al. 2006). The presence of abundant H_2S and HS^- in sediments or anoxic bottom waters can thus greatly enhance the preservation of OM. Reduced sulfur species (RSS) are formed during the process of bacterial sulfate reduction (BSR) and their concentrations are determined by sulfate concentrations and the relative rates of supply of organic matter and iron. When iron supply is low, such as when there is limited run-off from the land, the reaction between organic matter and RSS is the dominant process.

In the context of factors enhancing the preservation of labile sedimentary organic matter, it should be noted that during a significant part of Earth history both atmospheric oxygen (e.g. Holland 1999; Eriksson et al. 2004) and, consequently, marine sulfate levels were significantly lower than today. Furthermore, the assemblage of clay minerals supplied to the ocean was proposed to have been different before the rise of terrestrial plants (Kennedy et al. 2006) although it is important to note that there are alternative views (Tosca et al. 2010). Changes in the weathering rates of terrestrial rocks are an additional factor to consider (Berner 1991).

14.1.2.2 Kerogen Formation

Those organic compounds that evade microbial degradation polymerize and reorganize to form a macromolecular network termed kerogen (e.g. Gupta et al. 2007; Vandenbroucke and Largeau 2007; Fig. 14.1). Kerogen is less affected by microbial degradation than individual labile compounds and, provided that oxygen continues to be excluded and the temperature remains low, is relatively stable in the sedimentary environment. On the other hand, this relative stability also makes kerogen a difficult material to analyze. It does not dissolve in organic solvents and generally resists chemical attack. In its macromolecular state, one is limited to measurements of the bulk stable isotopic values and solid-state spectroscopy. In the absence of degradation techniques, no individual molecular fossils can be identified in kerogen – even if they are present, somewhere in this molecular network. Molecular analyses thus rely on the release of compounds from the kerogen (cracking), which occurs under elevated temperatures and pressures in nature, but can also be simulated in the laboratory.

14.1.2.3 Bitumen Expulsion

During progressive burial, sedimentary kerogen experiences increasingly high temperatures and pressures. To retain thermodynamic equilibrium, chemical bonds will be broken and individual compounds are cracked from the kerogen structure (Hunt 1995). These released compounds constitute the bitumen (Fig. 14.1). In contrast to the kerogen, bitumen is both mobile and soluble in organic solvents. Continuously increasing temperatures during burial will allow increasingly stronger bonds to be broken and this leads to a progressive shift in the chemical composition of both kerogen and bitumen. The content of bitumen continuously increases relative to kerogen and this cracking process, termed catagenesis, is generally accompanied by the formation of gaseous as well as liquid hydrocarbons.

Bitumen release is paralleled by a volumetric expansion, which entails a localized increase in internal pressure in the rock matrix and forces the bitumen to migrate away from the small, pressurized, pockets of generation. If sufficient bitumen is generated, which is commonly the case with highly OM-rich shales, the overpressuring will induce micro-fractures in the rock, and allow the bitumen to migrate away from its host sediment – a process termed primary migration or expulsion. Once expelled from the source rock, this bitumen becomes petroleum. Being commonly lighter than water, the petroleum starts a buoyancy-driven upward migration, termed secondary migration, until it is trapped by an impervious barrier or, alternatively, lost to the surface. Even after this petroleum has left its source rock, small amounts of bitumen will remain trapped. In this case, or when the volume of bitumen generated is insufficient to form a separate migrating liquid phase, it can be artificially recovered from the rock by extraction with organic solvents. Chemical analysis of this bitumen is the most common method for reconstructing the biotic input to sedimentary rocks.

14.1.2.4 Fluid Inclusions

Petroleum that migrates away from a source rock may eventually be trapped in a porous formation from which it cannot escape, or lost by seepage to the surface (Magoon and Dow 1994). At any point in this process, migrating petroleum can be occluded by secondary mineral growth to form fluid inclusions (Bhullar et al. 1999; Jensenius and Burruss 1990; George et al. 2004; Munz 2001; Volk et al. 2002). Being shielded from later fluid flow, such fluid inclusions are thought to form excellent time capsules, capable of preserving ancient bitumens uncontaminated by other sources. Consequently, much geochemical work has been dedicated to the study of their chemical compositions (e.g. Volk et al. 2005; Dutkiewicz et al. 2003, 2004, 2006, 2007). It should be noted however, that the OM content of fluid inclusions reflects the timing of mineral growth (primary inclusions) or mineral healing (secondary inclusions) and not the timing of bitumen generation and migration. This is distinct from the time of the original organic matter deposition. Even if one can reconstruct when the inclusion formed, the occluded bitumen must first be correlated to a sedimentary source unit for a greater understanding of the origin of the organic matter (Munz 2001).

14.2 Sampling Strategies

14.2.1 Precambrian Sedimentary Basins

Studies of ancient ecosystems through the analysis of preserved lipids are primarily limited by the conservation of Precambrian sediments. Plate tectonics run a constant recycling scheme and, with increasing age, the chances for survival of any sedimentary unit diminish. Marine basins and shelf sediments are especially prone to subduction as the underlying oceanic lithosphere continues its slow but relentless motion away from the spreading centers. Increasing density caused by the cooling of oceanic lithosphere will at some point induce spontaneous plate subduction and convert a passive continental margin into a subduction zone (Nicolas 1995). Oceanic crust is rarely older than 200 Ma and oceanic sediments have little chance for survival unless they become thrust up on to continental crust during ocean closure. Intracratonic basins sometimes escape orogenic events and are more readily preserved. Most Precambrian sediments that are presently available for study were originally deposited and preserved on stable cratons.

Being available for study merely by survival and escaping tectonic recycling is however only part of the story. The thermal overprint any given sedimentary stratum has experienced will affect the preservation state of the OM. Temperatures increase progressively with burial and lead to a increase in thermal maturity of the OM. More bitumen is generated from the kerogen and, at temperatures in excess of 200°C, much of this is eventually cracked to gas (Hunt 1995). In parallel, the conversion of higher hydrocarbons to methane is accompanied by loss of hydrogen and concomitant decrease in the hydrogen content (i.e. increasing aromaticity) of the residual kerogen (Hunt 1995). Highly mature bitumens are frequently dominated by low-molecular weight alkanes and some aromatics, whereas larger cyclic components such as steranes and hopanes have been largely destroyed (Brocks and Summons 2003, and references therein). It is therefore important to understand the regional tectonic and metamorphic history of any given region when sampling for biomarker analysis is envisaged.

Three aspects of the maturation process can be distinguished. Burial typically leads to higher temperature and pressure conditions and, as a very crude rule of thumb, one can assume deeper sediments in a simple tectonic setting to be more mature than shallower ones. Regional metamorphic events typically affect rocks on a larger scale, destroying the biomarker record of whole basins. Small-scale tectonic events and magmatic intrusions, on the contrary, affect small areas, whereas other parts of a basin might be unaffected. A thorough knowledge of the regional geology is thus an invaluable and important part of any sampling effort.

14.2.2 Outcrop and Core

Outcrop samples are usually easier to obtain but, with that benefit, there are also a wealth of problems. Most soils are known to be full of microorganisms and circulating meteoric waters can entrain their organic metabolites, as well as organic compounds generated from decaying soil organic matter. Also, deeper oxic waters can sustain an active microbial population and surface weathering can sometimes extend tens to hundreds of meters. Fungi can pervasively enter rocks (Sterflinger 2000) and, being eukaryotes, cause a modern addition of steroids (Summons et al. 2001). When outcrop samples are the only choice, they should optimally be taken from fresh exposures such as quarries, mines, road-cuts or from sites of recent exhumation. Only intact rocks should be sampled, and subjected to both inside/outside comparisons (Brocks et al. 2008) as well as comparisons of free- versus mineral-occluded bitumen (Sherman et al. 2007) as described below.

Core material is considered far superior to outcrop. When recovered from deep drill holes, no weathering or oxidation should have affected the rock. However, core samples pose their own set of problems. First of all, they are usually drilled for a purpose other than scientific research (e.g. mineral or petroleum exploration) and are typically more difficult to obtain from the preferred locations. When stored in a core repository, sampling can be requested but must be accompanied by a sound rationale. If sampling approval is granted, it is usually restricted to the taking of small pieces of the core so that the bulk is retained in the archive. Gaining insight into the purposes for which a core was drilled, what intervals were recovered, how the core was cut, where and how the rock material was originally stored and the availability of well logs and other metadata are important considerations. Storage for longer time periods might entail the deposition of organics onto the core, the biodegradation and oxidation of bitumen (Bennett and Larter 2000), or it may become brittle due to the oxidation of pyrite minerals. Apart from the effort to obtain core samples, one must also be aware that the equipment is lubricated in all drilling operations. Although saline water and synthetic lubricants are now quite common, petroleum products are still extensively used, especially in those cores drilled for mineral exploration purposes. This implies that one must assume that all core samples are affected to various degrees by contamination from drilling fluids. Some cores have only been superficially impregnated while others may have been pervasively soaked (Brocks et al. 2008). The degree to which this affects subsequent analysis depends on the lithology, porosity, permeability and fracturing, all of which must be evaluated on a case-by-case basis. However, even a contaminated core can still yield useful data when the spatial distribution of contaminants is carefully monitored and understood.

14.2.3 Interregional Approaches

Studying the geochemical characteristics of a stratigraphic profile in outcrop or in one core might yield very useful information on the past biosphere and environment. However, any such investigation can only produce a snapshot with respect to time and space. In terms of past environmental conditions it is of paramount importance to know whether anomalies or developments were of a global or local nature, and in the latter case, exactly how local. Although frequently very complicated, it is advisable to study stratigraphically correlated samples from multiple sedimentary settings (i.e. multiple facies), preferably progressing from shallow to deep water environments. Ideally, one would like to have information from multiple coeval sedimentary basins although this becomes more and more difficult with increasing age of the rocks. A solid chronostratigraphic framework is indispensable for such approaches and alludes to the importance of chemo-, and biostratigraphy, as well as absolute ages of ancient sediments.

14.3 Preparative Methods

14.3.1 Materials

Solvents, reagents, and laboratory ware must be scrupulously clean before use. All tools coming in contact with rocks should be rinsed with methanol (MeOH) and dichloromethane (DCM) before each application. Solvents should be purchased in the highest grade possible and should be tested before use by evaporating a sample, equal to the volume that will be used during the whole sample workup procedure, to ~50 μL and analyzing it with the same methodologies as applied to real samples. All recyclable laboratory glassware can be cleaned with water and soap, rinsed with deionized water, and baked (450–550°C, depending on glass type, thickness and shape) for a period of at least 8 h to combust any residual OM. Glass wool and aluminum foil should be baked at 550°C. Pipettes should be baked at 450°C. Sand and ceramics can be baked at 850°C to remove all organic remains.

Aqueous reagents should be extracted with organic solvents before use and tested for cleanliness. Water can be purchased as HPLC-grade, or cleaned by liquid-liquid extraction with organic solvents. While the latter is by far more economic it's production is labor intensive and one must evaluate cost versus convenience. Deionized water is cleaned by vigorous shaking with DCM in a ~4:1 volume ratio for 90 s in a separatory funnel. After phase separation, the DCM is discarded and the process is repeated 4 times. Subsequently the water should be boiled to remove residual DCM. Both HPLC-grade and laboratory-cleaned water should be tested on a 'per batch'-basis for cleanliness. For this purpose, ~400 mL water is extracted with 100 mL DCM for 90 s in a separatory funnel. After three extractions, the DCM fractions are combined, dried with sodium sulfate, spiked with internal standards, concentrated to a volume of ~50 μL and analyzed like a regular geochemical sample. Hydrochloric acid (HCl) can be cleaned in a similar fashion. After dilution of concentrated HCl (12 N) to ca. 6 N, the HCl is repeatedly extracted by shaking with DCM in a separatory funnel, and boiled to remove any remaining DCM.

Polytetrafluorethylene (PTFE, or Teflon) products (beakers, tubes) that are used for strong acid digestion of sedimentary rocks often absorb some of the residues from previous usage. Scrubbing with a brush does not fully remove this and therefore it is insufficient to clean Teflon-ware with organic solvents only. Teflon can even irreversibly adsorb organic molecules and is porous to a certain degree (Nagy and Bazsa 1991) possibly allowing it to act as a sponge for contaminants. Teflon items can be decontaminated by boiling in aqua regia (nitric acid [HNO_3]:hydrochloric acid [HCl], 1:3) for several hours. When combined, HNO_3 and HCl react to form Cl_2 and NOCl – strong oxidants that degrade organic carbon. Both NOCl and Cl_2 will evaporate during boiling and the aqua regia loses its potency and must be reactivated. Once clean, the mixture is kept boiling until all orange color disappears. Experience has shown that if Teflon-ware is removed from an aqua regia bath too early and allowed to dry at room temperature, it occasionally changes from being translucent to an undesirable opaque white color. Following this treatment, the items are then boiled in de-ionized water for approximately 1 h. Lastly, and just prior to use, the Teflon items are cleaned by ultrasonication in organic solvents. Alternatively, PTFE equipment can be cleaned by boiling in aqueous hydrogen peroxide (H_2O_2, ~10–15%) but extreme caution should be taken as H_2O_2 spontaneously disproportionates into water and dioxygen gas (O_2). Oxygen promotes combustion and its accumulation near sources of heat and organic matter can present an explosion hazard. Beakers with boiling H_2O_2 should thus never be covered and, if possible, the airflow in the hood should be increased.

14.3.2 Preparation of Rock Samples

Nitrile gloves should be worn at all times to protect the samples from fatty acids, squalene and other contaminants from skin. Rock samples, either core or outcrop, must be assumed to bear surficial contamination that might have penetrated to varying depths. An approach to evaluating and circumventing this problem is to divide each sample into layers, as with an onion skin, prior to extraction. Ideally the sample should be divided into three parts, but if this is precluded by small sample sizes, two parts will suffice. The outside, or combined outside and middle layers, should comprise at least 0.5 cm, preferably ~1 cm, of sediment. Before a sample is cut into pieces it should be cleaned by repeated ultrasonication in clean water, MeOH, and DCM. Sawing equipment should be cleaned with organic solvents and only clean water used as a lubricant, exchanging the water and cleaning the saw between samples.

After sawing, the individual sample pieces should again be repeatedly ultrasonicated in water, methanol, and DCM. The rock pieces are subsequently broken into <5 mm sized chips by wrapping them in aluminum foil and hammering. These chips can again be cleaned with water, MeOH, and DCM as described above. Although a small proportion of the extractable organic matter will be lost by doing so, it also reduces the chance of including drilling fluid that may have penetrated the core through fractures. In fact, brittle shale samples will often cleave along fractures during the first or second round of ultrasonication in solvents, making it extremely hard to remove edges by sawing. Rock chips are ground to a fine powder in a puck- or ball mill ideally constructed from stainless steel and small enough to completely immerse in solvent for cleaning. Prior to use and in between samples, the mill should be cleaned by scrubbing with water and soap (steel items can additionally be scrubbed with a wire brush), rinsing with deionized water, then methanol or acetone and, finally, by grinding and discarding fired sand at least three times. Blanks of combusted sand should be produced before grinding the first sample, and at regular intervals between samples to assess potential cross-contamination between samples.

14.3.3 Extraction

14.3.3.1 Free Bitumen (Bitumen 1)

The separation of solvent-extractable organic matter (EOM) from the rock powder can be achieved by ultrasonication of the powders in solvent, or by refluxing in a Soxhlet apparatus. Alternatively, one can use automated extraction equipment such as *Dionex* accelerated solvent extraction (ASE) or comparable microwave extraction devices. The solvents typically used are DCM and MeOH in a 9:1 mixture. If the sample is rich in polar compounds, as indicated by an intense dark coloration of the extract, a more polar mixture can be used. In Soxhlet extractions one should only use azeotropic solvent mixtures to avoid alteration of the mixing ratio by distillation. For the DCM/MeOH solvent system this corresponds to a weight ratio of 92.7:7.3. In any case, the sample should be exhaustively extracted, which is most efficiently achieved by refluxing in a Soxhlet apparatus for ~72 h. Very old and mature rocks typically contain very low amounts of EOM that are very lean in aromatic and polar compounds. As a consequence, bitumens and extracts are colorless, and the extraction efficiency cannot be visually estimated. In case of extraction by ultrasonication, the samples should be extracted at least three to four times at elevated temperature (~50–60°C) in a closed vessel and centrifuged before decanting the solvent to ensure a maximum solvent recovery during each extraction step.

14.3.3.2 Mineral-Occluded Bitumen (Bitumen 2)

It was found that even an exhaustive extraction of finely powdered sediments does not remove all extractable organic matter (e.g. Sherman et al. 2007; Waldbauer et al. 2009; Nabbefeld et al. 2010). After the removal of inorganic minerals, a second bitumen (denoted Bitumen 2) can be obtained, which is generally subtly different in composition to the first, free bitumen. The exact location of this Bitumen 2 is poorly defined but its presence in fluid inclusions can be excluded. Rather, it appears to represent bitumen that was not efficiently expelled from the kerogen matrix or which has become shielded within the layers of clay minerals. It also appears that, even when cores have been pervasively contaminated by drilling fluids, this second bitumen can be largely devoid of these contaminants and more likely to comprise syngenetic hydrocarbons (Hallmann, unpublished). Therefore the parallel analysis of both Bitumen 1 and Bitumen 2 is of paramount importance to understanding the contamination of samples and attributing select molecular fossils to the indigenous sedimentary organic matter.

The following steps should be conducted in a dedicated fume hood. Exhaustively extracted sediment powders are transferred to Teflon tubes (up to 10 g in a 60 mL tube) with caps, capable of being used in a centrifuge. Hydrochloric acid (HCl) is carefully added and allowed to react until the suspension exhibits no reaction even upon agitation. Tubes are then centrifuged, supernatant HCl is decanted and fresh HCl is added to ensure the removal of all carbonates. If a reaction is still observed, the last steps must be repeated. The remaining minerals are washed with clean, ideally warm, water to neutrality. Concentrated hydrofluoric acid (HF) is carefully added to the solid residue. Note that HF is an extremely dangerous reagent that can have lethal consequences if it contacts the skin. Specialized protective equipment and institutional health and safety training is a prerequisite for HF use. First, make dropwise additions to avoid any vigorous, delayed reaction. If no reaction is observed after a small addition, tubes should be gently agitated. If the reaction is mild or indiscernible, more HF can be added and the residue is left to react for 1–2 days, occasionally mixing the suspension with a clean Teflon rod. After the removal of the majority of minerals, the remaining kerogen and heavy minerals will form a dense slurry that cannot be easily agitated by hand. Tubes are centrifuged, the decanted HF is discarded, and a second aliquot of HF added. After additional ~24–48 h of reaction, the HF is discarded again and samples are washed with clean water until near neutrality. In our experience it is extremely hard to completely neutralize concentrated kerogen fractions. After drying at ~60°C, the kerogen is powdered and homogenized with a rounded glass rod in the same tube, and subsequently extracted by ultrasonication with hexane at ~50°C in the closed tube. The low specific gravity of kerogens makes extractions with DCM unfeasible as kerogen will not settle during centrifugation.

14.3.3.3 Elemental Sulfur

Solvent-extracted organic matter should be treated with HCl-activated copper powder, beads (shot) or turnings, to remove elemental sulfur. Elemental sulfur that is readily co-extracted with bitumen interferes with the subsequent analysis of saturated hydrocarbons and leads to rapid degradation of chromatographic columns. After activation, the copper should be washed with water to neutrality and ultrasonicated in methanol and DCM. Alternatively, activated copper can be precipitated from an acidic aqueous copper sulfate ($CuSO_4$) solution by adding elemental zinc powder under stirring. Brocks (2001) used 45 g $CuSO_4$ and 15 g zinc in 500 mL water and 20 mL HCl (10 N), and stored the precipitated copper in ice blocks until required.

The activated copper is added to the solvent extract and turns black through the formation of copper sulfides. Fresh copper should be added until no further color change is observed. Thereafter the extracts should be filtered, evaporated, and spiked with internal standards. If the EOM is allowed to come to complete dryness under elevated temperatures or under a stream of nitrogen gas, great care must be used in evaluating the relative abundances of volatile components. On the other hand, allowing an extract to gently come to dryness by leaving it to stand in a fume hood overnight ensures a consistent lower end (ca. C_{12}+) to the compound distribution. Under no circumstances should a bitumen extract be dried with the application of excessive heat or a vigorous stream of nitrogen.

14.3.3.4 Internal Standards

Internal standards, added in defined amounts, enable the quantitation of individual components. By adding standards at this early point, one can later ensure the quality of the preparative chromatographic separation. However, it is not uncommon for standards to be added just prior to analysis by gas chromatography (GC) and mass spectrometry (MS). The extremely low yield of most Precambrian bitumens precludes the gravimetric analysis of the EOM or individual fractions and compound concentrations are typically reported relative to TOC. Standards should be chemically similar to the analytes of interest, and we recommend using separate standards for alkanes (*anteiso*-docosane [C_{22}] or a deuterated alkane), cyclic isoprenoids (e.g. d_4-stigmastane) and aromatics (e.g. perdeuterated *para*-terphenyl). The concentration should be similar to those of the analytes. As a crude rule of thumb, we typically add 100, 10, and 100 ng, respectively, to very lean samples (e.g. Archean and Paleoproterozoic shales) that are dissolved in 25–40 μL solvent before analysis, and scale these numbers up when necessary.

14.3.4 Sample Fractionation by Liquid Chromatography

Intact bitumens are complex mixtures consisting of many thousands of individual components. In that state they can be analyzed for crude fingerprinting purposes, e.g. to see the distribution of *n*-alkanes and volatile, low molecular weight components. By conducting the equivalent of a 'whole oil GC' (using a flame ionizing detector [FID]; Sect. 14.3.2) one can gain an overview of the general concentration of the sample and the molecular weight distribution of its components. For more detailed and precise analyses, however, the mixture should be simplified by chromatographic separation into different compound classes. The compounds contained in bitumens cover a large polarity range and this can be exploited by liquid chromatography on a solid support such as silica or alumina powder. Following a general rule of thumb in organic chemistry (like dissolves like), solvents for chromatography are chosen to match the polarity of analytes that are desired to be collected within one fraction. The most typically used solvents in organic geochemistry are the very apolar and hydrophobic hexane, dichloromethane (DCM) of intermediate polarity, and the highly polar and hydrophilic solvent methanol.

14.3.4.1 Open Column Chromatography

Chromatography is a technique to separate mixtures of components by transporting them, dissolved in a mobile phase (solvent, gas stream), across a stationary phase. The separation is achieved by differing affinities of the individual analytes to the stationary phase. The standard technique to separate bitumens in organic geochemistry utilizes an open glass column filled with silica gel or aluminum oxide. The volume of the silica, and the size of the column, are chosen to match the size and composition of the bitumen sample. The length of the column positively affects the separation of compound classes. As a general rule, up to 10 mg of bitumen can be separated in a Pasteur pipette filled with ~500 mg of silica gel. If the bitumen is very lean in polar components, such as in the case of an oil-condensate, up to 50 mg or even more may be separated on a small column. However, when bitumens have an abundant polar content, the column might already be overloaded with a 5 mg charge.

Most chromatography columns resemble pipettes in shape. Typically, the larger ones are fitted with a stopcock allowing the user to pause the solvent flow at any time (although longer interruptions will reduce the quality of the separation). The base of the column is fitted with a short, loose plug of combusted, or solvent-washed glass wool forming a barrier to contain the stationary phase. The wool is then topped by a 1–2 mm layer of fired sand. This prevents wool strands protruding into the stationary phase and acting as a migration conduit. The stationary phase, most often activated silica gel, is then placed on top of the sand. Tapping the column while filling allows for an evenly packed adsorbent bed. The sample is then applied to the top of the bed in a minimal volume of DCM and allowed to come to complete dryness before proceeding. Individual fractions of the mixture are eluted by adding solvents of differing polarity to the column and allowing them to drain gravitationally. The separation of a bitumen into saturated hydrocarbon fraction, aromatic hydrocarbon fraction and polar fraction, as outlined above, is the most common and straight forward way to divide the sample into well-defined compound classes. This greatly simplifies their subsequent analysis by GC and GC-MS. Further, by modifying the polarity and eluting volume of the mobile phase, or the nature of the stationary phase, even more specialized fraction cuts can be obtained.

Medium and Large Columns

Columns are dry-filled as above and the sample, dissolved in DCM, is added dropwise to the top of the bed. The sample should form a concentrated, evenly distributed horizontal band at the top of the silica and not touch or be smeared against the column walls during application. If the sample slowly starts to migrate down the column for more than ~5 mm (medium columns) or 1 cm (large columns), the column is overloaded and less sample, or a larger column, should be employed. Large samples may be applied in a number of aliquots, each of a few drops, allowing the solvent to evaporate between each addition. All DCM should be evaporated before beginning the solvent sequence and overnight drying is an effective way to achieve this.

When adding solvent to the column to elute the different fractions, it should be done in small aliquots, rinsing the solvent circularly down the column walls. Once a small aliquot is added, the user should wait until the solvent level drains down to the top of the silica bed and repeat this process twice before carefully adding more solvent. This helps to retain the analytes in a concentrated band. Addition of excessive solvent at one time would disturb the bed of silica and, therefore, the localization of analytes. Starting with a dry column and carefully measuring the first fraction of solvent, in this case *n*-hexane, to the point at which the first drop of solvent exits, yields the dead-volume (DV) of the silica bed. Now, the saturated hydrocarbon fraction can be collected by eluting with 7/16 DV of *n*-hexane in a beaker or, preferably, a tared vial. Before switching to the next fraction, the tip of the column should be rinsed from the outside with a small volume of hexane to remove any remaining saturated hydrocarbons. Aromatic hydrocarbons can now be eluted with 4 DV of an 8:2 (vol./vol.) mixture of hexane and DCM. Again, it is recommended to start by rinsing small solvent aliquots down the column wall. After collecting the aromatic fraction, the polar fraction can be eluted with 2 DV of DCM/MeOH, 1:1 (vol./vol.). In the case of rich, colored bitumens, the aromatic and polar fraction can be seen as yellow to brownish bands, respectively, moving through the column. Saturated hydrocarbons, on the other hand, are colorless. If any colored fractions start to remobilize during the elution with hexane beyond ~1/5 of the length of the column (measured from the top of the silica), this also indicates that the column is overloaded, and the chromatographic separation will suffer to varying degrees.

Small Columns

A small-scale procedure was published by Bastow et al. (2007), who showed that a Pasteur-pipette (~1 mL)-based column, filled with ~600 mg activated (120°C for at least 8 h) silica gel (70–230 mesh) could be used for the rapid separation of small bitumen volumes into saturated hydrocarbons, aromatic hydrocarbons, and polar components. The packed column is made uniform by passing ca. 3 dead-volumes of *n*-pentane (can be substituted for *n*-hexane) through the column. The published method suggests pushing the solvent through the column by applying slight positive pressure with a pipette bulb on the top of the column. This is however discouraged since small amounts of organics can be leached or cross-contaminated by these bulbs. Just before the solvent reaches the top of the silica bed, the sample is applied on to the silica in a minimum volume of pentane. Saturated hydrocarbons can be eluted in 1.8 mL pentane, followed by recovery of the aromatic hydrocarbon fraction in 1.8 mL of pentane/DCM (7:3, vol./vol.) and the polar fraction in 1.8 mL DCM/MeOH (1:1, vol./vol.). Avoiding the drying of samples on the column allows even volatile compounds to be recovered and analyzed. Similar, but unpublished, methods have been commonly used in many other geochemistry laboratories.

Argentation Chromatography

Replacement of pure activated silica gel with silica that is impregnated with silver nitrate ($AgNO_3$) is the basis of a chromatographic procedure that will separate compounds according to the number, type, and position of double bonds (Morris 1966). In organic geochemistry, argentation chromatography is also useful for separating the total aromatic hydrocarbon fraction into mono-, di-, tri-, and poly-aromatic subfractions.

14.3.4.2 Molecular Sieve Separations

Shape-selective molecular sieves allow separation of compound classes based on size and geometry. Molecular sieves frequently consist of zeolites, or microporous aluminosilicates, that can be visualized as sponges with precisely defined pore diameters. In organic geochemistry, molecular sieves of the 5 Å-type (roughly 5 Å pore diameter) find a common application in the separation of the normal alkanes (i.e. *n*-alkanes) from all branched and cyclic alkanes in a saturated hydrocarbon fraction. This is most commonly used to obtain less complex alkane fractions for compound-specific isotope analysis, which requires compounds to be separated to the chromatographic base line, in order to exclude interference from neighboring compounds. Molecular sieves can also be used to prepare a more concentrated sterane and hopane fraction for increased sensitivity in GC-MS analysis. Molecular sieves with different pore diameters find application in organic geochemistry for enrichment of specific compound classes and have an additional advantage in that such separations, unlike chromatographic separations, are never accompanied by stable carbon isotopic fractionation (Ellis et al. 1992, 1994; Armanios et al. 1994; Kenig et al. 2000; Audino et al. 2004).

To separate *n*-alkanes from branched and cyclic alkanes, the saturated hydrocarbon fraction obtained from column chromatography must first be dissolved in cyclohexane – otherwise, the *n*-hexane that was used as eluant for column chromatography will occupy all the pores of the molecular sieve. Molecular sieve of the 5 Å type is 'activated' by heating to approximately 550°C in order to remove water and potential organic contaminants. Once treated in this way it can be maintained in a >100°C oven until use. An appropriate amount of sieve is placed into a gas-tight and pressure resistant vial to which the saturated hydrocarbon fraction is added in a small volume of cyclohexane. The capped vial is maintained at 80°C for at least 8 h with occasional monitoring of the solvent level to avoid leakage. After cooling to room temperature, the solvent, containing the branched and cyclic saturated hydrocarbons (often termed the molecular sieve non-adduct, MSNA), is separated from the sieve, which is further washed three to four times with small volumes of cyclohexane. A vortex mixer is useful for this purpose. After drying, the molecular sieve, which should now contain the occluded *n*-alkanes, is transferred to a Teflon test tube for digestion in HF. This reaction is carried out with occasional agitation at room temperature and with great care. Released *n*-alkanes are then extracted from the aqueous HF-phase by careful liquid-liquid extraction with *n*-pentane. Pentane will form the upper layer of the two-phase mixture and can be carefully removed with a pipette. The extraction should be repeated three times, the pentane fractions combined and then treated with a small amount of $NaSO_4$ to remove water. Finally, the pentane can be filtered through a short plug of silica gel to remove the $NaSO_4$ and neutralize any residual HF, thereby affording the molecular sieve adduct (MSA). After concentration, both adduct and non-adduct fractions can be analyzed by GC-MS to evaluate the outcome of the sieving process.

Grice et al. (2008) have shown that *n*-alkanes can be released from a molecular sieve without HF digestion and, instead, by heating with an excess of *n*-pentane.

Since the pentane displaces the larger, adducted alkanes, an almost-quantitative recovery should be possible, although it is not always the case in practice.

14.4 Analytical Methods

14.4.1 Gas Chromatography (GC)

GC or, more correctly, gas-liquid chromatography involves the separation of a complex mixture of volatile compounds by partition between a moving carrier gas stream (typically He or H_2) and a stationary liquid phase (Grob and Barry 2004). In its original form, the stationary phase was uniformly coated onto a solid support, such as diatomaceous earth, which was packed into a 1–2 m × 2–4 mm inner diameter (ID) glass or metal column. Subsequently it was found that much finer separations could be achieved with 10–60 m × 0.5–0.1 mm ID, open tubular fused silica capillary columns with a stationary phase coated or chemically bound to the column walls (wall-coated open tubular; WCOT). Retention of a particular compound on the column is a function of its boiling point and polarity relative to the temperature, thickness and polarity of the liquid film and the flow rate of the carrier gas. Precise control of the gas pressure (and flow) and the temperature of the oven housing the column results in uniform elution times (retention time) for each individual compound. Separations can be optimized by choice of column dimensions, polarity of the liquid phase and the program of temperature/pressure conditions for the oven. A variety of detectors are available for detecting and identifying components as they elute from the column (Fig. 14.3). Resulting data is typically shown as a chromatogram, plotting retention time on the x-axis versus response or intensity on the y-axis. Each eluting component will give rise to a discrete peak at the detector (Fig. 14.3). Depending on the operating conditions, different compounds can have similar retention times resulting in either partial or complete 'co-elution'. These can be 'de-convoluted' either by a different choice of operating parameters or, in the case of mass spectrometric detectors, by using numerical approaches.

14.4.1.1 GC Injection

The sample, dissolved in an organic solvent, can be introduced, via a syringe, to the GC system in different modes, each with their advantages and disadvantages. Injection (automatically or manually) can be done directly onto the column, or into a heated hollow open glass cylinder, or liner, of ca. 80 mm × 5 mm ID, which serves the purpose of vaporizing the solvent and analytes before they enter the GC column. The injector port temperature should be initially high enough to vaporize the solvent and all components of interest or programmed to differentially evaporate the solvent and analytes.

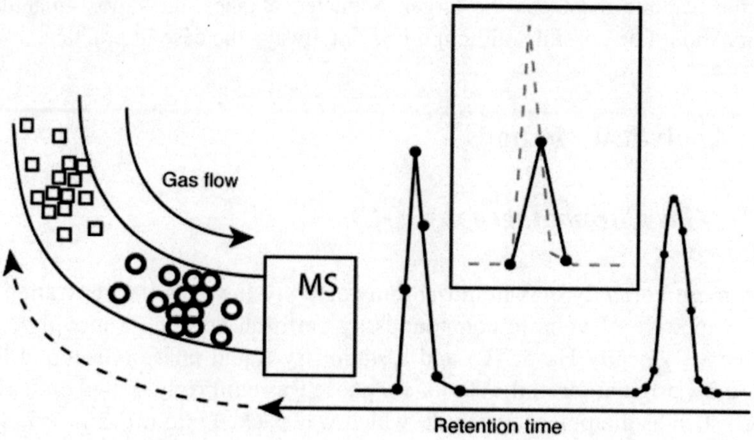

Fig. 14.3 Gas chromatography. Complex mixtures of compounds (illustrated here as *squares* and *circles*) can be separated by gas-liquid chromatography (GC). When combined with a mass spectrometer (MS) as the detecting device, it is important to match the sampling rate (*black dots* per time unit) to the period during which each component elutes. Insufficient data points (see *inset*) result in poorly defined peak shapes and inaccurate quantification

The *on-column* or *cool-on-column* injection method avoids use of a liner and the syringe needle directly enters the column to introduce the liquid sample-solvent mixture. The advantages of this technique lie in the introduction of thermally labile compounds directly onto the liquid phase. This minimizes possible degradation or selective adsorption onto the glass surface of the heated inner liner and ensures that no compositional fractionation occurs. With this method the compounds are simultaneously concentrated in a thin band at the front of the column, positively influencing the quality of subsequent chromatographic separation in the GC. The disadvantage is that all compounds in the sample will be introduced to the column. In the case of 'dirty' samples (that cannot be cleaned up by preparative techniques), unwanted molecules such as polar and involatile components will enter the column leading to its degradation over the longer term. This can however be overcome by regular trimming of the column (accompanied by loss of performance and a change of retention times) or by the use of a replaceable pre-column.

Splitless injection refers to a sample introduction protocol where both solvent and sample are volatilized in a heated liner. The temperature, pressure and valving conditions are selected in a manner that allows the analytes to concentrate in the liquid film at the head of the column in a cool oven. After 1–2 min, the most volatile components including the solvent are flushed from the injector at the same time as the oven temperature program begins. The advantage of using splitless injection lies in the afore-mentioned intended fractionation of unwanted compounds of high polarity or large molecular weight that would otherwise end up on the column. Unfortunately though, fractionation can also occur between analytes of differing molecular weight. Furthermore, volumetric expansion of the solvent in the heated liner limits the amount of solvent-sample mixture that can be injected.

A *split injection* similarly volatilizes all components of the sample but the valving protocol allows only a small fraction of both to enter the column; the remainder is vented and lost. It provides improved chromatographic resolution at the cost of lower overall sensitivity.

Programmable temperature vaporizing (PTV) injectors are now in common use and allow a wide choice of injection conditions optimized for different applications. For example, a PTV injector is capable of performing (a) multiple injections with solvent evaporation and purging between injections in order to concentrate analytes (b) using a temperature ramp that improves chromatography and (c) using a technique termed *pulsed splitless injection*, which uses carrier gas pressure pulses to increase sample transfer to the column. Different injector temperature ramps allow the user to monitor the fractionation between smaller versus larger compounds and experimentally establish the ideal conditions for the analysis of select molecules of interest.

14.4.1.2 GC Column and Oven Program Selection

The physical dimensions (length, diameter, film thickness), and chemical properties of the coating film allow the user to optimize both the length and resolution of a chromatographic analysis. For example, all else being equal, longer columns allow for a better separation whereas shorter columns are less expensive and allow more rapid analyses. Typical column lengths are 15–60 m but depending on the application, the quality-gain with a 60 m column might not be large enough to justify its price. The film thickness determines the amount of sample that can be separated. Injecting too much material onto a thin film will overload the column and cause 'fronting' of the peaks and decreased chromatographic quality. Using a thicker film increases the amount of sample that can be injected, but has slightly negative consequences for the quality of chromatography. The bore diameter determines gas flow rate that can be achieved under the same pressure conditions.

The polarity of the film coating should be chosen to match that of the analytes. The least polar columns use 100% methyl-substituted polysiloxane liquid phases (e.g. DB-1 or HP-1). Such columns are ideally suited for the separation of apolar analytes such as saturated hydrocarbon fractions. Aromatic hydrocarbon fractions are slightly more polar and preferably analyzed using a column with a somewhat higher polarity liquid phase (e.g. DB-5) where 5% of methyl substituents are replaced by phenyl groups. In practice, both saturated and aromatic hydrocarbon fractions can be successfully analyzed on 60 m columns with either DB-1 or DB-5 phases and without unduly compromising chromatographic quality. Chemically bonded liquid phases with these compositions are stable at oven temperatures up to 350°C. Compounds of higher polarity will exhibit peak tailing on DB-1 and DB-5 phases and should be analyzed using more polar columns. A vast range of columns and polarities exist and will not be described here. The most polar phases are based on a polyethylene-glycol (PEG) composition and can be used to analyze alcohols and fatty acids. The downside of these PEG-based columns is their lower temperature range (usually maximally ~260°C), compromising the analysis of components in the higher boiling range.

An oven temperature ramping program is typically used to improve chromatographic separations. Under isothermal conditions, all compounds would elute from the column at some point, but with increasing molecular weight the retention times become longer and the peaks broader. Use of an appropriate temperature ramp allows the user to maintain narrow peak widths and optimizes separation for most components. A rapid temperature ramp can be used to reduce analysis times but at the expense of peak resolution. Finally, at the end of every run, the oven is typically held at or near the maximum acceptable temperature for the liquid phase (typically 320–350°C for DB-1 and DB-5 columns) to allow time for any undesired high-molecular-weight compounds to elute. Typically employed temperature programs for biomarker analyses on DB-1 or DB-5 columns are: 60°C isothermal for 2 min, then programming to 150–180°C at 10°/min, followed by a slower ramp to 325°C at 3°/min and a final isothermal stage at 325°C for 15 min.

14.4.2 GC – Flame Ionization Detector (FID)

GC can be used for fingerprinting purposes when just fitted with a flame ionization detector (FID). Advantages of the FID are its high sensitivity to low carbon concentrations and a general lack of discrimination – i.e. peak areas closely approximate analyte abundances and the detector response does not vary with different compounds (unlike in mass spectrometry). However, no detector is perfectly linear for all compound classes and, for precise quantitation, standards should be used to establish calibration curves. Furthermore, an FID does not provide qualitative data. GC-FID is most useful to get an overview of the concentration and approximate composition of samples. Reliable quantification can only be performed on the most abundant compounds in a sample such as gasoline-range components and *n*-alkanes.

14.4.3 Mass Spectrometry (MS)

When coupled to GC, a mass spectrometer provides accurate qualitative data on analytes due to its mass selectivity. Compounds exiting the GC oven enter the *ion source* of the MS via a heated transfer line. To prevent peak broadening, this transfer line is ideally short and heated to a temperature just above the maximum column temperature. Fused silica capillary GC columns have the advantage of being non-conducting and can be placed directly inside the transfer line terminating at, or very close to, the ion source (Fig. 14.4). A very wide selection of ionization techniques exists (e.g. photo-ionization [PI] or chemical ionization [CI]) but the most commonly used in biomarker research is electron-impact ionization (EI). The energy of the electron beam can be varied to optimize preservation of intact molecular ions (low electron volts, eV) or to promote fragmentation of ionized

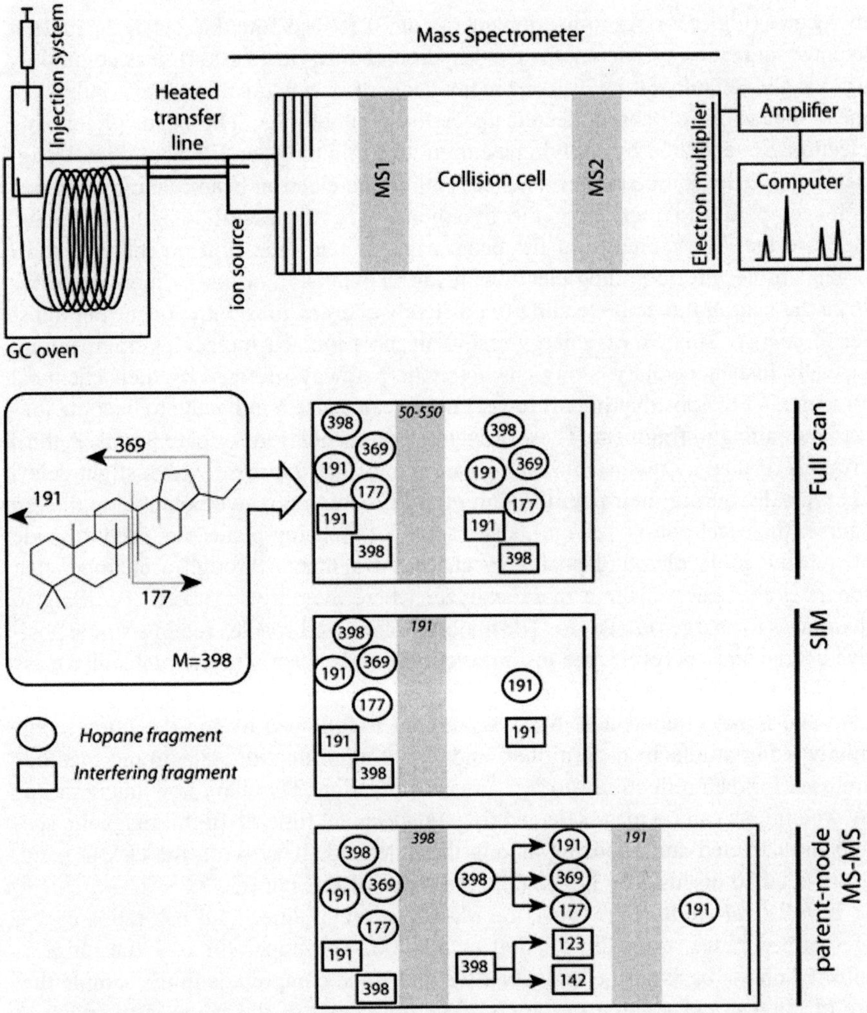

Fig. 14.4 Mass spectrometry. After the separation of compounds by GC, MS can be used for their identification and quantification. In *full scan* mode, all charged fragments that are generated from a compound in the ion source are analyzed within a selected mass range. Full mass spectra can be recorded and all eluting chromatographic peaks (Fig. 14.3) can be tentatively identified. By monitoring only one or a few select fragments in *SIM* mode, the sensitivity can be increased. Characteristic fragmentation patterns allow specific compounds of interest to be targeted. If, however, an analyte of interest (e.g. C_{29}-norhopane) coelutes with an interfering compound that produces the same characteristic molecular fragments (e.g. *m/z* 191), identification is impossible. This can be overcome in *multiple-reaction-monitoring* (MRM), or MS-MS modes, where precursor-product fragmentation reactions are monitored using two mass analyzers in tandem. Although sacrificing sensitivity, MRM and MS-MS allow to eliminate many interferences

molecules (high eV). An ionization energy of 70 eV was found to entail the highest average ionization efficiency for most hydrocarbons (Gross 2004). It is commonly used as the default setting as, over many years of data acquisition, most published mass spectra have been collected under these conditions. The beam of ionizing electrons is generated by applying a current to a thin tungsten filament. Small magnets outside the source confer a helical path to the electron beam, thereby creating a longer path and increasing the likelihood of a successful ionizing collision. Because the 70 eV energy of the beam exceeds the ionization potential of most organic molecules, colliding electrons result in expulsion of one or more electrons from the neutral molecule leading to positively charged ions with concurrent transfer of energy. This excess energy results in most ionized molecules fragmenting, virtually instantaneously, along characteristic pathways defined by their chemical structure. Collisions that impart lower energy can cause a molecule to become ionized but failing to fragment. These are termed molecular ions or parent ions. A third group is formed by metastable ions – parent ions that fragment with a slight delay, i.e. typically during their path through the MS. A high positive potential at the ion source (the accelerating voltage), and a series of focusing plates, are used to guide all ionized molecules and positively charged fragments through a chamber that separates the source from a mass analyzer where they are separated according to their mass to charge (*m*/*z*) ratio. The majority of ionized species retain a single positive charge and, therefore, the *m*/*z* ratio corresponds exactly to the molecular mass in dalton (Da). Species that evade ionizing electrons are pumped away by the vacuum and remain undetected. Mass separation is followed by ion detection, commonly using an electron multiplier, and signal amplification. Electronic methods are used for data reduction, storage and manipulation. There are now many means by which ions can be mass selected (e.g. quadrupole, time-of-flight, magnetic sector) and detected and choosing among them depends heavily on the task at hand. More detailed discussions are beyond the scope of this paper.

In *full scan* mode (Fig. 14.4), the MS focuses and detects all *m*/*z* ratios over a pre-defined range, typically 50–500 or 50–700. The upper limit of the range is selected on the basis of the molecular weight of the compounds in the sample that are of interest and volatile in the GC. The lower end of the range is the point at which the masses of fragment ions lose diagnostic value. Operating a mass spectrometer in the full scan mode maximizes the amount of information that can be gathered for an unknown sample. However, this also results in collection of a considerable amount of null data points because time is spent scanning when no ions are striking the detector. Effectively, this equates to a loss of sensitivity and can be overcome by concentrating the data acquisition time at narrower mass ranges, or even at selected *m*/*z* values to improve the signal to noise (S/N) ratio. Definition of a chromatographic peak and, consequently, the subtle aspects of its shape, is improved with an increasing number of data points. Minimally, six to eight data points are needed to define a chromatographic peak shape (Fig. 14.3) and therefore scan rates and scan ranges are chosen to match the chromatographic conditions. For example, a chromatographic peak that is 6 s wide demands that the user acquires at least six to eight samples (or spectra) during its elution so that its maximum height, shape and area are well defined. If the operator knows that none of the compounds of interest have *m/z* ratios higher than 350 or lower than 150, the scan range could be adjusted to 150–350 to maximize S/N. A lower scan range over the same time window allows the MS to dwell longer on each mass, resulting in more ions hitting the detector per collected data point. The summation of all ions detected in full scan acquisition, plotted versus the acquisition time is referred to as a total ion current (TIC). This is analogous to the FID trace in GC analysis and it often looks very similar with the main difference arising from the selective nature of the respective detectors. Alternatively, one can plot a specific mass or selection of masses versus acquisition time to generate ion chromatograms to visualize individual compound classes in a complex mixture.

The principle of concentrating acquisition times on compounds of interest can be taken to its next logical step by only collecting data for one or, more common, a few specific diagnostic ions. For example, in the case of hopanoid hydrocarbons, there is a prominent and diagnostic 191 Da fragment ion (Fig. 14.4). Use of a mass analyzer focused at *m/z* 191 for 100% of the time will maximize the chance of detecting hopanoids and measuring their abundances accurately. In this case, termed *selective ion monitoring* (SIM), the MS is asked to dwell on a 191 Da mass channel continuously. The ensuing improvement in S/N is significant and often enables detection of compounds that were not visible in full scan mode. Then why use full scan at all? While SIM analyses are useful for analyzing specific compounds that are present in low concentrations, they do not provide any information on whatever else may be present in the sample.

Most SIM analyses focus on more than one mass channel. Another gain in S/N can be achieved by bundling a number of m/z channels into groups. Instead of analyzing all ions of interest during the whole chromatographic run, groups of ions can be measured during pre-determined time windows if the elution time of analytes is known. In all cases, the idea is to maximize both the time spent detecting and counting ions of interest and the number of samples per unit of time so that the user gains the full benefit of the resolution afforded by the chromatographic system.

In practice, one finds that the chromatographic peaks of many compounds of interest overlap either partially or fully (Fig. 14.5). These 'co-elutions' compromise both identification and accurate quantification but can be minimized or completely overcome by using more exotic mass spectrometric tools. The most common way to reduce such interference is to use a linked combination of two mass spectrometers (i.e. MS-MS). To illustrate this we will take a closer look at the norhopane molecule. An intact C_{29} norhopane (Fig. 14.4) molecular has a parent ion of 398 Da and a number of diagnostic fragments including the ions at 191 Da (common to all hopanes) and 177 Da (specific to norhopanes). In GC-MS-MS analysis one can use the first mass analyzer (MS 1) to select for 398 Da precursor ions (aka 'parent' ions) and pass these through a collision chamber where collisions with an inert gas such as argon or helium promote fragmentation. This is followed by a second stage of mass selection (MS 2) to detect either 191 Da, 177 Da product (aka 'daughter') ions or both. In this technique, one is effectively generating specific reaction chromatograms (i.e. 398 $\rightarrow$ 191 and/or 398 $\rightarrow$ 177) and the resultant increases in selectivity overcome most flaws or deficiencies in chromatography.

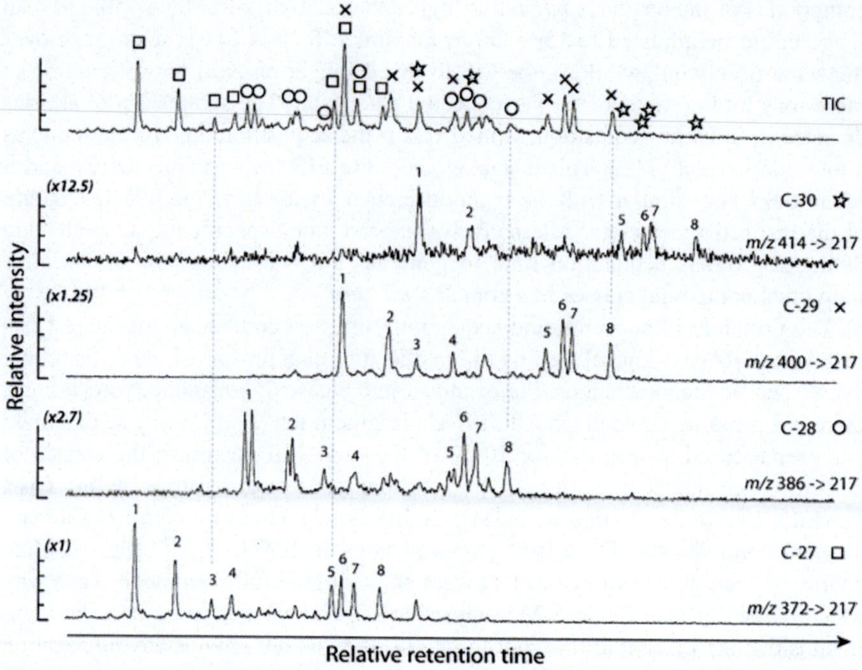

Fig. 14.5 SIM vs MRM mass spectrometry. Individual steranes exhibit strong coelution during GC, independent of the type of column used and SIM analyses of the *m/z* 217 typical sterane fragment (top chromatographic trace) do not allow for the quantification of individual steranes. MRM experiments (lower traces) overcome this problem by recording parent-to-daughter ion transitions. *1–4*: diasteranes, *5–8*: regular steranes. *1.* 13β,17α-diasterane 20*S*; *2.* 13β,17α-diasterane 20*R*; *3.* 13α,17β-diasterane 20*S*; *4.* 13α,17β-diasterane 20*R*; *5.* 14α,17α-sterane 20*S*; *6.* 14β,17β-sterane 20*R*; *7.* 14β,17β-sterane 20*S*; *8.* 14α,17α-sterane 20*R*

In practice one can combine different types of mass analyzers to accomplish MS-MS. Triple stage quadrupole instruments (TSQ) are exactly that with two mass analyzers separated by a third quadrupole acting as the collision chamber (Peters et al. 2005b). One can combine a single quadrupole with a time-of-flight analyzer (Q-TOF) and thereby take advantage of the TOFs high mass resolution capability. Ion trap systems can collect and 'store' molecular ions for short intervals of time followed by application of voltage patterns to fragment these and mass-analyze the products. Other, high resolution, mass spectrometers such as those with combined electrostatic (ESA) and magnetic sectors can be operated in a quasi-MS-MS fashion termed *metastable reaction monitoring* (MRM). So called 'metastable' ions are low energy species that leave the ion source as intact molecular ions but fragment during the time it takes to transit the space between the source and the ESA, that is, the first field-free region (FFR1) of the MS. Precise control of the accelerating voltage, the voltage on the ESA and the magnet current allow the user to select and measure multiple parent-to-product pairs during a chromatographic analysis (Fig. 14.5). Such MRM analyses have less specificity than that those afforded by TSQ instruments, but this is balanced by the greater inherent sensitivity and flexibility of the magnetic sector instruments. The ultimate form of MS-MS combines a magnetic sector 'front end', with its high resolution capability employed to precisely select molecular ions, with a quadrupole or TOF 'back end' for mass analysis of collision-induced decomposition (CID) products.

14.4.4 Pyrolysis

Pyrolysis is essentially the laboratory imitation of the subsurface process that generates bitumen from kerogen. Extracted sediments, or extracted kerogen concentrates are heated for variable amounts of time in a vacuum, or a neutral or reducing atmosphere, to induce cracking of macromolecules. Products resulting from the pyrolytic degradation of kerogen were covalently bound and thus syngenetic to it. Love et al. (2009) have successfully used a pyrolysis method to confirm that sponge biomarkers (24-isopropylcholestanes) were covalently bound to kerogens and, thereby, original components of the rocks they studied. However, pyrolysis is most useful in studies of kerogens below a certain maturity threshold. If bitumen has been largely generated in the subsurface, pyrolysis of such kerogens will afford mainly polyaromatic hydrocarbons of limited diagnostic value. Hydropyrolysis of a 2,500 Ma old kerogen from the Mt. McRae shale yielded only aromatic compounds and no alkanes or cyclic terpane biomarkers (Brocks et al. 2003a, b).

Generally, pyrolysis can be categorized as 'online' or 'offline'; the former is performed with the pyrolysis apparatus coupled to a GC-MS system, while the latter, which usually involves significantly larger samples, requires trapping of freshly-generated bitumen and additional cleanup steps prior to GC-MS. In *flash pyrolysis*, an online technique, the kerogen is rapidly heated to release products that directly enter the GC-MS. Compounds in these pyrolysates are typically functionalized and/or unsaturated. In hydrous pyrolysis (Lewan 1985), an aliquot of kerogen is heated with water in a high-pressure cell that is typically made of glass, gold or metal. The water at high pressures provides a source of hydrogen leading to a partial defunctionalization of compounds and reduction of double bonds. This yields a product with a higher proportion of saturated hydrocarbons and a composition comparable to the original bitumen. A similar rationale applies to hydropyrolysis or HyPy (e.g. Love et al. 1995; Rocha et al. 1997), the method that currently produces the most 'natural' bitumens. Micro-scale sealed vessel (MSSV) pyrolysis involves heating of small aliquots of kerogen; numerous analyses can be conducted on the same sample at different temperatures or timescales to determine the kinetics of hydrocarbon generation (e.g. Horsfield et al. 1989; Schenk and Horsfield 1993).

14.4.5 High Performance Liquid Chromatography (HPLC)

GC is limited to compounds that can be vaporized at temperatures below ~320°C (~400°C in specialized high-temperature GC [HT-GC]) and which are sufficiently stable at elevated temperatures to survive gas chromatographic analysis. Compounds that do not satisfy these requirements can be separated by high-performance liquid chromatography (HPLC) prior to MS analysis. The main classes of involatile compounds of geochemical interest are polar lipids such as triglycerides, phospholipids, glycolipids, diglycerol tetraether phospholipids, photosynthetic pigments and proteins. During diagenesis these large and labile compounds rapidly break down and are generally not preserved as intact molecules in sedimentary rocks beyond a few thousands to millions of years. HPLC therefore finds little application in Precambrian geochemistry. The exception is formed by porphyrin pigments and a few other types of resistant polar compounds. A potentially important, and little applied use for HPLC is in the preparative chromatographic separation of compound classes and individual components. Here, one can employ different combinations of solvent and stationary phases to purify a specific compound or class of compounds for isotopic or other kinds of spectral characterization.

14.4.6 Stable Isotope Analysis

The stable isotopic compositions of individual molecules can add valuable information (Fig. 14.6) that assists their attribution to particular source organisms or processes. This is especially true of compounds that can have many possible sources, such as *n*-alkanes and acyclic isoprenoids. The stable isotopic compositions of molecular fossils reflects a combination of processes including fixation and assimilation pathways, lipid biosynthetic pathways, and environmental influences on organisms such as e.g. pCO_2, pH and reservoir size (e.g. Hayes 1993; Freeman et al. 1990; Hayes 2001).

The most accessible and informative stable isotopes of are those of carbon, nitrogen, oxygen, sulfur and hydrogen whose natural abundances are: ^{13}C (~ 1.11%), ^{15}N (~0.37%), ^{18}O (~0.20%), ^{34}S (~4.20%) and ^{2}H (i.e. deuterium, ~0.015%). Isotopic fractionation processes generally operate in proportion to the difference in mass between the two isotopes and, therefore, one often finds that with increasing mass separation (N: 15/14 = 1.07, C: 13/12 = 1.08, H: 2/1 = 2), the isotopic differences between the precursor and product of a reaction tend to be more pronounced. Isotopic fractionation processes are often described as being under either 'equilibrium' or 'kinetic' control. The former include fractionations associated with phase changes, diffusion or ionic equilibria. With the latter, we are mostly concerned with biological reactions catalyzed by enzymes. Examples include the fractionation of carbon isotopes during assimilation (House et al. 2003) and biosynthesis (Hayes 2001) (Fig. 14.6).

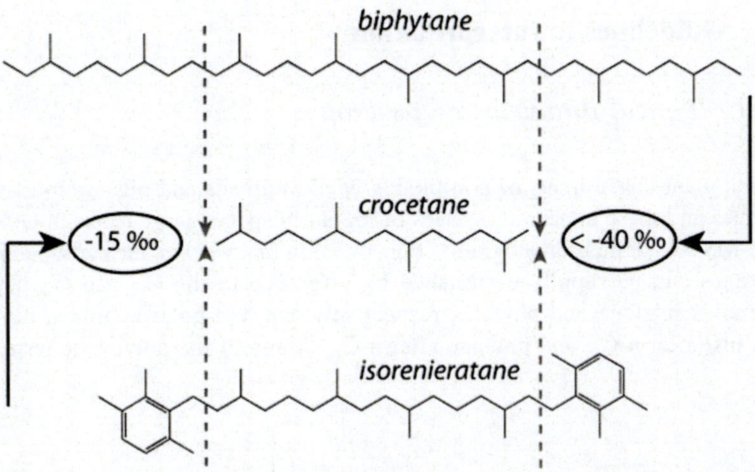

Fig. 14.6 Stable carbon isotopes can aid in identifying the biosynthetic origin of select compounds. Crocetane can be a thermally-cleaved fragment of either archaeal biphytane or isorenieratane from green sulfur bacteria (Maslen et al. 2009). The stable carbon isotopic composition of crocetane can reveal whether it stems from a precursor that was biosynthesized by green sulfur bacteria that use the reverse tricarboxylic acid (TCA) cycle (Summons and Powell 1986), or by archaea engaging in the anaerobic oxidation of methane (AOM; Orphan et al. 2001)

Stable isotopic compositions are typically measured and expressed as ratios in parts per thousand (‰). Data are collected and reported with reference to international standards which, by convention, are assigned a value of 0‰. For carbon and oxygen in carbonate rocks, the original standard was a belemnite (extinct group of marine cephalopods) from the Cretaceous Pee Dee Formation in South Carolina (Pee Dee belemnite, PDB). For nitrogen, the standard is the largest terrestrial reservoir, that is, the atmosphere. For hydrogen and oxygen in water the standard is 'standard mean ocean water' (SMOW). The sulfur standard is a meteorite, the Canyon Diablo Troilite (CDT). For current usage, these standard names are preceded by a 'V' for Vienna (e.g. VPDB), the location of the International Atomic Energy Agency laboratory that performed most recent calibrations and retains the current standards. The true $^{13}C/^{12}C$ ratio of PDB is 0.0112372 (Craig 1957).

The measurement of the stable isotopic compositions of individual organic compounds can be performed by the on-line combusting of bulk samples to CO_2 and the measurement of the C-isotopic composition of that gas. Alternatively, a GC can be coupled to an isotope ratio monitoring mass spectrometer (irm-MS) via a combustion oven enabling the isotopic analysis of the individual components in a continuously flowing gas stream. This method is referred to as compound-specific isotope analysis (CSIA; Matthews and Hayes 1978; Hayes et al. 1990; Freeman et al. 1990; Meritt and Hayes 1994a; Ricci et al. 1994; Merrit et al. 1995) and recent advances have made all the light elements (C, H, N, O and S) accessible to continuous flow techniques (e.g. Meritt and Hayes 1994b) with particular advantages accruing from the isotopic analysis of multiple elements in the same compound (Hinrichs et al. 2001).

14.5 Guidelines to Interpretation

14.5.1 Typical Bitumen Composition

The dominant constituents of non-biodegraded bitumens and oils are most often *n*-alkanes and these appear as a series of regularly spaced large peaks in GC-FID or GC-MS total ion chromatograms (Fig. 14.7). In many cases, the carbon number of alkanes can be rapidly established by reference to the C_{19} and C_{20} acyclic isoprenoids pristane and phytane, respectively. On non-polar columns, pristane elutes just after *n*-C_{17} and phytane after *n*-C_{18}. Many of the polycyclic terpenoid

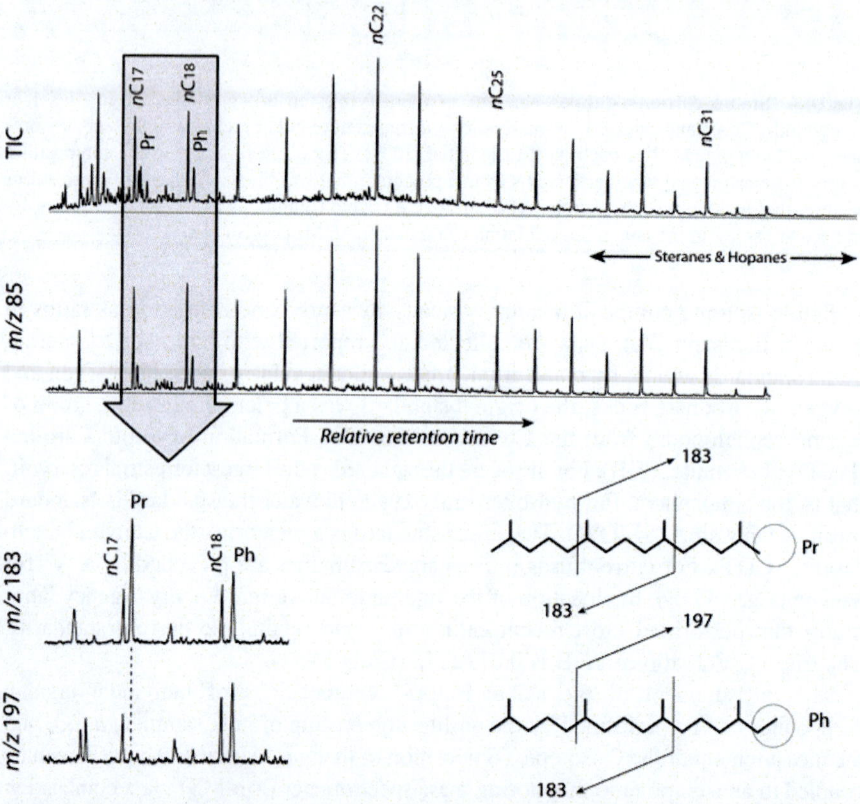

Fig. 14.7 Saturated hydrocarbons. The saturated hydrocarbon fraction of bitumens is typically dominated by a homologous series of linear alkanes, which can be visualized with reduced interference by monitoring their characteristic *m/z* 85 molecular fragment. On apolar GC columns (DB-1 and DB-5), the isoprenoid lipids pristane (Pr) and phytane (Ph) (Fig. 14.2) elute shortly after *n*-C17 and *n*-C18, respectively. Pristane and phytane can be identified by monitoring characteristic molecular fragments at *m/z* 183 and *m/z* 197, the latter of which is only generated from phytane. Steranes and hopanes elute in the C_{27}–C_{37} alkane range but their concentration is typically far below that of the *n*-alkanes

biomarkers, such as tricyclic terpanes, hopanes and steranes, elute in the vicinity of their corresponding carbon-numbered *n*-alkanes. The combined effects of structure, stereochemistry and conformation lead most steranes to elute earlier than corresponding carbon-numbered hopanes. The lower concentration of terpanes in relation to the *n*-alkanes makes them hard to spot in GC-FID traces and full-scan total ion chromatograms (TIC). Accordingly, the specific ion-chromatograms for 191 Da or 217 Da, or MRM chromatograms, are most often used to study these compounds. For an extensive description of various biomarkers, their masses, fragmentation pattern, elution profiles, characteristics and interpretations, the reader is referred to *The Biomarker Guide* (Peters et al. 2005a, b) and the book chapter *Sedimentary Hydrocarbons, Biomarkers for early Life* (Brocks and Summons 2003).

14.5.2 Compound Identification

The identity of molecular fossils can be established in multiple ways. Well-known compounds are typically analyzed with methods similar to those described in the literature and compared to published data. When the mass spectra of molecules are known to the analyst, or can be found in the scientific literature or one of the many web-based resources (e.g. http://lipidlibrary.aocs.org/), these can be specifically monitored during GC-MS analyses by employing SIM or MRM approaches that minimize background noise and interferences. In some cases, the mass spectra of individual compounds are simple enough to deduce structural features (e.g. pristane and phytane in Fig. 14.7), which can then be checked by reference to the published spectra of standards. While there is generally little doubt about the identities of common components such as e.g. *n*-alkanes, other molecules that are present in very low concentrations, or whose distributions or mass spectra are ambiguous, typically have to be confirmed by co-injection/co-elution experiments. Many compounds of biogeochemical interest are commercially available and can be added to an aliquot of the sample as a 'spike'. If the tentatively identified peak in the chromatogram becomes more intense but is not broadened by the addition of the standard, this suggests positive identification. Such co-elution experiments should be repeated on chromatographic columns of differing polarity, as the co-elution on only one column could be a coincidence. When switching from, for example, a DB-1 to a DB-17 column, the relative elution positions of most compounds changes; the probability that two compounds will exactly co-elute on different columns is low. If pure, certified, standards are unavailable, an alternative approach is to use an oil or bitumen that has been reported to contain the compound of interest. In instances where this approach fails, geochemists have used preparative LC or GC to obtain sufficient for structural identification by nuclear magnetic resonance (NMR) (e.g. Sinninghe-Damsté et al. 2002), X-ray crystallography (e.g. Moldowan et al. 1991a, b) or chemical synthesis (e.g. Moldowan et al. 1991a, b; Summons and Capon 1991).

14.5.3 Contamination

Considerable concern surrounds the integrity of bitumens isolated from ancient rock samples (e.g., McKirdy 1974; Hayes et al. 1983; Hoering and Navale 1987; Summons and Walter 1990; Brocks et al. 2008; Waldbauer et al. 2009). The problem is particularly acute in studies of very old rocks, but applies in one form or another to all analyses of sediments that are either lean in organic matter and/or characterized by a high thermal maturity.

Contamination can generally be classified as three types: (a) laboratory workup contamination, (b) drilling fluid contamination, and (c) ancient or modern subsurface contamination. Any contaminants that are introduced during laboratory procedures can be controlled by awareness of contaminant sources together with consistent use of procedural blanks. Contamination of samples during their recovery as drill core is considerably more complicated. Drilling fluids almost always contain at least small amounts of hydrocarbons, including biomarker steranes and terpanes, that are deliberately or inadvertently added or leached from overlying sediments. Core samples must thus always be assumed to be contaminated. Brocks et al. (2008) have suggested that, by comparing the bitumen composition of inside versus the outside parts of cores, the extent of contamination can be estimated, thereby deeming select parts of the core either useful or otherwise. However, frequently cores have been pervasively impregnated with traces of drilling fluids. Sherman et al. (2007) have shown that even when this is the case, bitumens can sometimes be recovered by extracting the mineral-occluded bitumen fraction of core insides. These bitumens, which often have higher biomarker contents than freely extractable bitumens, and subtly different characteristics, are less likely to have been corrupted by drilling contaminants. Currently, efforts are underway to drill Archean cores for biomarker analysis with drilling fluids that have been chemically spiked to assess the role of drilling fluids in core contamination.

Subsurface contamination is hard to reconstruct as is staining by migrating petroleum fluids and hydrocarbon-containing basinal brines. Biomarkers are poorly soluble in water under standard conditions but this might change when aqueous fluids contain dissolved organics such as phenols, benzene, and toluene (Love et al. 2003; Zhang et al. 2005) as well as gases such as methane and CO_2 under elevated temperatures and pressures. A thorough understanding of the local geology is thus essential, and aspects such as the presence of other organic rich units in the basin, local hydrology, rock fracturing, and the permeability of adjoining units must be taken into account

14.5.3.1 Branched Alkanes with Quaternary Carbon Atoms (BAQCs)

Branched alkanes with quaternary carbon atoms (BAQC), such as n,n-dimethyl and n,n-diethyl alkanes, have been reported to occur in ancient sediments and were initially attributed to microbial sources of unknown affinities (e.g. Kenig et al. 2003; Greenwood et al. 2004). However, Grosjean and Logan (2007) have since shown that BAQC, amongst other compounds, are non-natural contaminants associated with the manufacture of plastics. Due to their moderate volatility, they can leach from polyethylene and polypropylene containers. Being anthropogenic materials, they are a most obvious sign of sample contamination and very easily picked up by sediment samples during storage. While BAQCs and plasticizers are easily recognized and do not interfere with the identification and quantitation of diagnostic terpenoids, plastics may also leach traces of petroleum-derived hydrocarbons (Aziz et al. 2008; Brocks et al. 2008). Based on the fact that anthropogenic hydrocarbons such as BAQCs have similar adsorption and diffusion properties to petrogenic hydrocarbons, the distribution of BAQCs has been proposed as a means to gauge contaminant penetration potential (Brocks et al. 2008). A spatial gradient in BAQCs, e.g. BAQC-free organic extracts from rock fractions inside the core or from mineral-occluded bitumen contrasting BAQC-rich extracts from core edges may also be used to infer the indigenous or contaminant nature of other hydrocarbons in the core. While core edges are mostly enriched in BAQCs, we find that BAQC-free fractions can often be recovered from mineral occluded bitumens.

Other sources of potential contamination include sunscreen, cosmetics and common lubricants (Grosjean and Logan 2007). These contaminants included fatty acid amides, chemical antioxidants, alkyl cyclohexanes and alkyl cyclopentanes (Grosjean and Logan 2007; Brocks et al. 2008). In any case, the presence of BAQCs or other contaminants does not necessarily render a sample void but, rather, can be a valuable tool for monitoring the spatial distribution of contaminants versus indigenous components.

14.5.3.2 Patterns of *n*-Alkane Hydrocarbons

The pattern of *n*-alkanes in an uncontaminated sample should be consistent with what is known about the prevailing biology and thermal history of a sediment sample. Linear alkanes are made *de novo* by algae, bacteria, and land plants. An odd over even preference in the abundances of *n*-alkanes, particularly *n*-C_{27}, *n*-C_{29} and *n*-C_{31}, is a universal feature of leaf wax and, therefore, suggests the presence of land-plants in a sedimentary environment (Peters et al. 2005b). Data from fossil spores and pollen suggest that land plants originated in the mid-Ordovician (Qiu et al. 2006); therefore an odd over even preferences in C_{27+} alkanes may be seen as contamination in samples thought to pre-date the evolution of land plants. One has to be careful here, however, because some green algae also contain long-chain hydrocarbons with an odd over even carbon number preference (e.g. Knights et al. 1970). Further, at the high end of the maturity spectrum, most carbon number preferences disappear as the result of hydrocarbon cracking and the *n*-alkanes progressively assume a unimodal distribution. A bimodal pattern, though not necessarily indicative of contamination, creates the possibility that one distribution is of a more recent contribution to the bitumen. Lastly, biodegradation is, along with other factors, indicated by loss of light *n*-alkanes and generation of an unresolved complex mixture (UCM) in a gas chromatogram (Peters et al. 2005a, b). Biodegradation is the microbial attack on petroleum hydrocarbons and thus reflects the activity of organisms that post-date the original organics. This signal of decay is an additional warning sign that a sample may contain several generations of hydrocarbon biomarkers.

14.5.3.3 Thermal Maturity Parameters and Their Use

During the maturation of sedimentary organic matter, the macromolecular, insoluble and immobile kerogen responds by cracking to smaller and more mobile molecules (Fig. 14.1). The effects of thermal stress also result in modifications to individual molecules, seen in changes at the level of stereochemical configuration (Figs. 14.8 and 14.9). All biomarkers are formed as the result of sequences of enzymatic reactions that recognize, confer and preserve the three-dimensional characteristics of molecules (i.e. their stereochemistry). Originally formed lipids have a unique combination of spatial patterns that is commonly referred to as the 'biological configuration'. This is the particular arrangement of atoms in the molecule, akin to the homochirality in amino acids and sugars, that confers optimal functionality in a living organism. Very often, however, the 'biological configuration' is the least thermodynamically stable one and there is, inevitably, a drive to revert to a more stable form. Almost all biomarkers, therefore, undergo stereochemical changes during diagenesis and catagenesis that lead to their conversion from the single, unique 'biological configuration' to a combination of more thermodynamically stable forms, that is, 'geological configurations'. This may include some molecules retaining the original stereochemical form. Stereochemical changes are progressive, along with the overall transformation of kerogen, and aligned to the time and temperature of burial under the cover of a sedimentary pile. The term 'increasing maturity' is used describe the process (Killops and Killops 2005; Peters et al. 2005b) of thermal evolution. Maturity refers to both the overall level of thermal alteration of kerogen (e.g. a darkening in color and increasing reflectivity) as well as the extent of stereochemical change in individual molecules.

In steroidal hydrocarbons, the stereocenters that are most prone to alteration during burial are those at C-5, C-14, C-17 and C-20 (Fig. 14.8). In the former three, which are at ring-junctions in the tetracyclic ring system, the hydrogen can be below or above the plane formed by the four rings and these are referred to as the alpha (α) and beta (β) positions respectively (Eliel 1994). The stereocenter at C-20 is not a ring junction and is described according to the *S* and *R* convention. Therefore, the biological configuration is formally described as 5α, 14α, 17α, 20*R*. All sterols are exclusively in the 20*R* from but this can change to a mixture of 20*S* and 20*R* isomers in a fossilized steroid and all its diagenetic products (Fig. 14.8). The 20*S*/(20*S*+20*R*) ratio in steranes is therefore a common means to assess thermal maturity in bitumen and oil. Because of its great age, Precambrian organic matter is expected to be 'thermally mature'. Values at or near 0.55 for the 20*S*/(20*S*+20*R*) ratio in steranes indicate sediment samples (Seifert and Moldowan 1986) that are

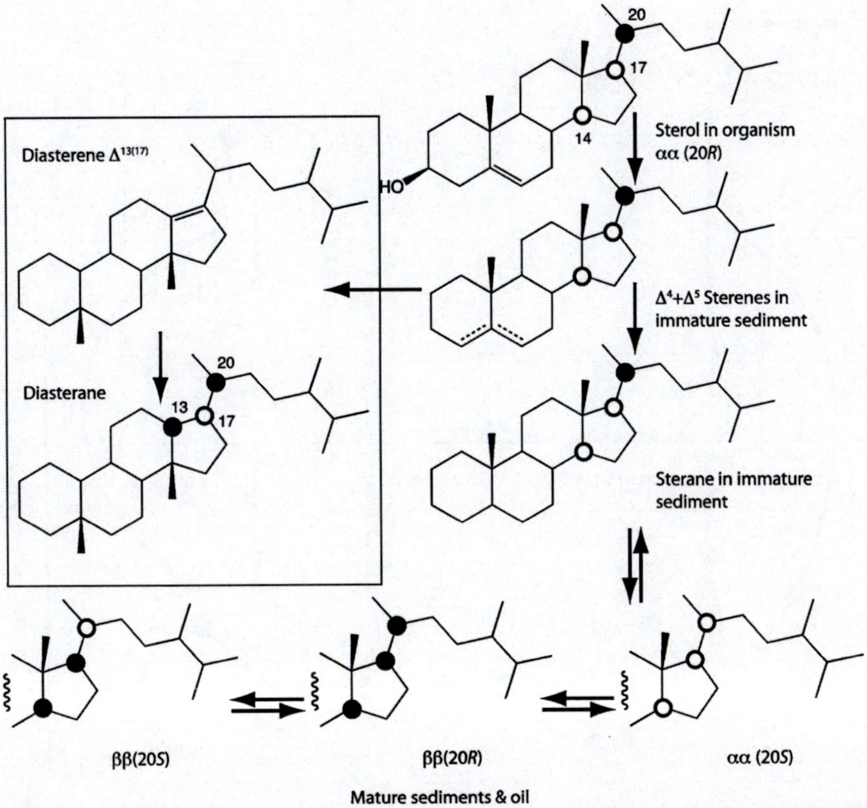

Fig. 14.8 Steroid stereochemistry during thermal maturation. Steroids are biosynthesized as unsaturated (Δ^5) alcohols (sterols) with a 5α, 14α, 17α, 20*R* stereochemical configuration. The position of hydrogen atoms at ring junctions (C-5, C-14, C-17) is described as being below (α) or above (β) the plane of the molecule and depicted by an open (α) or a filled circle (β). Stereogenic centers that are not part of a cyclic system (C-20) are either in an *S* (*open circle*) or *R* (*filled circle*) configuration. A more complete description of stereochemical rules can be found in Killops and Killops (2005), Peters et al. (2005b) or on the IUPAC website (http://www.chem.qmul.ac.uk/iupac/stereo/). Sterols are defunctionalized during early diagenesis and give rise to sterenes and, subsequently, steranes without affecting the stereochemical configuration. Elevated temperature and pressure conditions induced by sedimentary burial will drive the progressive conversion of the biological sterane configuration to more thermodynamically stable forms: 20*R* is changed to a mixture of 20*S* and 20*R*, and 14α, 17α to a mixture of αα and ββ. The 20*S*/(20*S*+20*R*) and ββ/(αα+ββ) ratios can thus be used to asses the level of thermal maturation that a bitumen has experienced. Immature sterenes can undergo rearrangement to form diasterenes and diasteranes. The conversion is thought to be catalyzed by acidic sites on clays (Rubinstein et al. 1975; Sieskind et al. 1979; van Kaam-Peters et al. 1998). Due to the methyl-migration, the stereocenter at C-14 shifts to C-13 and thermal maturation gives rise to a mixture showing 13β, 17α 20*S* and 20*R* (major isomers) and 13α, 17β 20*S* and 20*R* (minor isomers) stereochemistries (Peters et al. 2005b)

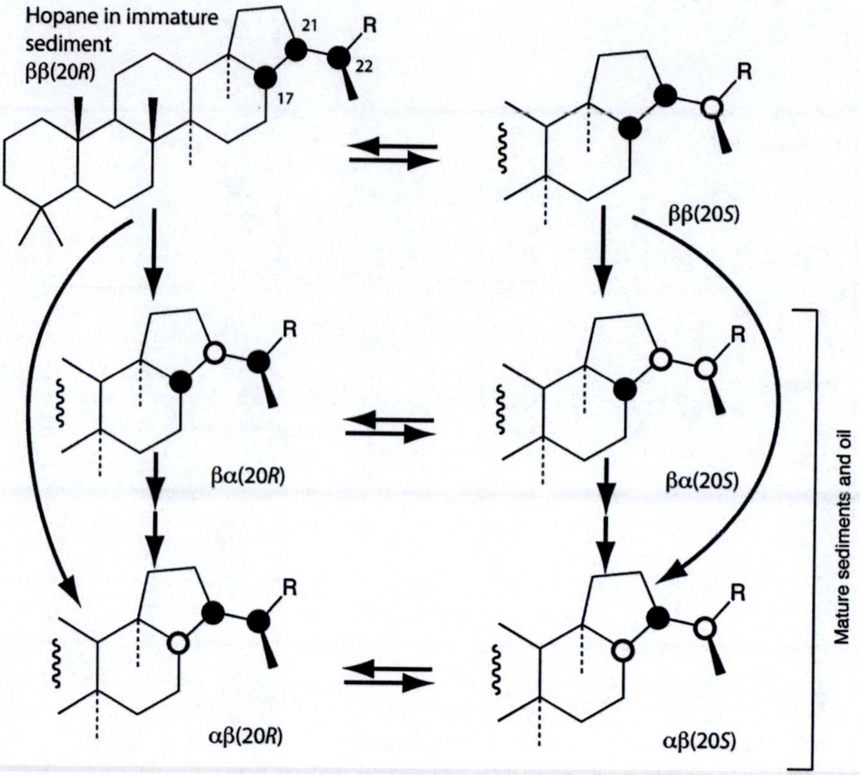

Fig. 14.9 Hopanoid stereochemistry during thermal maturation. After defunctionalization of the biological precursor, hopanes in immature sediments carry the 17β, 21β stereochemical configuration. Hopanes with 31 or more carbon atoms (i.e. with a more intact side chain) carry an additional stereocenter at C-22 that is biosynthesized in the *R* form. During thermal maturation, isomerization at C-22 leads to a mixture of 22*R* and 22*S* hopanes. Concurrently, the ββ configuration is converted to an αβ and βα mixture (Farrimond et al. 1998), with an initial (under slightly elevated temperature and pressure conditions) predominance of βα hopanes due to a lower energy barrier from ββ to βα than from ββ to αβ. At higher temperatures, a further conversion of βα to αβ hopanes will take place through a ββ intermediate. The αβ/(αβ+βα) and 22*S*/(22*S*+22*R*) ratios are used as molecular maturity indicators (Seifert and Moldowan 1980)

mature enough to have expelled much of their potential petroleum. Because oils contain steranes with different carbon numbers, an important test of consistency is to check that the values for the 20*S*/(20*S*+20*R*) ratio are identical or very similar at each carbon number (i.e. C_{27}, C_{28} and C_{29}). If these biomarkers are all indigenous to the sample, the values of all their maturity ratios should be very close, as they will have seen the same thermal history. A large difference in the two ratios suggests multiple sources of different maturities, and likely different age. If, for example, the C_{28} steranes have a lower ratio, then it is likely that they are younger than the indigenous bitumen and that the sample has been compromised.

The hopanoid triterpanes have maturity-sensitive stereocenters at C-17, C-21 in the ring system and at C-22 in the side-chain of those hopanes with 31 or more carbon atoms (homohopanes) (Fig. 14.9). The biologically unique form is termed

17β, 21β, 22*R* while the geologically stable forms are 17α, 21β (also called hopanes) and 17β, 21α (now called moretanes). All fossilized homohopanes comprise mixtures of 22*S* and 22*R* isomers. A hopane with αβ stereochemistry is thermodynamically more stable than one with βα configuration and this is, in turn, more stable than the ββ form (Fig. 14.9). In practice, the ββ form completely disappears during the early stages of burial and the abundance of the βα (i.e. moretane) is the next best index of low maturity. Empirical studies suggest that high αβ/βα (i.e. hopane/moretane) and 22S/(22S + 22R) ratios are also indicative of sediment samples (Seifert and Moldowan 1980) that are mature for petroleum generation. Therefore, one would not expect to find Precambrian samples with low values for these parameters. For a more comprehensive treatment of maturity parameters, the reader is referred to texts such as the *Biomarker Guide* (Peters et al. 2005a, b) or *Introduction to organic geochemistry* (Killops and Killops 2005).

14.5.3.4 Age-Sensitive Biomarker Distributions

Paleontology informs us that plants, animals and algae have evolved through geological time. This evolution has been accompanied by the appearance of new biosynthetic pathways and an ever-changing prevalence of certain types of molecular fossils in sedimentary rocks through the ages. A particularly well-established example is oleanane, a higher plant-derived pentacyclic terpenoid whose functionalized precursor, β-amyrin, is found in many classes of flowering plants (ten Haven and Rullkötter 1988; Das and Mahato 1983). Its secular distribution in sediments and petroleum has been shown to follow the Cenozoic diversification of angiosperm families as measured by pollen occurrences (Moldowan et al. 1994). Typically, little or no oleanane is found in sediments pre-dating the Late Cretaceous rise of angiosperms (Fig. 14.10a). When found in older rocks, oleanane likely represents contamination from a younger generation of hydrocarbons.

Other examples are molecules that are produced by algae of relatively recent origin. Molecular phylogenetic and culture studies have shown that C_{25} highly-branched isoprenoids (HBI) are biosynthesized only by two classes of diatoms (Sinninghe-Damsté et al. 2004). The sedimentary distribution of HBI can be shown to follow the rise of the older of the two classes, the rhizosolenoid diatoms, during the Late Cretaceous. HBI can thus be used as an age-dependent marker and contaminant tracer in ancient sediments, similar to oleanane (Fig. 14.10b).

A third example involves the distribution patterns of the ubiquitous but low abundance steranes with 26 carbon atoms. Two isomers (24-nor and 27-nor) of the C_{26} sterol precursors to these molecular fossils are known to occur in marine invertebrates and some algae (Schmitz 1978). A third isomer (21-nor) is found in sediments but has no known natural precursor. The structures of the chemical fossils were established by Moldowan and co-workers (Moldowan et al. 1991a) through rigorous chemical synthesis. A study of evolving petroleum composition (Holba et al. 1998) showed that the ratio of 24-norcholestanes/ (24-norcholestanes + 27-norcholestanes), and their rearranged sterane (diasterane) isomers, roughly follows the

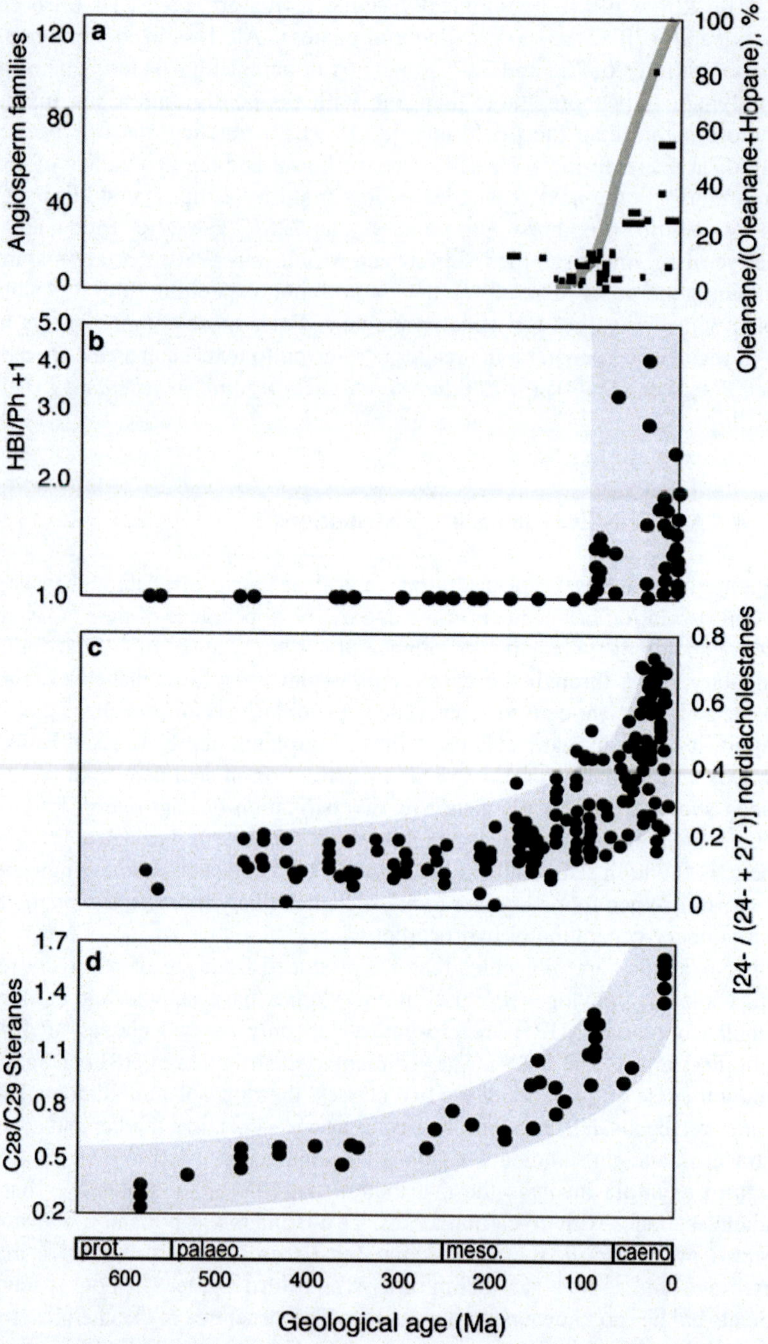

Fig. 14.10 Age-diagnostic markers. A selection of molecules which show secular changes that follow the radiation or emergence of their source organisms. In some cases they may be used to recognize contamination in old samples. After: **(a)** Moldowan et al. (1994), **(b)** Sinninghe-Damsté et al. (2004), **(c)** Holba et al. (1998), **(d)** Grantham and Wakefield (1988)

rise of diatom fossils in sedimentary rocks. It was therefore hypothesized that some kinds of diatoms were the predominant source of 24-norsterols in geological samples. This hypothesis was further supported by the identification of 24-norsterols in a cold-water diatom and in select dinoflagellates by Rampen et al. (2007). These authors therefore suggested that the ratio is affected by the evolution of both diatoms and dinoflagellates and is supported by other findings of 24-norsterols in dinoflagellates (Goad and Withers 1982; Kokke and Spero 1987; Thomson et al. 2004). Because diatoms and dinoflagellates both rose to prominence during the Mesozoic Era, it is expected that the 24-nor/(24-nor+27-nor) sterane ratios should be very low prior this. In Ediacaran and Cambrian sediments, for example, this ratio is expected to be less than 0.25 (Holba et al. 1998; Fig. 14.10c). It is essential, however, to remember that this ratio is based on data gathered from oils, which drain from large volumes of sedimentary rocks and therefore represent an averaged signal that integrates material deposited over large areas and intervals of time. It may be the case that diatoms and dinoflagellates, or their ancestors, proliferated for short intervals of time, or under unusual environmental conditions prior to their subsequent, large-scale radiation in the Mesozoic. Accordingly, while high values of some biomarker parameters may be used as guides to contamination, especially in early Paleozoic or Neoproterozoic rocks, one must not automatically exclude the possibility that such data are a real signal of paleobiological significance.

As an age-sensitive geochemical parameter, the ratio of $C_{28}/(C_{28}+C_{29})$ steranes (Fig. 14.10d) was originally shown to vary systematically with the geological age of petroleum source rocks (Grantham and Wakefield 1988). An increase in this ratio through geological time was thought to be associated with the diversification of phytoplankton assemblages through the Mesozoic. As with the 24-nor/(24-nor+27-nor) sterane ratios, a high ratio (e.g. >~0.4) for bitumen from a putative Precambrian rock may be an indicator of contaminated samples or an incorrect age assignment (Grantham and Wakefield 1988; Schwark and Empt 2006). Knoll et al. (2007) attribute the trend to the evolution of the so-called 'modern plankton' that is diatoms, dinoflagellates and coccolithophorids that produce both chlorophylls a and c. The issue of 'signal averaging', discussed above, is well illustrated by some studies of Paleozoic rock samples, as opposed to oils, where abundances of C_{28} steranes have been observed to be elevated during the Kacak, and Upper and Lower Kellwasser extinction events of the Devonian and Carboniferous (Schwark and Empt 2006). These 'events' manifest themselves as spikes over the average background level for that time. One interpretation for observed high values for the $C_{28}/(C_{28}+C_{29})$ steranes ratio in some ancient sediments may be proliferation of a taxon of green algae, the *Prasinophyceae*, which appear to be resilient under oxygen depleted environments (Schwark and Empt 2006). Some prasinophytes are known to biosynthesize an abundance of C_{28} sterols (Volkman 1986; Kodner et al. 2008). Little is yet known about the behavior of the $C_{28}/(C_{28}+C_{29})$ ratio during the Precambrian. A high value may be a sign that a bitumen sample contains a component of younger hydrocarbons. Alternatively, one has to consider that the ratio marks a real 'event' and conduct appropriate experiments to distinguish these possibilities.

14.5.3.5 Early or Late Eukaryotes?

The emergence of the eukaryotic lineage was a significant evolutionary event that paved the way for all the subsequent increase in biotic complexity. Knowledge of the first eukaryotes is important for the calibration of molecular clocks and our understanding of fundamental aspects of evolution and the role of environment as a driver of that. The age at which eukaryotes first appeared on Earth is however strongly debated. While the oldest likely eukaryotic body fossils are known from the early Mesoproterozoic (Javaux et al. 2003; Knoll et al. 2006), steroid molecular fossils have been reported from Neoarchean sediments (Brocks et al. 1999; Eigenbrode 2004; Sherman et al. 2007; Waldbauer et al. 2009). Steroids are essential for physiological functioning in all eukaryotes, and the presence of syngenetic sedimentary steranes is thus taken as evidence for the presence eukaryotes during deposition. Some criticism arose based on the fact that a number of bacteria are also capable of biosynthesizing primitive steroids (e.g. Pearson et al. 2003) and it was argued that Archean steroids could have a bacterial source (e.g. Cavalier-Smith 2006). However, no bacterium has yet been shown to induce downstream modifications and make steroids that are alkylated at position C-24; the extended steroids ergostane (C_{28}) and stigmastane (C_{29}) can thus be regarded as firm eukaryotic markers when found in sediments. Both ergostane and stigmastane were part of the Archean sedimentary lipid suites reported by Brocks et al. (1999), Eigenbrode (2004), Dutkiewicz et al. 2006; Sherman et al. (2007) and Waldbauer et al. (2009).

A bigger point of concern regards sources of contamination. Brocks et al. (2003a, b) argued that the "*arguments for syngeneity evidently outweigh the objections*". Over the years, however, the awareness of contamination-related problems has risen (e.g. Grosjean and Logan 2007; Brocks et al. 2008; George et al. 2010). More recently, the results of micro-scale in-situ measurements of $\delta^{13}C$ on kerogen and pyrobitumen in 2.7 Ga old rocks from the Pilbara craton lead Rasmussen et al. (2008) to believe that "*the biomarkers...*" [reported by Brocks et al. 1999] "*...must have entered the rock after peak metamorphism ~2.2 Gyr ago and thus do not provide evidence for the existence of eukaryotes and cyanobacteria in the Archaean eon*". While extreme prudence in the interpretation of such old sedimentary lipids is certainly warranted, Rasmussen et al. (2008) do not identify the exact stratigraphic location of the analyzed samples and use stable carbon isotopic values of alkanes and isoprenoids reported by Brocks et al. (1999) to exclude the syngeneity of hopanoids or steroids, which makes the reported 1:1 comparison of their analyzed samples and those studied by Brocks et al. (1999) not fully justifiable. More importantly, one should keep in mind that by now multiple studies have reported steroids in Archean sediments and fluid inclusions (see Sect. 14.1.2.4) from different basins, depositional ages and lithologies (Eigenbrode 2004; Dutkiewicz et al. 2006; Sherman et al. 2007; Waldbauer et al. 2009).

Even so, it is important to remind ourselves of the potential sources of contamination and the evidence with which we may disprove their presence. The separate analysis of core outsides and insides (Brocks et al. 2008) or of free and mineral-occluded bitumen (Sherman et al. 2007) can yield valuable insight to the penetration depth of contaminants during and after drilling. We should also expect to find sharp and reproducible transitions in certain biomarker parameters at lithological boundaries. Nevertheless, these approaches cannot offer absolute confidence in the nature of encountered lipids. It will be up for future studies, employing 'ultra-clean' drilling operations, to resolve the syngeneity of Archean steroids. While clean aqueous drilling fluids can be spiked with water-soluble hydrocarbons to assess the degree of impregnation of drilling fluids into the recovered core samples, staining-contamination by subsurface fluid flow can only be evaluated by a thorough knowledge of the geological setting. Additionally, preferential migration conduits of aqueous brines should be targeted for core recovery and analysis. While entrained organics can be expected to preferentially accumulate where aquifers border with lithologies that are prone to adsorb organics, such as clay- and/or organic-rich strata, hydrocarbon expulsion from shale units preferentially takes place at their edges (Leythaeuser and Schaefer 1984). Any significant enrichment of bitumen and biomarkers at the edge of shale units, or at clay-sandstone boundaries, relative to the center of these units, should consequently be a warning sign and remind us of the uncertainties and difficulties involved in the analysis of trace analytes in such ancient material.

14.5.4 Blanks

To asses potential contamination induced during the laboratory workup procedure, procedural blanks should be processed parallel to the actual samples. We suggest processing of both procedural blanks that undergo the full preparative methodological procedure, as well as blanks for individual processes such as grinding, extraction and acid-digestion. In the case of a specific contaminating step in the rock treatment, these process-blanks enable easy recognition of the sources of contamination.

In our experience, a common type of contamination involves cross-contamination within a set of samples between ones that have abundant bitumen and others that are largely devoid of hydrocarbons. A potential source of this cross-contamination is suspected to be the use of pipettes. Small amounts of sample material might be drawn up into the pipette bulb, and enter subsequent samples when the bulb is re-used. On the other hand, the use of a new pipette bulb for every pipette is highly uneconomic. If pipettes are used to transfer samples, we advise plugging its top with a small amount of combusted glass wool, minimizing any potential sample transfer from pipette to bulb. Ideally, however, we prefer to use syringes with a cemented tip that can be rapidly and efficiently cleaned with organic solvents. Despite taking these precautions, absolutely clean procedural blanks are an exception rather than the rule. The analyst must evaluate whether the signal in the blanks is significant enough to suspect the same contents in the true samples to be a laboratory contamination and not original constituents of the bitumen. Brocks et al. (2003a, b) suggested to calculate a blank/sample ratio for individual components, setting a cutoff at a value of 20 (i.e. a component is 20× more abundant in the sample than in the blank), which we endorse.

14.6 Concluding Remarks and Future Directions

The methodological overview given in this chapter is far from complete and, especially in regard to gas chromatography and mass spectrometry, only touches on the basics of the subject. Regardless of all the problematic issues that surround the reconstruction of ancient environments and past organismic diversity with sedimentary lipids, the field of organic geochemistry has come a long way since Treibs' (1936) recognition of the biogenic origin of petroleum. Over the years, awareness of potential contamination has been growing, and methodological approaches have continuously been improved. A prognosis based on past and current analytical improvements gives hope that current limitations in terms of sample size will be overcome at some point, and thereby allow a more precise evaluation of the originality of sedimentary lipids and increase our understanding of the evolution of early life on Earth.

Acknowledgments Early life research in the Geobiology Laboratory at MIT is supported by grants from the NASA Astrobiology Institute, the NASA Exobiology Program, the National Science Foundation and the Agouron Institute to RES. CH thanks the Agouron Institute for a Geobiology Postdoctoral Fellowship. The authors are grateful to Marc Laflamme, James Schiffbauer and Stephen Dornbos for the opportunity to write this chapter and to Herbert Volk and Srinath Krishnan for helpful reviews.

References

Armanios C, Alexander R, Kagi RI, Sosrowidjojo IB (1994) Fractionation of sedimentary higher-plant derived pentacyclic triterpanes using molecular sieves. Org Geochem 21:531–543

Audino M, Grice K, Alexander R, Kagi R (2004) Macrocyclic alkanes in crude oils and sediment extracts: enrichment using molecular sieves. Org Geochem 35:661–663

Aziz N, Greenwood PF, Grice K et al (2008) Chemical fingerprinting of adhesive tapes by GCMS detection of petroleum hydrocarbon products. J Forensic Sci 53:1130–1137

Bastow TP, van Aarssen BGK, Lang D (2007) Rapid small-scale separation of saturate, aromatic and polar components in petroleum. Org Geochem 38:1235–1250

Bennett B, Larter SR (2000) The isolation, occurrence and origin of fluorenones in crude oils and rock extracts. Org Geochem 31:117–126

Berner RA (1991) A model for atmospheric CO_2 over Phanerozoic time. Am J Sci 291:339–376

Bhullar AG, Karlsen DA, Backer-Owe K et al (1999) Dating reservoir filling – a case history from the North Sea. Mar Petrol Geol 16:581–603

Biddle JF, Lipp JS, Lever MA et al (2006) Heterotrophic Archaea dominate sedimentary subsurface ecosystems off Peru. Proc Natl Acad Sci USA 103:3846–3851

Briggs DEG (2003) The role of decay and mineralization in the preservation of soft-bodied fossils. Annu Rev Earth Planet Sci 31:275–301

Brocks JJ, Logan GA, Buick R, Summons RE (1999) Archean molecular fossils and the early rise of eukaryotes. Science 285:1033–1036

Brocks JJ (2001) Molecular fossils in Archean rocks. Dissertation, University of Sydney

Brocks JJ, Love GD, Snape CE et al (2003a) Release of bound aromatic hydrocarbons from late Archean and Mesoproterozoic kerogens via hydropyrolysis. Geochim Cosmochim Acta 67:1521–1530

Brocks JJ, Buick R, Logan GA, Summons RE (2003b) Composition and syngeneity of molecular fossils from the 2.78 to 2.45 billion-year-old Mount Bruce Supergroup, Pilbara Craton, Western Australia. Geochim Cosmochim Acta 67:4289–4319

Brocks JJ, Summons RE (2003) Sedimentary hydrocarbons, biomarkers for early life. In: Schlesinger WH (ed) Biogeochemistry, vol 8, Treatise on geochemistry. Elsevier, Amsterdam

Brocks JJ, Grosjean E, Logan GA (2008) Assessing biomarker syngeneity using branched alkanes with quaternary carbon (BAQCs) and other plastic contaminants. Geochim Cosmochim Acta 72:871–888

Canfield DE (1994) Factors influencing organic carbon preservation in marine sediments. Chem Geol 114:315–329

Canfield DE, Teske A (1996) Late Proterozoic rise in atmospheric oxygen concentration inferred from phylogenetic and sulfur-isotope studies. Nature 382:127–132

Cavalier-Smith T (2006) Cell evolution and Earth history: stasis and revolution. Phil Trans R Soc B 361:969–1006

Coolen MJL, Overmann J (2007) 217, 000-year-old DNA sequences of green sulfur bacteria in Mediterranean sapropropels and their implications for the reconstruction of the paleoenvironment. Environ Microbiol 9:238–249

Craig H (1957) Isotopic standards for carbon and oxygen and correction factors for mass-spectrometric analysis of carbon dioxide. Geochim Cosmochim Acta 12:133–149

Das MC, Mahato SB (1983) Triterpenoids. Phytochemistry 5:1071–1095

Des Marais DJ, Strauss H, Summons RE, Hayes JM (1992) Carbon isotope evidence for the stepwise oxidation of the Proterozoic environment. Nature 359:605–609

de Leeuw JW, Versteegh GJM, van Bergen PF (2006) Biomacromolecules of algae and plants and their fossil analogues. Plant Ecol 182:209–233

Didyk BM, Simoneit BRT, Brassel SC, Eglinton G (1978) Organic geochemical indicators of palaeoenvironmental conditions of sedimentation. Nature 272:216–222

Dutkiewicz A, Volk H, Ridley J, George S (2003) Biomarkers, brines, and oil in the Mesoproterozoic, Roper Superbasin, Australia. Geology 31:981–984

Dutkiewicz A, Volk H, Ridley J, George S (2004) Geochemistry of oil in fluid inclusions in a middle Proterozoic igneous intrusion: implications for the source of hydrocarbons in crystalline rocks. Org Geochem 35:937–957

Dutkiewicz A, Volk H, George SC et al (2006) Biomarkers from Huronian oil-bearing fluid inclusions: an uncontaminated record of life before the Great Oxidation Event. Geology 34:437–440

Dutkiewicz A, George SC, Mossman DJ et al (2007) Oil and its biomarkers associated with the Palaeoproterozoic Oklo natural fission reactors, Gabon. Chem Geol 244:130–154

Eigenbrode JL (2004) Late Archean microbial ecology: an integration of isotopic, molecular, and lithologic studies. Dissertation, The Pennsylvania State University

Eliel EL (1994) The stereochemistry of organic compounds. Wiley, New York

Ellis L, Kagi RI, Alexander R (1992) Separation of petroleum hydrocarbons using dealuminated mordenite molecular sieve I. Monoaromatic hydrocarbons. Org Geochem 18:587–593

Ellis L, Alexander R, Kagi RI (1994) Separation of petroleum hydrocarbons using dealuminated mordenite molecular sieve II. Alkylnaphthalenes and alkylphenanthrenes. Org Geochem 21:849–855

Eriksson PG, Altermann W, Nelson DR et al (eds) (2004) The precambrian Earth: tempos and events, vol 12, Developments in Precambrian geology. Elsevier, New York

Farrimond P, Taylor A, Telnaes N (1998) Biomarker maturity parameters: the role of generation and thermal degradation. Org Geochem 29:1181–1197

Freeman KH, Hayes JM, Trendel J-M, Albrecht P (1990) Evidence from carbon isotope measurements for diverse origins of sedimentary hydrocarbons. Nature 343:254–256

Gaines SM, Eglinton G, Rullkötter J (2009) Echoes of life: what fossil molecules reveal about earth history. Oxford University Press, New York

George SC, Ahmed M, Liu K, Volk H (2004) The analysis of oil trapped during secondary migration. Org Geochem 35:1489–1511

George SC, Volk H, Romero-Sarmiento M-F, Dutkiewicz A, Mossman DJ (2010) Diisopropylnaphthalenes: environmental contaminants of increasing importance for organic geochemical studies. Org Geochem 41:901–904

Goad LJ, Wither N (1982) 27-Nor-(24R)-24-methylcholesta-5, 22-dien-3/3-ol and brassicasterol as the major sterols of the marine dinoflagellate Gymnodinium simplex 1. Lipids 17:853–858

Gold T (1992) The deep, hot biosphere. Proc Natl Acad Sci USA 89:6045–6049

Grantham PJ, Wakefield LL (1988) Variations in the sterane carbon number distributions of marine source rock derived crude oils through geological time. Org Geochem 12:61–73

Greenwood PF, Arouri KR, Logan GA, Summons RE (2004) Abundance and geochemical significance of C_{2n} dialkylalkanes and highly branched C_{3n} alkanes in diverse Meso- and Neoproterozoic sediments. Org Geochem 35:331–346

Grice K, de Mesmay R, Glucina A, Wang S (2008) An improved and rapid 5A molecular sieve method for gas chromatography isotope ratio mass spectrometry of n-alkanes (C8-C30+). Org Geochem 39:284–288

Grob RL, Barry EF (eds) (2004) Modern practice of gas chromatography, 4th edn. Hoboken, Wiley

Grosjean E, Logan GA (2007) Incorporation of organic contaminants into geochemical samples and an assessment of potential sources: examples from Geoscience Australia marine survey S282. Org Geochem 38:853–869

Gross JH (2004) Mass spectrometry: a textbook. Springer, Berlin

Gupta NS, Michels R, Briggs DEG et al (2007) Experimental evidence for the formation of geomacromolecules from plant leaf lipids. Org Geochem 38:28–36

Hallmann C, Schwark L, Grice K (2008) Community dynamics of anaerobic bacteria in deep petroleum reservoirs. Nat Geosci 1:588–591

Hayes JM, Kaplan IR, Wedeking KW (1983) Precambrian organic geochemistry; preservation of the record. In: Schopf JW (ed) Origin and evolution of the Earth's earliest biosphere. Princeton University Press, Princeton, pp 93–132

Hayes JM, Freeman KH, Popp BN, Hoham CH (1990) Compound-specific isotope analyses: a novel tool for reconstruction of ancient biogeochemical processes. Org Geochem 16:1115–1128

Hayes JM (1993) Factors controlling ^{13}C contents of sedimentary organic compounds: principles and evidene. Mar Geol 113:111–125

Hayes JM (2001) Fractionation of the isotopes of carbon and hydrogen in biosynthetic processes. Rev Mineral Geochem 43:225–277

Head IM, Jones DM, Larter SR (2003) Biological activity in the deep subsurface and the origin of heavy oil. Nature 426:344–352

Hebting Y, Schaeffer P, Behrens A et al (2006) Biomarker evidence for a major preservation pathway of sedimentary organic carbon. Science 312:1627–1631

Hedges JI, Keil RG (1995) Sedimentary organic matter preservation: an assessment and speculative synthesis. Mar Chem 49:81–115

Hedges JI, Baldock JA, Gellnas Y et al (2001) Evidence for non-selective preservation of organic matter in sinking marine particles. Nature 409:801–804

Hinrichs K-U, Eglinton G, Engel MH, Summons RE (2001) Exploiting the multivariate isotopic nature of organic compounds. Geochem Geophy Geosy 2:8. doi:10.1029/2001GC000142

Hoering TC, Navale V (1987) A search for molecular fossils in the kerogen of Precambrian sedimentary rocks. Precambrian Res 34:247–267

Holba AG, Tegelaar EW, Huizinga BJ et al (1998) 24-Norcholestanes as age-sensitive molecular fossils. Geology 26:783–786

Holland HD (1999) When did Earth's atmosphere become oxic? A reply. Geochem News 100:20–23

Horsfield B, Disko U, Leistner F (1989) The micro-scale simulation of maturation: outline of a new technique and its potential applications. Geol Rundsch 78:361–374

House CH, Schopf JW, Stetter KO (2003) Carbon isotopic fractionation by Archaeans and other thermophilic prokaryotes. Org Geochem 34:345–356

Hunt JM (1995) Petroleum geochemistry and geology, 2nd edn. Freeman, New York

Javaux EJ, Knoll AH, Walter MR (2003) Recognizing and interpreting the fossils of early eukaryotes. Orig Life Evol B 33:75–94

Jensenius J, Burruss RC (1990) Hydrocarbon-water interactions during brine migration: evidence from hydrocarbon inclusions in calcite cements from Danish North Sea oil fields. Geochim Cosmochim Acta 54:705–713

Keil RG, Montlucon DB, Prahl FG, Hedges JI (1994) Sorptive preservation of labile organic matter in marine sediments. Nature 370:549–552

Kenig F, Popp BN, Summons RE (2000) Preparative HPLC with ultrastable-Y zeolite for compound-specific carbon isotope analyses. Org Geochem 31:1087–1094

Kenig F, Simons DH, Crich D et al (2003) Branched aliphatic alkanes with quaternary substituted carbon atoms in modern and ancient geological samples. Proc Natl Acad Sci USA 100:12554–12558

Kennedy M, Droser M, Mayer L et al (2006) Late Precambrian oxygenation; Inception of the clay mineral factory. Science 311:1446–1449

Killops SD, Killops VJ (2005) Introduction to organic geochemistry, 2nd edn. Blackwell, Oxford

Knights BA, Brown AC, Conway E, Middleditch BS (1970) Hydrocarbons from the green form of the freshwater alga *Botryococcus braunii*. Phytochemistry 9:1317–1324

Knoll AH, Javaux EJ, Hewitt D, Cohen P (2006) Eukaryotic organisms in Proterozoic oceans. Phil Trans R Soc B 361:1023–1038

Knoll AH, Summons RE, Waldbauer J, Zumberge J (2007) The geological succession of primary producers in the oceans. In: Falkowski P, Knoll AH (eds) The evolution of primary producers in the sea. Elsevier, Burlington, pp 133–163

Kodner RB, Pearson A, Summons RE, Knoll AH (2008) Sterols in red and green algae: quantification, phylogeny, and relevance for the interpretation of geologic steranes. Geobiology 6:411–420

Kok MD, Schouten S, Sinninghe-Damste JS (2000) Formation of insoluble, nonhydrolyzable, sulfur-rich macromolecules via incorporation of inorganic sulfur species into algal carbohydrates. Geochim Cosmochim Acta 64:2689–2699

Kokke WCMC, Spero HJ (1987) Sterol pattern determination as a probe for new species of zooxanthellae in marine invertebrates: application to the dinoflagellate symbiont of the foraminifer *Orbulina universa*. Biochem Syst Ecol 15:475–478

Konhauser K (2007) Introduction to geomicrobiology. Blackwell, Oxford

Lewan MD (1985) Evaluation of petroleum generation by hydrous pyrolysis experimentation. Phil Trans Roy Soc A 315:123–134

Leythaeuser D, Schaefer RG (1984) Effects of hydrocarbon expulsion from shale source rocks of high maturity in Upper Carboniferous strata of the Ruhr area, Federal Republic of Germany. Org Geochem 6:671–681

Love GD, Aplin AC, Larter SR, Taylor G (2003) Determination of stable carbon ($\delta^{13}C$) isotope systematics for alkylphenols and light aromatic hydrocarbons (BTEX) in petroleum formation waters and co-produced oils. J Geochem Explor 78–79:465–467

Love GD, Snape CE, Carr AD, Houghton RC (1995) Release of covalently-bound alkane biomarkers in high yields from kerogen via catalytic hydropyrolysis. Org Geochem 23:981–986

Love GD, Grosjean E, Stalvies C et al (2009) Fossil steroids record the appearance of Demospongiae during the Cryogenian period. Nature 457:718–721

Magoon LB, Dow WG (1994) The petroleum system – from source to trap. American Association of Petroleum Geologists, Tulsa

Maslen E, Grice K, Gale JD et al (2009) Crocetane: a potential marker of photic zone euxinia in thermally mature sediments and crude oils of Devonian age. Org Geochem 40:1–11

Matthews DE, Hayes JM (1978) Isotope-ratio monitoring gas chromatography-mass spectrometry. Anal Chem 50:1465–1473

McKirdy DM (1974) Organic chemistry in Precambrian research. Precambrian Res 1:75–137

McMurry J (2003) Fundamentals of organic chemistry, 5th edn. Thomson, Pacific Grove

Meritt DA, Hayes JM (1994a) Factors controlling precision and accuracy in isotope-ratio-monitoring mass-spectrometry. Anal Chem 66:2336–2347

Meritt DA, Hayes JM (1994b) Nitrogen isotopic analyses by isotope-ratio-monitoring gas-chromatography and mass-spectrometry. J Am Soc Mass Spectrom 5:387–397

Merrit DA, Freeman KH, Ricci MP, Studley SA, Hayes JM (1995) Performance and optimization of a combustion interface for isotope ratio monitoring gas-chromatography mass-spectrometry. Anal Chem 67:2461–2473
Moldowan JM, Lee CY, Watt DS et al (1991a) Analysis and occurence of C_{26}-steranes in petroleum and source rocks. Geochim Cosmochim Acta 55:1065–1081
Moldowan JM, Fago FJ, Carlson RMK et al (1991b) Rearranged hopanes in sediments and petroleum. Geochim Cosmochim Acta 55:3333–3353
Moldowan JM, Dahl J, Huizinga BJ et al (1994) The molecular fossil record of oleanane and its relation to angiosperms. Science 265:768–771
Morris LJ (1966) Separations of lipids by silver ion chromatography. J Lipid Res 7:717–732
Munz IA (2001) Petroleum inclusions in sedimentary basins: systematics, analytical methods and applications. Lithos 55:195–212
Nabbefeld B, Grice K, Schimmelmann A et al (2010) A comparison of thermal maturity parameters between freely extracted hydrocarbons (Bitumen I) and a second extract (Bitumen II) from within the kerogen matrix of Permian and Triassic sedimentary rocks. Org Geochem 41:78–87
Nagy IP, Bazsa G (1991) Artifacts caused by plastics in chemical experiments: absorption, degradation, catalysis and irreproducible behavior. React Kinet Catal Lett 45:15–25
Nicolas A (1995) The mid-oceanic ridges: mountains below sea level. Springer, Berlin
Orphan VJ, House CH, Hinrichs K-U et al (2001) Methane-consuming archaea revealed by directly coupled isotopic and phylogenetic analysis. Science 293:484–487
Pearson A, Budin M, Brocks JJ (2003) Phylogenetic and biochemical evidence for sterol synthesis in the bacterium *Gemmata obscuriglobus*. Proc Natl Acad Sci USA 101:15352–15357
Peters KE, Walter CC, Moldowan JM (2005a) The biomarker guide, vol 1: Biomarkers and isotopes in the environment and human history. Cambridge University Press, Cambridge
Peters KE, Walter CC, Moldowan JM (2005b) The biomarker guide, vol 2: Biomarkers and isotopes in petroleum systems and earth history. Cambridge University Press, Cambridge
Peterson KJ, Summons RE, Donoghue PCJ (2007) Molecular palaeobiology. Palaeontology 50:775–809
Qiu Y-L, Li L, Wang B et al (2006) The deepest divergences in land plants inferred from phylogenomic evidence. Proc Natl Acad Sci USA 103:15511–15516
Rampen SW, Schouten S, Abbas B et al (2007) On the origin of 24-norcholestanes and their use as age-diagnostic biomarkers. Geology 35:419–422
Rasmussen B, Fletcher IR, Brocks JJ, Kilburn MR (2008) Reassessing the first appearance of eukaryotes and cyanobacteria. Nature 455:1101–1104
Revsbech NP, Sorensen J, Blackburn TH (1980) Distribution of oxygen in marine sediments measured with microelectrodes. Limnol Oceanogr 25:403–411
Ricci MP, Merritt DA, Freeman KH, Hayes JM (1994) Acquisition and processing of data for isotope-ratio-monitoring mass-spectrometry. Org Geochem 21:561–571
Rocha JD, Brown SD, Love GD, Snape CE (1997) Hydropyrolysis: a versatile technique for solid fuel liquefaction, sulphur speciation and biomarker release. J Anal Appl Pyrol 40:91–103
Rubinstein I, Sieskind O, Albrecht P (1975) Rearranged steranes in a shale: occurence and simulated formation. J Chem Soc Perkin Trans 1:1833–1836
Schenk HJ, Horsfield B (1993) Kinetics of petroleum generation by programmed-temperature close- versus open-system pyrolysis. Geochim Cosmochim Acta 57:623–630
Schippers A, Neretin LN, Kallmeyer J et al (2005) Prokaryotic cells of the deep sub-seafloor biosphere identified as living bacteria. Nature 433:861–864
Schmitz FJ (1978) Uncommon marine steroids. In: Scheuer PJ (ed) Marine natural products, vol I, Chemical and biological perspectives. Academic, New York, pp 241–297
Schouten S, de Graaf W, Sinninghe-Damste JS et al (1993) Laboratory simulation of natural sulphurization: II Reaction of multi-functionalized lipids with inorganic polysulphides at low temperatures. Org Geochem 22:825–834
Schwark L, Empt P (2006) Sterane biomarkers as indicators of palaeozoic algal evolution and extinction events. Palaeogeogr Palaeoclimatol Palaeoecol 240:225–236
Seifert WK, Moldowan JM (1980) The effect of thermal stress on source-rock quality as measured by hopane stereochemistry. Phys Chem Earth 12:229–237

Seifert WK, Moldowan JM (1986) Use of biological markers in petroleum exploration. In: Johns RB (ed) Methods in geochemistry and geophysics, vol 24. Elsevier, Amsterdam, pp 261–290
Sherman LS, Waldbauer JR, Summons RE (2007) Improved methods for isolating and validating indigenous biomarkers in Precambrian rocks. Org Geochem 38:1987–2000
Sieskind O, Joly G, Albrecht P (1979) Simulation of the geochemical transformation of sterols: superacid effects on clay minerals. Geochim Cosmochim Acta 43:1675–1679
Sinninghe-Damsté JS, Strous M, Rijpstra WIC et al (2002) Linearly concatenated cyclobutane lipids form a dense bacterial membrane. Nature 419:708–712
Sinninghe-Damsté JS, Muyzer G, Abbas B et al (2004) The rize of the rhizosolenoid diatoms. Science 304:584–587
Sterflinger K (2000) Fungi as geological agents. Geomicrobiol J 17:97–124
Summons RE, Powell TG (1986) Chlorobiaceae in Palaeozoic seas revealed by biological markers, isotopes and geology. Nature 319:763–765
Summons RE, Capon RJ (1991) Identification and significance of 3β-ethyl steranes in sediments and petroleum. Geochim Cosmochim Acta 55:2391–2395
Summons RE, Jahnke LL, Cullings KW et al. (2001) Cyanobacterial biomarkers: triterpenoids plus steroids? EOS Trans AGU Fall Meet Suppl 82. Abstract B22D–0184
Summons RE, Walter MR (1990) Molecular fossils and microfossils of prokaryotes and protists from Proterozoic sediments. Am J Sci 290A:212–244
ten Haven HL, Rullkötter J (1988) The diagenetic fate of taraxer-14-ene and oleanene isomers. Geochim Cosmochim Acta 52:2543–2548
Thomson PG, Wright SW, Bolch CJS et al (2004) Antarctic distribution, pigment and lipid composition, and molecular identification of the brine dinoflagellate *Polarella glacialis* (Dinophyceae). J Phycol 40:867–873
Tosca NJ, Johnston DT, Mushegian A et al (2010) Clay mineralogy, organic carbon burial, and redox evolution in Proterozoic oceans. Geochim Cosmochim Acta 74:1579–1592
Treibs A (1936) Chlorophyll- und Haemderivate in organischen mineralstoffen. Angew Chem 39:682–686
Vandenbroucke M, Largeau C (2007) Kerogen origin, evolution and structure. Org Geochem 38:719–833
van Kaam-Peters HME, Koester J, van der Gaast SJ et al (1998) The effect of clay minerals on diasterane/sterane ratios. Geochim Cosmochim Acta 62:2923–2929
Volk H, George SC, Dutkiewicz A, Ridley J (2005) Characterisation of fluid inclusion oil in a mid-Proterozoic sandstone and dolerite (Roper Superbasin, Australia). Chem Geol 223:109–135
Volk H, Horsfield B, Mann U, Suchy V (2002) Variability of petroleum inclusions in vein, fossil and vug cements – a geochemical study of the Barrandian Basin (Lower Palaeozoic, Czech Republic). Org Geochem 33:1319–1341
Volkman JK (1986) A review of sterol markers for marine and terrigenous organic matter. Org Geochem 9:83–99
Waldbauer JR, Sherman LS, Sumner DY, Summons RE (2009) Late Archean molecular fossils from the transvaal supergroup record the antiquity of microbial diversity and aerobiosis. Precambrian Res 169:28–47
Wilhelms A, Larter SR, Head I et al (2001) Biodegradation of oil in uplifted basins prevented by deep-burial sterilization. Nature 411:1034–1037
Woese CR, Kandler O, Wheelis ML (1990) Towards a natural system of organisms: proposal for the domains archaea, bacteria, and eucarya. Proc Natl Acad Sci USA 87:4576–4579
Zhang Y, Person M, Merino E (2005) Hydrologic and geochemical conrols on soluble benzene migration in sedimentary basins. Geofluids 5:83–105

Chapter 15
Carbon and Sulfur Stable Isotopic Systems and Their Application in Paleoenvironmental Analysis

Kathleen A. McFadden and Amy E. Kelly

Contents

K.A. McFadden (✉)
ConocoPhillips, Houston, TX 77079-1100, USA
e-mail: meikailing@yahoo.com

A.E. Kelly
School of Earth and Space Exploration, Arizona State University,
85281-1404, Tempe, AZ, USA
e-mail: kellya@alum.mit.edu

M. Laflamme et al. (eds.), *Quantifying the Evolution of Early Life*, Topics in Geobiology 36,
DOI 10.1007/978-94-007-0680-4_15, © Springer Science+Business Media B.V. 2011

Abstract Stable isotope geochemistry has important applications for understanding past environments and should be incorporated, where appropriate, into paleontological research. This chapter aims to provide a basic introduction to the field of stable isotope geochemistry, highlight fundamental isotopic systems used in paleoenvironmental analysis, outline methods used to acquire isotopic data, explain common primary and secondary processes that can affect the interpretation of that data, and discuss the secular trends of carbon and sulfur isotopes through Earth history. In order to properly interpret stable isotope data, it is critical to understand the origin of the sample used in the geochemical analysis and use consistent sampling methodologies and preparation techniques. Careful attention should be made to assess the sample for diagenetic alteration and contamination prior to preparation for geochemical analysis, including the use of petrographic analysis, CL and SEM petrography, and spot sampling obvious diagenetic zones to compare against samples considered less diagenetically altered. Secular variation of carbon and sulfur isotopes has varied dramatically over geological history. Most isotopic excursions in the geological record appear to manifest from sudden imbalances carbon and sulfur cycles. Long-term variation has important implications on productivity, redox state of the atmosphere and oceans, and potentially the extinction and radiation of organisms. However, geochemists still struggle to understand the mechanisms, duration, and triggers of isotopic excursions. In many case studies, such as the Shuram anomaly, there are multiple scenarios that can produce isotopic excursions. The challenge for future research is to better address stable isotopic variation by integrating datasets across disciplines to improve our understanding of the complex interactions between environmental and biological systems.

Keywords Stable isotopes • Carbon • Sulfur • Carbonate diagenesis • Shuram anomaly

15.1 Introduction

Stable isotope geochemistry has become a leading tool in paleobiology to characterize a wide variety of biological processes and environmental conditions in the geological record. The variation of carbon, sulfur, and oxygen isotopes has been used to understand short-term change in redox stability, salinity, nutrient flux, productivity, and temperature of the global oceans (Epstein et al. 1951; Kump 1991; Scholle and Arthur 1980; Zachos et al. 2001). Likewise, long-term isotopic trends in the carbon and sulfur records have provided valuable insight into the fluctuation of atmospheric O_2 and its relationship to continental weathering as well as the burial of organic carbon and sedimentary sulfides (Berner 2006; Canfield 2001; Hayes et al. 1999; Kump 1989). Development of integrated techniques that analyze the temporal and stratigraphic distribution of geochemical signatures together within a strong depositional and diagenetic framework is critical to evaluate paleoenvironmental conditions and resolve the complex interactions between environmental change and the evolution of life. This chapter is not intended to be a full review of all known geochemical systems nor their applications to geology – there is a considerable amount of literature dedicated to stable isotope geochemistry (e.g. Hoefs 2008; Sharp 2007; Valley and Cole 2001 and references therein). Instead, we give a brief summary concerning conventional isotopic systems used in paleoenvironmental analysis.

The basic principles of stable isotope geochemistry are presented first, in order to review what isotopes are, how samples are prepared, how isotopic ratios are measured, and what processes contribute to variation in isotopic values in the geological record. We discuss two major isotopic systems, carbon and sulfur, that reflect secular variations in the isotopic mass balance of the world oceans, which in turn reflect long-term variability in global oceanic redox conditions and productivity. We summarize standard sampling protocols and extraction techniques that allow for the best reproducibility and confidence in the interpretation of geochemical data. Diagenetic overprinting, which occurs from time of deposition through burial, is an important process that must be considered for the accurate interpretation of any isotopic data. Therefore, we illustrate how rock-fluid interactions can potentially alter isotopic signatures and should be evaluated using rigorous petrographic analysis. Finally, we summarize the secular carbon and sulfur isotopic trends through geologic history and focus on one of the largest negative carbon isotopic excursions, called the Shuram anomaly ($\delta^{13}C$ minima near −15‰). Recent geochemical studies have suggested that the Shuram anomaly reflects pervasive and permanent ventilation of the deep oceans, which may have played a role in the evolution of early animals (Canfield and Teske 1996; Canfield et al. 2007; Fike et al. 2006; McFadden et al. 2008; Rothman et al. 2003). We discuss the timing, duration, and mechanisms of the Shuram anomaly as it is related to our current understanding of the carbon and sulfur cycle during the Neoproterozoic.

15.2 Introduction to Stable Isotopes

15.2.1 Basic Principles and Reference Standards

Isotopes are defined as atoms with the same number of protons, but a different number of neutrons. Isotopes can be either radioactive or stable; radioactive isotopes breakdown, or decay, into other isotopic forms or elements, whereas stable isotopes do not. Both types of isotopes are broadly applied in the geosciences. Radioactive isotopes are used to determine the ages of sedimentary rocks and can also act as chemical tracers in the geological record. Stable isotopes are commonly used to indirectly characterize surface and near-surface environmental conditions. The majority of stable isotopic literature attempts to describe past environmental conditions by focusing on a small number of light elements (Table 15.1). These elements are commonly analyzed by geochemists because they are (1) relatively common constituents in the Earth atmosphere and crust and (2) consist of at least two stable isotopic forms which exhibit large variation and are potentially related to both geological and biological processes.

Table 15.1 Stable isotopic abundances, gases commonly measured in gas source isotopic ratio mass spectrometers, isotopologues measured during analysis, and standards for five biogeochemically relevant elements (Buffle and van Leeuwen 1992)

Element	Stable isotope	Relative abundance %	Delta notation	Gas	Common isotopologues	Isotopologue ratios measured	Standard(s)
Hydrogen	^{1}H	99.999	δD	H_2	2 and 3	3/2	SMOW – Standard Mean Ocean Water
	^{2}H or D	0.015					
Carbon[a]	^{12}C	98.900	$\delta^{13}C$	CO_2	44, 45, and 46	45/44	PDB – Cretaceous Pee Dee Belemnite
	^{13}C	1.100					
Nitrogen[b]	^{14}N	99.630	$\delta^{15}N$	N_2	28, 29, and 30	29/28 and 30/28	NBS-14 – National Bureau of Standards
	^{15}N	0.370					AIR – Atmospheric N_2
Oxygen	^{16}O	99.762	$\delta^{18}O$	CO_2	44, 45, and 46	46/44	SMOW – Standard Mean Ocean Water
	^{17}O	0.038		O_2 (Fluorination)	32, 33, and 34	34/32	PDB – Cretaceous Pee Dee Belemnite
	^{18}O	0.200		CO (pyrolysis)	28 and 30	30/28	
Sulfur[c]	^{32}S	95.020	$\delta^{34}S$	SO_2	64 and 66	66/64	CDT – Canyon Diablo Triolite
	^{33}S	0.750					
	^{34}S	4.210	$\delta^{36}S$	SF_6	127, 128, 129 and 131	Mono-isotopic	
	^{36}S	0.014					

[a] Carbon also has several radioactive isotopes (^{10}C, ^{11}C, ^{14}C, and ^{16}C). ^{4}C is most commonly referred to as "radiocarbon" with a half-life of 5,730 ± 40 years (Godwin 1962)

[b] Nitrogen also has several radioactive isotopes

[c] Sulfur has 25 known isotopes, 4 of which are stable. The radioactive forms have half-lives on the order of microseconds to days (Holden 2010). The isotopologues for SF_6 are monoisotopic, which means that different masses of SF_6 represent variations in the sulfur

The absolute abundance of an individual isotope cannot be measured accurately because the amount of the rare, usually heavier, isotope, such as ^{18}O and ^{13}C, is much less than their familiar "light" isotopic counterparts, ^{16}O and ^{12}C. However, relative differences in the isotopic ratios ($^{18}O/^{16}O$ and $^{13}C/^{12}C$) between two samples can be measured with great precision. Therefore, the isotopic ratio of a sample is compared to a predefined standard and expressed using the delta (δ) notation:

$$\delta = (\mathrm{Rspl} - \mathrm{Rstd})/\mathrm{Rstd} \times 1000 \tag{15.1}$$

where Rspl represents the isotopic ratio of the sample and Rstd is the isotopic ratio of the standard. Delta values are reported in permil (‰), or parts per thousand. Positive δ values indicate that the isotopic ratio of the sample is greater than the isotopic ratio of the standard, while a negative δ value signifies the opposite. Common isotopic notations and references used in geochemical analyses for light elements are included in Table 15.1.

Global standards are important for calibration of instruments and meaningful comparison of isotopic compositions of samples between laboratories. The history of most reference standards is long and complex (e.g. Beaudoin et al. 1994; Coplen et al. 1983; Coplen and Krouse 1998; Craig 1953; Mariotti 1983). Early calibration of extraction techniques used in mass spectrometer analysis involved the development of standards from an outcrop or laboratory-derived sources. Over time, supply of most international standards has been exhausted, and carefully calibrated secondary and tertiary standards have replaced the originals.

Some samples are commonly described using different standards, although the use of reference standards is almost always restricted to specific applications in geochemistry. For example, $\delta^{18}O$ values extracted from water are typically reported with respect to Standard Mean Ocean Water (SMOW), whereas $\delta^{18}O$ values extracted from carbonates are almost always reported as Pee Dee Belemnite (PDB) (Coplen 1996). It is important to note that the isotopic values of standards often do not equal each other and must be mathematically converted in order to be compared effectively (Craig 1957).

15.2.2 Isotopic Fractionation and the Fractionation Factor

Isotopic fractionation is the process by which isotopes partition during biological, chemical, and physiological reactions. Early laboratory work first calibrated the isotopic fractionation of calcite to estimate the paleotemperature of seawater from calcite in fossils (Craig 1965; Epstein et al. 1951; McCrea 1950). Fractionation has also been used to describe biological processes associated with partitioning isotopes during metabolic reactions like photosynthesis, which in turn is used at gross scales to model productivity (Schidlowski et al. 1983).

Isotopic fractionation is controlled by a number of factors; mainly temperature, reaction rate, and atomic mass. Secondary controls include pressure, composition, and atomic structure, which will not be discussed (see Sharp 2007; Hoefs 2008 for in-depth review). Isotopic exchange reactions are classified as kinetic or equilibrium reactions. Kinetic reactions are unidirectional and include both physical and chemical processes that partition isotopes as a function of reaction rate. The fractionation effects are primarily determined by binding energies. Isotopically lighter molecules have higher velocities, react more rapidly, and have weaker bond strengths than their heavy isotopic equivalents, which lead to comparatively lower binding energies. Equilibrium reactions, on the other hand, are reversible and involve redistribution of isotopes, or isotopic exchange, between two chemical substances, phases, or individual molecules during a chemical reaction. The fractionation process during these reactions is largely a function of temperature.

The process of isotopic fractionation can be mathematically described by comparing the isotopic ratios between two substances before and after a reaction

$$aA + bB \leftrightarrow bA + aB \tag{15.2a}$$

where A and B represent chemical species, and a and b refer to the heavy and light isotopes, respectively. One such example is given by the fractionation between CO_2 and bicarbonate during the dissolution of CO_2 into water:

$$^{13}CO_2 + H^{12}CO_{3^-} \leftrightarrow {}^{12}CO_2 + H^{13}CO_{3^-} \tag{15.2b}$$

The isotope fractionation factor (α) is then expressed as the ratio of the isotopic ratios of the two species (R_A and R_B)

$$\alpha = R_A / R_B \tag{15.3}$$

The fractionation factor for equilibrium reactions changes when the reactants are not present in a one to one ratio, in which case an equilibrium constant (K) must be used. However, most applications in geochemistry assume that K is approximately equal to α because partitioning functions and other parameters such as temperature are not known and must be estimated. For most stable isotopic systems, $\alpha \approx 1$; therefore, deviation from unity becomes important, particularly in kinetic reactions associated with metabolic processes. This deviation is also called fractionation (ε), and is defined as:

$$\varepsilon = (\alpha - 1) * 1000 \tag{15.4}$$

Fractionation is a small number and is expressed in ‰, even though α is unitless. More often in geochemical studies, α is approximated to capital delta (Δ), which is defined as the difference in the isotopic values between two substances in ‰

$$\alpha \approx \Delta_{A-B} = \delta_A - \delta_B = 10^3 \ln \alpha_{A-B} \tag{15.5}$$

There are several qualitative rules to consider when evaluating fractionation factors in a suite of geochemical data (Bigeleisen and Mayer 1947; Schauble 2004):

1. Isotopic fractionation is greater at low temperature, proportional to $1/T$ at low temperatures and $1/T^2$ at high temperatures. This is because bonds vibrate faster at high temperatures and consequently break more readily for both light and heavy isotopes during a reaction. Thus, at high temperatures, isotopic fractionation is weaker. For example, the isotopic fractionation of carbon during the precipitation of calcite from CO_2 at 25°C is 10.9‰. However, at 40°C fractionation decreases to 7.4‰ (Romanek et al. 1992).
2. Isotopic fractionation is large when the mass ratio of an element is large (defined as $(m_h-m_l)/(m_h*m_l)$, where m_h and m_l represent the heavy and light isotopic species, respectively). For example, hydrogen isotopic fractionation is very large because the isotopic mass of ^2H (deuterium – also denoted as D) is double the isotopic mass of ^1H. However, for carbon isotopes, the isotopic mass of ^{13}C is only 8.3% heavier than the isotope ^{12}C; therefore, the isotopic fractionation of carbon is relatively smaller than the isotopic fractionation of hydrogen.
3. Heavy isotopes have greater bond strength than light isotopes because differences in the vibrational frequency of a molecule affect how readily a molecular bond will break. For example, during evaporation of water, ^{16}O has a weaker bond than ^{18}O. As a result, ^{16}O will more readily volatilize into the vapor phase over ^{18}O. Heavy isotopes are also preferentially incorporated into sites with a stronger bond, which is commonly associated with molecules of higher oxidation state. Likewise, the lighter isotopes tend to be incorporated into reduced molecular forms.
4. As reaction rate increases, fractionation decreases. Much like temperature, as particles react faster, isotopic partitioning is reduced. Further, in closed systems where the abundance of reactant is limited, fractionation will decrease because all of the substance is being reacted, thus all of the light and heavy isotopes are exchanged.

15.3 The Carbon Cycle and Isotopic System

The carbon cycle is fundamental to all life on Earth, and has been extensively studied from ecological time-scales (days to thousands of years) to geological time-scales (millions of years) (see Bolin et al. 1979; Holser et al. 1988; Jacobson et al. 2000 for review). Stable carbon isotopic studies can be used to help estimate the relative sizes and fluxes into and out of various carbon reservoirs through time, as well as understand how these fluxes interact with biological and environmental systems, such as productivity, metabolic pathways, and climate. Carbon exists in several forms and has oxidation states ranging from +IV to –IV. Elemental forms include amorphous carbon, graphite, and diamond. Oxidized forms can occur (1) as gases in the atmosphere such as carbon dioxide (CO_2) and carbon monoxide (CO); (2) as dissolved ions in the hydrosphere in the form of carbonic acid ($H_2CO_{3(aq)}$), bicarbonate ($HCO_3^-{}_{(aq)}$), and carbonate ($CO_3^{2-}{}_{(aq)}$); and (3) as solid carbonate minerals in the lithosphere such as calcite ($CaCO_3$), dolomite ($CaMg(CO_3)_2$), and siderite

($FeCO_3$). Reduced forms of carbon include organic matter composed of complex chains of carbon and hydrogen with trace amounts of sulfur, phosphorous, nitrogen, oxygen, and metals. The most reduced form of carbon is methane (CH_4).

15.3.1 Carbon Fractionation and the Vital Affect

Fractionation of oxidized carbon by inorganic processes is relatively small and controlled by equilibrium thermodynamics. However, carbon exhibits very large fractionation between ^{12}C and ^{13}C during certain metabolic processes (Fig. 15.1a). For example, oxygenic photosynthesis transforms, or "fixes", inorganic CO_2 into organic matter by oxidizing water into free oxygen through a photochemical reaction. During this process, ^{12}C is preferentially incorporated into the organic matter over ^{13}C because the bond strength in ^{12}C is weaker relative to ^{13}C due to its lighter mass. In other words, it takes less energy to break a covalent bond in $^{12}CO_2$ compared to $^{13}CO_2$. The degree of fractionation depends on the metabolic pathway, type of enzyme the organism utilizes, and reaction rate.

Autotrophic organisms that use CO_2 as their sole source of carbon create large isotopic fractionations that can be useful to understand productivity and organic

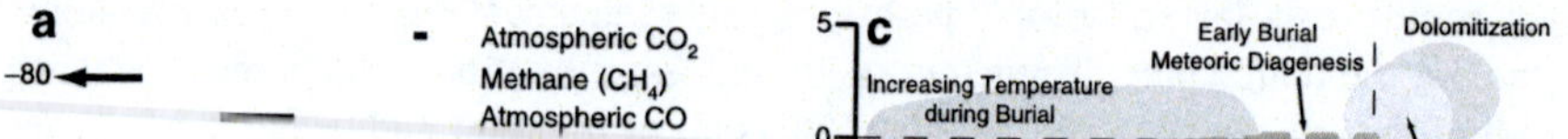

Fig. 15.1 The carbon cycle. (**a**) General isotopic range for the typical sources in the carbon cycle; (**b**) Schematic diagram of the carbon cycle outlining the primary reservoirs and their respective sizes; (**c**) Cross-plot of $\delta^{13}C$ with $\delta^{18}O$ depicting how diagenetic factors impact the original isotopic values of a sample

matter burial in the rock record (e.g. Hayes et al. 1999; Hayes 1993; Meyer 1994; Schidlowski et al. 1983). There are many forms of autotrophs, including green plants, algae, various chemoautotrophs, and photosynthetic bacteria. These organisms use five known metabolic pathways to fix inorganic carbon in nature, the best studied being the pentose phosphate cycle (Calvin cycle), the reverse tricarboxylic (rTCA) cycle, and the acetyl-CoA pathway (White 2000). Oxygenic photosynthetic organisms, like cyanobacteria and algae, fix carbon through the Calvin Cycle, which has a characteristic fractionation range of ($\delta^{13}C$ = −24‰ to −37‰) (Schidlowski 2001). Other organisms fix carbon chemoautotrophically by utilizing reduced compounds as electron donors, such as organic sulfur compounds and hydrogen gas. Methanogens, for example, reduce CO_2 into CH_4 anaerobically via the acetyl-CoA pathway which can result in extremely large isotopic fractionation ($\delta^{13}C$ = −40‰ to −80‰) (Summons et al. 1998). Other metabolic pathways can also produce exceedingly low $\delta^{13}C$ values, such as the serine pathway used by methanotrophs in which CH_4 is oxidized to CO_2, resulting in low $\delta^{13}C$ values in the CO_2 end product (Jahnke et al. 1999).

In general, carbonate minerals directly precipitated in the ocean are close to isotopic equilibrium with ambient waters, and the isotopic composition of the carbonate is mainly controlled by temperature. However, much of the carbonate precipitated in the ocean today is biologically controlled by both autotrophic and heterotrophic organisms. The secretion of carbonate shells can impose isotopic fractionation because of kinetic processes that result in an isotopic disequilibrium between the secreted carbonate shell and the ambient waters, known as the "vital effect" (Epstein et al. 1951). When the organism secretes its shell, ^{12}C is preferentially incorporated into the shell lattice, resulting in low $\delta^{13}C$ values of the shell relative to the ambient water. Much like autotrophic CO_2 fixation, the degree of fractionation varies among organisms by several permil (Anderson and Arthur 1983; Wefer and Berger 1991) and is strongly dependent on growth rate, such that slow growth will lead to carbonate isotopic values closer to equilibrium (McConnaughey 1989a, b).

15.3.2 Carbon Reservoirs

Carbon exists in four primary reservoirs: the lithosphere, atmosphere, hydrosphere, and biosphere. Carbon in each reservoir has its own residence time, and the flux of carbon between reservoirs differs. Table 15.2 and Fig. 15.1b illustrate the main reservoirs, abundances, fluxes, residence times, and average carbon isotopic compositions. Simplistically, on geological time scales (>10^5 years), it can be assumed that the amount of carbon input to the atmosphere, hydrosphere, and biosphere is balanced by the amount of carbon buried in the lithosphere, and that the fluxes into and out of each reservoir are roughly equivalent (Holser et al. 1988). This equilibrium is known as steady state. Imbalances in the carbon cycle on geological time scales via changes in the fluxes of carbon into or out of various carbon reservoirs ultimately control secular isotopic variation recorded in the geological record.

Table 15.2 Main carbon reservoirs, residence time, and average isotopic composition in the carbon cycle

Reservoir	Mass (Pg C)	Residence time (years)	Average $\delta^{13}C$ value (‰ PDB)
Atmosphere			
CO_2	600	5	−7
CH_4	3	9	−50
CO	0.2	0.2	−32 to −22
Hydrosphere			
Dissolved inorganic carbon (DIC)	38,000	50,000	0
Dissolved organic carbon (DOC)	1,000	3,400	−20
Biosphere			
Terrestrial biomass[a]	480–1,600	50	−25 to −17
Soils (humus)	3,000	1,000	−25 to −12
Lithosphere and mantle			
Sedimentary inorganic carbonates	65,000,000	1×10^8	0
Organic carbon	16,000,000	1×10^8	−25
Continental silicic crust	7,000,000	1×10^8	−5
Mantle	342,000,000	–	−5

Abundance for each reservoir from Anderson and Arthur (1983), Bazilevich et al. (1970), Blake and Rowland (1988), Deines (2002), Des Marais et al. (1992), Garrels et al. (1971), Li (1972), Mopper and Degens (1979), Petit et al. (1999), Schlesinger (1997), and Takahashi et al. (1981). Residence times from Anderson and Arthur (1983), Broecker (1973), Holser et al. (1988), Mopper and Degens (1979), and Schlesinger (1997). Isotopic values for each reservoir are from Anderson and Arthur (1983), Craig et al. (1988), Deines (2002), Holser et al. (1988), Schidlowski et al. (1983), and Stevens et al. (1972), 1 Pg C = 10^{15} g C

[a]Terrestrial biomass isotopic composition dependent on the dominant primary producers. For example, C4 plants have $\delta^{13}C$ values between −10‰ and −20‰, where as C3 plants have $\delta^{13}C$ values between −20‰ and −32‰ (Schidlowski et al. 1983)

The mantle and lithosphere account for the largest carbon reservoir in the carbon cycle, together estimated around 424 × 10^6 Pg C (1 Pg C = 10^{15} g C). The vast majority of carbon is stored in the mantle (342 × 10^6 Pg C), and has an average isotopic value around −6‰ (Deines 2002). Carbon, once trapped in the lithosphere, also has the longest residence time of hundreds of millions of years (Broecker 1973). Fluxes into the atmosphere and hydrosphere occur through volcanism and rock weathering (Berner et al. 1993; Resing et al. 2004), whereas fluxes into the lithosphere occur through burial of carbonates, shales, and organic matter. The isotopic composition of sedimentary carbonates grossly reflects the isotopic composition of the dissolved inorganic carbon subreservoir in the hydrosphere, but is also a function of vital affects associated with biologically mediated growth of carbonate shells (McConnaughey 1989a, b). About 15 × 10^6 Pg C is found in organic-rich rocks and the average $\delta^{13}C$ value is around −23‰.

Carbon in the atmosphere dominantly occurs as CO_2, with lesser amounts of CO, CH_4, and other organic gases. Carbon dioxide has been the most studied compound because of its potential affect on the climate, where modern-day concentrations are around 365 ppmv (or 750 Pg C) (IPCC 1990). Pre-industrial concentrations of CO_2

have been estimated to be 280 ppmv (or 600 Pg C) (Petit et al. 1999). However, CO_2 levels during the Cretaceous may have been 3–12 times greater than modern pre-industrial levels (Berner 1992). The average global isotopic composition of atmospheric CO_2 is around –6‰. Methane constitutes approximately 1% of the carbon budget in the atmosphere, but can have a dramatic impact on climate as a greenhouse gas. Modern levels of methane are around 1.7 ppmv, or 3 Pg C (Blake and Rowland 1988). The isotopic composition of methane is significantly lower than the isotopic composition of CO_2 (average atmospheric $\delta^{13}C = -50‰$) (Craig et al. 1988). Methane is typically oxidized in the atmosphere by the hydroxyl radical to yield CO, so the resulting isotopic composition for CO is also low ($\delta^{13}C = -32‰$ to –22‰) relative to CO_2 (Stevens et al. 1972). Carbon monoxide concentrations range from 0.05 to 0.20 ppmv, or roughly 0.2 Pg C.

Carbon in the oceans is characterized by two primary subreservoirs: (1) dissolved inorganic carbon (DIC) and (2) dissolved organic carbon (DOC) (Jacobson et al. 2000). The DIC reservoir is by far the largest store of carbon in the ocean at 38,000 Pg C (Takahashi et al. 1981), with a residence time of about 50,000 years (Holser et al. 1988). The fluxes between the atmosphere and the DIC reservoirs are large, and as a result DIC is isotopically buffered with respect to the atmosphere because the volume of carbonate entering and leaving the DIC reservoir at the atmosphere-water contact is large. There is approximately ~8‰ ^{13}C enrichment that occurs when atmospheric CO_2 dissolves into water (Jacobson et al. 2000), and the average $\delta^{13}C$ value of the modern DIC reservoir is near 0‰ (Anderson and Arthur 1983). The DOC reservoir accounts for approximately 1,000 Pg C in the oceans with a residence time of approximately 3,400 years (Mopper and Degens 1979). The average $\delta^{13}C$ value of the DOC is around –20‰ (Anderson and Arthur 1983).

The biosphere reservoir can be subdivided into two subreservoirs: the marine biosphere and terrestrial biosphere. The marine biosphere encompasses both particulate organic carbon (POC) and marine biota., which account for a minor amount of carbon in the oceans, roughly 30 Pg C and 3 Pg C, respectively (De Vooys 1979; Mopper and Degens 1979). For most of geological history, the marine biosphere has played a more important role in the carbon cycle, however, the terrestrial biosphere has become increasingly significant since the Silurian (~350 Ma), when vascular plants became dominant on land (Kenrick and Crane 1997). Modern estimated terrestrial biomass varies widely from 480 Pg C to 1,600 Pg C (Anderson and Arthur 1983; Bazilevich et al. 1970; Garrels et al. 1971). The average residence time is short, on the order 50 years, and the mean $\delta^{13}C$ value is between –25‰ and –17‰ (Anderson and Arthur 1983).

15.4 The Sulfur Cycle and Isotopic System

Sulfur is an important component in the chemical cycling of the Earth and as such, stable sulfur isotopes have become an important research tool to characterize past depositional environments and oxidation of the ancient atmosphere (Berner 1999;

Canfield et al. 2000; Farquhar et al. 2000), as well as to better understand the evolution of key biological clades (Canfield 2001; Canfield and Teske 1996). Sulfur exists in several oxidation states ranging from +VI to –II. Major sources of sulfur include sulfate (SO_4^{2-}) in the ocean, evaporates such as gypsum ($CaSO_4{\bullet}2(H_2O)$), anhydrite ($CaSO_4$), and barite ($BaSO_4$), sulfide minerals such as pyrite (FeS_2), and sulfuric acid in the atmosphere (H_2SO_4). Elemental sulfur can be found in hot springs and igneous ore deposits. Major sources of sulfides occur in igneous and metamorphic rocks, volcanic ash, and organic sulfur compounds. The most reduced form of sulfur is hydrogen sulfide (H_2S).

15.4.1 Sulfur Fractionation

Sulfur isotopic fractionation can be very large, but typically only at high temperatures (between 300°C and 750°C) (Rye 2005). Low temperature sulfur isotopic fractionation occurs through two major processes: exchange reactions associated with evaporation that result in the concentration of ^{34}S in the most oxidized sulfur compounds and reduction of sulfate to hydrogen sulfide (H_2S) through metabolic processes associated with anaerobic bacteria. Each source of sulfur has characteristic $\delta^{34}S$ values (Fig. 15.2a), which range from –40‰ to +30‰.

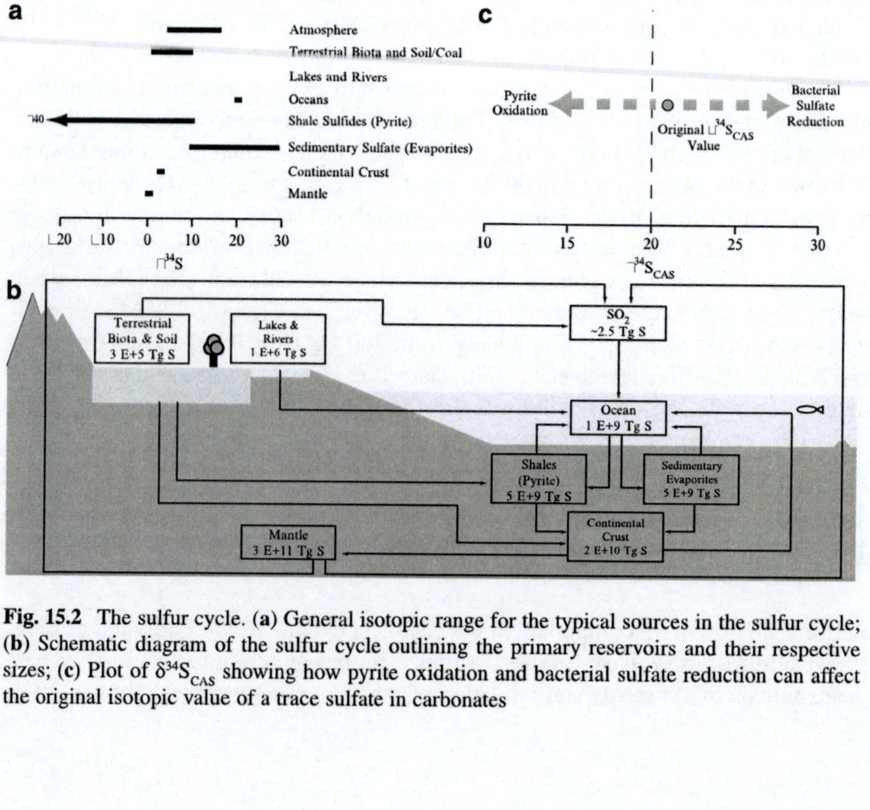

Fig. 15.2 The sulfur cycle. (**a**) General isotopic range for the typical sources in the sulfur cycle; (**b**) Schematic diagram of the sulfur cycle outlining the primary reservoirs and their respective sizes; (**c**) Plot of $\delta^{34}S_{CAS}$ showing how pyrite oxidation and bacterial sulfate reduction can affect the original isotopic value of a trace sulfate in carbonates

The isotopic fractionation associated with evaporation is small, between 1‰ and 2‰. During evaporation, ^{34}S is preferentially incorporated into gypsum, anhydrite, and barite (Payton et al. 1998; Thode and Monster 1965). The relative isotopic similarity and minor fractionation between dissolved oceanic sulfate and sulfate minerals has allowed the isotopic composition of trace sulfate in carbonates and evaporites to serve as a proxy for the isotopic value of oceanic sulfate.

Several organisms can reduce sulfate into sulfides through metabolic processes with varying degrees of fractionation (see Canfield 2001; Canfield and Raiswell 1999 for summary). The resulting sulfides have lower $\delta^{34}S$ values relative to the dissolved $\delta^{34}S$ values of oceanic sulfate. Assimilatory sulfate reduction is the most common metabolic process with a large sulfur isotope fractionation, ranging from –4.5‰ to 0.5‰ (Kaplan 1983). Plants, algae, and various prokaryotes reduce sulfate into an organic sulfur compound called sulfhydryl, which is a thiol used to synthesize proteins and amino acids (White 2000).

A second common metabolic process called dissimilatory sulfate reduction, or bacterial sulfate reduction, is utilized by the delta subgroup of Proteobacteria such as *Desulfovibrio* and *Desulfotomaculum*, in addition to a few thermophiles, nitrogen-oxidizers, diazotrophs, and Archaea. These organisms anaerobically oxidize organic matter ($2CH_2O + SO_4^{2-} \rightarrow 2HCO_3^- + H_2S$) and impose the largest isotopic fractionation (up to 70‰) (Brunner and Bernasconi 2005). When sulfate is reduced to H_2S, the sulfide becomes enriched in ^{32}S relative to sulfate because (much like with carbon) the bond strength in $^{32}SO_4^{2-}$ is weaker than the bond strength in $^{34}SO_4^{2-}$. The degree of fractionation during sulfate reduction is dependent on reaction rate and the availability of sulfate. When the reaction rate is high, the fractionation will be small (Goldhaber and Kaplan 1974; Kaplan and Rittenberg 1964). Also, fractionation will be minimal if sulfate concentrations are below 200 μmol (Habicht et al. 2005). In closed systems, where all pore-water sulfate is converted to sulfide, the isotopic value of the bulk pyrite will be equal to the isotopic value of the oceanic sulfate. However, in most oceanic sedimentary rocks, there is constant flux of oceanic sulfate into the pore-water. Therefore, the isotopic values of the corresponding sulfide is more dependent on the reaction rate, availability of organic matter, and presence of metal cations (like iron) that can react with H_2S (Kempe and Thode 1968).

15.4.2 Sulfur Reservoirs

Sulfur occurs in the lithosphere, atmosphere, hydrosphere, and biosphere (Fig. 15.2b; Table 15.3). The vast majority of sulfur is stored within the lithosphere and hydrosphere as sulfate ($CaSO_4{\cdot}H_2O$) and pyrite (FeS_2), which record very different isotopic values. Although the primary fluxes of sulfur occur in the atmosphere and terrestrial biosphere, these fluxes have a negligible impact on the sulfur isotopic composition of sulfide and sulfate-rich sedimentary rocks of the geologic record and therefore will not be discussed in detail.

Table 15.3 Main sulfur reservoirs, residence time, and average isotopic composition in the sulfur cycle

Compound	Abundance (Tg S)	Residence time (years)	Average $\delta^{34}S$ value (‰ CTD)
Atmosphere			
Continental atmosphere	1.6	0.004	3.2
Marine atmosphere	3.2	0.015	15.6
Hydrosphere			
Oceans	1.3×10^9	2×10^7	20–21[a]
Lakes and rivers	1.3×10^6	<1–0,000s	0–20
Biosphere			
Soils and land biota	2.4×10^4	<1–1,000	1–17[b]
Lithosphere and mantle			
Shales	4.74×10^9	1.4×10^8	−17
Evaporites	4.86×10^9	1.4×10^8	16
Sulfur in crust	2×10^{10}	2.4×10^8	2–7
Mantle (above 1000 km)	3×10^{11}	–	0

Abundances for each reservoir are from Chaussidon et al. (1989), Holser et al. (1988), Reeburgh (1997), and Schlesinger (1997). Residence times are from Bottrell and Newton (2006), Charlson et al. (1992), Garrels and MacKenzie (1972), Mayer et al. (1995), Reeburgh (1997), Rees et al. (1978), and Toon et al. (1987). Isotopic values for each reservoir from Claypool et al. (1980), Jensen and Nakai (1961), Nielsen (1970), Kampschulte and Strauss (2004), Kaplan (1983), Mizota and Sasaki (1996), Thode et al. (1961). 1 Tg S = 10^{12} g S

[a] 20‰ value calculated from SF_6 gas extraction techniques (Rees et al. 1978). 21‰ value from SO_2 extraction techniques (Kampschulte and Strauss 2004; Thode et al. 1961)

[b] Isotopic composition of soils depends on distance from coastline; soils near the coastline have a greater proportion of sulfides from sea spray (~20‰), while soils inland will have locally isotopic composition of sulfate from locally derived igneous/volcanic deposits, evaporites, or oxidized pyrite (Mizota and Sasaki 1996)

Sulfur in the lithosphere is primarily stored as sulfides in shale and sulfates in evaporites. Both reservoirs are larger than the oceanic sulfate reservoir (4.74×10^9 Tg S sulfides and 4.86×10^9 Tg S sulfates in shale and evaporite deposits, respectively) (Holser et al. 1988), and also have long residence times, sequestered in sedimentary rocks on the order of hundreds of millions of years (Charlson et al. 1992). The isotopic composition of evaporites range from +10‰ to +30‰, but generally reflect the isotopic composition of oceanic sulfate at any given time (Claypool et al. 1980). Conversely, sulfides in shales have isotopic values between −40‰ and +10‰, with an average value around −17‰ (Claypool et al. 1980). As a whole, sulfur stored in the crust is about 2×10^{10} Tg S with a residence time of 1×10^8 years (Reeburgh 1997). The $\delta^{34}S$ values for sulfur in the crust range from 2‰ in oceanic crust to 7‰ in continental crust (Chaussidon et al. 1989). The mantle stores 3×10^{11} Tg S (Chaussidon et al. 1989) with an average $\delta^{34}S$ value of 0‰ (Holser et al. 1988).

The amount of sulfur in the atmosphere at any given time is very small and the residence time of sulfuric gases is on the order of days because the fluxes of sulfur into and out of the atmosphere via volcanic outgassing (3.2–9.6 Tg S year^{-1}; 1 Tg S = 10^{12} g S) and sulfuric acid in rain, respectively, are relatively large (Toon et al.

1987). There are a number of chemical transformations that occur in the atmosphere (Charlson et al. 1985). Natural sources of sulfur into the atmosphere include biomass burning (SO_2), volcanic outgassing (H_2S and SO_2), and biological processes that release reduced sulfur-containing gases (H_2S, CS_2, OCS, SCH_3SH, DMS, and DMDS) (Charlson et al. 1992). The isotopic composition and abundance of atmospheric sulfur measured in snow and rainwater range from 3.2‰ to 15.6‰ and 1.6 to 3.2 Tg S in continental and marine environments, respectively (Jensen and Nakai 1961).

Nearly all dissolved sulfur in the oceans today occurs as sulfate and constitutes a major reservoir (1.3×10^9 Tg S) (Schlesinger 1997). Continental lakes and rivers comprise a much smaller reservoir (1.3×10^6 Tg S), but have a large flux into the oceans. Sulfur enters the oceans via continental weathering of sulfur-bearing sedimentary rocks (0.6 Tg S $year^{-1}$) or from the atmosphere via volcanic outgassing that then dissolves in the surface ocean (Canfield 2004). Sulfur is removed from the ocean by precipitation and burial of sulfate and sulfide minerals. The reduction of sulfate into sulfides in the oceans is almost exclusively biologically mediated. The isotopic composition of oceanic sulfate averages around 21‰ (Kampschulte and Strauss 2004; Thode et al. 1961) and has a residence time around of 20 Ma (Rees et al. 1978). The long term sulfur isotopic variability of the ocean is controlled by the amount of sulfides stored in sedimentary rocks and the fluxes into and out of the ocean due to weathering or burial of sulfides.

The residence time for sulfur in the terrestrial hydrosphere is quite variable, ranging from <1 to 10,000s of years (Bottrell and Newton 2006) and has an average $\delta^{34}S$ of 0–20‰ (Nielsen 1970). Soils and land biota account for 2.4×10^4 Tg of sulfur (Schlesinger 1997), with a residence time of <1–1,000 years (Mayer et al. 1995) and $\delta^{34}S$ values ranging from 1‰ to 17‰ (Mizota and Sasaki 1996).

15.5 Stable Isotopes Mass Spectrometry and Sampling Techniques

15.5.1 Isotope Ratio Mass Spectrometry

Standard sampling protocols and extraction techniques have been developed over the decades to help calibrate mass spectrometer analyses and to create globally comparable sampling strategies between laboratories (Coplen and Krouse 1998; Craig 1953, 1957; Epstein et al. 1951; Macnamara and Thode 1950; Mariotti 1983; McCrea 1950). Stable isotopes are measured on gas-source isotope ratio mass spectrometers that require the sample to be in a pure gaseous phase. Therefore, protocols have been developed to concentrate the desired gas through chemical trapping, combustion, and gas chromatographic feeds. Table 15.1 lists the typical gases and their isotopic masses used in gas source mass spectrometry for the major light elements.

All gas-sourced isotope ratio mass spectrometers consist of five components: (1) an inlet system, (2) ion source, (3) flight tube, (4) ion collector assembly, and

(5) recording system. The gas that enters the mass spectrometer is not purely elemental gas, but rather molecules that can have several potential isotopic compositions, called isotopologues. For example, CO_2 gas is used for isotopic analysis of carbon and can have three isotopologues characterized by their molecular masses, 44 ($^{16}O^{12}C^{16}O$), 45 ($^{16}O^{13}C^{16}O$, $^{17}O^{12}C^{16}O$, and $^{16}O^{12}C^{17}O$) and 46 ($^{17}O^{13}C^{16}O$, $^{16}O^{13}C^{17}O$, $^{17}O^{12}C^{17}O$, $^{18}O^{12}C^{16}O$, and $^{16}O^{12}C^{18}O$). The mass ratio of the isotopologues (45/44) can be directly related to the isotopic ratio of the elements ($^{13}C/^{12}C$) themselves.

The most common type of mass spectrometer used today is the gas chromatograph isotope ratio mass spectrometer (GC/IRMS). This machine uses a continuous flow of gas from several inlets and combusts a powered sample directly into a gas (Matthews and Hayes 1978). There are several advantages to using GC/IRMS over a conventional mass spectrometer: (1) the sample can be prepared just before analysis, whereas in conventional IRMS, samples must be isolated in a pure gas form prior to injection; (2) a much smaller sample is required for analysis (only 0.35 μg of sample is required rather than 0.5 g of sample for extraction via conventional mass spectrometers); and (3) the analysis is much more rapid than conventional mass spectrometers. However, the entire sample is used during analysis. The machine is also less precise than conventional mass spectrometers, although careful calibration of the machine will yield accurate and reproducible values.

Recent development of laser-based microanalytical methods can now precisely measure isotopic compositions of samples *in situ*. Laser extraction techniques have been calibrated to analyze the isotopic ratio in carbonates, silicates, oxides, and sulfides (Chaussidon et al. 1989; Franchi et al. 1986; Sharp 1990, 1992). There are numerous kinds of ion microprobes calibrated to analyze light stable isotopes (C, O, and S), such as the secondary ion mass spectrometer (SIMS), as well as heavy isotopes (U-Pb-Th geochronology), including the sensitive high-resolution ion microprobe (SHRIMP) (Lovering 1975; Riciputi et al. 1998). The ion microprobe uses a highly focused ion beam of oxygen or cesium to ionize the targeted material. The ions released by the beam then enter a mass spectrometer and the isotopic ratios are measured. One chief advantage of the ion microprobe is that the technique can analyze very small samples (nm-μm diameter pits and several tens of nm deep, depending on the microprobe used), allowing fine resolution spot sampling to avoid diagenetically altered zones. Unfortunately, the ion microprobe is not as precise as conventional mass spectrometry with an average reproducibility within 1‰ (Leshin et al. 1997).

15.5.2 Gases Used in Carbon and Sulfur Isotopic Extraction

The gas used in all carbon isotopic measurements is CO_2. As mentioned earlier, the mass spectrometer measures the 45/44 mass ratio of the CO_2 gas in order to calculate the isotopic value of a sample against a standard. Carbon isotopic values for inorganic carbon are typically reported as $\delta^{13}C_{carb}$ or simply $\delta^{13}C$ and organic carbon as $\delta^{13}C_{org}$. All extraction techniques convert the inorganic or organic carbon into a gas form and then isolate the CO_2 through a vacuum. Preparation and extraction methodologies have evolved from an "offline" carbon combustion apparatus (Craig 1953; McCrea 1950) to automatic combustion methods (Kaufman et al. 1991; Ostermann and Curry 2000). Careful calibration of the mass spectrometers yields reproducible values, ±0.05‰ for carbonates and ±0.5‰ for organic carbon.

Much like carbon, sample preparation and extraction of sulfur isotopes from sulfides and trace sulfate has a long history (Burdett et al. 1989; Canfield et al. 1986; Gellatly and Lyons 2005; Kaufman et al. 2007b; Puchelt et al. 1971; Thode et al. 1961). In general, sulfur isotopic values for trace sulfate is reported as $\delta^{34}S_{CAS}$ or $\delta^{34}S_{SO4}$, while isotopic composition of pyrite is reported as $\delta^{34}S_{py}$. Two gases can be used for sulfur isotopic measurements: SO_2 and SF_6. SO_2-based extraction techniques are more commonly used in paleobiological research today because of the rapidity and simplicity of the procedure. However, SO_2 has several major drawbacks; namely it sticks in the mass spectrometer because of its polarity, which ultimately reduces accuracy. SO_2 can also react easily with trace amounts of water in the vacuum line of the mass spectrometer to form sulfuric acid, and thus the line must be cleaned on a regular basis. Another problem with SO_2 is distinguishing 66/64 mass ratios between $^{34}S^{16}O^{16}O$ and $^{32}S^{18}O^{16}O$ (Halas 1985). There is no way to directly distinguish the ^{18}O contribution to the mass 66 between these isotopologues because $^{34}S/^{32}S$ and $^{18}O/^{16}O$ are not correlated. Therefore, a correction must be made by either independently measuring the $\delta^{18}O$ value from the SO_2 gas (Rees 1978) or by rigorous comparison against sulfate standards.

In contrast, SF_6 is a clean, inert gas with a very high ionization potential relative to SO_2 (Halas 1985). Fluorine is monospecific, which means that the different masses measured in the mass spectrometer for the ionized species SF_5+ (127, 128, 129, 131) are uniquely related to a single isotope of sulfur. As a result, measurements using SF_6 are more accurate relative to SO_2; however, the use of SF_6 is considerably more dangerous because the extraction procedure requires fluorination of sulfur. The extraction technique is also more elaborate and time consuming than SO_2-extraction (Beaudoin and Taylor 1994).

15.5.3 Common Preparation Techniques and Quality Control

Consistent sampling methods allow for greater reproducibility and confidence in the interpretation of geochemical data. Careful attention should be made to assess the quality of hand samples and core fragments for diagenetic alteration and contamination while preparing a suite of samples for geochemical analysis. Several tests can help identify diagenetically altered zones in a sample. Petrographic analysis provides a first pass characterization of the primary and secondary cements in the host rock. For example, void-filling cements, veins, styolites, and authigenic clay minerals represent post depositional alteration of the host rock and should be avoided (Fig. 15.3a–d) (see McIlreath and Morrow (1990) and Moore (2001) and

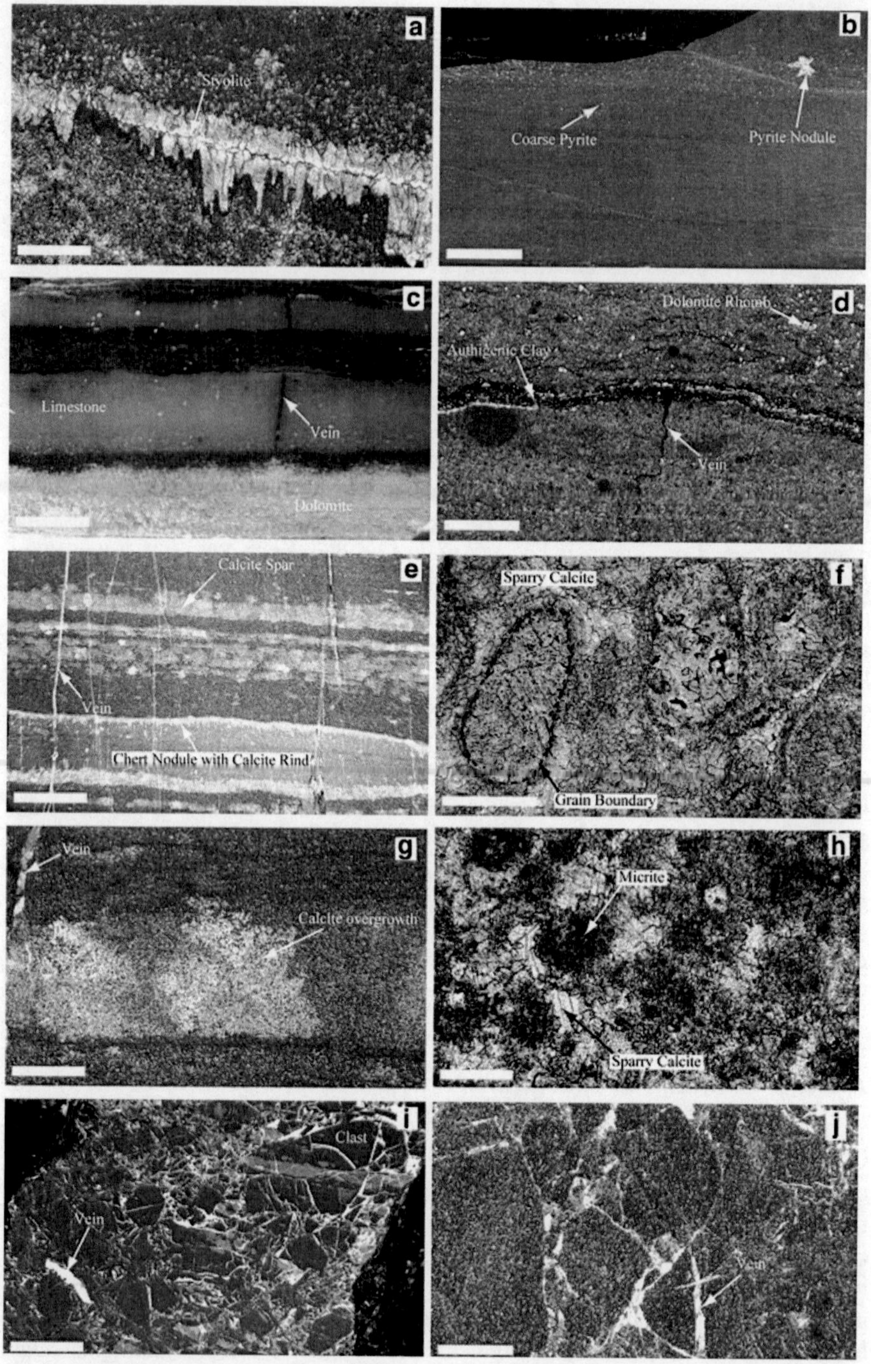

Fig. 15.3 Hand sample and photomicrographs from the Doushantuo and Denying Formations, South China. (**a**) Photomicrograph showing a styolite filled with coarse crystalline calcite; scale 200 μm. (**b**) Hand sample and photomicrograph showing coarse pyrite disseminated in

references therein for an in-depth review of diagenetic cements in carbonates). Cathodoluminescence (CL) petrography of polished thin sections can also identify zones of alteration that might not be immediately apparent during petrographic screening (Hemming et al. 1989; Savard et al. 1995). Manganese and iron are common cations in meteoric and burial fluids incorporated into calcite cements and veins. These cements fluoresce and can be useful for understanding redox and pH conditions of diagenetic fluids (Machel 1985). SEM is also useful in identifying fine-scale diagenetic cements and textures for micro-sampling carbonates and *in situ* isotopic analysis (Bruckschen et al. 1995; Bruhn et al. 1995).

Figure 15.4 outlines common procedures for the preparation and extraction of carbon and sulfur isotopes. For inorganic carbon isotopic analysis, powdered samples are usually acquired via micro-drilling on a mirror image of the thin section so that sampling is precise (Kaufman and Knoll 1995). Bulk samples for organic carbon, trace sulfate, and pyrite are acquired by crushing bulk samples into a very fine powder (80–200 μm) in order to increase the surface area, which aids in complete dissolution during preparation and release of any sedimentary residues, fluid inclusions, and organically bound minerals that might contaminate the sample.

It is very important to make sure that bulk samples are pre-treated and cleaned prior to preparation for geochemical analysis. For example, the preparation for trace sulfate suggests washing powdered samples with both ultra-pure Milli-Q water and 5.25% bleach (NaOCl) for 24 h each. These two steps are critical in order to avoid contamination associated with sedimentary sulfate such as anhydrite or gypsum, dissolved sulfate in fluid inclusions, sulfate formed through oxidation of pyrite or metastable sulfides, or release of organic-bound sulfur. Samples that are particularly organic rich or pyritic (>1 wt% pyrite), can be additionally treated with hydrogen peroxide (30% H_2O_2) for 48 h (Gellatly and Lyons 2005); however, samples with a high abundance of pyrite should be avoided because of the likelihood of sulfide oxidation during acidification (Morenco et al. 2008). By thoroughly pre-treating the sample to remove potential contamination *a priori*, the isotopic data will more likely reflect the trace sulfate bound in the carbonate and not incorporate contaminants.

Fig. 15.3 (continued) dolo-mudstone (scale 1 cm). (**c**) Hand sample micritic limestone interbedded with dolostone. (**d**) Photomicrograph of micrite cross-cut by a vein and styolite filled with authigenic clay; scale 200 μm. (**e**) Hand sample and photomicrograph of cherty micrite overprinted by coarse calcite and cross-cut by veins; scale 1 cm. (**f**) Recrystallized grainstone with micritic rings along grain boundaries from middle Doushantuo Formation, Yangjiaping section South China; scale 100 μm; image courtesy of Ganqing Jiang. (**g**) Photomicrograph under cross-polar light illustrating micrite overprinted with calcite spar and cross-cutting veins; scale 200 μm. (**h**) Photomicrograph of strongly altered micritic carbonate with sparry calcite voids from middle Doushantuo Formation, Yangjiaping section South China; scale 150 μm; geochemical analysis of micrite ($\delta^{13}C = -0.6‰$; $\delta^{18}O =$ 4.5‰) and surrounding cements ($\delta^{13}C = -3.3‰$; $\delta^{18}O = -9.9‰$) show differing isotopic values, indicating that the sparitic cements formed in pore fluids isotopically different relative to the micrite; image courtesy of Ganqing Jiang. (**i, j**) Hand sample and photomicrograph of a breccia with clasts composed of micrite cross-cut by sparitic calcite veins; hand samples scale 1 cm; photomicrograph scale 200 μm; geochemical analysis of the micritic clast ($\delta^{13}C = -1.6‰$; $\delta^{18}O = -3.2‰$) and surrounding veins ($\delta^{13}C = -1.0‰$; $\delta^{18}O = -6.7‰$) show differing isotopic values, indicating that the sparitic veins formed in pore fluids isotopically different relative to the micritic clasts

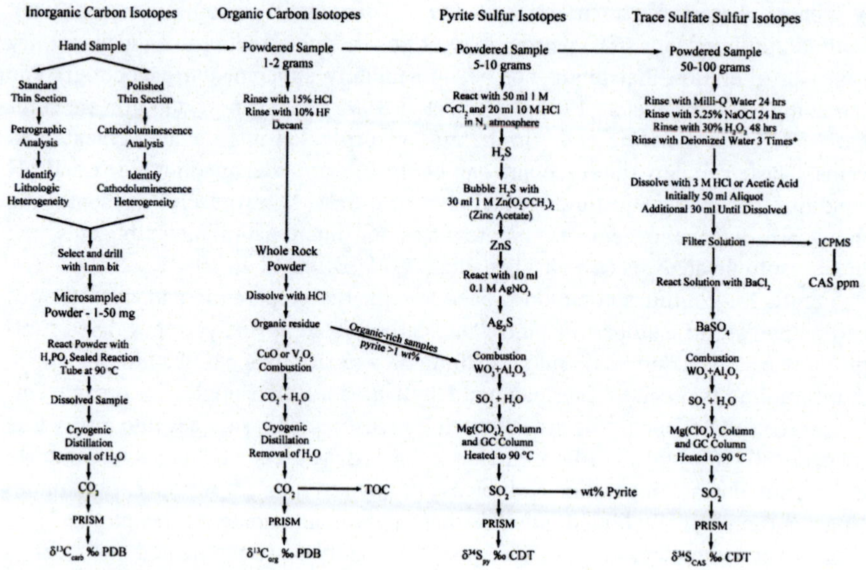

Fig. 15.4 Summary of common preparation and extraction procedures for carbon and sulfur isotopes. Inorganic carbon isotopic preparation and extraction techniques are from Ostermann and Curry (2000). Organic carbon isotope preparation and extraction methods are described in detail by Craig (1953), Hoefs and Schidlowski (1967) and Kaufman et al. (2007a, b). Chromium reduction method for sulfur extraction from pyrites is summarized by Canfield et al. (1986). Sulfur extraction techniques for pyrites via combustion summarized by Studley et al. (2002) and Kaufman et al. (2007a, b). Trace sulfate preparation and extraction techniques from Burdett et al. (1989) and Gellatly and Lyons (2005) (Modified from Kah et al. 1999, Kaufman and Knoll 1995)

15.6 Diagenetic Alteration of Stable Isotopes

Many geochemical studies assert that the carbon and sulfur isotopic values in the rock record represent the primary isotopic composition of the global oceans based on careful petrographic evaluation and comparison against global secular trends (Claypool et al. 1980; Kampschulte and Strauss 2004; Veizer et al. 1999). However, interpretation of any isotopic data requires an appreciation of the depositional environment as well as the burial history of the sedimentary rock. Alteration of sediment after deposition is referred to as diagenesis (Larson and Chilingar 1979) and includes physical deformation, chemical remineralization, and microbial processes that can potentially alter the mineral structure and/or composition of the host rock. Diagenesis occurs from just after deposition to burial kilometers in the subsurface.

Carbonate precipitating in the ocean today, including low-magnesium calcite, high-magnesium calcite, and aragonite, is metastable and will eventually dissolve and reprecipitate into coarser, interlocking crystals of diagenetic low magnesium calcite. This process is termed "stabilization" and represents the diagenetic conversion of originally precipitated carbonate particles into limestone or dolostone. The majority of isotopic "resetting" of $\delta^{13}C$ and $\delta^{34}S$ values in carbonate is thought to occur during stabilization because most or all of the metastable carbonate particles are reprecipitated *in situ* (Fairchild et al. 1990; Fairchild and Spiro 1987). Therefore, the resulting isotopic values of the limestone will represent the original carbonate particles and the diagenetic fluids present during stabilization. If the diagenetic fluids are different relative to the original carbonate, then the isotopic value of the carbonate rock will shift toward the isotopic value of pore fluids (Given and Lohmann 1985). For example, marginal marine carbonates exposed to meteoric fluids during stabilization will have isotopic values that reflect both the primary marine environment and the meteoric diagenetic fluids (Lohmann 1988). On the other hand, if stabilization of marine carbonates occurs at the seafloor in contact with penecontemporaneous seawater with similar isotopic values, then the isotopic values in the carbonate rock will not be markedly different relative to the isotopic value of the original marine carbonate (Melim et al. 2002).

Late burial alteration by hydrothermal fluids also influences the isotopic value of sedimentary carbonates. However, unlike early diagenetic stabilization, burial diagenesis typically does not entirely replace the carbonate rock and is characterized by coarse cements, veins, styolites, and other authigenic minerals that fill remaining pore space or cross cut primary fabrics (Choquette and James 1990). Late burial complete replacement can occur during late burial dolomitization; however, this process tends to be fabric destructive and can be avoided with careful petrographic analysis and regional mapping. Determining the diagenetic history of a given sample requires a strong understanding of the depositional environment and cement chronology. As discussed previously, to minimize the contamination of diagenetic products in a given sample during geochemical analysis, interpretations should be based on careful petrography, cathodoluminescence, and SEM microscopy. Literature aimed at recognizing diagenetic cements and their affects on stable isotopes can be reviewed for further information (Mackenzie 2005; McIlreath and Morrow 1990; Melim et al. 2002; Moore 2001).

Geochemical analysis of diagenetic phases is also recommended, in that isotopic analysis of obvious veins, cements, and unusual textures in addition to ideal spot samples can help identify diagenetic alteration. For example, some sparitic cements completely replace pore space in a mosaic of sparitic cement with zones of relic carbonate precipitates (Fig. 15.3e–h). Figure 15.3j–h, carbon and oxygen isotopic analyses were taken from micritic clasts and surrounding sparitic calcite veins in a breccia. The isotopic values of the veins ($\delta^{13}C = -1.0‰$; $\delta^{18}O = -6.7‰$) are different relative to the isotopic values in the micritic clasts ($\delta^{13}C = 1.6$; $\delta^{18}O = -3.2‰$), indicating that these two carbonates precipitated from fluids of differing isotopic compositions and likely at different times. Comparison of geochemical analyses between the sparitic zones and relic micrite may give insight as to the relative timing of cementation.

15.6.1 Diagenetic Alteration of Carbon Isotopes

There are several general isotopic trends associated with diagenesis of carbon isotopes (Fig. 15.1c). Negative carbon isotopic excursions can result from either variation in the isotopic composition of seawater, an increase in temperature, or the interaction of isotopically light carbon with the host rock at any point during diagenesis. As previously mentioned, shallow marine carbonates can undergo radical isotopic transformation during early diagenesis if the carbonate is exposed to meteoric waters, because the $\delta^{13}C$ and $\delta^{18}O$ values of soil bicarbonate and meteoric water are lower relative to those of oceanic bicarbonate (Lohmann 1983). Microbial and abiotic processes can also alter the $\delta^{13}C$ value of carbonate, especially in argillaceous sedimentary rocks with abundant organic matter deposited under anoxic conditions. For example, sulfate reduction ($\delta^{13}C = -25‰$), fermentation ($\delta^{13}C = +15‰$), and thermal release of CO_2 ($\delta^{13}C = -20‰$) can all impose an isotopic shift (Irwin et al. 1977).

Cross-plots of $\delta^{18}O$ versus $\delta^{13}C$ can be useful to characterize diagenetic alteration of marine carbonate and cements as well as estimate the original isotopic value of oceanic DIC (Lohmann 1983, 1988). One common isotopic trend observed in $\delta^{18}O$–$\delta^{13}C$ cross-plots is the "J-curve," which illustrates covariation between $\delta^{18}O$ and $\delta^{13}C$ toward low values associated with progressive meteoric diagenesis (Lohmann 1988). Covariation can also reflect kinetic reactions associated with vital effects (McConnaughey 1989a, b), as ^{12}C and ^{16}O diffuse faster, relative to ^{13}C and ^{18}O, into the carbonate lattice during secretion of shells. Covariation is most pronounced in certain groups of benthic algae (Rhodophycaea and Chlorophycaea), corals (Hermatypic Schleractinia), foraminifera, and echinoderms (Asteroidea and Holothuroidea) (Anderson and Arthur 1983; Spero et al. 1997; Wefer and Berger 1991). However, the exhibited isotopic spread tends to be limited over several permil, whereas covariation associated with meteoric diagenesis can create an isotopic spread over tens of permil in both oxygen and carbon isotopes (Lohmann 1988).

A second common isotopic trend observed in $\delta^{18}O$–$\delta^{13}C$ cross-plots is a progressive trend towards lower $\delta^{18}O$ values during deep burial due to increasing temperature with depth (Choquette and James 1990). As a limestone is buried, the $\delta^{13}C$ values become isotopically buffered against further diagenetic alteration because of increased rock-water interaction, such that the evolving isotopic composition eventually reflects the isotopic value of the sediment during burial (Choquette and James 1990; Moore and Druckman 1981). During deep burial and metamorphism, carbonates can also undergo decarbonation when in the presence of silicates, where carbonate and silicates, such as quartz or feldspar, react to produce Ca- and Mg-silicates and CO_2. ^{13}C is preferentially incorporated in the CO_2, resulting in lower $\delta^{13}C$ in the remaining carbonate (Shieh and Taylor 1969).

Dolomitization is a process by which the original carbonate particles are replaced with a stoichiometric carbonate called dolomite ($MgCa(CO_3)_2$) during very early to late diagenesis (Morrow 1990). The pore fluids associated with dolomitization are typically hypersaline with an elevated Mg^{2+}/Ca^{2+} ratio relative to normal marine waters, alkaline environments, or hydrothermal environments (temperatures >50°C) (Machel and Mountjoy 1986). As a result, the $\delta^{18}O$ values of dolomite can be enriched by as much as 4–7‰ (Sheppard and Schwarcz 1970). The $\delta^{13}C$ value of dolomite depends on both the $\delta^{13}C$ value of the precursor carbonate and the $\delta^{13}C$ value of the diagenetic fluids (Tucker and Wright 1990).

It should be noted that while $\delta^{18}O$–$\delta^{13}C$ cross-plots are useful for qualitatively evaluating diagenetic alteration, the range of isotopic values for ancient seawater must be known in order to fully understand the effects of diagenesis in a given sample. The isotopic composition of the oceans has changed over time; therefore the starting isotopic value for a given carbonate will be different at any given time. Unfortunately, it is difficult to ascertain the original isotopic value of the DIC because most of the original carbonate has been replaced during stabilization. For this reason, comparison against established secular trends in addition to careful petrographic analysis is required to estimate the isotopic composition of the "primary" carbonate and delineate subsequent diagenetic alteration.

15.6.2 Diagenetic Alteration of Sulfur Isotopes

Carbonate associated sulfate (CAS) is structurally boundary sulfate incorporated into the calcite lattice via substitution of the carbonate ion (Burdett et al. 1989). CAS concentrations in modern carbonates range from approximately 1,000 to 24,000 ppm (Staudt and Schoonen 1995). However, ancient limestone and dolomite often have lower sulfate concentrations relative to modern carbonates because of preferential uptake of sulfate into aragonite relative to low-magnesium calcite (Lyons et al. 2004; Staudt and Schoonen 1995), diagenetic replacement of carbonate particles with low sulfate-bearing cements (Gellatly and Lyons 2005), and lower marine sulfate concentrations in the geological past (Canfield 2004; Hurtgen et al. 2002, 2006; Shen et al. 2002).

Although the absolute CAS concentration in ancient sedimentary carbonates may be compromised by diagenetic processes, the sulfur isotopic values of CAS, in general, have been shown to be good proxies for the coeval isotopic value of oceanic sulfate (Burdett et al. 1989; Busenberg and Plummer 1985; Claypool et al. 1980; Kampschulte and Strauss 2004; Pingitore et al. 1995; Rees 1978; Staudt et al. 1993). Still, diagenetic alteration of sulfur isotopes can occur as a positive or a negative shift in $\delta^{34}S_{CAS}$ values (Fig. 15.2c) (Kampschulte and Strauss 2004). A positive isotopic shift can reflect continuous bacterial sulfate reduction (BSR) in the sediment under limited sulfate conditions, which will result in ^{34}S enrichment of the remaining dissolved sulfate in the pore fluids (Pierre 1985; Strauss 1997). Such conditions are commonly associated with deposition in anoxic settings with a high influx of organic carbon and iron (Raiswell and Berner 1985). Conversely, a negative isotopic shift in $\delta^{34}S_{CAS}$ values can occur if sedimentary carbonate is exposed to ^{34}S-depleted sulfate in the pore-fluids. Influx of ^{34}S-depleted sulfate is often associated with the oxidation of sedimentary pyrite previously formed via BSR during deposition or burial (Chambers and Trudinger 1979). The degree of isotopic overprinting can be assessed by analyzing the heterogeneity of $\delta^{34}S_{CAS}$ values obtained from multiple analyses in the same time interval, where large heterogeneity indicates a greater likelihood of diagenetic alteration (Kampschulte and Strauss 2004).

Under anoxic-euxinic conditions, sulfate reduction can result in the formation of pyrite in the water-column and sediments. Pyrite ultimately forms when dissolved sulfide (HS^-) reacts with iron derived from ferric oxides in detrital clays, resulting in a metastable intermediate iron sulfide and eventually sedimentary iron pyrite (Fe_2S) (Berner 1970). Pyrite formation in the water column forms microcrystalline framboidal pyrite that aggregate as laminae on the sediment surface (Love 1967). Early sedimentary pyrite can occur as fine crystalline pyrite randomly disseminated throughout the carbonate and can be later overprinted during burial by large (μm-mm scale) euhedral crystals (Mackenzie 2005). Late burial pyrite formation is often associated with other late diagenetic cements textures such as styolites and veins.

Since the bulk of pyrite formation is a consequence of bacterial sulfate reduction, the sulfide becomes enriched with ^{32}S (Goldhaber and Kaplan 1974; Trudinger et al. 1985). However, with the exception of pyrite formation in the water column, formation of sedimentary pyrite requires the mobilization of sulfur in pore-fluids. As a result, the origin and isotopic composition of the sulfide might not necessarily originate from seawater sulfate. A wide range of sulfur variability in a given suite of geochemical samples can be associated with the isotopic differences between pyrite size fractions that formed during different stages of diagenesis (Canfield et al. 1992; Goldhaber and Kaplan 1974; Kohn et al. 1998). Although early pyrite formation might occur in the water column or shallow subsurface with open circulation of seawater sulfate, most late diagenetic pyrite occurs in a closed system with limited sulfate availability (Goldhaber and Kaplan 1980). As dissolved sulfate becomes depleted through continuous sulfate reduction, the isotopic composition of pyrite formed late in diagenesis will have higher $\delta^{34}S$ values relative to pyrite formed earlier (Canfield et al. 1992). However, recall that the degree of fractionation is also largely a function of reaction rate, availability of reactive iron, sulfate, and organic matter (Berner 1970; Kempe and Thode 1968). Therefore, rapid sulfate reduction rates can also produce isotopically heavy sedimentary pyrite (Goldhaber and Kaplan 1974). Careful petrographic analysis of pyrite crystal size, location relative to veins and styolites, and multiple geochemical analyses should be considered to evaluate the timing of pyrite formation and source of sulfur.

15.7 Mass Balance, Mechanisms, and Secular Trends Through Geologic Time

15.7.1 Carbon

On geological time scales, the $\delta^{13}C$ value obtained from marine carbonate reflect the average isotopic composition of the oceans (or at least the well-mixed surface waters of the oceans), which has varied widely over geologic time (Fig. 15.5) (Berner and Canfield 1989; Des Marais 2001; Kaufman and Knoll 1995; Prokoph et al. 2008; Shields and Veizer 2002; Veizer et al. 1999). Secular variation of carbon isotopes is controlled by several factors: (1) the burial of organic matter, which removes isotopically light carbon from the carbon cycle and increases the $\delta^{13}C$ value in coeval sedimentary carbonates; (2) chemical weathering of organic matter, carbonates, and silicates that reintroduces isotopically light carbon back into the DIC reservoir, potentially resulting in low $\delta^{13}C$ values in coeval sedimentary carbonates; and (3) atmospheric CO_2 concentrations that affect isotopic fractionation during photosynthesis, where greater CO_2 concentrations allow greater isotopic fractionation, enriched isotopic values in the DIC reservoir and coeval sedimentary carbonates, and potentially enhanced organic carbon burial (Hayes et al. 1999; Kump and Arthur 1999; Kump and Garrels 1986).

The isotopic values of sedimentary carbonate and organic carbon are also indirect measurements of atmospheric CO_2 and consequently O_2. For every one mole of CO_2 consumed by oxygenic photosynthetic organisms, one mole of O_2 is produced and released to the atmosphere, represented by the reversible reaction $CO_2 + H_2O \rightarrow CH_2O + O_2$ where CH_2O is an estimate of organic matter. Therefore, increased productivity by photosynthetic organisms will decrease atmospheric CO_2 and increase O_2 if the organic carbon is sequestered in sediments. The burial of light organic carbon would also lead to higher $\delta^{13}C$ values of the coeval carbonates. Silicate weathering also plays a large role in the consumption CO_2, where CO_2 + $(Ca,Mg)SiO_3 \rightarrow (Ca,Mg)CO_3 + SiO_2$ sequesters CO_2 during carbonate and silica precipitation (Berner 2006).

Since the isotopic value of sedimentary carbonate can be indirectly tied to atmospheric CO_2, it is not surprising that atmospheric CO_2 concentrations have also varied over geologic time (Berner 2006; Pearson and Palmer 2000). However, direct measurements of atmospheric CO_2 in the ancient record are difficult to establish, except for fluid inclusions in ice cores, which can be used (based on the chemical equilibrium between the atmosphere and oceans) to estimate the isotopic values and concentration of CO_2 in the ancient atmosphere (Petit et al. 1999). Instead, geologists use the isotopic values from sedimentary carbonate and organic matter, together with some basic assumptions of organic burial rate, volcanic input, and other fluxes to model the concentration and isotopic values for atmospheric CO_2. Although there are a geochemical models that vary in complexity, they all assume the carbon cycle is at steady state such that (1) the fluxes into the system equal the fluxes out of the system and (2) the isotopic composition of carbon entering the system is the same as the isotopic composition of carbon leaving the system (Hayes et al. 1999). This concept is called mass balance, which is generally accurate for a closed system at steady state. The combination of these two rules can be mathematically expressed as:

$$\delta_i = f_o\delta_o + (1 - f_o) / \delta_c \tag{15.6}$$

Where δ_i represents the isotopic composition of the carbon input, δ_o represents the isotopic composition of the organic carbon in organic-rich rocks, and δ_c represents the isotopic composition of the sedimentary carbonate. The fraction of carbon buried in organic matter is denoted as f_o. For any instance in time, the isotopic value of

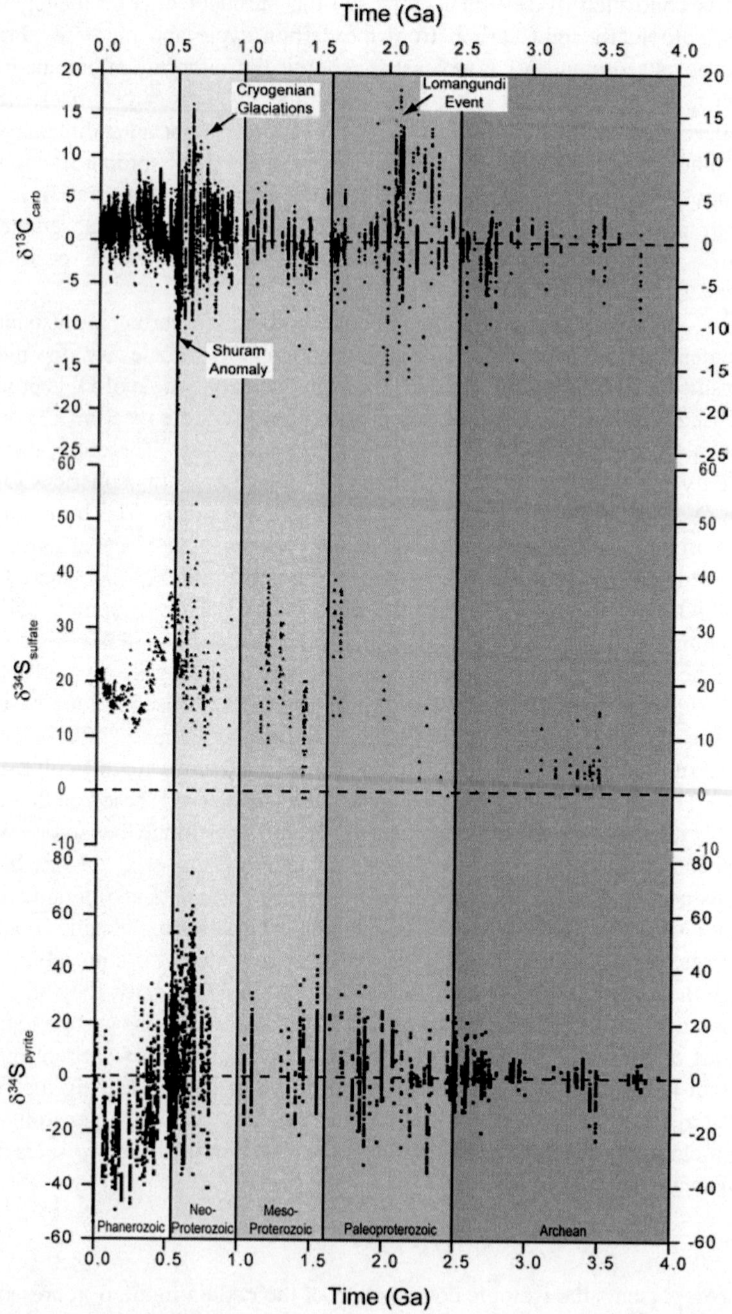

Fig. 15.5 Secular trends of stable isotopes ($\delta^{13}C_{carb}$, $\delta^{34}S_{sulfate}$, $\delta^{34}S_{pyrite}$) through geologic time. Note the sulfur isotopes have a much greater range of values. As a whole, for any given period of time, there is a large range of variability in the data because all isotopic data reflect both

sedimentary carbonate is linearly related to the fraction of organic carbon buried, assuming that the isotopic fractionation between the organic carbon and sedimentary carbonate ($\Delta \approx \delta_c - \delta_o$) remains constant:

$$\delta_c = \delta_i + f_o\Delta \qquad (15.7)$$

In other words, if the isotopic values of the sedimentary carbonate and coeval organic matter can be measured, then the isotopic composition of the incoming carbon via riverine inputs and the fraction of organic carbon buried can be estimated. The assumption that fractionation has remained constant through time as well as robust reconstructions of δ_i are two of the primary challenges with this model (Hayes et al. 1983; Kump 1989; Raymo 1997). Several studies have demonstrated that fractionation has varied greatly over geological time, particularly during the Proterozoic (Des Marais 2001; Hayes et al. 1999; Rothman et al. 2003). Variation in organic carbon burial appears to be related to a combination of climate, CO_2 concentrations, and primary production in the oceans, specifically the dominance of chemoautotrophy (Canfield and Teske 1996; Hayes et al. 1999).

Abrupt perturbations in the carbon isotopic record reflect periods of sudden change in the carbon cycle. Events that can result in sudden shifts to low $\delta^{13}C$ values in coeval sedimentary carbonates include an influx of riverine carbon from enhanced chemical weathering of exposed organic-rich sediments (Higgins and Schrag 2006), remineralization of organic matter associated with overturning of anoxic, organic-rich bottom waters (Knoll et al. 1996; Kump 1991), mass extinction events (Hsü and McKenzie 1985), and input of methane into the atmosphere (Dickens et al. 1995). Likewise, enhanced productivity due to increased nutrient supply from upwelling and terrestrial weathering and/or enhanced preservation of organic carbon associated with local or global bottom water anoxia can result in high $\delta^{13}C$ values in coeval sedimentary carbonates (Scholle and Arthur 1980). Determining the duration and triggers of carbon isotopic anomalies is a challenge in part because the isotopic value and amount of incoming carbon must be estimated. Both values are typically modeled by resolving differential equations to account for fluxes of carbon such as riverine input and volcanic outgassing (see Hayes et al. 1999; Hayes and Waldbauer 2006; Kump 1991 for an in-depth summary).

Fig. 15.5 (continued) diagenetic and depositional factors. Sulfate data prior to 1.0 Ga is sparse because sulfate concentrations were likely very low and sulfate minerals are rare. The $\delta^{13}C_{carb}$ data are fairly stable around 0‰ with major exceptions being the Lomagundi positive excursion in the Paleoproterozoic and the Shuram anomaly in the Neoproterozoic. The $\delta^{34}S_{sulfate}$ data generally increase through time, with a significant drop in values over the first half of the Phanerozoic, indicating a decrease in the fractional burial of sedimentary pyrite. The $\delta^{34}S_{pyrite}$ data become more variable after the Great Oxidation Event (2.3 Ga) with the most positive values in the Neoproterozoic and most negative values in the Phanerozoic. Carbon isotopic data compiled from Prokoph et al. (2008), Shields and Veizer (2002), Veizer et al. (1999) and references therein. Sulfur Isotopic data compiled from Canfield and Farquhar (2009), Johnston et al. (2005), Kampschulte and Strauss (2004), Prokoph et al. (2008), Strauss (1993), Thomazo et al. (2009) and references therein

Carbonate $\delta^{13}C$ values in the Archean (4.0–2.5 Ga) were fairly stable around −0.1‰ until the Lomagundi event (2.22–2.06 Ga), characterized by a prolonged positive carbon isotopic excursion ($\delta^{13}C$ > 10‰) likely related to the rise of atmospheric oxygen (Bekker et al. 2006). Karhu and Holland (1996) proposed that increased organic carbon burial between 2.22 and 2.06 Ga resulted in high $\delta^{13}C$ values in sedimentary carbonates and contributed to a net production of oxygen in the atmosphere. Although the global nature of the Lomagundi event has become accepted, the mechanism for the excursion has been challenged. Alternative mechanisms for the positive excursion include methanogenic diagenesis and active methanogenesis, by which with the signal being recorded in sediments after oxidants reached a certain threshold and ceasing to be recorded in sediments as oxidant concentrations increased further (e.g. Hayes and Waldbauer 2006). By the Paleoproterozoic (2.5–1.6 Ga), the carbonate isotopic record returned to around 0‰ and remained invariant, likely due to a high concentration of atmospheric CO_2 buffered by a large DIC reservoir (Bartley and Kah 2004; Grotzinger and Kasting 1993), which is evidenced by extensive abiotic crystalline carbonate deposits in peritidal to deep-shelf environments (Grotzinger 1989).

The isotopic record during the Mesoproterozoic (1.6–1.0 Ga) can be subdivided into three isotopic intervals: (1) stable isotopic interval with $\delta^{13}C$ values around 0‰; (2) subdued variation with slightly positive isotopic values and minor negative excursions ($\delta^{13}C$ between −1‰ and ~3.5‰); and (3) short-term, high amplitude variation exceeding 4–5‰ and becomes enhanced in the Neoproterozoic (Bartley et al. 2001; Kah et al. 1999; Kaufman and Knoll 1995). The transition from invariant to high amplitude variation may be explained by a reduction in the size of the DIC reservoir, such that the DIC surface oceans sensitive to perturbations (Bartley and Kah 2004). Overall positive $\delta^{13}C$ values may be attributed to a combination of strong stratification of the oceans with oxic surface waters and anoxic/euxinic conditions in the deep ocean (Shields et al. 2002), enhanced burial rates of organic carbon ($f$org = 0.5–1) than modern burial rates ($f$org = 0.2–0.3) (Des Marais 2001), and a very high concentration of DOC, allowing the ^{12}C pool to be dominated by the large store of organic carbon in the deep, anoxic oceans (Rothman et al. 2003).

Strongly positive $\delta^{13}C$ values ($\delta^{13}C$ = 4–8‰) continue into the Neoproterozoic Era (1,000–542 Ma) and are punctuated by high amplitude negative excursions (Halverson et al. 2005; Kaufman and Knoll 1995). Many of these negative excursions ($\delta^{13}C$ minima near −10‰) have been linked to extreme climatic fluctuations, known as Snowball Earth, which reflect periods when much of the Earth was potentially covered in ice (Hoffman et al. 1998; Hoffman and Schrag 2002; Kirschvink 1992). Two glaciations, the Sturtian (746–713 Ma) (Allen et al. 2002; Hoffman et al. 1996) and Marinoan (663–635 Ma) (Condon et al. 2005; Hoffmann et al. 2004; Zhou et al. 2004), have been recognized globally based on diamictites capped by carbonates with low $\delta^{13}C$ values and composed of unusual cements and sedimentary structures (e.g. Allen and Hoffman 2005; James et al. 2001; Jiang et al. 2006a; Kennedy 1996; Kennedy et al. 2001; Nogeira et al. 2003). A third, younger glacial deposit called the Gaskiers glaciation (~580 Ma) has been observed in successions in Newfoundland (Bowring et al. 2003), Australia (Calver et al. 2004), Brazil (de Alvarenga et al. 2007), and North China (Xiao 2004). The origin of the cap carbonates and their respective negative carbon isotopic excursions is still highly controversial, although several hypotheses exist: (1) high rates of siliciclastic weathering (mantle derived $\delta^{13}C$ values around –6‰) during deglaciation increased the alkalinity of the surface ocean (Higgins and Schrag 2003; Hoffman and Schrag 2002); (2) post-glacial upwelling of ^{13}C-depleted alkaline anoxic bottom waters during transgression (Grotzinger and Knoll 1995; Kaufman et al. 1997; Ridwell et al. 2003; Shields 2005); and (3) oxidation of methane clathrates ($\delta^{13}C$ values –50‰ to –80‰) during deglaciation (Jiang et al. 2003; Kennedy et al. 2001). These excursions were followed by the large negative carbon isotope excursion known as the Shuram anomaly, which has been linked to a rise in atmospheric oxygen and subsequent ventilation of the oceans (Fike et al. 2006; McFadden et al. 2008).

Phanerozoic secular variation shows a return to relatively stable carbon isotopes near 0‰ with moderate excursions with amplitudes ±4‰ associated with periods of biotic crisis (Saltzman et al. 1995). These negative excursions are interpreted as periods of decreased productivity while positive excursions may have been associated with enhanced carbon burial. Interestingly, many Phanerozoic isotopic excursions coincide with mass extinctions (Magaritz et al. 1986), including positive excursions in the early Dresbachian (SPICE), early Franconian (Late Cambrian), Late Ordovician, late Frasnian, late Famennian, and late Cenomanian, and negative excursions at the early Toarcian, Cretaceous- Tertiary, and Late Eocene (Bambach 2006). Oceanic anoxic events also are associated with negative excursions at the Permian-Triassic boundary (Cao et al. 2002) and throughout the Jurassic and Cretaceous (Cohen et al. 2007; Herrle et al. 2004; Schouten et al. 2000).

15.7.2 Sulfur

The secular variation of sulfur isotopes over geological time is controlled by terrestrial weathering and BSR (Canfield 2004; Claypool et al. 1980; Fike and Grotzinger 2008; Garrels and Lerman 1981; Hurtgen et al. 2002; Kampschulte and Strauss 2004; Pelechaty 1998; Strauss 1997, 1999). During intervals of intense weathering of shales and organic matter, which contain abundant sulfides, the isotopic value of oceanic sulfate will decrease. Conversely, when BSR is high, there will be an increased burial of sulfides in the lithosphere, and the isotopic value of oceanic sulfate will increase (Claypool et al. 1980) Similar to the burial/oxidation of organic carbon, the burial/exhumation and oxidative weathering of sulfides can also have an affect on (and be affected by) the redox state of the atmosphere and oceans (Canfield 2004; Kah et al. 2004; Kump and Garrels 1986; Petsch and Berner 1998). Various mass balance models similar to equation 1.6 have also been used to estimate the fraction of sulfide burial over geological time, which is indirectly tied to atmospheric O_2 concentrations (Berner 1999; Canfield and Teske 1996; Holland 2006).

Secular trends of the respective sulfur isotopic reservoirs, extracted from trace sulfate in sedimentary carbonates ($\delta^{34}S_{CAS}$) and corresponding pyrite ($\delta^{34}S_{pyrite}$) in organic rich shales, suggest that sulfur fractionation has also varied considerably over geological time (Fig. 15.5) (Canfield and Raiswell 1999). Unlike carbon isotopes, sulfur isotopic data is often interpreted by analyzing the difference in sulfur isotopic composition between CAS and sulfide, or $\Delta^{34}S$ ($\delta^{34}S_{CAS} - \delta^{34}S_{pyrite}$). During the Archean, the primary source of oceanic sulfate was through photochemical oxidation of volcanogenic sulfur compounds in the oxygen-poor atmosphere which induces a mass independent fractionation of sulfur isotopes (Farquhar et al. 2000). Archean $\Delta^{34}S$ fractionations can be explained by abiological processes alone, though this does not mean BSR was inactive since biological fractionation is minimal if there is less than 1 mM sulfate present (Shen et al. 2001), which was likely the case up to around 2.3 Ga (Canfield and Teske 1996). By 2.4–2.1 Ga the mass independent signature was replaced by a mass dependent one like that of the Phanerozoic, demonstrating that enough oxygen had built up in the atmosphere to allow for weathering to be the dominant source of oceanic sulfate (Farquhar et al. 2000). Although evidence for BSR is present in older rocks from isolated areas, there is no global oceanic BSR signature until around 2.4 Ga (Anbar and Knoll 2002; Canfield and Teske 1996; Canfield 1998).

During the early Proterozoic, $\Delta^{34}S$ values increased to greater than 20‰, likely associated with BSR and an increase in sulfate availability. The origin of the oxidative sulfur cycle, which includes the oxidation of hydrogen sulfide to S^0 or sulfate and bacterial sulfur disproportionation (BSD), is debated. BSD is when bacteria disproportionate (both oxidize and reduce) S^0, or some other intermediate form of sulfur, into sulfide and sulfate (Canfield and Teske 1996). The oxidation of sulfides has no affect on the isotopic value of the residual sulfur, and the repeated recycling of ^{34}S-depleted sedimentary sulfides into sulfates over time leads to a significant increase in the isotopic fractionation of sulfur (Canfield 2004). Using the more sensitive $\Delta^{33}S$ ($\delta^{33}S - [(\delta^{34}S/1000 + 1)^{0.515} - 1] \times 1000$), sulfur disproportionation has been detected by 1.3 Ga (Johnston et al. 2005). There is, however, little global evidence for the values of $\Delta^{34}S$ to be consistently greater than 45‰ (the maximum fractionation produced by sulfate reducing bacteria) until the Neoproterozoic.

The stable isotopic record during the Neoproterozoic Era (1,000–542 Ma) is characterized by low sulfate concentrations. During the Neoproterozoic, the radiation of non-photosynthetic sulfide-oxidizing bacteria dramatically increased the total sulfur isotopic fractionation from 46‰ to >70‰ (Canfield and Teske 1996), suggesting that at least the surface oceans were likely well oxygenated by the end of the Neoproterozoic. Fike et al. (2006) constrained the timing for BSD to be significant after ~548 Ma, which is consistent with $\Delta^{34}S$ values greater than 45‰ by the early Cambrian. It is, however, possible that BSD occurred throughout the Proterozoic and that $\Delta^{34}S$ values remained relatively low due to low sulfate concentrations (Hurtgen et al. 2005). The average isotopic value measured from evaporites and CAS suggest that the isotopic composition of oceanic sulfate during the Neoproterozoic was similar to the Cenozoic ($\delta^{34}S_{CAS}$ = ~21‰) but varied considerably

(Hurtgen et al. 2002; Strauss et al. 2001). However, $\Delta^{34}S$ was narrow (<46‰) and the burial rate of sulfides ($f$py=0.6–0.7) was greater relative to the Cenozoic ($\Delta^{34}S$ = ~24–71‰; $f$py=0.3) (Canfield 2001; Canfield and Teske 1996; Hurtgen et al. 2005; Hurtgen et al. 2002). The $\delta^{34}S_{py}$ was also unusually enriched, near Cenozoic $\delta^{34}S_{CAS}$ values (Canfield 2004). Low sulfur fractionation, high rates sulfide deposition, $\delta^{34}S_{CAS}$ variability, and enriched $\delta^{34}S_{py}$ values suggest that sulfate concentrations in the Neoproterozoic were generally low. Some estimates of sulfate concentrations are as low as 3 mM (Hurtgen et al. 2004), which may lend to greater sensitivity of the sulfur isotopic record during this time because a small sulfur reservoir would lead to greater isotopic heterogeneity in the $\delta^{34}S_{CAS}$ and $\delta^{34}S_{py}$ records (Hurtgen et al. 2006).

Throughout the Phanerozoic, the ocean has had significant concentrations of sulfate, allowing both BSR and BSD to flourish. Average $\delta^{34}S_{CAS}$ values range from ~30‰ in the Cambrian to ~10‰ at the Permian-Triassic boundary before settling around 20‰ today (Strauss 1997). High $\delta^{34}S_{CAS}$ values during the late Cambrian signifies a return to low sulfate concentrations and fluctuating redox (Hurtgen et al. 2009). Low $\delta^{34}S_{CAS}$ values during the Permian have been explained by a variety of mechanisms, including reduced continental weathering and nutrients in the ocean (Kramm and Wedepohl 1991), a bolide impact that released mantle sulfur (Kaiho et al. 2001), and fluctuating stratified surface waters that lead to the burial of sedimentary sulfides which were then oxidized as the chemocline deepened (Riccardi et al. 2006).

15.8 Case Study: The Terminal Neoproterozoic Shuram Anomaly

Thus far in the chapter, we have broadly reviewed how to measure and interpret secular carbon and sulfur isotopic variations in the geological record. Despite the great volume of research over past decades, geochemists still struggle to understand how many of the major isotopic excursions were generated and speculate as to the mechanisms, duration, triggers, and affects on both the global environmental and biological systems. Integration of paleoenvironmental and paleontological proxies is necessary in order to better understand changes in secular variation of the carbon and sulfur isotopic record because the isotopic composition of most sedimentary carbonates, in part, reflects both biological activity and environmental change.

One of the largest negative carbon isotopic excursions in Earth history is known as the Shuram anomaly ($\delta^{13}C$ minima near −15‰), which culminated during the terminal Neoproterozoic Era. As discussed earlier, the Neoproterozoic Era is characterized by extreme climatic fluctuations (Hoffman et al. 1998), major carbon and sulfur isotopic variation, ventilation of the deep oceans (Canfield and Teske 1996; Canfield et al. 2007; Fike et al. 2006), and radiation of early animals (Knoll and Carroll 1999). Although these environmental and biological events have been speculatively linked, their temporal relationships are debated. Recently, it has been proposed that the pervasive and permanent ventilation of the oceans may have played a role in the increase in taxonomic diversity and complexity throughout the Neoproterozoic (Canfield et al. 2007; Fike et al. 2006; McFadden et al. 2008; Rothman et al. 2003), which is evidenced by the earliest molecular evidence of demosponges (700–635 Ma) (Love et al. 2009), proliferation of acanthomorphic acritarchs and macroscopic algae (635–550 Ma) (Ding et al. 1996; Grey 2005; Knoll 1992; Moczydlowska 2005; Tiwari and Knoll 1994; Zhou et al. 2007), appearance of complex macro fauna known as the Ediacara biota (575–542 Ma) (Waggoner 2003), motile bilaterian eumetazoans (~550 Ma) (Weber et al. 2007), and first occurrence biomineralizing animals including *Cloudina* and *Sinotubulites* (Grotzinger et al. 2000; Hofmann and Mountjoy 2001; Hua et al. 2005).

The Shuram anomaly, roughly bracketed between the last known occurrence of acanthomorphic acritarchs and the first global appearance of Ediacara Biota (Grey 2005; Zhou et al. 2007), is thought to represent the ventilation of the deep oceans. Yet, the timing, duration, and mechanisms of the anomaly are still poorly understood. For example, although it is recognized in many locations, there is internal isotopic variation from region to region, which has lead some to question whether the Shuram can be correlated globally and whether the isotopic anomaly is truly representative of Neoproterozoic seawater. There are multiple hypotheses about the origin and interpretation of the Shuram event. As with most scientific queries, more than one geochemical scenario or model is possible. Here we discuss the characteristics, correlation, and potential mechanisms for the anomaly.

15.8.1 Characteristics and Correlation of the Shuram Anomaly

The Shuram anomaly was an unusual isotopic event. The timing and onset of the negative excursion is not well constrained. It is roughly the same age as the Gaskiers glaciation, which has led some to suggest that the excursion may be similar to isotopic excursions observed within cap carbonates overlying the Cryogenian diamictites (Condon et al. 2005). However, the excursion does not directly overly a glacial diamictite, nor is there a cap carbonate present. Additionally, the $\delta^{13}C$ excursion is unusually long-lived (1–50 Ma) (Condon et al. 2005; Le Guerroue et al. 2006) relative to negative excursions associated with cap carbonates (<1 Ma) (Hoffman et al. 2007; Kennedy et al. 2001). The termination of the anomaly around 551–547 Ma is more definitively based on U-Pb age-constraints in South China (Condon et al. 2005), Namibia (Grotzinger et al. 1995), and Oman (Bowring et al. 2007).

The Shuram anomaly is characterized by three distinct isotopic trends. The most obvious is the sharp negative $\delta^{13}C$ excursion to −12‰ between 580 and 550 Ma. The second distinctive isotopic trend is an overall increase in $\Delta^{34}S$ values from <30‰ to >50‰ (Fike et al. 2006; McFadden et al. 2008), indicating that sulfate concentrations increased throughout the Neoproterozoic. The third, less obvious, characteristic is the decoupling between inorganic carbon isotopes ($\delta^{13}C_{carb}$) and organic carbon isotopes ($\delta^{13}C_{org}$). Although $\delta^{13}C$ values distinctly vary throughout the Neoproterozoic, $\delta^{13}C_{org}$ values remained constant. In the modern oceans, $\delta^{13}C_{carb}$ and $\delta^{13}C_{org}$ co-vary because the primary autotrophic organisms creating the fractionation between inorganic and organic carbon utilize specific enzymes with fixed fractionation factors that are controlled by temperature and reaction rate (Hayes et al. 1999). Decoupling of $\delta^{13}C_{org}$ values from $\delta^{13}C_{carb}$ values indicates that the DOC was likely isotopically buffered by a large DOC pool (Rothman et al. 2003).

The negative $\delta^{13}C$ excursion has been observed in Oman (Fike et al. 2006; Le Guerroue and Cozzi 2006), Namibia (Grotzinger et al. 1995), China (Zhou and Xiao 2007), Australia (Calver 2000), SW United States (Kaufman et al. 2007a), India (Kaufman et al. 2006), Brazil (Misi et al. 2007), and Siberia (Melezhik et al. 2005), and an increase in sulfur fractionation has been observed in Oman (Fike et al. 2006), and China (McFadden et al. 2008). The recognition at many disparate localities suggests that the Shuram anomaly is a global event, however, intra-basinal isotopic heterogeneity has been observed. For example, there are several negative $\delta^{13}C$ excursions (EN1, EN2, and EN3) recognized in the Doushantuo Formation in South China (Fig. 15.6a), (Zhou and Xiao 2007), which is not the observation in Oman (Fig. 15.6b). EN1 has been attributed to methane destabilization during post-Marinoan glacial warming (Jiang et al. 2006b). EN2, on the other hand, occurs toward the top of a regionally recognized sedimentary cycle in the middle of the Doushantuo Formation, which has previously been interpreted to represent an unconformity associated with the Gaskiers glaciation (Condon et al. 2005). There is no record of glacial diamictites in the Doushantuo Formation, and a 599 ± 4 Ma Pb–Pb age above the EN2 excursion in other areas of South China are inconsistent with a 580 Ma Gaskiers event (Barfod et al. 2002). Therefore, in the absence of a glacial deposit, EN2 has since been reinterpreted to represent an oxidation event prior to the main Shuram anomaly (McFadden et al. 2008). Pre-Shuram negative $\delta^{13}C$ excursions have also been observed in the India (Kaufman et al. 2006) and SW United states (Corsetti and Kaufman 2003; Kaufman et al. 2007a), which suggests that ventilation event(s) prior to the Shuram may have been local or regional in origin. Alternatively, negative isotopic excursions may not be recognized in other localities due to poor sampling resolution.

Spatial variability of $\delta^{13}C$ excursions also differs between Oman and China. In Oman, the negative $\delta^{13}C$ excursion has been recognized at more than 30 localities with little variation in duration or magnitude (Burns and Matter 1993; Le Guerroue and Cozzi 2006; McCarron 2000). In China, there is a large isotopic shift toward negative $\delta^{13}C$ values from shelf into basin, such that successions on the shallow shelf are characterized by positive $\delta^{13}C$ values near +6‰ punctuated by large negative excursions, while successions in the deep basin are represented by $\delta^{13}C$ values near −2‰ and subdued negative excursions (Jiang et al. 2007). Isotopic variation in $\Delta^{34}S$ values are also observed from shelf to basin in China. Shelf successions exhibit an increase in $\Delta^{34}S$ values to >30‰ by 551 Ma and deep basin successions record suppressed $\Delta^{34}S$ values around 15‰ (Li et al. 2010).

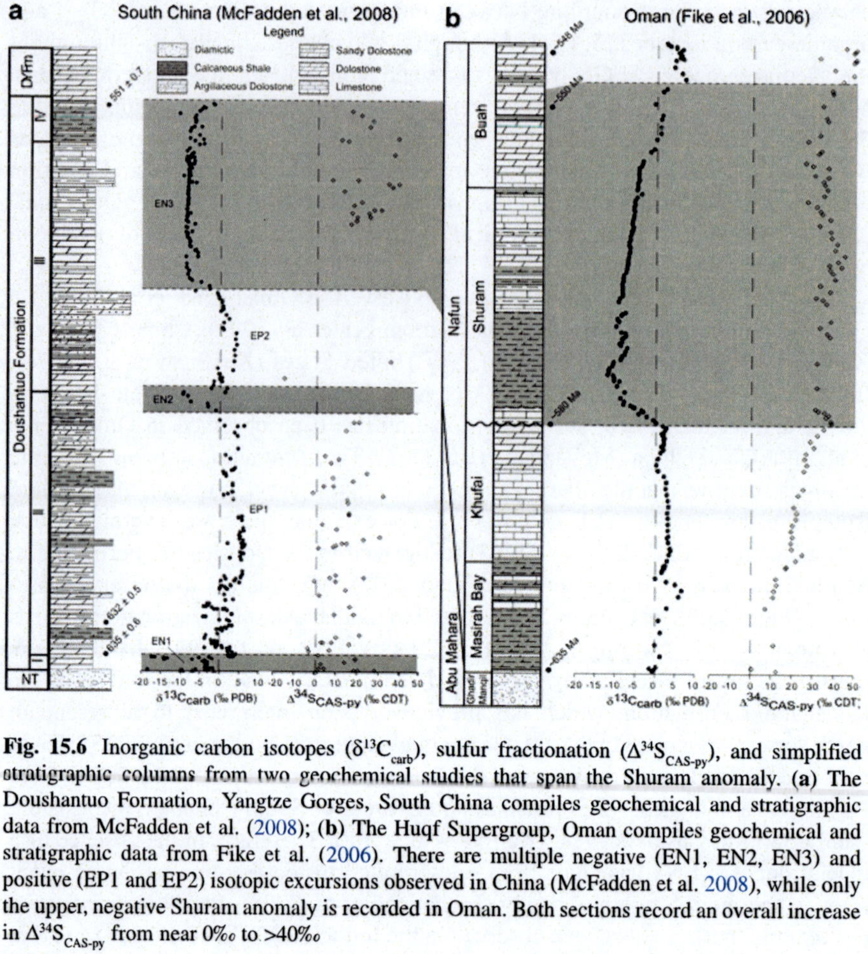

Fig. 15.6 Inorganic carbon isotopes ($\delta^{13}C_{carb}$), sulfur fractionation ($\Delta^{34}S_{CAS\text{-}py}$), and simplified stratigraphic columns from two geochemical studies that span the Shuram anomaly. (**a**) The Doushantuo Formation, Yangtze Gorges, South China compiles geochemical and stratigraphic data from McFadden et al. (2008); (**b**) The Huqf Supergroup, Oman compiles geochemical and stratigraphic data from Fike et al. (2006). There are multiple negative (EN1, EN2, EN3) and positive (EP1 and EP2) isotopic excursions observed in China (McFadden et al. 2008), while only the upper, negative Shuram anomaly is recorded in Oman. Both sections record an overall increase in $\Delta^{34}S_{CAS\text{-}py}$ from near 0‰ to >40‰

This shift from strongly positive carbon isotopic values on the shelf to negative carbon isotopic values in the basin has been interpreted to represent a strong chemical gradient, whereby the shallow shelf was positioned within oxic waters relative to anoxic/euxinic/ferruginous bottom waters in the deep basin separated by a chemocline (Jiang et al. 2007).

15.8.2 Mechanisms for the Shuram Anomaly

Mechanisms for isotopic anomaly remain poorly understood. The magnitude and proposed duration of the Shuram $\delta^{13}C$ excursion challenge our current understanding of the carbon cycle. It is difficult to explain isotopic values significantly lower than the mantle ($\delta^{13}C$ = <–6‰) over prolonged periods of time (>0.5 Ma) because of the short residence time of carbon in the modern oceans (10^5 years) and the large oceanic DIC reservoir that isotopically buffers $\delta^{13}C$ values in carbonates (Holser et al. 1988). As mentioned previously, short term negative isotopic perturbations require the sudden input of large quantities of ^{13}C-depleted carbon such as the oxidative decay of organic-rich sedimentary rocks, overturning of stratified oceans, or oxidation of methane. These sources of carbon are generally exhausted rapidly (Berner 2002) and the associated carbon isotopic excursions typically do not drop below the mantle value (Kump and Arthur 1999).

Due to these challenges, the Shuram anomaly has been suggested to be simply the result of diagenetic alteration (Burns and Matter 1993; Derry 2010 , McCarron 2000). In some localities, such as the SW United States (Kaufman et al. 2007a), the onset of the negative excursion occurs near an erosional unconformity and is co-variant with oxygen isotopes. As a result, some studies conclude that the Shuram excursion was caused by subaerial exposure and diagenetic alteration of the carbonates during meteoric diagenesis (Burns and Matter 1993; Knauth and Kennedy 2005). Derry (2010) suggested rock-water interactions during deep burial may have created the Shuram anomaly. The negative isotopic excursion has also been interpreted to represent strong stratification under restricted marine to lacustrine conditions with no connection to the open ocean (Calver 2000). Restricted basins have a smaller carbon reservoir (DIC and DOC) relative to the global ocean, and therefore, local inputs of oxidants can potentially increase sensitivity to isotopic perturbations. If any of the above is correct, then the Shuram excursion may not represent a global, or primary, signal at all.

In addition to the extreme magnitude and duration, the amount of oxidants necessary to create the excursion on a global scale has been difficult to quantify (Bristow and Kennedy 2008; Halverson and Hurtgen 2007); however there are hypotheses that address all of these challenges. If the Shuram anomaly represents primary isotopic values of the Neoproterozoic oceans, then it is likely that the fluxes of carbon and the sizes of the carbon reservoirs (particularly the DIC and DOC reservoirs) were different relative to our understanding of the modern carbon cycle. Rothman et al. (2003) derived a model depicting deep anoxic oceans with a DOC inventory 100–1,000 times larger than the modern DOC. Such a large DOC pool would have resulted in decoupling of the oceanic DOC and DIC reservoirs, in turn buffering the isotopic composition of the DOC reservoir, and increasing the residence time of organic carbon in the oceans. Therefore, minor changes in the DOC pool via periodic inputs of oxidants would have had a dramatic impact on the isotopic composition of the DIC reservoir, leading to high magnitude $\delta^{13}C$ excursions below mantle values in coeval carbonate sedimentary rocks. Complete depletion of the DOC reservoir to a modern size would have occurred through prolonged exposure of oxidants introduced to the deep oceans, creating a protracted isotopic excursion(s). Decoupling of the DIC and DOC reservoirs indicates that the carbon cycle was not at steady state and evolved dynamically through quantitative reduction of the DOC reservoir to near modern values by the Early Cambrian.

15.9 Summary

The purpose of this chapter was to provide a basic introduction to the field of stable isotope geochemistry, highlight fundamental isotopic systems used in paleoenvironmental analysis, outline methods used to acquire isotopic data, explain common primary and secondary processes that can affect the interpretation of that data, and discuss the secular trends of carbon and sulfur isotopes through Earth history. Stable isotope geochemistry has important applications for understanding past environments and should be incorporated, where appropriate, into paleontological research. The primary take away messages are as follows:

1. Consistent sampling methodologies and preparation techniques are critical in order to produce meaningful results. Careful attention should be made to assess the quality of the sample for diagenetic alteration and contamination prior to preparation for geochemical analysis, including petrographic analysis, CL and SEM petrography, and spot sampling obvious diagenetic zones to compare against samples considered less diagenetically altered.
2. In order to properly interpret stable isotope data, it is critical to understand the origin of the sample used in the geochemical analysis. The stable isotopic composition of any sample is a combination of the primary depositional environment and its diagenetic history. Therefore, the design of a geochemical suite should incorporate all available data and approach the isotopic results with multiple hypotheses and models.
3. The secular trends of carbon and sulfur have varied dramatically over geological history and reflect major changes in the carbon and sulfur cycle. Most isotopic excursions in the geological record appear to manifest from sudden imbalances in the influx/outflow of carbon and sulfur into/out of specific reservoirs. The long-term variation has important implications on productivity, redox state of the atmosphere and oceans, and potentially the extinction and radiation of organisms.
4. Geochemists still struggle to understand the mechanisms, duration, and triggers of isotopic excursions. In many case studies, such as the Shuram anomaly, there are multiple scenarios that produce isotopic excursions. The challenge for future research is to better address stable isotopic variation by integrating datasets across disciplines to improve our understanding of the complex interactions between environmental and biological systems.

Acknowledgments We appreciate Peir Pufahl and Linda Kah for their thorough review and comments that helped to greatly improve the quality of this manuscript. Special thanks also to the editors of this book for their comments and discussion, which helped in refining the flow of the manuscript. Finally, we also thank Dr. Ganqing Jiang for his contribution of several photomicrographs of diagenetic cements.

References

Allen PA, Hoffman PF (2005) Extreme winds and waves in the aftermath of a Neoproterozoic glaciation. Nature 433:123–127

Allen PA, Bowring SA, Leather J, Brasier M, Cozzi A, Grotzinger JP, McCarron G, Amthor J (2002) Chronology of Neoproterozoic glaciations: new insights from Oman. In: Abstract volume, 16th international sedimentological congress, Johannesburg, pp 7–8. International Association of Sedimentologists

Anbar AD, Knoll AH (2002) Proterozoic ocean chemistry and evolution: a bioinorganic bridge? Science 297:1137–1142

Anderson TF, Arthur MA (1983) Stable isotopes of oxygen and carbon and their application to sediementologic and paleoenvironmental problems. In: Arthur MA, Anderson TF, Kaplan IR, Veiser J, Land LS (eds) Stable isotopes in sedimentary geology, vol 10. SEPM, Tulsa, pp 1–151

Bambach RK (2006) Phanerozoic biodiversity mass extinctions. Annu Rev Earth Planet Sci 34:127–155

Barfod GH, Albaréde F, Knoll AH, Xiao S, Frei R, Baker J (2002) Implications for the Neoproterozoic biological and climatic history from dating of the doushantuo phosphorites, S. China. EOS Trans AGU Fall Meeting 83(47)

Bartley JK, Kah LC (2004) Marine carbon reservoir, C_{org}-C_{carb} coupling, and the evolution of the Proterozoic carbon cycle. Geology 32(2):129–132

Bartley J, Semikhatov M, Kaufman A, Knoll A, Pope M, Jacobsen S (2001) Global events across the Mesoproterozoic-Neoproterozoic boundary: C and Sr isotopic evidence from Siberia. Precambrian Res 111(1–4):165–202

Bazilevich NI, Rodin L, Roznov NN (1970) Geographical aspects of biological productivity. Sov Geogr Rev Transl 12(5):293–317

Beaudoin G, Taylor BE (1994) High precision and spatial resolution sulfur isotope analysis using MILES laser microprobe. Geochim Cosmochim Acta 58(22):5055–5063

Beaudoin G, Taylor BE, Rumble D III, Thiemens M (1994) Variations in the sulfur isotope composition of troilite from the Canon Diablo iron meteorite. Geochim Cosmochim Acta 58(19):4253–4255

Bekker A, Karhu JA, Kaufman AJ (2006) Carbon isotope record for the onset of the Lomagundi carbon isotope excursion in the Great Lakes area, North America. Precambrian Res 148:145–180

Berner RA (1970) Sedimentary pyrite formation. Am J Sci 268:1–23

Berner RA (1992) Paleo-CO_2 and climate. Nature 358:114–114

Berner RA (1999) Atmospheric oxygen over Phanerozoic time. Proc Natl Acad Sci USA 96:10955–10957

Berner RA (2002) Examination of hypotheses for the Permo–Triassic boundary extinction by carbon cycle modeling. Proc Natl Acad Sci USA 99:4172–4177

Berner RA (2006) GEOCARBSULF: a combined model for Phanerozoic atmospheric O_2 and CO_2. Geochim Cosmochim Acta 70:5653–5664

Berner RA, Canfield D (1989) A model for atmospheric oxygen over Phanerozoic time. Am J Sci 289:333–361

Berner RA, Lasaga AC, Garrels RM (1993) The carbonate-silicate geochemical cycle and its effect on atmospheric carbon dioxide over the past 100 million years. Am J Sci 283:641–683

Bigeleisen J, Mayer MG (1947) Calculation of equilibrium constancts for isotopic exchange reactions. J Chem Phys 15:261–267

Blake DR, Rowland FS (1988) Continuing worldwide increase in tropospheric methane 1978 to 1987. Science 239:1129–1131

Bolin B, Degens ET, Kempe S, Ketner P (1979) Global carbon cycle. Wiley, New York

Bottrell SH, Newton RJ (2006) Reconstruction of changes in global sulfur cycling from marine sulfate isotopes. Earth Sci Rev 75:59–83

Bowring S, Myrow P, Landing E, Ramezani J, Grotzinger J (2003) Geochronological constraints on terminal Neoproterozoic events and the rise of metazoans. Geophys Res Abstr 5:13219

Bowring SA, Grotzinger JP, Condon DJ, Ramezani J, Newall MJ, Allen PA (2007) Geochronologic constraints on the chronostratigraphic framework of the Neoproterozoic Huqf Supergroup, Sultanate of Oman. Am J Sci 307:1097–1145

Bristow TF, Kennedy MJ (2008) Carbon isotope excursions and the oxidant budget of the Ediacaran atmosphere and ocean. Geology 36(11):863–866

Broecker WS (1973) Factors controlling CO2 content in the oceans and atmosphere. In: Woodwell GM, Pecan EV (eds) Carbon and the biosphere. United States Atomic Energy Commission, Washington, DC, pp 32–50

Bruckschen P, Bruhn F, Meijer J, Stephan A, Veizer J (1995) Diagenetic alteration of calcitic fossil shells: proton microprobe (PIXE) as a trace element tool. Nucl Instrum Methods Phys Res Sect B 104(1–4):427–431

Bruhn F, Bruckschen P, Richter DK, Meijer J, Stephan A, Veizer J (1995) Diagenetic history of sedimentary carbonates: constraints from combined cathodoluminescence and trace element analyses by micro-PIXE. Nucl Instrum Methods Phys Res Sect B 104(1–4):409–414

Brunner B, Bernasconi SM (2005) A revised isotope fractionation model for dissimilatory sulfate reduction in sulfate reducing bacteria. Geochim Cosmochim Acta 69:4759–4771

Buffle J, van Leeuwen HP (1992) Environmental particles, series on analytical and physical chemistry of environmental systems, vol 1. Environmental analytical and physical chemistry series. Lewis Publishers Boca Raton, FL

Burdett ST, Arthur MA, Richardson M (1989) A Neogene seawater isotope age curve from calcareous pelagic microfossils. Earth Planet Sci Lett 94:189–198

Burns SJ, Matter A (1993) Carbon isotopic record of the latest Proterozoic from Oman. Eclog Geol Helv 86:595–607

Busenberg E, Plummer LN (1985) Kinetic and thermodynamic factors controlling the distribution of SO^4 and Na in calcites and selected aragonites. Geochim Cosmochim Acta 49:713–725

Calver CR (2000) Isotope stratigraphy of the Ediacarian (Neoproterozoic III) of the Adelaide Rift Complex, Australia, and the overprint of water column stratification. Precambrian Res 100:121–150

Calver CR, Black LP, Everard JL, Seymour DB (2004) U-Pb zircon age constraints on late Neoproterozoic glaciation in Tasmania. Geology 10:893–896

Canfield DE (1998) A new model for Proterozoic ocean chemistry. Nature 396:450–453

Canfield D, Canfield D (2001) Biogeochemistry of sulfur isotopes. In: Valley JW, Cole DR, Valley JW, Cole DR (eds) Stable isotope geochemistry. Reviews in mineralogy and geochemistry, vol 43. The Mineralogical Society of America, Washington, DC, pp 607–636

Canfield D (2004) The evolution of the earth surface sulfur reservoir. Am J Sci 304:839–861

Canfield D, Farquhar J (2009) Animal evolution, bioturbation, and the sulfate concentration of the oceans. Proc Natl Acad Sci USA 106(20):8123–8127

Canfield D, Raiswell R (1999) The evolution of the sulfur cycle. Am J Sci 299:697–723

Canfield D, Teske A (1996) Late Proterozoic rise in atmospheric oxygen concentration inferred from phylogenetic and sulphur-isotope studies. Nature 382:127–132

Canfield DE, Raiswell R, Westrich JT, Reaves CM, Berner RA (1986) The use of chromium reduction in the analysis of reduced inorganic sulfur in sediments and shale. Chem Geol 54:149–155

Canfield D, Raiswell R, Bottrell SH (1992) The reactivity of sedimentary iron minerals toward sulfide. Am J Sci 292:659–683

Canfield DE, Habicht KS, Thamdrup B (2000) The Archean sulfur cycle and the early history of atmospheric oxygen. Science 288:658–661

Canfield DE, Poulton SW, Narbonne GM (2007) Late-Neoproterozoic deep-ocean oxygenation and the rise of animal life. Science 315(5808):92–95

Cao C, Wang W, Jin Y (2002) Carbon isotope excursions across the Permian-Triassic boundary in the Meishan section, Zhejiang Province, China. Chin Sci Bull 47:1125–1129

Chambers LA, Trudinger PA (1979) Microbiological fractionation of stable sulfur isotopes: a review and critique. Geomicrobiol J 1:249–293

Charlson RJ, Chameides WL, Kley D (1985) The transformations of sulfur and nitrogen in the remote atmosphere. In: Galloway JN, Charlson RJ, Andreae MO (eds) The biogeochemical cycling of sulfur and nitrogen in the remote atmosphere. Reidel, Dordrecht, pp 67–80
Charlson RJ, Anderson TL, McDuff RE (1992) The sulfur cycle. In: Butcher SS, Charlson RJ, Orian GH, Wolfe GV (eds) Global biogeochemical cycles. Academic, San Diego, pp 285–299
Chaussidon M, Albarede F, Sheppard SMF (1989) Sulphur isotope variations in the mantle from ion microprobe analyses of micro-sulphide inclusions. Earth Planet Sci Lett 92:144–156
Choquette PW, James NP (1990) Limestone: the burial diagenetic environment. In: McIlreath IA, Morrow DW (eds) Diagenesis, vol 4, Geoscience Canada. Runge, Ottawa, pp 75–112
Claypool GE, Holser WT, Kaplan IR, Sakai H, Zak I (1980) The age curves of sulfur and oxygen isotopes in marine sulfate and their mutual interpretation. Chem Geol 29:199–260
Cohen AS, Coe AL, Kemp DB (2007) The Late Palaeocene-Early Eocene and Toarcian (Early Jurassic) carbon isotope excursions: a comparison of their time scales, associated environmental changes, causes and consequences, J Geol Soc Lond 164:1093–1108
Condon DJ, Zhu M, Bowring SA, Wang W, Yang A, Jin Y (2005) U-Pb ages from the neoproterozoic Doushantuo Formation China. Science 308:95–98
Coplen TB (1996) New Guidelines for reporting stable hydrogen, carbon, and oxygen isotope-ratio data. Geochim Cosmochim Acta 60:3359–3360
Coplen TB, Krouse HR (1998) Sulphur isotope data consistency improved. Nature 392:32
Coplen TB, Kendall C, Hopple J (1983) Comparison of stable isotope reference samples. Nature 302:236–238
Corsetti FA, Kaufman AJ (2003) Stratigraphic investigations of carbon isotope anomalies and Neoproterozoic ice ages in Death Valley California. GSA Bull 115:916–932
Craig H (1953) The geochemistry of the stable carbon isotopes. Geochim Cosmochim Acta 3:53–92
Craig H (1957) Isotopic standards for carbon and oxygen and correction factors for mass-spectrometric analysis of carbon dioxide. Geochim Cosmochim Acta 12(1–2):133–149
Craig H (1965) The measurement of oxygen isotope paleotemperatures. In: Tongiorgi E (ed) Stable isotopes in oceanographic studies and paleotemperatures. Consiglio Nazionale selle Ricerche, Laboratorio de Geologia Nucleare, Pisa, pp 161–182
Craig H, Chou C, Welhan J, Stevens CM, Engelkemeir A (1988) The isotopic composition of methane in polar ice cores. Science 242:1535–1539
de Alvarenga CJS, Figueiredo MF, Babinski M, Pinho FEC (2007) Glacial diamictites of Serra Azul Formation (Ediacaran, Paraguay belt): evidence of the Gaskiers glacial event in Brazil. J S Am Earth Sci 23(2–3):236–241
De Vooys CGN (1979) Primary production in aquatic environments. In: Bolin B, Degens ET, Kempe S, Ketner S (eds) The global carbon cycle. Wiley, New York, pp 259–292
Deines P (2002) The carbon isotope geochemistry of mantle xenoliths. Earth Planet Sci Lett 58:247–278
Derry LA (2010) A burial diagenesis origin for the Ediacaran Shuram–Wonoka carbon isotope anomaly. Earth Planet Sci Lett 294:152–162
Des Marais DJ (2001) Isotopic evolution of the biogeochemical carbon cycle during the Precambrian. Stable Isot Geochem Rev Miner Geochem 43:555–578
Des Marais DJ, Strauss H, Summons RE, Hayes JM (1992) Carbon isotope evidence for the stepwise oxidation of the Proterozoic environment. Nature 359:605–609
Dickens GR, O'Neil JR, Rea DK, Owen RM (1995) Dissociation of oceanic methane hydrate as a cause of the carbon isotope excursion at the end of the Paleocene. Paleoceanography 10:965–971
Ding L, Li Y, Hu X, Xiao Y, Su C, Huang J (1996) Sinian miaohe biota. Geological Publishing House, Beijing
Epstein S, Buchsbaum R, Lowenstam H, Urey HC (1951) Carbonate-Water isotopic temperature scale. GSA Bull 62:417–426
Fairchild IJ, Spiro B (1987) Petrological and isotopic implications of some contrasting Late Precambrian carbonates, NE Spitsbergen. Sedimentology 34:973–989

Fairchild IJ, Marshall JD, Bertrand-Sarfati J (1990) Stratigraphic shifts in carbon isotopes from Proterozoic stromatolitic carbonates (Mauritania): influences of primary mineralogy and diagenesis. In: Knoll AH, Ostrom JH (eds) American Journal of Science, Cloud Volume, Proterozoic Evolution and Environments 290–A:46–79

Farquhar J, Bao H, Thiemens M (2000) Atmospheric influence of Earth's earliest sulfur cycle. Science 289:756–758

Fike D, Grotzinger J (2008) A paired sulfate-pyrite $\delta^{34}S$ approach to understanding the evolution of the Ediacaran-Cambrian sulfur cycle. Geochim Cosmochim Acta 72(11):2636–2648

Fike DA, Grotzinger JP, Pratt LM, Summons RE (2006) Oxidation of the Ediacaran Ocean. Nature 444(7120):744–747

Franchi IA, Wright IP, Gibson JEK, Pillinger CT (1986) The laser microprobe: a technique for extracting carbon, nitrogen, and oxygen from solid samples for isotopic measurements. J Geophys Res 91:D514–D524

Garrels RM, Lerman A (1981) Phanerozoic cycles of sedimentary carbon and sulfur. Proc Natl Acad Sci USA 78:4652–4656

Garrels RM, Mackenzie FT (1972) A quantitative model for the sedimentary rock cycle. Mar Chem 1(1):27–41

Garrels RM, Mackenzie FT, Hunt C (1971) Chemical cycles and the global environment. Kaufmann, Los Altos

Gellatly AM, Lyons TW (2005) Trace sulfate in mid-Proterozoic carbonates and the sulfur isotope record of biospheric evolution. Geochim Cosmochim Acta 69(15):3813–3829

Given R, Lohmann KC (1985) Derivation of the original isotopic composition of Permian marine cements. J Sediment Petrol 55:430–439

Godwin H (1962) Radiocarbon dating. Nature 195(4845):943–945

Goldhaber MB, Kaplan IR (1974) The sulfur cycle, vol 5, The sea. Wiley, New York

Goldhaber MB, Kaplan IR (1980) Mechanisms of sulfur incorporation and isotope fractionation during early diagenesis in sediments of the Gulf of California. Mar Chem 9:95–143

Grey K (2005) Ediacaran palynology of Australia. Mem Assoc Aust Palaeontol 31:1–439

Grotzinger JP (1989) Facies and evolution of Precambrian carbonate depositional systems: emergence of the modern platform archetype. In: Crevello PD, Wilson JL, Sarg JF, Read JF (eds) Controls on carbonate platform and basin development, vol 44, SEPM special publication. SEPM, Tulsa, pp 79–106

Grotzinger JP, Kasting JF (1993) New constraints on Precambrian ocean composition. J Geol 101:235–243

Grotzinger JP, Knoll AH (1995) Anomalous carbonate precipitates: is the Precambrian the key to the Permian? Palaios 10:578–596

Grotzinger JP, Bowring SA, Saylor BZ, Kaufman AJ (1995) Biostratigraphic and geochronologic constraints on early animal evolution. Science 270:598–604

Grotzinger JP, Watters WA, Knoll AH (2000) Calcified metazoans in thrombolite-stromatolite reefs of the terminal Proterozoic Nama Group, Namibia. Paleobiology 26(3):334–359

Habicht KS, Salling L, Thamdrup B, Canfield D (2005) Effect of low sulfate concentration on lactate oxidation and isotope fractionation during sulfate reduction by *Archaeoglobus fulgidus* strain Z. Appl Environ Microbiol 71(7):3770–3777

Halas S (1985) On bias in $^{34}S/^{32}S$ data obtained using SO_2 gas in mass spectrometry. In: Studies on sulphur isotope variations in nature. International Atomic Energy Agency, Vienna, pp 105–111

Halverson GP, Hurtgen MT (2007) Ediacaran growth of the marine sulfate reservoir. Earth Planet Sci Lett 263(1–2):32–44

Halverson GP, Hoffman PF, Schrag DP, Maloof AC, Rice AHN (2005) Toward a Neoproterozoic composite carbon-isotope record. GSA Bull 117(9–10):1181–1207

Hayes JM (1993) Factors controlling ^{13}C contents of sedimentary organic compounds: principles and evidence. Mar Geol 113:111–125

Hayes JM, Waldbauer JR (2006) The carbon cycle and associated redox processes through time. Philos Trans R Soc Lond B Biol Sci 361:931–950

Hayes JM, Kaplan IR, Wedeking KW (1983) Precambrian organic geochemistry: preservation of the record. In: Schopf JW (ed) Earth's earliest biosphere. Princeton University Press, Princeton, pp 93–134

Hayes J, Strauss H, Kaufman A (1999) The abundance of ^{C}-13 in marine organic matter and isotopic fractionation in the global biogeochemical cycle of carbon during the past 800 Ma. Chem Geol 16(1–3):103–125

Hemming NG, Meyers WJ, Grams JC (1989) Cathodoluminescence in diagenetic calcites: the roles of Fe and Mn as deduced from electron probe and spectrophotometric measurments. J Sediment Petrol 59:401–411

Herrle JO, Kößler P, Friedrich O, Erlenkeuser H, Hemleben C (2004) High-resolution carbon isotope records of the Aptian to Lower Albian from SE France and the Mazagan Plateau (DSDP Site 545): a stratigraphic tool for paleoceanographic and paleobiologic reconstruction. Earth Planet Sci Lett 218:149–161

Higgins JA, Schrag DP (2003) Aftermath of a snowball Earth. Geochem Geophys Geosyst (G^3) 4:Article Number 1028. DOI 1010.1029/2002GC000403

Higgins JA, Schrag DP (2006) Beyond methane: towards a theory for the Paleocene-Eocene thermal maximum. Earth Planet Sci Lett 245:523–537

Hoefs J (2008) Stable isotope geochemistry, 6th edn. Springer, Berlin/Heidelberg

Hoefs J, Schidlowski M (1967) Carbon isotope composition of carbonaceous matter from the precambrian of the witwatersrand system. Science 155:1096

Hoffman PF, Schrag DP (2002) The snowball Earth hypothesis: testing the limits of global change. Terra Nova 14:129–155

Hoffman PF, Hawkins DP, Isachsen CE, Bowring SA (1996) Precise U-Pb zircon ages for early Damaran magmatism in the Summas Mountains and Welwitschia Inlier, norther Namara belt, Namibia. Commun Geol Surv Namibia 11:47–52

Hoffman PF, Kaufman AJ, Halverson GP, Schrag DP (1998) A Neoproterozoic snowball Earth. Science 281:1342–1346

Hoffman PF, Halverson GP, Domack EW, Husson JM, Higgins JA, Schrag DP (2007) Are basal Ediacaran (635 Ma) post-glacial "cap dolostones" diachronous? Earth Planet Sci Lett 258(1–2):114–131

Hoffmann K-H, Condon DJ, Bowring SA, Crowley JL (2004) U-Pb zircon date from the Neoproterozoic Ghaub Formation, Namibia: constraints on Marinoan glaciation. Geology 32:817–820

Hofmann HJ, Mountjoy EW (2001) *Namacalathus-Cloudina* assemblage in Neoproterozoic Miette Group (Byng Formation), British Columbia: Canada's oldest shelly fossils. Geology 29(12):1091–1094

Holden NE (2010), Section 11. Table of the Isotopes.In: Lide D(ed) CRC handbook of chemistry and physics, 90th edn (Internet Version). CRC Press/Taylor & Francis, Boca Raton

Holland HD (2006) The oxygenation of the atmosphere and oceans. Philos Trans R Soc Lond B Biol Sci 361(1470):903–915

Holser WT, Schidlowski M, Mackenzie FT, Maynard JB (1988) Biogeochemical cycles of carbon and sulfur. In: Gregor CB, Garrels RM, Mackenzie FT, Maynard JB (eds) Chemical cycles in the evolution of the earth. Wiley, New York, pp 105–173

Hsü KJ, McKenzie JA (1985) A "Stranglove" Ocean in the earliest tertiary. In: Sundquist ET, Broacker WS (eds) The carbon cycle and atmospheric co_2: Natural variations archean to present. American Geophysical Union, Wahsington, D.C., pp 487–492

Hua H, Chen Z, Yuan X, Zhang L, Xiao S (2005) Skeletogenesis and asexual reproduction in the earliest biomineralizing animal *Cloudina*. Geology 33(4):277–280

Hurtgen MT, Arthur MA, Suits NS, Kaufman AJ (2002) The sulfur isotopic composition of Neoproterozoic seawater sulfate: implications for a snowball Earth? Earth Planet Sci Lett 203(1):413–429

Hurtgen MT, Arthur MA, Prave AR (2004) The sulfur isotopic composition of carbonate-associated sulfate in Mesoproterozoic to Neoproterozoic carbonates from Death Valley, CA. In: Lyons TW (ed) Sulfur biogeochemistry – past and present, Geological Society of America Special Publication, vol 379. Geological Society of America, Boulder, pp 177–194

Hurtgen MT, Arthur MA, Halverson GP (2005) Neoproterozoic sulfur isotopes, the evolution of microbial sulfur species, and the burial efficiency of sulfide as sedimentary pyrite. Geology 33(1):41–44

Hurtgen MT, Halverson GP, Arthur MA, Hoffman PF (2006) Sulfur cycling in the aftermath of a 635-Ma snowball glaciation: evidence for a syn-glacial sulfidic deep ocean. Earth Planet Sci Lett 245(3–4):551–570

Hurtgen MT, Pruss SB, Knoll AH (2009) Evaluating the relationship between the carbon and sulfur cycles in the later Cambrian ocean: an example from the Port au Port Group, western Newfoundland, Canada. Earth Planet Sci Lett 281:288–297

IPCC (1990) Climate change. Cambridge University Press, Cambridge

Irwin H, Curtis C, Coleman M (1977) Isotopic evidence for source of diagenetic carbonates formed during burial of organic-rich sediments. Nature 269:209–213

Jacobson MC, Charlson RJ, Rodhe H, Orians GH (2000) Earth system science: from biogeochemical cycles to global change, vol 72, International geophysics series. Elsevier, Oxford

Jahnke LJ, Summons RE, Hope JM, Des Marais DJ (1999) Carbon isotopic fractionation in lipids from methanotrophic bacteria II: the effects of physiology and environmental parameters on the biosynthesis and isotopic signatures of biomarkers. Geochim Cosmochim Acta 63(1):79–93

James NP, Narbonne GM, Kyser TK (2001) Late Neoproterozoic cap carbonates, Mackenzie Mountains, northwestern Canada: precipitation and global glacial meltdown. Can J Earth Sci 38:1229–1262

Jensen ML, Nakai N (1961) Sources and isotopic composition of atmospheric sulfur. Science 134(3496):2102–2104

Jiang G, Kennedy MJ, Christie-Blick N (2003) Stable isotopic evidence for methane seeps in Neoproterozoic postglacial cap carbonates. Nature 426:822–826

Jiang GQ, Kennedy MJ, Christie-Blick N, Wu HC, Zhang SH (2006a) Stratigraphy, sedimentary structures, and textures of the late neoproterozoic doushantuo cap carbonate in south China. J Sediment Res 76(7–8):978–995

Jiang GQ, Shi XY, Zhang SH (2006b) Methane seeps, methane hydrate destabilization, and the late Neoproterozoic postglacial cap carbonates. Chinese Sci Bull 51(10):1152–1173

Jiang GQ, Kaufman AJ, Christie-Blick N, Zhang SH, Wu HC (2007) Carbon isotope variability across the Ediacaran Yangtze platform in South China: implications for a large surface-to-deep ocean delta C-13 gradient. Earth Planet Sci Lett 261(1–2):303–320

Johnston DT, Wing BA, Farquhar J, Kaufman AJ, Strauss H, Lyons TW, Kah LC, Canfield DE (2005) Active microbial sulfur disproportionation in the Mesoproterozoic. Science 310(5753):1477–1479

Kah LC, Sherman AG, Narbonne GM, Knoll AH, Kaufman AJ (1999) $\delta^{13}C$ stratigraphy of the Proterozoic Bylot Supergroup, Baffin Island, Canada: implications for regional lithostratigraphic correlations. Can J Earth Sci 36:313–332

Kah LC, Lyons TW, Frank TD (2004) Low marine sulphate and protracted oxygenation of the proterozoic biosphere. Nature 431(7010):834–838

Kaiho K, Kajiwara Y, Nakano T, Miura Y, Kawahata H, Tazaki K, Ueshima M, Chen Z, Shi GR (2001) End-Permian catastrophe by a bolide impact: evidence of a gigantic release of sulfur from the mantle. Geology 29:815–818

Kampschulte A, Strauss H (2004) The sulfur isotopic evolution of Phanerozoic seawater based on the analysis of structurally substituted sulfate in carbonates. Chem Geol 204:255–286

Kaplan IR (1983) Stable isotopes of sulfur, nitrogen and deuterium in recent margin environments. In: Arthur MA, Anderson TF, Kaplan IR, Veizer J, Land LS (eds) Stable isotopes in sedimentary geology, vol 10, SEPM short course. SEPM, Tulsa, pp 2.1–2.108, Columbia, SC

Kaplan IR, Rittenberg SC (1964) Microbiological fractionation of sulphur isotopes. J Gen Microbiol 34:195–212

Karhu J, Holland H (1996) Carbon isotopes and the rise of atmospheric oxygen. Geology 24:867–870

Kaufman AJ, Knoll AH (1995) Neoproterozoic variations in the C-isotope composition of sea water: stratigraphic and biogeochemical implications. Precambrian Res 73(3–4):27–49

Kaufman AJ, Hayes JM, Knoll AH, Germs GJB (1991) Isotopic compositions of carbonates and organic carbon from upper Proterozoic successions in Namibia: stratigraphic variation and the effects of diagenesis and metamorphism. Precambrian Res 49:301–327

Kaufman AJ, Knoll AH, Narbonne GM (1997) Isotopes, ice ages, and terminal Proterozoic earth history. Proc Natl Acad Sci USA 94:6600–6605

Kaufman AJ, Jiang GQ, Christie-Blick N, Banerjee DM, Rai V (2006) Stable isotope record of the terminal Neoproterozoic Krol platform in the Lesser Himalayas of northern India. Precambrian Res 147(1–2):156–185

Kaufman AJ, Corsetti FA, Varni MA (2007a) The effect of rising atmospheric oxygen on carbon and sulfur isotope anomalies in the Neoproterozoic Johnnie Formation, Death Valley, USA. Chem Geol 237(1–2):47–63

Kaufman AJ, Johnston DT, Farquhar J, Masterson AL, Lyons TW, Bates S, Anbar AD, Arnold GL, Garvin J, Buick R (2007b) Late Archean biospheric oxygenation and atmospheric evolution. Science 317(5846):1900–1903

Kempe A, Thode HG (1968) The mechanism of the bacterial reduction of sulphate and of sulphite from isotope fractionation studies. Geochim Cosmochim Acta 32:71–91

Kennedy MJ (1996) Stratigraphy, sedimentology, and isotopic geochemistry of Australian Neoproterozoic postglacial cap dolostones; deglaciation, $d^{13}C$ excursions, and carbonate precipitation. J Sediment Res 66:1050–1064

Kennedy MJ, Christie-Blick N, Sohl LE (2001) Are Proterozoic cap carbonates and isotopic excursions a record of gas hydrate destabilization following Earth's coldest intervals? Geology 29(5):443–446

Kenrick P, Crane PR (1997) The origin and early diversification of land plants: a cladistic study. Smithsonian Institution Press, Washington, DC

Kirschvink JL (1992) Late Proterozoic low-latitude global glaciation: the snowball Earth. In: Schopf JW, Klein C (eds) The proterozoic biosphere: a multidisciplinary study. Cambridge University Press, Cambridge, pp 51–52

Knauth LP, Kennedy MJ (2005) An alternative view of C isotope variations in Neoproterozoic carbonates. Geol Soc Am Annu Meet Abstr Prog 37(7):43

Knoll AH (1992) Vendian microfossils in metasedimentary cherts of the Scotia Group, Prins Karls Forland, Svalbard. Palaeontology 35:751–774

Knoll AH, Carroll SB (1999) Early animal evolution: emerging views from comparative biology and geology. Science 284(5423):2129–2137

Knoll AH, Bambach RK, Canfield D, Grotzinger J (1996) Comparative Earth history and late Permian mass extinction. Science 273(5274):452–457

Kohn MJ, Riciputi LR, Stakes DS, Orange D (1998) Sulfur isotope variability in biogenic pyrite: reflections of heterogeneous bacterial colonization? Am Miner 83:1454–1468

Kramm U, Wedepohl KH (1991) The isotopic composition of strontium and sulfur in seawater of Late Permian (Zechstein) age. Chem Geol 90:253–262

Kump LR (1989) Alternative modeling approaches to the geochemical cycles of carbon, sulfur, and strontium isotopes. Am J Sci 289:390–410

Kump LR (1991) Interpreting carbon-isotope excursions: strangelove oceans. Geology 19:299–302

Kump LR, Arthur MA (1999) Interpreting carbon-isotope excursions: carbonates and organic matter. Chem Geol 161:181–198

Kump LR, Garrels RM (1986) Modeling atmospheric O_2 in the global sedimentary redox cycle. Am J Sci 286:337–360

Larson G, Chilingar GV (1979) Diagenesis in sediments and sedimentary rocks. In: Developments in sedimentology, vol 25A. Elsevier, Amsterdam

Le Guerroue E, Cozzi A (2006) A chemostratigraphic and sedimentological framework of the largest negative carbon isotopic excursion in Earth history: the Neoprotoerozoic Shuram Formation (Nafun Group, Oman). Precambrian Res 146:69–92

Le Guerroue E, Allen PA, Cozzi A, Etienne JL, Fanning M (2006) 50 Myr recovery from the largest negative delta C-13 excursion in the Ediacaran ocean. Terra Nova 18(2):147–153

Leshin LA, Rubin AE, McKeegan KD (1997) The oxygen isotopic composition of olivine and pyroxene from CI chondrites. Geochim Cosmochim Acta 61(4):835–845

Li YH (1972) Geochemical mass balance among lithosphere, hydrosphere, and atmosphere. Am J Sci 272:119–137

Li C, Love GD, Lyons TW, Fike D, Sessions AL, Chu X (2010) A stratified redox model for the Ediacaran Ocean. Science 328:80–83

Lohmann KC (1983) Unraveling the diagenetic history of carbonate reservoirs. In: Wilson JL, Wilkinson BH, Lohmann KC (eds) New ideas and methods for exploration of carbonate reservoirs – notes for a short course. Dallas Geological Society, Dallas

Lohmann KC (1988) Geochemical patterns in meteoric diagenetic systems and their application to studies of paleokarsts. In: James NP, Choquette PW (eds) Paleokarst, vol 58–80. Springer, New York

Love LG (1967) Early diagenetic iron sulphide in recent sediments of the wash (England). Sedimentology 9:327–352

Love GD, Grosjean E, Stalvies C, Fike DA, Grotzinger JP, Bradley AS, Kelly AE, Bhatia M, Meredith W, Snape CE, Bowring SA, Condon DJ, Summons RE (2009) Fossil steroids record the appearance of Demospongiae during the Cryogenian period. Nature 457:718–722

Lovering JF (1975) Application of SIMS microanalysis techniques of trace element and isotopic studies in geochemistry and cosmochemistry. In: US National Bureau of Standards, Special Publication, vol 427, pp 135–178

Lyons TW, Gill BC, Shim MJ, Frank TD, Hurtgen MT, Saltzman MR, Gellatly AM, Kah LC (2004) Carbonate-associated sulfate as a paleoceanographic proxy: an update. Geochim Cosmochim Acta 68(11):A337–A337

Machel HG (1985) Cathodoluminescence in calcite and dolomite and its chemical interpretation. Geosci Can 12:139–147

Machel HG, Mountjoy EW (1986) Chemistry and environments of dolomitization: a reappraisal. Earth Sci Rev 23(3):175–222

Mackenzie FT (2005) Sediments, diagenesis, and sedimentary rocks. In: Treatise on geochemistry, vol 7. Elsevier, Amsterdam

Macnamara J, Thode HG (1950) Comparison of the isotopic constitution of terrestrial and meteoritic sulfur. Phys Rev 78:307–308

Magaritz M, Holser WT, Kirschvink JL (1986) Carbon-isotope events across the Precambrian/Cambrian boundary on the Siberian Platform. Nature 320:258–259

Mariotti A (1983) Atmospheric nitrogen is a reliable standard for natural ^{15}N abundance measurements. Nature 303:685–687

Matthews JM, Hayes JM (1978) Isotope-ratio-monitoring gas chromatography-mass spectrometry. Anal Chem 50:1465–1473

Mayer B, Fritz P, Prietzel HR, Krouse HR (1995) The use of stable sulfur and oxygen isotope ratios for interpreting the mobility of sulfate in aerobic forest soils. Appl Geochem 10:161–173

McCarron G (2000) The sedimentology and chemostratigraphy of the Nafun Group, Huqf Supergroup, Oman. University of Oxford, Oxford

McConnaughey TA (1989a) ^{13}C and ^{18}O isotopic disequilibrium in biological carbonates: I. Patterns. Geochim Cosmochim Acta 53:151–163

McConnaughey TA (1989b) ^{13}C and 18O disequilibrium in biological carbonates: II in vitro simulation of kinetic isotope effects. Geochim Cosmochim Acta 53:163–171

McCrea JM (1950) On the isotope geochemistry of carbonates and a paleotemperature scale. J Chem Phys 18:849–857

McFadden K, Huang J, Chu XL, Jiang GQ, Kaufman AJ, Zhou CM, Yuan SL, Xiao SH (2008) Redox instability and biological evolution in the Ediacaran Doushantuo formation. Proc Natl Acad Sci USA 105(9):3197–3202

McIlreath IA, Morrow DW (1990) Diagenesis, vol 4, Geocience Canada reprint series. Runge, Ottawa

Melezhik VA, Fallick AE, Pokrovsky BG (2005) Enigmatic nature of thick sedimentary carbonates depleted in ^{13}C beyond the canonical mantle value: the challenges to our understanding of the terrestrial carbon cycle. Precambrian Res 137(3–4):131–165

Melim LA, Westphal H, Swart PK, Eberli GP, Munnecke A (2002) Questioning carbonate diagenetic paradigms: evidence from the Neogene of the Bahamas. Mar Geol 185:27–53
Meyer PA (1994) Preservation of elemental and isotopic course identification of sedimentary organic matter. Chem Geol 114:289–302
Misi A, Kaufman AJ, Veizer J, Powis K, Azmy K, Boggiani PC, Gaucher C, Teixeira JBG, Sanches AL, Iyer SSS (2007) Chemostratigraphic correlation of neoproterozoic successions in South America. Chem Geol 237(1–2):143–167
Mizota C, Sasaki A (1996) Sulfur isotope composition of soils and fertilizers: differences between northern and southern hemispheres. Geoderma 71(1–2):77–93
Moczydlowska M (2005) Taxonomic review of some Ediacaran acritarchs from the Siberian Platform. Precambrian Res 136(3–4):283–307
Moore CH (2001) Carbonate reservoirs: porosity evolution and diagenesis in a sequence stratigraphic framework. In: Developments in sedimentology. Elsevier, Amsterdam
Moore CH, Druckman Y (1981) Burial diagenesis and porosity evolution, Upper Jurassic Smackover, Arkansas and Louisiana. AAPG Bull 65(4):597–628
Mopper K, Degens ET (1979) Organic carbon in the ocean: nature and cycling. In: Bolin B, Degens ET, Kempe S, Ketner P (eds) The global carbon cycle. Wiley, New York,pp 365–385
Morenco PJ, Corsetti FA, Hammond DE, Kaufman AJ, Bottjer DJ (2008) Oxidation of pyrite during extraction of carbonate associated sulfate. Chem Geol 247:124–132
Morrow DW (1990) Dolomite part II: dolomitization models and ancient dolostones. In: McIlreath IA, Morrow DW (eds) Diagenesis, vol 4, Geoscience Canada. Runge, Ottawa, pp 125–140
Nielsen H (1970) Sulfur 16-B. Isotopes in nature. In: Handbook of geochemistry, vol 2. Springer, New York
Nogeira ACR, Riccomini C, Sial AN, Moura CAV, Fairchild TR (2003) Soft-sediment deformation at the base of the Neoproterozoic Puga cap carbonate (southwestern Amazon craton, Brazil): confirmation of rapid icehouse to greenhouse transition in snowball Earth. Geology 31:613–616
Ostermann DR, Curry WB (2000) Calibration of stable isotopic data: an enriched $\delta^{18}O$ standard used for source gas mixing detection and correction. Paleoceanography 15:353–360
Payton A, Kastner M, Campbell D, Thiemens MH (1998) Sulfur isotopic composition of Cenozoic seawater sulfate. Science 282:1459–1462
Pearson PN, Palmer MR (2000) Atmospheric carbon dioxide concentrations over the past 60 million years. Nature 406:695–699
Pelechaty SM (1998) Integrated chronostratigraphy of the Vendian System of Siberia: implications for a global stratigraphy. J Geol Soc Lond 155(6):957–973
Petit JR, Jouzel J, Raynaud D, Barkov NI, Barnola JM, Basile I, Bender M, Chappellaz J, Davis M, Delaygue G, Delmotte M, Kotlyakov VM, Legrand M, Lipenkov VY, Lorius C, Pepin L, Ritz C, Saltzman E, Stievenard M (1999) Climate and atmospheric history of the past 420, 000 years from the Vostok ice core, Antarctica. Nature 399:429–436
Petsch ST, Berner RA (1998) Coupling the geochemical cycles of C, P, Fe, and S: the effect of atmospheric O_2 and the isotopic records of carbon and sulfur. Am J Sci 298:246–262
Pierre C (1985) Isotopic evidence for the dynamic reduc cycle of dissolved sulphur compounds between free and interstitial solutions in amrine salt pans. Chem Geol 53:191–196
Pingitore JNE, Meitzner G, Love KM (1995) Identification of sulfate in natural carbonates by X-ray absorption spectroscopy. Geochim Cosmochim Acta 59:2477–2483
Prokoph A, Shields G, Veizer J (2008) Compilation and time-series analysis of a marine carbonate δ18O, δ13C, 87Sr/86Sr and δ34S database through Earth history. Earth-Sci Rev 87(3–4): 113–133
Puchelt H, Sabels BR, Hoering TC (1971) Preparation of sulfur hexafluoride for isotope geochemical analysis. Geochim Cosmochim Acta 35:625
Raiswell R, Berner RA (1985) Pyrite formation in euxinic and semi-euxinic sediments. Am J Sci 285:710–724
Raymo ME (1997) Carbon cycle models: how strong are the constraints? In: Ruddiman W (ed) Tectonics, uplift and climate change. Plenum, New York, pp 367–381

Reeburgh W (1997) Figures summarizing the global cycles of biogeochemically important elements. Bull Ecol Soc Am 78:260–267

Rees CE (1978) Sulfur isotope measurements using SO_2 and SF_6. Geochim Cosmochim Acta 42:383–390

Rees CE, Jenkins WJ, Monster J (1978) The sulphur isotopic composition of ocean water sulphate. Geochim Cosmochim Acta 42:377–382

Resing JA, Lupon JE, Feely RA, Lilley MD (2004) CO_2 and ^3He in hydrothermal plumes: implications for mid-ocean ridge CO2 flux. Earth Planet Sci Lett 226:449–464

Riccardi AL, Arthur MA, Kump LR (2006) Sulfur isotopic evidence for chemocline upward excursions during the end-Permian mass extinction. Geochim Cosmochim Acta 70:5740–5752

Riciputi LR, Paterson BA, Ripperdan RL (1998) Measurement of light stable isotope ratios by SIMS: matrix effects for oxygen, carbon, and sulfur isotopes in minerals. Int J Mass Spectrom 178:81–112

Ridwell AJ, Kennedy MJ, Caldeira K (2003) Carbonate deposition, climate stability, and Neoproterozoic ice ages. Science 302:859–862

Romanek CS, Grossman EL, Morse JW (1992) Carbon isotopic fractionation in synthetic aragonite and calcite: effects of temperature and precipitation rate. Geochim Cosmochim Acta 56:419–430

Rothman DH, Hayes JM, Summons R (2003) Dynamics of the Neoproterozoic carbon cycle. Proc Natl Acad Sci USA 100(14):8124–8129

Rye RO (2005) A review of the stable-isotope geochemistry of sulfate minerals in selected igneous environments and related hydrothermal systems. Chem Geol 215:5–36

Saltzman E, Davidson JP, Holden P, Runnegar B, Lohmann KC (1995) Sea level-driven changes in ocean chemistry at an Upper Cambrian extinction horizon. Geology 23:893–896

Savard MM, Veizer J, Hinton R (1995) Cathodoluminescence at low Fe and Mn concentrations: a SIMS study of zones in natural calcite. J Sediment Petrol A65:208–213

Schauble EA (2004) Applying stable isotope fractionation theory to new systems, vol 55, Geochemistry of non-traditional stable isotopes. Mineralogical Society of America, Washington, DC

Schidlowski M (2001) Carbon isotopes as biogeochemical recorders of life over 3.8 Ga of Earth history: evolution of a concept. Precambrian Res 106:117–134

Schidlowski M, Hayes JM, Kaplan IR (1983) Isotopic inferences of ancient biochemistries: carbon, sulfur, hydrogen, and nitrogen. In: Schopf JW (ed) Earth's earliest biosphere: its origin and evolution. Princeton University Press, Princeton, pp 149–186

Schlesinger WH (1997) Biogeochemistry: an analysis of global change, 2nd edn. Academic, San Diego

Scholle PA, Arthur MA (1980) Carbon isotope fluctuations in Cretaceous pelagic limestones: potential stratigraphic and petroleum exploration tool. AAPG Bull 64(1):67–87

Schouten S, Van Kaam-Peters HME, Rijpstra WIC, Schoell M, Sinninghe Damste JS (2000) Effects of an oceanic anoxic event on the stable carbon isotopic composition of early Toarcian carbon. Am J Sci 200:1–22

Sharp ZD (1990) A laser-based microanalytical method for the insitu determination of oxygen isotope ratios in silicates and oxides. Geochim Cosmochim Acta 54:1353–1357

Sharp ZD (1992) In situ laser microprobe techniques for stable isotope analysis. Chem Geol Isotope Geosci Sect 101:3–19

Sharp Z (2007) Principles and stable isotope geochemistry. Prentice Hall, Upper Saddle River

Shen Y, Buick R, Canfield D (2001) Isotopic evidence for microbial sulphate reduction in the early Archaean era. Nature 410:77–81

Shen Y, Canfield DE, Knoll AH (2002) Middle Proterozoic ocean chemistry: evidence from the McArthur Basin, northern Australia. Am J Sci 302:81–109

Sheppard SMF, Schwarcz HP (1970) Fractionation of carbon and oxygen isotopes and magnesium between coexisting metamorphic calcite and dolomite. Contrib Mineralog Petrol 26(3):161–198

Shieh YN, Taylor HP (1969) Oxygen and carbon isotope studies of contact metamorphism of carbonate rocks. J Petrol 10:307–331

Shields G (2005) Neoproterozoic cap carbonates: a critical appraisal of existing models and the plumeworld hypothesis. Terra Nova 17:299–310

Shields G, Veizer J (2002) Precambrian marine carbonate isotope database: version 1.1. Geochem Geophys Geosyst 3:1–12

Shields G, Brasier M, Stille P, Dorjnamjaa D (2002) Factors contributing to high $\delta^{13}C$ values in cryogenian limestones of western mongolia. Earth Planet Sci Lett 196:99–111

Spero HJ, Bijma J, Lea D, Bemis BE (1997) Effect of seawater carbonate concentration on foraminiferal carbon and oxygen isotopes. Nature 390:497–500

Staudt WJ, Schoonen MAA (1995) Sulfate incorporation into sedimentary carbonates. Am Chem Soc Symp 612:332–347

Staudt WJ, Oswald EJ, Schoonen MAA (1993) Determination of sodium, chloride and sulfate in dolomites: a new technique to constrain the composition of dolomitizing fluids. Chem Geol 107:97–109

Stevens CM, Krout L, Walling D, Venters A, Engelkemeir S, Ross LE (1972) The isotopic composition of atmospheric carbon monoxide. Earth Planet Sci Lett 16:147–165

Strauss H (1993) The sulfur isotopic record of Precambrian sulfates: new data and a critical evaluation of the existing record. Precambrian Res 63(3–4):225–246

Strauss H (1997) The isotopic composition of sedimentary sulfur through time. Palaeogeogr Palaeoclimatol Palaeoecol 132:97–118

Strauss H (1999) Geological evolution from isotope proxy signals – sulfur. Chem Geol 161(1–3):89–101

Strauss H, Banerjee DM, Kumar V (2001) The sulfur isotopic composition of Neoproterozoic to early Cambrian seawater – evidence from the cyclic Hanseran evaporites, NW India. Chem Geol 175(1–2):17–28

Studley SA, Ripley EM, Elswich MS, Dorais MD, Fong J, Finkelstein D, Pratt LM (2002) Analysis of sulfides in whole rock matrices by elemental analyzer-continuous flow isotope ratio mass spectrometry. Chem Geol 192(1–2):141–148

Summons RE, Franzmann PD, Nichols PD (1998) Carbon isotopic fractionation associated with methylotrophic methanogenesis. Org Geochem 28(7–8):465–475

Takahashi M, Broacker WS, Bainbridge AE (1981) The alkalinity and total carbon dioxide concentration in the world oceans. In: Bolin B (ed) Carbon dioxide modeling. Wiley, New York, pp 271–286

Thode HG, Monster J (1965) Sulfur-isotope geochemistry of petroleum, evaporites, and ancient seas. Am Assoc Pet Geol Mem 4:367–377

Thode HG, Monster J, Dunford M (1961) Sulfur isotope geochemistry. Geochim Cosmochim Acta 25:159–174

Thomazo C, Ader M, Farquhar J, Philippot P (2009) Methanotrophs regulated atmospheric sulfur isotope anomalies during the Mesoarchean (Tumbiana Formation, Western Australia). Earth Planet Sci Lett 279(1–2):65–75

Tiwari M, Knoll AH (1994) Large acanthomorphic acritarchs from the Infra Krol Formation of the Lesser Himalaya and their stratigraphic significance. J Himal Geol 5:193–201

Toon OB, Kasting JF, Turco RP, Liu MS (1987) The sulfur cycle in the marine atmosphere. J Geophys Res 92:943–963

Trudinger PA, Chambers LA, Smith JW (1985) Low temperature sulphate reduction: biological versus abiological. Can J Earth Sci 22:1910–1918

Tucker ME, Wright VP (1990) Carbonate sedimentology. Blackwell Scientific, Oxford

Valley JW, Cole DR (2001) Stable isotope geochemistry. Reviews in mineralogy and geochemistry, vol 43. Mineralogical Society of America, Washington, DC

Veizer J, Ala D, Azmy K, Bruckschen P, Buhl D, Bruhn F, Carden GAF, Diener A, Ebneth S, Godderis Y, Jasper T, Korte C, Pawellek F, Podlaha OG, Strauss H (1999) $^{87}Sr/^{86}Sr$, $\delta^{13}C$ and $\delta^{18}O$ evolution of Phanerozoic seawater. Chem Geol 161:59–88

Waggoner B (2003) The Ediacaran biotas in space and time. Integr Comp Biol 43:104–113

Weber B, Steiner M, Zhu MY (2007) Precambrian-Cambrian trace fossils from the Yangtze Platform (South China) and the early evolution of bilaterian lifestyles. Palaeogeogr Palaeoclimatol Palaeoecol 254(1–2):328–349

Wefer G, Berger WH (1991) Isotope paleontology: growth and composition of extant calcareous species. Mar Geol 100:207–248

White D (2000) The physiology and biochemistry of prokaryotes. Oxford University Press, New York, p 565

Xiao S (2004) Neoproterozoic glaciations and biological evolution: data from the Tarim and Yangtze blocks, China. In: Reimold WU, Hofmann A (eds) Abstract Volume of GSA/GSSA Field Forum "Processes on the Early Earth". p 99

Zachos J, Pagani M, Sloan L, Thomas E, Billups K (2001) Trends, rhythms, and aberrations in global climate 65 Ma to present. Science 272:686–693

Zhou C, Xiao S (2007) Ediacaran $\delta^{13}C$ chemostratigraphy of South China. Chem Geol 237(1–2):89–108

Zhou C, Tucker R, Xiao S, Peng Z, Yuan X, Chen Z (2004) New constraints on the ages of Neoproterozoic glaciations in South China. Geology 32:437–440

Zhou CM, Xie GW, McFadden K, Xiao SH, Yuan XL (2007) The diversification and extinction of Doushantuo-Pertatataka acritarchs in South China: causes and biostratigraphic significance. Geol J 42(3–4):229–262

Index

M. Laflamme et al. (eds.), *Quantifying the Evolution of Early Life*, Topics in Geobiology 36,
DOI 10.1007/978-94-007-0680-4, © Springer Science+Business Media B.V. 2011

C

D